FUNDAMENTALS
with Elements of Algebra

KAUFMANN, *Intermediate Algebra for College Students, Fourth Edition*
KAUFMANN, *Elementary and Intermediate Algebra: A Combined Approach*
KAUFMANN, *Algebra for College Students, Fourth Edition*
KAUFMANN, *Algebra with Trigonometry for College Students, Third Edition*
KAUFMANN, *College Algebra, Third Edition*
KAUFMANN, *Trigonometry, Second Edition*
KAUFMANN, *College Algebra and Trigonometry, Third Edition*
KAUFMANN, *Precalculus, Second Edition*
LAVOIE, *Discovering Mathematics*
MCCOWN/SEQUEIRA, *Patterns in Mathematics: From Counting to Chaos*
RICE/STRANGE, *Plane Trigonometry, Sixth Edition*
RIDDLE, *Analytic Geometry, Fifth Edition*
RUUD/SHELL, *Prelude to Calculus, Second Edition*
SGROI/SGROI, *Mathematics for Elementary School Teachers*
SWOKOWSKI/COLE, *Fundamentals of College Algebra, Eighth Edition*
SWOKOWSKI/COLE, *Fundamentals of Algebra and Trigonometry, Eighth Edition*
SWOKOWSKI/COLE, *Fundamentals of Trigonometry, Eighth Edition*
SWOKOWSKI/COLE, *Algebra and Trigonometry with Analytic Geometry, Eighth Edition*
SWOKOWSKI/COLE, *Precalculus: Functions and Graphs, Seventh Edition*
WELTMAN/PEREZ, *Beginning Algebra, Second Edition*
WELTMAN/PEREZ, *Intermediate Algebra, Third Edition*

THE PWS SERIES IN CALCULUS AND UPPER-DIVISION MATHEMATICS

ALTHOEN/BUMCROT, *Introduction to Discrete Mathematics*
ANDRILLI/HECKER, *Linear Algebra*
BURDEN/FAIRES, *Numerical Analysis, Fifth Edition*
CROOKE/RATCLIFFE, *A Guidebook to Calculus with Mathematica*
CULLEN, *An Introduction to Numerical Linear Algebra*
CULLEN, *Linear Algebra and Differential Equations, Second Edition*
DENTON/NASBY, *Finite Mathematics, Preliminary Edition*
DICK/PATTON, *Calculus*
DICK/PATTON, *Single Variable Calculus*
DICK/PATTON, *Technology in Calculus: A Sourcebook of Activities*
EDGAR, *A First Course in Number Theory*
EVES, *In Mathematical Circles*
EVES, *Mathematical Circles Revisited*
EVES, *Mathematical Circles Squared*
EVES, *Return to Mathematical Circles*
FAIRES/BURDEN, *Numerical Methods*
FINIZIO/LADAS, *Introduction to Differential Equations*
FINIZIO/LADAS, *Ordinary Differential Equations with Modern Applications, Third Edition*
FLETCHER/HOYLE/PATTY, *Foundations of Discrete Mathematics*
FLETCHER/PATTY, *Foundations of Higher Mathematics, Second Edition*
GILBERT/GILBERT, *Elements of Modern Algebra, Third Edition*
GORDON, *Calculus and the Computer*
HARTFIEL/HOBBS, *Elementary Linear Algebra*
HARTIG, *Guidebook to Linear Algebra for Theorist*
HILL/ELLIS/LODI, *Calculus Illustrated*
HILLMAN/ALEXANDERSON, *Abstract Algebra: A First Undergraduate Course, Fifth Edition*
HUMI/MILLER, *Boundary-Value Problems and Partial Differential Equations*

LAUFER, *Discrete Mathematics and Applied Modern Algebra*

LEINBACH, *Calculus Laboratories Using Derive*

MARON/LOPEZ, *Numerical Analysis, Third Edition*

MIECH, *Calculus with Mathcad*

MIZRAHI/SULLIVAN, *Calculus with Analytic Geometry, Third Edition*

MOLLUZZO/BUCKLEY, *A First Course in Discrete Mathematics*

NICHOLSON, *Elementary Linear Algebra with Applications, Third Edition*

NICHOLSON, *Introduction to Abstract Algebra*

O'NEIL, *Advanced Engineering Mathematics, Third Edition*

PENCE, *Calculus Activities for Graphic Calculators*

PENCE, *Calculus Activities for the TI-81 Graphic Calculator, Second Edition*

PLYBON, *An Introduction to Applied Numerical Analysis*

POWERS, *Elementary Differential Equations, Brief Edition*

POWERS, *Elementary Differential Equations with Boundary-Value Problems*

PRESCIENCE CORPORATION, *The Student Edition of Theorist*

RIDDLE, *Calculus and Analytic Geometry, Fourth Edition*

SCHELIN/BANGE, *Mathematical Analysis for Business and Economics, Second Edition*

SENTILLES, *Applying Calculus in Economics and Life Science*

SWOKOWSKI, OLINICK, AND PENCE, *Calculus, Sixth Edition*

SWOKOWSKI, OLINICK, AND PENCE, *Calculus of a Single Variable, Second Edition*

SWOKOWSKI, *Calculus, Fifth Edition (Late Trigonometry Version)*

SWOKOWSKI, *Elements of Calculus with Analytic Geometry: High School Edition*

TAN, *Applied Finite Mathematics, Fourth Edition*

TAN, *Calculus for the Managerial, Life, and Social Sciences, Third Edition*

TAN, *Applied Calculus, Third Edition*

TAN, *College Mathematics, Third Edition*

TRIM, *Applied Partial Differential Equations*

VENIT/BISHOP, *Elementary Linear Algebra, Third Edition*

VENIT/BISHOP, *Elementary Linear Algebra, Alternate Second Edition*

WATTENBERG, *Calculus in a Real and Complex World*

WIGGINS, *Problem Solver for Finite Mathematics and Calculus*

ZILL, *Calculus, Third Edition*

ZILL, *A First Course in Differential Equations, Fifth Edition*

ZILL/CULLEN, *Differential Equations with Boundary-Value Problems, Third Edition*

ZILL/CULLEN, *Advanced Engineering Mathematics*

THE PWS SERIES IN ADVANCED MATHEMATICS

EHRLICH, *Fundamental Concepts of Abstract Algebra*

JUDSON, *Abstract Algebra: Theory and Applications*

KIRKWOOD, *An Introduction to Analysis*

PATTY, *Foundations of Topology*

RUCKLE, *Modern Analysis: Measure Theory and Functional Analysis with Applications*

SIERADSKI, *An Introduction to Topology and Homotopy*

STEINBERGER, *Algebra*

STRAYER, *Elementary Number Theory*

TROUTMAN, *Boundary-Value Problems of Applied Mathematics*

FUNDAMENTALS
with Elements of Algebra
SECOND EDITION

Patricia J. Cass
Columbus State Community College

Elizabeth R. O'Connor
Columbus State Community College

PWS PUBLISHING COMPANY
Boston

PWS PUBLISHING COMPANY
20 Park Plaza, Boston, MA 02116-4324

PWS Publishing Company is a division of Wadsworth, Inc.

International Thomson Publishing
The trademark ITP is used under license

 This book is printed on recycled, acid-free paper.

Library of Congress Cataloging-in-Publication Data
Cass, Patricia J.
    Fundamentals with Elements of Algebra, Second Edition / Patricia J. Cass, Elizabeth R. O'Connor.
        p.   cm.
    Includes index.
    ISBN 0-534-93477-3
    1. Algebra. I. O'Connor, Elizabeth, R. II. Title.
    QA107.C39   1993   93-25645
    513—dc20   CIP

Cover image copyright © Jeff Hunter/The Image Bank.

Sponsoring Editor: Al Bruckner
Developmental Editor: Susan M. Gay
Production Coordinator: Kirby Lozyniak
Editorial Assistant: Mary Boulger
Interior Cover Designer: Kirby Lozyniak
Interior Illustrator: Beacon Graphics Corporation
Marketing Manager: Marianne C. P. Rutter
Manufacturing Coordinator: Lisa Flanagan
Compositor: Beacon Graphics Corporation
Cover Printer: Henry N. Sawyer Company, Inc.
Text Printer and Binder: Courier Companies Inc./Westford

Printed and bound in the United States of America.

93 94 95 96 97 98—10 9 8 7 6 5 4 3 2 1

We would like to dedicate this text to our husbands and children who gave up their wives and mothers for so very long so that we could fulfill this dream. Without their support, encouragement, and love, this project would never have been possible.

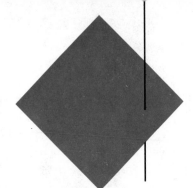

Contents

9   The Why and What of Algebra   237

10   Equations   259

11   Word Problems   289

12 Signed Numbers   313

Cumulative Review, Chapters 1–12   339

13 Expressions, Equations, and Inequalities   341

14 More Word Problems   365

15 Polynomial Operations 397

16 Introduction to Factoring 433

17 ◆ Introduction to Graphing   461

Cumulative Review, Chapters 1–17   505

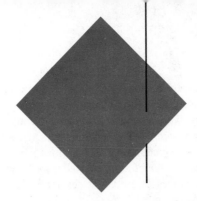

Preface

Our initial intent in writing this book was to give returning adult students a chance to be successful with basic math and elementary algebra. However, in our teaching experience, we have found that the same strategies that work for our older, nontraditional students, also work very well for our younger, more traditional college students. What many of these students have in common is a less-than-successful history with math. A majority have anxiety about math, especially algebra. They see no real purpose in studying it, and find many reasons for avoiding it. Older students return to school with new educational and career goals, but they are still intimidated by math and the people who teach it, and many are self-programmed for failure. While most of these students don't desire to become whizzes at high-level math, they now know that some math skills are necessary to succeed, not only in school, but also in life.

As developmental math teachers, we can help these students work through their difficulties. We believe this text will enable them to have a positive experience with mathematics and provide the skills, tools, and confidence to go on in math. We also hope that this text will enable other instructors to help students achieve that positive experience.

Since we understand students' fear and dislike of math, we have written the text in an almost conversational style. We did this without sacrificing any mathematical truths. We focused on the need for students to advance to other math courses and we have presented methods and materials which would best prepare them for subsequent courses.

In keeping with the NCTM Standards, this revised edition includes more emphasis on estimation, interpreting data, and critical thinking. A unique feature, located at the end of each chapter, are the Thought-Provoking Problems. These involve written responses, analysis activities, calculator experimentation, and skill-stretching problem-solving. An instructor wanting to expand these areas might use these Thought-Provoking Problems as a starting point for class discussion questions, small group activities, extra credit questions, or other creative activities.

The arithmetic chapters form a basis for students' future work in algebra. Little time was spent working with whole numbers, but topics such as the properties of whole numbers, prime factorization, Least Common Multiple and Greatest Common Factor were developed in depth because of their application to algebra. The fraction and decimal chapters (Chapters 3, 4, and 5) are included to help students brush up on their basic skills so they will be able to solve and check some algebraic equations. The treatment of percents in Chapter 7 is formula oriented. While many instructors may find this to be a new approach, students will need to be able to work with formulas in their future courses, and this has proven to be a good way to introduce them to these concepts. The text is arranged so that those instructors wishing to teach percents by the proportion method can skip Section 7.4 and use the methods in Section 7.5. Chapter 8, "Practical Geometry," could be studied as a separate chapter if the students' needs warrant it. We have, however, tried to incorporate geometry throughout the text whenever the topics could reasonably be included.

The algebra topics are presented in a nontraditional order. We introduce the ideas of formulas, evaluating, equation solving, and word problems before introducing signed numbers. This approach gives the students success with algebra concepts before complicating problems with signed numbers. There is ample practice in, and after, Chapter 12 to perfect the signed-number operations.

Word problems are presented throughout the text. Since critical thinking and decision-making processes are being developed through this course, extensive practice with word problems is essential. We have included a wide variety of problems, knowing that teachers will pick and choose the ones most appropriate for their students.

The presentation of polynomial operations is fairly standard, which includes a chapter on factoring, gives the students a good foundation, using both trial-and-error and product-sum methods. The graphing chapter provides explanation and practice not only in graphing equations and inequalities, but also in writing equations from given information. Should time not allow coverage of all chapters in this text, the later chapters are a good resource for students going on to a typical Introduction to Algebra course. We believe that once students have completed our material, they can progress into more traditional algebra courses and successfully build on their new-found skills.

This text is an excellent resource for students. It includes extensive examples with complete solutions. Chapter tests at the end of each chapter can be used as practice. The cumulative review exercises are grouped according to type so a student can see the connection among, for example, addition of whole numbers, fractions, decimals, signed numbers, and expressions. The continuous emphasis on word problems and critical thinking carries into the Thought-Provoking Problems.

Supplements

This text is accompanied by a comprehensive set of supplements available to all adopters:

1. *Instructor's Manual*: The Instructor's Manual includes notes and approaches for each chapter, one readiness and three mastery tests for each chapter, and answers to all even-numbered problems in the text.

2. *Transparency Masters*: The Transparency Masters have been extensively class-tested. They correspond with the topics covered in each chapter and provide full explanations of key concepts, and problems for the students to work during the instructor's presentation.

3. *EXPTest*: The EXPTest (available for IBM-PCs and compatibles on both 5 1/4" and 3 1/2" disks) is a computerized test bank containing hundreds of questions keyed specifically to the text from which instructors can choose, delete, and edit multiple choice, true-false, and open-ended questions. Additional questions may also be added for a more customized test.

4. *EXAMbuilder*: The EXAMbuilder is a computerized test bank similar to EXPTest, available for Macintosh users.

5. *Developmental Mathematics Video Series*: These videos are available for qualified adopters. Through the departmental or college audiovisual library, students can check out these videos and use them to review material when they need additional help.

6. *Tutorial Software*: This text-specific, interactive tutorial software in Windows and Macintosh formats allows students to practice the skills taught in the textbook. For each topic in the text, the tutorial provides multiple-choice, true-false, and short-answer exercises. The student is presented with a step-by-step solution to incorrectly answered problems. The program keeps track of right and wrong responses, and can report to the instructor on the students' progress.

Acknowledgments

We would like to thank our reviewers, who have been so encouraging and generous in giving their time and expertise to suggest changes that improved these materials. These educators have played an important part in helping to develop this text:

Marilyn Burton
Columbus State Community College

Lucinda Schweller
Wright State University

Linda Duttlinger
Purdue University, North Central

Joan Spivey
Normandale Community College

Gladys M. Kerr
Wingate College

Lana Taylor
Siena Heights College

Peg Pankowski
Community College of Allegheny, South Campus

William T. Wheeler
Abraham Baldwin College

We would also like to thank our editors, Al Bruckner, and Tim Anderson, Susan Gay, Kirby Lozyniak, and the staff at PWS. They have all been helpful and supportive during this very interesting experience. Their expertise has added to the quality of our efforts and we greatly appreciate their time and talent.

We hope that you enjoy working with this text. Our faculty and students have. Many students have said that it is the first time that they really understood a math book. We hope that your experiences with this text are at least as good as ours have been.

Patricia J. Cass
Elizabeth R. O'Connor

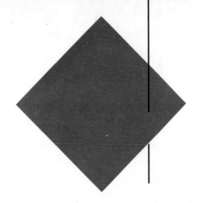

To the Student

In order to succeed in this or any math course, you must start from the same point as every other student in your class and use the same vocabulary and the same rules. Since each of your past math experiences has been different, we need to begin with a few basics to insure that everyone is starting with a solid foundation. It is likely that many of the topics in these first few chapters will be familiar to you. Nevertheless, please read each section as if the material was new to you. Pay attention to details. Underline the important statements and outline each topic. If you study all of the sections, you will begin to fill in the gaps which exist in your own background. You will see how topics are related and you can progress with confidence. The topics which are truly review will bring quick success. The weaker areas will be repaired and reinforced and will soon become sources of confidence as well.

In spite of your past experiences with math texts, we urge you to READ this one. It is written with you, the adult learner, in mind. It doesn't matter if you are an adult of 17 or an adult of 45. You can pick things up quickly if they are explained clearly. You don't want to waste your time. You want to learn the facts and how to apply them so you can move on to a successful working knowledge of math. To help you accomplish this, we have included many example problems with detailed explanations. Chapter objectives are stated clearly at the beginning of each chapter and are summarized in the chapter review. Many practice problems in a variety of types and real-life applications are included, with the odd-numbered answers at the back of the book. Ask your instructor if the videotape and computer software supplements are available on your campus.

Whether or not you have had difficulty with mathematics courses in the past, we urge you to read and use the Appendix materials. They are not in the back of the book because we thought that they were unimportant, but rather because they are support sections for anyone learning math. You will find help in the Appendices to develop good study skills and some suggestions for overcoming math anxiety.

We wish you success in your mathematical endeavors, and hope that this text helps you to accomplish your goals.

Patricia J. Cass
Elizabeth R. O'Connor

FUNDAMENTALS
with Elements of Algebra

About Whole Numbers

1.1 Introduction

In this chapter, we will cover many topics that are very elementary (maybe even boring). Adding, subtracting, and multiplying whole numbers are skills that most students take for granted. Here you are expected only to review and rebuild these skills. Long division, however, is one topic that a number of adult students find difficult. Consequently, we will spend time going through division in detail. The official rules, called **properties**, that govern our dealings with numbers will also be discussed. You use these properties all the time but may not be familiar with their names. In this chapter also, your work with word problems will begin. These application problems are designed to make mathematics real. The step-by-step approach introduced in this chapter will be used for word problems throughout the text.

Learning Objectives

When you have completed this chapter, you should be able to:

1. Read, write, and round whole numbers.
2. Perform the four basic mathematical operations (addition, subtraction, multiplication, and division) using whole numbers.
3. Identify and make use of the following number properties:
   a. Commutative Properties of Addition and Multiplication
   b. Associative Properties of Addition and Multiplication
   c. Properties of Zero and One
   d. The Distributive Principle
4. Estimate the answers to some calculation problems.
5. Estimate answers to word problems.
6. Read word problems and solve them by using one or more of the four basic operations.
7. Find the mean, median, mode, and range of given data.
8. Interpret data represented on a graph.

1.2   Reading, Writing, and Rounding

It is not often that one writes a check for an amount as large as $4,208. Check writing, however, is one important activity where we write numbers in both word form and number form. Look at the following sample check and note where the word description and the numeric (number) designation appear on it.

No. _263_

_7/4_ 19 _90_   25-2/440

PAY TO THE
ORDER OF   _Tri-City Car Sales_ $ | 4,208.00 |

        Four Thousand Two Hundred Eight and °°/100   DOLLARS

**THE FIRST
NATIONAL BANK**
Columbus, Ohio 43260

MEMO _____   _Jane L. Edwards_

⑆044000024⑆   03898851946⑈

The ability to translate from word to number form is directly related to **place value,** so that is where we must begin.

Look at the chart in Figure 1. In this chapter, you will use only whole numbers. Whole numbers are often thought of as the counting numbers, or the natural numbers and zero. The **whole numbers,** then, are 0, 1, 2, 3, 4, and so on. Fractions, decimals, and all numbers less than zero are not included among the whole numbers. In the example of the check, there were no cents involved, and only zeros followed the decimal point. For now, we will look only at digits that fall to the left side of the decimal point: only whole dollars in the check example, only whole numbers in the other problems.

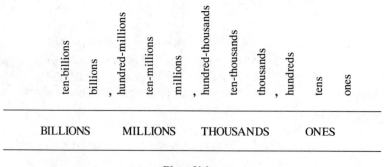

BILLIONS   MILLIONS   THOUSANDS   ONES

Place Value

FIGURE 1

Note that when we read from the right side of the number, the names for place value fall naturally into groups of three. Each of these groups of three is called a **period.** Each period is separated from the other groups by a comma. The three place values in the thousands period, for example, are all grouped together. It is the same for the millions, the billions, and so on. This is how knowledge of place value makes writing numbers in numeric form possible.

E X A M P L E **1** Write eight thousand, two hundred sixteen in digits.

Solution Because you know you need eight thousand, write 8 in the thousands place and list the remaining three spaces.

8, ⎯ ⎯ ⎯

Fill in two hundred sixteen, 216, in the blank spaces.

8,216 ◆

Note that the commas in the verbal statement and the commas in the number statement are written in the same place. Counting from right to left, commas are placed after each group of three digits. The commas separate the periods, or blocks of numbers, in both word and numeric form. Figure 2, where each digit is written under the correct column in the place value chart, also shows how the place value names correspond to placement of the digits in the number.

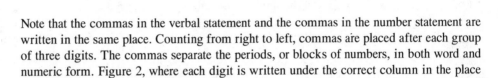

F I G U R E **2**

E X A M P L E **2** Write four million, twelve thousand, seven in digits.

Solution You need two periods of three spaces each to the right of the millions place. Write 4 in the millions place and indicate the other places that need to be filled.

4, ⎯ ⎯ ⎯ , ⎯ ⎯ ⎯

Twelve thousand must be positioned in the thousands period. You have only two digits plus an extra space, so you will fill that space with a place-holding zero. This zero needs to go before the 12. If the zero were placed between the 1 and the 2, the number in the thousands period would be 102. If the zero were placed after the 12, it would be 120. Position of the place-holding zeros is very important.

4,012, ⎯ ⎯ ⎯

The seven, 7, must be placed in the ones space. Fill in the blank spaces with place-holding zeros. You cannot write 070 because that would be seventy. And you cannot put both zeros after the 7 because that would yield seven hundred.

4,012,007

Therefore, four million, twelve thousand, seven is written 4,012,007. Figure 3 shows how place value and the words and the digits all fit together for this number. ◆

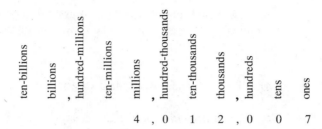

FIGURE 3

Always go back and read the number as you have written it to make sure that you wrote what was asked for in the problem. That is an easy way to check your answer.

EXAMPLE 3   Write fourteen thousand, eighty in digits.

Solution   Write 14 in the thousands period and follow it with the remaining three blanks.

14, __ __ __

Put 80 into the remaining period, filling in the blank space with a place-holding zero.

14,080

Therefore, fourteen thousand, eighty is written 14,080. ◆

EXAMPLE 4   Write thirty-seven billion, four thousand, seventeen in digits.

Solution   Write 37 in the billions period and list the spaces for millions, thousands, and ones.

37, __ __ __ , __ __ __ , __ __ __

Write 4 in the thousands place and 17 in the ones period.

37, __ __ __ , __ __ 4, __ 17

Fill in the blanks with place-holding zeros.

37,000,004,017

Therefore, thirty-seven billion, four thousand, seventeen is written 37,000,004,017. ◆

Most numbers, of course, are not so large, but the length of the number actually makes no difference. Now that you understand how the periods are used, you can break every number down into groups of at most three digits.

The reverse process—going from numbers to words—is also possible if you break the number down into groups of three digits and use the period (block of three) idea. But before you can correctly write the numbers in words, you need to know which numbers are hyphenated. All numbers between twenty and one hundred that are read as a combination of two single numbers are written with a hyphen. For instance, twenty-one, forty-three, and ninety-six are all hyphenated. Study the following examples to see how this works.

EXAMPLE 5   Write 3,264,581 in words.

Solution   The first period on the left, millions, has only one digit, a 3, so it is read

"three million"

The next period, thousands, uses all three spaces and is read

"two hundred sixty-four thousand"

The third period, ones, uses all three spaces and is read

"five hundred eighty-one"

Therefore, 3,264,581 is written

"three million, two hundred sixty-four thousand, five hundred eighty-one" ◆

Note that we do not use the word *and* in reading whole numbers. *And* stands for the decimal point, which separates the whole numbers from values that are less than one. For example, $26.17 is read "twenty-six dollars *and* seventeen cents." Because we are interested only in whole numbers right now, *and* has no place here.

E X A M P L E **6**   Write 75,009 in words.

Solution   The first period on the left side, thousands, uses only two digits, 7 and 5. It is read "seventy-five thousand." The second period, ones, has only one digit, 9, and that digit is located in the ones place. It is read "nine." Therefore, 75,009 is read "seventy-five thousand nine." Remember, no *and*. ◆

E X A M P L E **7**   Write 203,014,085 in words.

Solution   Two hundred three million, fourteen thousand, eighty-five ◆

Rounding numbers also requires knowledge of place value. When we round in everyday situations, $19.89 becomes $20, and a gallon of gasoline for 92⁹ becomes 93 cents a gallon. When we round in a mathematical sense, however, we must follow certain rules. Consider this problem.

Round 24,325 to the nearest *thousand*.

The question really is asking, "Is 24,325 closer to 24,000 or to 25,000?" Asked that way, it is easy to answer. It is closer to 24,000.

To determine the answer to this type of problem, it is helpful to follow these steps:

1. Mark the digit that occupies the place to which you are to round.
2. Underline the number to its right.
3. If the underlined number is less than 5, leave the digit in question unchanged, and replace all the digits that follow it with zeros.
4. If the underlined number is 5 or more, increase the digit in question by one and replace all the digits that follow it with zeros.

E X A M P L E **8**   Round 24,325 to the nearest *thousand*.

Solution   24,325   4 is the digit in the thousands place. Mark it.

24,325   3 is the digit to the right of that 4. Underline it.

24,000   Because 3 is less than 5, leave the digit in the thousands place as it is and replace all the remaining digits with zeros.

The answer you found by following the steps agrees with the answer you determined before "by inspection." ◆

E X A M P L E   **9**   Round 22,876 to the nearest *hundred*.

Solution

22,876	8 is the digit in the hundreds place. Mark it.
22,876	7 is the digit to the right of that 8. Underline it.
22,900	Because 7 is greater than 5, increase the 8 by one and replace all the remaining digits with zeros. ◆

E X A M P L E   **10**   Round 1,054,839 to the nearest *hundred-thousand*.

Solution

1,054,839	0 is the digit in the hundred-thousands place. Mark it.
1,054,839	5 is the digit to the right of that 0. Underline it.
1,100,000	Because the underlined number is 5, increase the 0 by one and replace all the remaining digits with zeros. ◆

E X A M P L E   **11**   Round 186,925 to the nearest *hundred*.

Solution

186,925	9 is the digit in the hundreds place. Mark it.
186,925	2 is the digit to the right of that 9. Underline it.
186,900	Because 2 is less than 5, leave the 9 as it is and replace all the remaining digits with zeros. ◆

E X A M P L E   **12**   Round 19,832 to the nearest *thousand*.

Solution

19,832	9 is the digit in the thousands place. Mark it.
19,832	8 is the digit to the right of that 9. Underline it.
20,000	Because 8 is more than 5, increase the 19 by one and replace all the remaining digits with zeros. **Note that when a 9 is the number in question, and it needs to be increased, the change also involves the digit to the left of the 9.** ◆

E X A M P L E   **13**   Round 23,971 to the nearest *hundred*.

Solution

23,971	9 is the digit in the hundreds place. Mark it.
23,971	7 is the digit to the right of that 9. Underline it.
24,000	Because 7 is more than 5, increase the 239 by one to 240 and replace all the remaining digits with zeros. ◆

E X A M P L E   **14**   Round $27,392,086 to the nearest *ten-thousand*.

Solution   Rounding dollar amounts is no different from any other rounding.

$27,392,086	9 is the digit in the ten-thousands place. Mark it.
$27,392,086	2 is the digit to the right of that 9. Underline it.
$27,390,000	Because 2 is less than 5, leave the 9 as it is and replace all the remaining digits with zeros. ◆

1.2 ◆ Practice Problems

A. Write each of the following expressions in numeric form.

1. eight thousand, two hundred sixty-five

2. nine thousand, four hundred seventy-two

3. six million, fourteen thousand, three hundred eighty-five

4. eight million, twenty-six thousand, one hundred seventeen

5. forty-eight million, two thousand, eighty-five

6. seventy-three million, nineteen thousand, forty-seven

7. eight billion, four thousand, seven

8. nine billion, seven million, twelve

9. twenty-nine thousand, four hundred sixty

10. fifty-six thousand, twenty-nine

11. forty-six million, nine hundred eighty thousand, two

12. ninety million, thirty-six thousand, seven hundred four

13. seventy-two million, four hundred eighty thousand

14. fourteen billion, six hundred twenty-four thousand

15. eight million, six thousand, forty-three

16. twenty million, five thousand, two hundred twenty-four

17. two billion, forty-five thousand, one hundred fifty-two

18. three million, twelve thousand, fifteen

19. two hundred million, eighty-five thousand, two hundred six

20. three thousand, two hundred fifteen

B. Write each of the following numbers in words.

21. 16,205

22. 34,509

23. 8,914

24. 5,423

25. 2,009

26. 8,017

27. 70,070

28. 56,060

29. 1,234,567

30. 9,876,543

31. 14,008

32. 51,007

33. 12,014,235,006

34. 21,415,006,140

35. 35,000,000,008

36. 47,000,080,000

37. 9,081,060

38. 6,170,003

39. 76,008

40. 45,030

C. Round each of these numbers to the place requested.

41. 14,276 to the nearest hundred

42. 37,581 to the nearest hundred

43. 1,526 to the nearest thousand

44. 8,568 to the nearest thousand

45. 18,623 to the nearest hundred

46. 37,419 to the nearest hundred

47. 32,999 to the nearest hundred

48. 57,999 to the nearest hundred

49. 42,396 to the nearest ten

50. 92,598 to the nearest ten

51. 24,311 to the nearest thousand

52. 72,141 to the nearest thousand

53. 19,876 to the nearest thousand

54. 39,712 to the nearest thousand

55. 114,912 to the nearest ten-thousand

56. 315,014 to the nearest ten-thousand

57. 37,065,214 to the nearest ten-thousand

58. 86,396,021 to the nearest ten-thousand

59. 219,875 to the nearest hundred-thousand

60. 476,810 to the nearest hundred-thousand

D. Round each of these dollar amounts to the place requested.

61. $4,376 to the nearest hundred

62. $1,263,545 to the nearest thousand

63. $36,264 to the nearest thousand

64. $915,337 to the nearest hundred

65. $27,543,260 to the nearest million

66. $19,620,249 to the nearest million

1.3 ◆ Three Whole Number Operations

The use of calculators has made doing computations "in your head" almost a thing of the past. Many people have become so dependent on their calculators that they have almost lost their own computational skills. You should do the calculations in this section using only your head and paper and pencil—not by using a calculator. You may certainly use your fingers to count on and to "carry," but don't use them to push buttons! Once you have found the answer to a problem, go ahead and check your answers in the answer key at the back of the text. And if your instructor agrees, it is appropriate to check your work using your calculator. Most of the problems in this text are not so difficult that a calculator will be necessary. Furthermore, there are many situations—working with fractions, for one—where calculators are of little benefit, and your personal skills must be used. If you have not used your own calculation skills in a long time, you will be amazed at how quickly you can work the rust off them with a little serious effort.

Addition   Given enough fingers and toes, anyone can do an addition problem successfully. Addition is the most basic of the mathematical operations. There is nothing wrong with using your fingers to help you add, but it is easier and faster to do it in your head. The only requirement for addition is that quantities must be combined with *like quantities:* thousands with thousands, hundreds with hundreds, ones with ones, and so on. If you always line up the numbers to be added (called **addends**) so that the ones values are under each other, the rest is easy. Study these examples for your review.

E X A M P L E   1   Add 2,304, 87, and 426.

Solution   A good approach in problem solving is first to estimate the answer by combining easy-to-add numbers such as 2,300, 100, and 400 that are close to the actual numbers. When you do this addition, you get an estimate of 2,800. Your actual answer should be close to that.

$$
\begin{array}{r}
2,304 \\
87 \\
+\quad 426 \\
\end{array}
$$
Line up the numbers vertically so that the digits in the ones place (4, 7, and 6) are under each other.

$$
\begin{array}{r}
{\scriptstyle 1\,1} \\
2,304 \\
87 \\
+\quad 426 \\
\hline
2,817 \\
\end{array}
$$
Add each column, beginning with the ones and carrying numbers over to the next column as required.

The estimate of 2,800 was very close to the actual answer of 2,817.

Check
$$
\begin{array}{r}
{\scriptstyle 1\,1} \\
2,304 \\
426 \\
+\quad 87 \\
\hline
2,817 \checkmark \\
\end{array}
$$
You can check an addition problem that has three or more addends by arranging them in a different order. You can rewrite the problem, as is shown here. Or, if you normally add from top to bottom, you can add from the bottom to the top.   ◆

E X A M P L E   2   Find the sum of 91, and 384, and 78.

Solution   An estimate might be 100 + 400 + 100, or 600.

$$\begin{array}{r} {\scriptstyle 2\,1} \\ 91 \\ 384 \\ +\ \ 78 \\ \hline 553 \end{array}$$

The word *sum* indicates addition. Line up the digits in the ones place: 1, 4, and 8. Add, and carry to the next place as necessary.

Check
$$\begin{array}{r} 384 \\ 78 \\ +\ \ 91 \\ \hline 553 \ \checkmark \end{array}$$

Check by rearranging the addends and adding again.

◆

E X A M P L E **3** Find the total of 19, 3,804, and 198.

Solution
$$\begin{array}{r} {\scriptstyle 1\,1\,2} \\ 19 \\ 3{,}804 \\ +\ \ 198 \\ \hline 4{,}021 \end{array}$$

The word *total* indicates addition. Line up the 9, 4, and 8. Add, and carry as required.

Check
$$\begin{array}{r} 3{,}804 \\ 198 \\ +\ \ \ 19 \\ \hline 4{,}021 \ \checkmark \end{array}$$

Check by rearranging the addends and adding again.

◆

You could have first estimated the answer by adding 20, 3,800, and 200. The result would have been 4,020, which is very close to the actual answer, 4,021.

Subtraction Subtraction also requires keeping the place values lined up properly. Begin with the ones place and line up the digits. Often when subtracting, you will need to **borrow**. Borrowing will be discussed in the examples where it is necessary. You can check subtraction problems by adding, because addition and subtraction are opposite operations.

E X A M P L E **4** Subtract 623 from 8,897.

Solution Estimate an answer by subtracting 600 from 9,000.

$$\begin{array}{r} 8{,}897 \\ -\ \ 623 \end{array}$$

The number to be subtracted *from* goes on top. Line up the 7 and the 3.

$$\begin{array}{r} 8{,}897 \\ -\ \ 623 \\ \hline 8{,}274 \end{array}$$

Subtract, beginning with the ones column and moving to the left.
3 from 7 is 4.
2 from 9 is 7.
6 from 8 is 2.
Nothing from 8 is 8.

Check
$$\begin{array}{r} 8{,}897 \\ -\ \ 623 \\ \hline +\ 8{,}274 \\ \hline 8{,}897 \ \checkmark \end{array}$$

Check the solution by adding 623 and 8,274. 8,897 is the same as 8,897, so the answer is right.

◆

E X A M P L E **5**   Find the difference between 17,821 and 9,460.

Solution

$$\begin{array}{r} 17,821 \\ -\ 9,460 \end{array}$$

The word *difference* indicates subtraction. Line up the 1 and the 0.

$$\begin{array}{r} 17,821 \\ -\ 9,460 \\ \hline 1 \end{array}$$

Subtract from right to left. 0 from 1 is 1.

$$\begin{array}{r} {}^{7\,1}\\ 17,\!821 \\ -\ 9,460 \\ \hline 61 \end{array}$$

Because 6 cannot be taken from 2, borrow 1 from the column to the left, making the 8 a 7, and creating a 12 in the column in which you are working. 6 from 12 is 6.

$$\begin{array}{r} {}^{7\,1}\\ 17,\!821 \\ -\ 9,460 \\ \hline 8,361 \end{array}$$

Continue to subtract.
4 from 7 is 3.
9 from 17 is 8.

Check

$$\begin{array}{r} 17,821 \\ -\ 9,460 \\ +\ 8,361 \\ \hline 17,821\ \checkmark \end{array}$$

Check the subtraction answer by adding.

◆

E X A M P L E **6**   Subtract 438 from 2,004.

Solution

You could estimate an answer by subtracting 400 from 2,000. The actual answer should be close to 1,600.

$$\begin{array}{r} 2,004 \\ -\ 438 \end{array}$$

Line up the digits in the ones column, 4 and 8.

$$\begin{array}{r} {}^{1\ 9\,9\,1}\\ 2,\!004 \\ -\ 438 \end{array}$$

You cannot subtract 8 from 4, and you must borrow 1 from 200 because you cannot borrow from 0. When you borrow 1 from 200, 199 remains.

$$\begin{array}{r} {}^{1\ 9\,9\,1}\\ 2,\!004 \\ -\ 438 \\ \hline 1,566 \end{array}$$

Continue to subtract.
8 from 14 is 6.
3 from 9 is 6.
4 from 9 is 5.
Nothing from 1 is 1.

Check

$$\begin{array}{r} 2,004 \\ -\ 438 \\ +\ 1,566 \\ \hline 2,004\ \checkmark \end{array}$$

Check by adding.

◆

E X A M P L E **7**   Subtract 2,684 from 30,000.

Solution

$$\begin{array}{r} 30,000 \\ -\ 2,684 \end{array}$$

Line up the digits in the ones column, 0 and 4.

$$\begin{array}{r} {}^{2\,9\ 9\,9\,1}\\ 30,\!000 \\ -\ 2,684 \end{array}$$

You cannot subtract 4 from 0, and you must borrow 1 from 3,000 because you cannot borrow 1 from 0. When you borrow 1 from 3,000, the number 2,999 remains.

$$\begin{array}{r} {\scriptstyle 29\ 991} \\ \cancel{30,000} \\ -\ \ 2,684 \\ \hline 27,316 \end{array}$$

Continue to subtract.
4 from 10 is 6.
8 from 9 is 1.
6 from 9 is 3.
2 from 9 is 7.
Nothing from 2 is 2.

Check

$$\begin{array}{r} 30,000 \\ -\ \ 2,684 \\ +\ 27,316 \\ \hline 30,000\ \sqrt{} \end{array}$$

Check by adding.

◆

E X A M P L E **8**   Find the difference between 70,165 and 18,476.

Solution and Check

$$\begin{array}{r} 70,165 \\ -\ 18,476 \\ +\ 51,689 \\ \hline 70,165\ \sqrt{} \end{array}$$

◆

Multiplication   Multiplication of whole numbers demands a firm knowledge of the basic multiplication facts. You must know without hesitation that $6 \times 7 = 42$ and that $9 \times 4 = 36$. You may not have used these basic facts for a while. Take the short multiplication review check that follows and see if you can answer these problems easily, quickly, and correctly. If you cannot, go to the Skills Check (Appendix C of this text) and do the exercises suggested to help you rebuild this skill. It will not take a lot of time, but the effort you invest *now* will pay off in the rest of your math work.

Try to do all the problems in 2 minutes, and then check your answers against the answers that follow. Be honest with yourself and work on this skill now if you need to brush up.

$4 \times 7 =$	$9 \times 3 =$	$0 \times 5 =$	$11 \times 10 =$	$8 \times 5 =$
$6 \times 4 =$	$7 \times 8 =$	$4 \times 8 =$	$3 \times 9 =$	$7 \times 6 =$
$12 \times 6 =$	$2 \times 7 =$	$2 \times 6 =$	$4 \times 1 =$	$11 \times 1 =$
$0 \times 4 =$	$12 \times 9 =$	$10 \times 3 =$	$7 \times 3 =$	$4 \times 0 =$
$7 \times 9 =$	$5 \times 7 =$	$5 \times 9 =$	$11 \times 12 =$	$8 \times 3 =$
$6 \times 3 =$	$12 \times 8 =$	$10 \times 7 =$	$1 \times 7 =$	$8 \times 9 =$
$3 \times 5 =$	$6 \times 7 =$	$4 \times 3 =$	$9 \times 5 =$	$0 \times 9 =$
$12 \times 5 =$	$6 \times 0 =$	$11 \times 4 =$	$6 \times 8 =$	$4 \times 9 =$

Multiplication answers:

28	27	0	110	40
24	56	32	27	42
72	14	12	4	11
0	108	30	21	0
63	35	45	132	24
18	96	70	7	72
15	42	12	45	0
60	0	44	48	36

Multiplication, like addition, requires paying careful attention to place value and to carrying. The **factors** (the numbers that are multiplied) are lined up vertically so that the ones values are under each other, just as in addition. Most students prefer to put the larger factor on top, but that is not necessary. It usually makes the multiplication easier, though, because you are multiplying by fewer digits. Study these examples.

E X A M P L E **9**   Find the product of 473 and 85.

Solution

473
× 85

The word *product* indicates multiplication. Line up the ones values, 3 and 5.

3 1
473
× 85
2365

Multiply by the 5. Because 5 is located in the first column, the first "partial product" begins in the first column. Carry as required.
$5 \times 3 = 15$. Write 5, carry 1.
$5 \times 7 = 35 + 1 = 36$. Write 6, carry 3.
$5 \times 4 = 20 + 3 = 23$. Write 23.

5 2
473
× 85
2365
3784

Now multiply by the 8. It is located in the second column, so this partial product begins in the second column. Carry as needed.
$8 \times 3 = 24$. Write 4, carry 2.
$8 \times 7 = 56 + 2 = 58$. Write 8, carry 5.
$8 \times 4 = 32 + 5 = 37$. Write 37.

473
× 85
2365
+3784
40205

Finally, add the partial products. Carry in the addition as required.
$5 + \text{nothing} = 5$.
$6 + 4 = 10$. Write 0, carry 1.
$1 + 3 + 8 = 12$. Write 2, carry 1.
$1 + 2 + 7 = 10$. Write 0, carry 1.
$1 + 3 = 4$. Write 4.

$473 \times 85 = 40{,}205$ ◆

There are two ways to check a multiplication problem. One is to rearrange the factors and multiply again. The partial products will be different, but the final product will be the same. The second way to check multiplication is to divide the final product by one of the factors. The **quotient** (answer) in the division should be the other factor. This division should be easier than most long division problems, because you already have an idea of which numbers should appear in the quotient. Both methods are shown here. Of course, you could use your calculator, instead of paper and pencil, to do these checking calculations.

```
      85                    473 √
×    473              85)40205
     255                − 340
     595                  620
     340                − 595
   40205 √                255
                        − 255
```

E X A M P L E **10**   Multiply 2483 by 79.

Solution   You could estimate the answer by multiplying 2,500 by 80 and know that your answer should be close to 200,000.

4 7 2
2483
×   79
22347

Line up the digits in the ones column, 3 and 9. Multiply by 9, beginning the answer in the first column. Carry as required.
$9 \times 3 = 27$. Write 7, carry 2.
$9 \times 8 = 72 + 2 = 74$. Write 4, carry 7.
$9 \times 4 = 36 + 7 = 43$. Write 3, carry 4.
$9 \times 2 = 18 + 4 = 22$. Write 22.

$$
\begin{array}{r}
{}^{3\,5\,2}\\
2483\\
\times \quad 79\\
\hline
22347\\
+\ 17381\\
\hline
196157
\end{array}
$$

Now multiply by 7. Since 7 is in the second column, the answer begins in that column.

$7 \times 3 = 21$. Write 1, carry 2.

$7 \times 8 = 56 + 2 = 58$. Write 8, carry 5.

$7 \times 4 = 28 + 5 = 33$. Write 3, carry 3.

$7 \times 2 = 14 + 3 = 17$. Write 17.

Add the partial products. Carry as required.

$2,483 \times 79 = 196,157$

The actual answer of 196,157 is very close to the estimate of 200,000.

Check Check your answer. Two methods are shown. The checks should be the same if you are using a calculator.

$$
\begin{array}{r}
79\\
\times \quad 2483\\
\hline
237\\
632\\
316\\
158\\
\hline
196157\ \checkmark
\end{array}
\qquad
\begin{array}{r}
2483\ \checkmark\\
79\overline{)196157}\\
-158\\
\hline
381\\
-316\\
\hline
655\\
-632\\
\hline
237\\
-237
\end{array}
$$

◆

It is time-consuming to check multiplication problems, so you should do so only if you are not sure you solved the problem correctly.

E X A M P L E 11 Find the product of 9,371 and 407.

Solution

$$
\begin{array}{r}
{}^{2\,4}\\
9371\\
\times \quad 407\\
\hline
65597
\end{array}
$$

The word *product* indicates multiplication. Line up the factors that occupy the ones place. Multiply by 7 first.

When you multiply by 0, the result is always 0. You can indicate this product in two different ways. You can insert a whole row of zeros, or you can just put a zero in the second column and then multiply by the 4, beginning your answer in the *third* column. Both ways are shown.

$$
\begin{array}{r}
{}^{1\,2}\\
{}^{2\,4}\\
9371\\
\times \quad 407\\
\hline
65597\\
0000\\
37484\\
\hline
3813997
\end{array}
\qquad
\begin{array}{r}
{}^{1\,2}\\
{}^{2\,4}\\
9371\\
\times \quad 407\\
\hline
65597\\
374840\\
\hline
3813997
\end{array}
$$

$9,371 \times 407 = 3,813,997$

◆

E X A M P L E   **12**   Find the product of 48 and 206.

Solution   You could estimate your answer by finding the product of 50 and 200. The approximate answer is 10,000.

$$\begin{array}{r} \overset{2}{}\overset{4}{} \\ 206 \\ \times\quad 48 \\ \hline 1648 \\ 824 \\ \hline 9888 \end{array}$$

The word *product* indicates multiplication. As the checking process in Examples 9 and 10 shows, you can line up the factors with either number on top, but it is easier to multiply twice instead of three times. Remember, $8 \times 0 = 0$.

$48 \times 206 = 9{,}888$ ◆

E X A M P L E   **13**   Find the product of 2,615 and 3,009.

Solution   You could estimate the answer to be 7,800,000 by multiplying 2,600 by 3,000. In this problem, you need to be careful to place the partial products correctly when multiplying by the zeros.

$$\begin{array}{r} \overset{1\ \ 1}{\overset{5\,1\,4}{2615}} \\ \times\quad 3009 \\ \hline 23535 \\ 0000 \\ 0000 \\ 7845 \\ \hline 7868535 \end{array} \qquad \begin{array}{r} \overset{1\ \ 1}{\overset{5\,1\,4}{2615}} \\ \times\quad 3009 \\ \hline 23535 \\ 7845000 \\ \hline 7868535 \end{array}$$

$2{,}615 \times 3{,}009 = 7{,}868{,}535$ ◆

E X A M P L E   **14**   From the sum of 1,289 and 3,046 subtract 986.

Solution   You need to add the numbers 1,289 and 3,046 and then subtract 986 from that answer.

$$\begin{array}{r} 1{,}289 \\ +3{,}046 \\ \hline 4{,}335 \end{array} \qquad \begin{array}{r} 4{,}335 \\ -\ 986 \\ \hline 3{,}349 \end{array}$$ ◆

1.3 ◆ Practice Problems

A. Perform the indicated operations. Whenever practical, check your answers by the methods suggested in this section. Use a calculator to check your answers if your instructor suggests that you do so.

1. $\begin{array}{r} 245 \\ +566 \\ \hline \end{array}$

2. $\begin{array}{r} 372 \\ 29 \\ +495 \\ \hline \end{array}$

3. $\begin{array}{r} 59 \\ 178 \\ +\ 35 \\ \hline \end{array}$

4. $\begin{array}{r} 614 \\ +\ 92 \\ \hline \end{array}$

5. $\begin{array}{r} 3{,}125 \\ -\ 256 \\ \hline \end{array}$

6. $\begin{array}{r} 801 \\ -\ 45 \\ \hline \end{array}$

7. $\begin{array}{r} 776 \\ -125 \\ \hline \end{array}$

8. $\begin{array}{r} 1{,}254 \\ -\ 363 \\ \hline \end{array}$

9. $\begin{array}{r} 856 \\ \times\ 37 \\ \hline \end{array}$

10. $\begin{array}{r} 579 \\ \times\ 58 \\ \hline \end{array}$

11. $\begin{array}{r} 1{,}892 \\ \times\ 92 \\ \hline \end{array}$

12. $\begin{array}{r} 297 \\ \times\ 46 \\ \hline \end{array}$

13. $725 - 88$

14. $8{,}921 + 726$

15. 149×8

16. 497×6

17. $\begin{array}{r} 20{,}007 \\ -\ 8{,}564 \\ \hline \end{array}$

18. $\begin{array}{r} 4{,}098 \\ \times\ 306 \\ \hline \end{array}$

19. 6,078 847 +4,651	**20.** 2,056 −1,778	**21.** 4,003 − 875
22. 486 × 39	**23.** 1,756 326 + 28	**24.** 895 3,046 + 57
25. 7,631 − 987	**26.** 15,080 − 8,795	**27.** 348 49 +806
28. 1,006 × 278		

29. Find the difference between 1,587 and 849.

30. What is the product of 2,016 and 83?

31. Find the total of 2,985, 384, and 1,912.

32. From the sum of 3,548 and 2,095 take 3,296.

33. Multiply the sum of 1,483 and 982 by 14.

34. From the product of 48 and 32 take 893.

35. Find the difference between 54,008 and 27,519.

36. Find the product of 431, 5, and 22.

37. Subtract 578 from 2,113.

38. Find the difference between 62,000 and 8,747.

39. Find the product of 387 and 78.

40. Subtract 5,376 from 10,002.

B. **Use your calculator to find the answers to the following problems.**

41. 204,965 371,508 + 734,918	**42.** 16,243,002 − 9,875,799
43. 17,395 × 2,645	**44.** 20,034 × 857
45. 40,000,031 − 29,998,754	**46.** 1,836,752 39,098 + 157,489

47. Find the sum of 17,384 and 209,195.

48. Find the difference between 198,000 and 99,493.

49. Find the product of 16,453 and 975.

50. From the product of 2,493 and 26,548 subtract 43,821.

1.4 Whole Number Division

Division can be indicated in one of four ways:

$$6\overline{)4{,}521} \qquad 4{,}521 \div 6 \qquad \frac{4{,}521}{6} \qquad \text{“Find the quotient of 4,521 divided by 6.”}$$

The number that does the dividing is called the **divisor**. The number into which the divisor goes is called the **dividend**. The answer that you get when you divide is called the **quotient**. If the division does not come out evenly, there is a **remainder**.

$$\overset{\text{quotient} \quad \text{remainder}}{\text{divisor}\overline{)\text{dividend}}} \qquad \text{dividend} \div \text{divisor} \qquad \frac{\text{dividend}}{\text{divisor}}$$

Dividing by a single digit is the easiest to do. Carefully studying one of these example problems will serve as a review of the basics of division.

E X A M P L E **1** Divide 4,521 by 6.

Solution You could estimate an answer by dividing 4,800 by 6, because 6 divides easily into 48. The actual answer should be close to 800.

$$\begin{array}{r} 753 \\ 6\overline{)4521} \\ -42 \\ \hline 32 \\ -30 \\ \hline 21 \\ -18 \\ \hline 3 \end{array}$$

6 will not go into 4, but 6 will go into 45.

6 × 7 = 42. Place the 7 over the 5.

45 − 42 = 3. Bring down the next digit, 2.

6 × 5 = 30. Place the 5 over the 2.

32 − 30 = 2. Bring down the next digit, 1.

6 × 3 = 18. Place the 3 over the 1.

There are no more digits, so there is a remainder of 3. Whenever there is a remainder, it must be written as part of the quotient. Otherwise, someone looking at the answer might think that the number divided exactly (with a zero remainder).

$$4{,}521 \div 6 = 753 \text{ R}3$$

The actual answer is, in fact, close to the estimate, 800.

Check   To check the division, multiply 753 by 6 and add 3.

$$753 \times 6 = 4{,}518 \qquad 4{,}518 + 3 = 4{,}521 \checkmark$$

◆

E X A M P L E 2   Find the quotient of 3,652 divided by 8.

Solution

```
      456   R4
  8)3652
   − 32
     45
    − 40
      52
     − 48
       4
```

Check

```
      456
  ×     8
     3648
  +     4
     3652 √
```

◆

E X A M P L E 3 $21{,}341 \div 7 = ?$

Solution

```
     3048   R5
 7)21341
  − 21
    03
   − 00
     34
    − 28
      61
     − 56
       5
```

This problem has a new situation. When you bring the 3 down, the divisor, 7, will not divide into 3. Before you can continue, you must show that 03 ÷ 7 has no natural number answer. Place a zero in the quotient above the 3, multiply 0 by 7, and write down the product, 00. Subtract, getting 3. Then you can bring down the next digit, 4, and continue to divide 7 into 34 and so on. When you have run out of numbers in the dividend in a whole number division problem, you are finished. If the final remainder is not zero, your answer must include that remainder.

Check

```
     3048
 ×      7
    21336
 +      5
    21341 √
```

◆

Dividing by a number with more than one digit follows the same pattern as dividing by a single digit, but the calculations are longer. Try to do the following example on your own paper first. Then follow the worked-out steps and see if your process agrees with these steps.

E X A M P L E **4**   Divide 3,815 by 26.

Solution   You could estimate an answer by dividing 3,900 by 30. This shows that the answer in the original problem should be close to 1,300.

$$
\begin{array}{r}
1 \\
26\overline{)3815} \\
-\ 26 \\
\hline
121
\end{array}
$$

26 goes into 38 one time. Write 1 over the 8.
$1 \times 26 = 26$.
$38 - 26 = 12$. 12 is less than 26. This step is extremely important. Each time you subtract, be sure to check to see that *the number you get is less than the divisor.* If it is not, erase the number in the quotient and try a larger number. 12 is less than 26. OK. Bring down the next digit, 1.

$$
\begin{array}{r}
14 \\
26\overline{)3815} \\
-\ 26 \\
\hline
121 \\
-\ 104 \\
\hline
17
\end{array}
$$

You can guess that 26 goes into 121 four or five times because $30 \times 4 = 120$ and $30 \times 5 = 150$.
Try 5. $5 \times 26 = 130$. Too big.
Try 4. $4 \times 26 = 104$. $121 - 104 = 17$.
17 is less than 26. OK.

$$
\begin{array}{r}
146 \quad \text{R19} \\
26\overline{)3815} \\
-\ 26 \\
\hline
121 \\
-\ 104 \\
\hline
175 \\
-\ 156 \\
\hline
19
\end{array}
$$

Bring down the 5.
You know that $6 \times 30 = 180$, which is close to 175, so try $6 \times 26 = 156$. Place 6 above the 5 and see if, when you subtract $175 - 156$, your remainder is less than the divisor, 26. $175 - 156 = 19$. OK. Write the remainder in the quotient. The answer, 146 R19, is close to the estimate of 130.

Check
$$
\begin{array}{r}
146 \\
\times\ \ 26 \\
\hline
876 \\
292 \\
\hline
3796 \\
+\ \ 19 \\
\hline
3815 \ \sqrt{}
\end{array}
$$

Check the exact answer by multiplying. Remember to add on the remainder, 19, in the form R19.

Therefore, 3,815 divided by 26 equals 146 R19.   ◆

E X A M P L E **5**   Find the quotient of 6,204 divided by 73.

Solution   Estimate an answer: $6,300 \div 70 = 90$

$$
\begin{array}{r}
8 \\
73\overline{)6204} \\
-\ 584 \\
\hline
36
\end{array}
$$

The word *quotient* indicates division. 73 will not go into 6. 73 will not go into 62. But 73 will go into 620.
$70 \times 8 = 560$, so we can try 8.
$8 \times 73 = 584$, which is less than 620.
$620 - 584 = 36$. 36 is less than 73. OK.

$$\begin{array}{r} 84 \quad R72 \\ 73\overline{)6204} \\ -\ 584 \\ \hline 364 \\ -\ 292 \\ \hline 72 \end{array}$$

Bring down the 4.
$70 \times 5 = 350$, so we can try 5.
$73 \times 5 = 365$. 365 is bigger than 364.
$73 \times 4 = 292$. $364 - 292 = 72$.
72 is less than 73. OK.
Write the remainder in the answer. The estimate is close, so 84 R72 is a reasonable answer to the problem.

Check

$$\begin{array}{r} 84 \\ \times\ 73 \\ \hline 252 \\ 588 \\ \hline 6132 \\ +\ 72 \\ \hline 6204 \ \checkmark \end{array}$$

Check by multiplication. Add the remainder.

Therefore, the quotient of 6,204 divided by 73 is 84 R72. ◆

As you do the four exercises that follow, try estimating the answer before you divide. Check your answers against those given below the exercises. If you have made an error, see if you can find it and correct it. It may help to know that the four most common errors in long division problems are

1. placing the first answer over the wrong digit

2. subtracting incorrectly

3. not spotting a remainder that is larger than the divisor (correcting this error requires that you increase the digit in the quotient and multiply again; be sure to check your remainder each time)

4. not placing a zero in the quotient *once you have brought a digit down* and realized that the divisor is larger than the number created

Exercises

1. $9{,}876 \div 42$ **2.** $56\overline{)61{,}152}$ **3.** $\dfrac{68{,}452}{18}$

4. Find the quotient of 147,729 divided by 68.

Answers:

1. 235 R6   **2.** 1,092   **3.** 3,802 R16

4. 2,172 R33

Hint: When you do division problems on a test, it is always a good idea to check your answer by multiplying the divisor by the quotient and adding the remainder if there is one.

1.4 ◆ Practice Problems

A. Divide as indicated in each of the following problems. Check your answers by multiplication by hand or by calculator.

1. $2{,}136 \div 8$ **2.** $3{,}213 \div 7$

3. $4{,}032 \div 6$ **4.** $17{,}810 \div 5$

5. $42{,}728 \div 7$ **6.** $74{,}754 \div 9$

7. $8{,}246 \div 38$ **8.** $15{,}687 \div 27$

9. $26{,}187 \div 43$ **10.** $1{,}435 \div 8$

11. $4{,}735 \div 9$ **12.** $6{,}381 \div 7$

13. $\dfrac{63{,}015}{61}$ **14.** $\dfrac{82{,}196}{34}$

15. $\dfrac{77{,}458}{29}$ **16.** $62\overline{)24{,}159}$

17. $77\overline{)53{,}215}$ **18.** $36\overline{)11{,}614}$

19. Find the quotient of 116,248 divided by 287.

20. Find the quotient of 320,064 divided by 215.

21. What is the result when 3,458 is divided by 88?

22. What is the result when 7,083 is divided by 45?

23. Divide 5,092 by 41.

24. Divide 8,176 by 64.

25. Find the quotient of 76,368 divided by 37.

26. What is the result when 51,815 is divided by 43?

27. Divide 8,020 by 26.

28. Find the quotient when 65,060 is divided by 13.

 B. Use your calculator to find the answers to the following problems.

29. $72{,}430{,}140 \div 20{,}036$

30. $6{,}368{,}670 \div 16{,}542$

31. $\dfrac{41{,}382{,}302}{2{,}351}$

32. $\dfrac{39{,}135{,}200}{5{,}680}$

33. $7{,}895\overline{)15{,}813{,}685}$

34. $3{,}006\overline{)26{,}975{,}844}$

35. Find the quotient of 18,293,658 divided by 5,001.

36. What is the result when 2,118,825 is divided by 3,225?

37. What is the result when 2,950,992 is divided by 8,019?

38. Find the quotient of 48,650,625 divided by 6,975.

39. Divide 1,056,784 by 1,028.

40. Divide 17,538,458 by 8,743.

1.5 Some Common Properties of Whole Numbers

When you are beginning any new project, it is important to review the basics and be sure you know the rules. It is difficult to follow any process if you don't understand what is going on from the very beginning. In this section, you are going to study some properties of numbers. These properties will not be new to you. You use them all the time in solving arithmetic problems. Now, however, you need to make sure you know what their names are and can identify and use them correctly.

Does it make the answer any different if you add $2 + 3$ or $3 + 2$? Of course not. You get 5 in either case. If you add $7 + 12$ and $12 + 7$, you get the same answer: 19. The fact that numbers can be added in any order reflects a number property called the **Commutative Property of Addition.** In formal, mathematical language, it is stated in this way:

Commutative Property of Addition

For all numbers represented by a and b,

$$a + b = b + a$$

You know from your own experience that $4 + 6 = 6 + 4$ and that $120 + 18 = 18 + 120$. The statement of the property uses a and b instead of numbers to show that any values would work, not just specific values such as 4 and 6.

The next property is like the first except that it applies to multiplication problems. What is 6×7 equal to? What is 7×6 equal to? Both answers are 42. What about 12×5 and 5×12? You get 60 each time. It doesn't matter which factor comes first. That is what this next property says.

Commutative Property of Multiplication

For all numbers represented by a and b,

$$a \times b = b \times a$$

Does this same property work for subtraction and for division? You can discover the answer to that question by finding only one example where it does not work. If you find such an example, the property cannot be true for *all* numbers.

If you have $20 in your checking account and write a check for $8, how much money is left in your account? $20 − $8 = $12. If, however, you have only $8 in your checking account and write a check for $20, what happens? $8 − $20 = an overdrawn account. You are $12 in the red, in debt. From this example, you know that $20 − $8 is not the same as $8 − $20. Therefore, subtraction is *not* commutative.

The Commutative Property is easy to disprove for division as well. Suppose you have made 8 pies for Thanksgiving and you are splitting them up equally for your two daughters to take home to their families. How many pies does each daughter get? Eight pies divided by 2 daughters gives 4 pies to each family: $8 \div 2 = 4$. However, if you only have two pies, and there are 8 people at your dinner table, how much pie does each person get? $2 \div 8 = 1/4$. By this example, you can tell that $8 \div 2 = 4$ but $2 \div 8 = 1/4$. The results 4 and 1/4 are obviously not the same; the Commutative Property does not work for division.

When you are trying to learn about something as unexciting as "properties of numbers," it is a good idea to try to get a practical or even funny picture in your mind to help you remember the fact. It makes the property more interesting and easier to remember.

The next two properties, the **Associative Property of Addition** and the **Associative Property of Multiplication,** are easiest to show by example. Properties cannot be proved; they are just accepted as fact by everyone who uses them. Associates are people who group together to do something. Barnes and Associates might be a law firm or a consulting firm or a construction company. The Associative Properties of Addition and Multiplication talk about the way numbers are grouped together.

The way in which addends are grouped together does not change the answer in an addition problem. You know from your own experience that if you are doing a column addition problem, it does not matter whether you start from the top and add each succeeding number going down the column, or if you begin at the bottom of the column, getting sums as you go up the column (the Commutative Property). You can also *combine* certain numbers in the middle of a column to form 10s or 9s if you choose. The grouping does not matter in an addition problem. Look at this example. The parentheses, ( ), join quantities in a special way. The operations that appear inside the parentheses are always done *first*.

$$3 + (7 + 9) = (3 + 7) + 9$$
$$3 + (16) = (10) + 9$$
$$19 = 19$$

The order of the addends is the same, but they are grouped in different ways. However, both arrangements produce the same answer.

Associative Property of Addition

For all numbers represented by a, b, and c,

$$(a + b) + c = a + (b + c)$$

The same property holds true for multiplication. Look at this example:

$$(5 \times 4) \times 8 = 5 \times (4 \times 8)$$
$$(20) \times 8 = 5 \times (32)$$
$$160 = 160$$

The order of the numbers is the same but the groupings are different.

Stated formally, this grouping property is:

Associative Property of Multiplication

For all numbers represented by a, b, and c,

$$(a \times b) \times c = a \times (b \times c)$$

Is there an Associative Property for Subtraction? Consider this example.

Is $12 - (8 - 2)$ the same as $(12 - 8) - 2$?

$12 - (8 - 2)$	?	$(12 - 8) - 2$
$12 - 6$	?	$4 - 2$
6	$\neq$	2

In order to disprove a property, you need to find only one example that will not work. This example proves that subtraction is not associative. See if you can disprove an Associative Property for Division. All you need is one situation where the property fails to be true. Neither subtraction nor division is associative.

Now let's consider zero. Zero is one digit that has some special properties that are not true for other numbers. What happens when you add zero to something? Absolutely nothing changes. You add $0 to your checking account and nothing happens to your balance. $5 + 0 = 5$. $38 + 0 = 38$. The **Addition Property of Zero** is just that simple.

Addition Property of Zero

> For any number represented by a,
>
> $$a + 0 = a$$

The **Multiplication Property of Zero** is easy to understand and to remember. You know from your own experience that zero times any number is equal to zero. $0 \times 8 = 0$, $12 \times 0 = 0$, and so on. That is all the property says.

Multiplication Property of Zero

> For any number represented by a,
>
> $$a \times 0 = 0$$

What about division and zero? What happens there? Look at the following examples of some division problems that involve zero. See if you can figure out what the properties would be.

$$6\overline{)0}^{\,?} \qquad 35\overline{)0}^{\,?} \qquad 114\overline{)0}^{\,?}$$

Remember, to check division, you multiply the answer times the divisor. The foregoing division problems might be worded another way:

"What number times 6 = 0?"

"What number times 35 = 0?"

"What number times 114 = 0?"

The answer to all these questions is, of course, zero. According to the Multiplication Property of Zero, $0 \times a$ (any number) $= 0$. Therefore, *whenever you divide zero by any number you get zero.* This is not a formal property, but it is always true nonetheless.

What happens when you divide *by zero*? Look at these examples.

$$0\overline{)7}^{\,?} \qquad 0\overline{)42}^{\,?} \qquad 0\overline{)215}^{\,?}$$

Put another way,

"What number times 0 equals 7?"

"What number times 0 equals 42?"

"What number times 0 equals 215?"

The answer is, of course, that you cannot multiply 0 by any number and get anything but zero. There is no number that will answer these questions. Division by zero cannot be done. It is **undefined**. When you encounter a problem that involves division by zero, your answer should be "Undefined" or "No Solution." This is because division by zero has no answer.

What about one? Does it have a special rule? You know that multiplying a number by one doesn't change a thing. $7 \times 1 = 7, 25 \times 1 = 25$, and so on. That is exactly what the **Multiplication Property of One** says.

Multiplication Property of One

For any number represented by a,

$$a \times 1 = a$$

The last property we will study at this time is one that combines addition and subtraction operations with multiplication. The **Distributive Principle** is used throughout mathematics, and it is not difficult to understand. Consider the example that follows.

$6(5 + 4) = ?$ Because a number written just outside a set of parentheses means to multiply, the problem says to multiply 6 times the result when you add 5 and 4.

$6(9) = 54$ In order to do that, first add 5 and 4 to get 9, and then multiply 9 by 6.

Now, do the problem by finding the products first and then adding.

$6(5) + 6(4) = ?$ This time, find the product of 6 and 5 and the product of 6 and 4 and then combine those results.

$30 + 24 = 54$ You get 54 both times, so the two expressions must be equal.

$$6(5 + 4) = 6(5) + 6(4)$$

$$6(9) = 30 + 24$$

$$54 = 54$$

This example suggests that it doesn't matter whether you add first and then multiply the sum by the factor or you find the two products first and then add those results. This is, in fact, always true.

Distributive Principle

For all numbers represented by a, b, and c,

$$a(b + c) = a(b) + a(c)$$

and

$$a(b - c) = a(b) - a(c)$$

The following examples show that you get the same answer whether you add (or subtract) first and then multiply or you multiply first and then add (or subtract). See if you can understand each step of the "proof."

E X A M P L E **1** $5(9 + 3) = 5(9) + 5(3)$

$5(12) = \quad 45 \; + \; 15$

$60 = 60$ ◆

E X A M P L E **2** $8(7 - 5) = 8(7) - 8(5)$

$8(2) = \quad 56 \; - \; 40$

$16 = 16$ ◆

The Distributive Principle is a very powerful tool. To multiply 7 by 15 mentally, you could use the Distributive Principle and turn the problem into $7(10) + 7(5)$. That equals $70 + 35$, which equals 105. If you wanted to multiply 23×14, you could do the multiplication in long form, using paper and pencil, or you could do the problem partly (or wholly) in your head by using the Distributive Principle.

$$23 \times 14 = 20(14) + 3(14)$$
$$= 280 + 42$$
$$= 322$$

In this case, the multiplication by 14 is *distributed over* the sum of 20 and 3.

Sometimes in class, instructors amaze students by their ability to do complicated calculations in their heads. Perhaps they are using the Distributive Principle. Practice it and see if you can develop your own mental skills.

All of these properties are very useful for you to know. You have, in fact, been using most of them all along and may never have been aware of them. It is important for you to know each property and to be able to recognize it when it is used.

1.5 ◆ Practice Problems

A. Match each of the statements in Column I with the appropriate property in Column II.

Column I

1. $7 \times 0 = 0$
2. $1 \times 9 = 9 \times 1$
3. $1 + (2 + 3) = (1 + 2) + 3$
4. $6(7 - 5) = 6(7) - 6(5)$
5. $4 \times (3 \times 2) = (4 \times 3) \times 2$
6. $0 + 1 = 1 + 0$
7. $1 + 0 = 1$
8. $19 \times 1 = 19$

Column II

a. Distributive Principle
b. Multiplication Property of One
c. Addition Property of Zero
d. Multiplication Property of Zero
e. Commutative Property of Addition
f. Associative Property of Addition
g. Commutative Property of Multiplication
h. Associative Property of Multiplication

B. Verify that each of the following statements is true by using the Distributive Principle. (See Examples 1 and 2.)

9. $8(12 + 2) = 8(12) + 8(2)$
10. $25(10 - 7) = 25(10) - 25(7)$
11. $7(9 - 3) = 7(9) - 7(3)$
12. $13(8 + 3) = 13(8) + 13(3)$
13. $4(5) + 4(7) = 4(5 + 7)$
14. $6(9) - 6(4) = 6(9 - 4)$

15. $14(10) - 14(6) = 14(10 - 6)$
16. $8(9) + 8(7) = 8(9 + 7)$
17. $(12 - 5)22 = 12(22) - 5(22)$
18. $104(38) - 104(23) = 104(38 - 23)$
19. $68(33) + 22(33) = (68 + 22)33$
20. $47(53 + 27) = 47(53) + 47(27)$

1.6   Word Problems!

There is probably no more dreaded phrase in a mathematics classroom than "word problems." Students, as a whole, dislike them and are uncomfortable trying to solve them. This really is strange because word problems, which are also called application problems, are much more practical than plain calculation problems. They are the real-world part of math for most people. What probably makes them so feared is that they force students to combine two kinds of thinking. You need to *read and interpret* the words and then apply the *mathematical operation* necessary to reach a solution. Having to do two things in one problem hardly seems fair, does it? But once you understand the cause of your dislike and discomfort, word problems will be easier for you to do.

This is the approach that you should follow when faced with a word problem.

Suggested Approach to Word Problems

1. *Read* the problem carefully once to see what is going on in the problem. What's it about?

2. *Read* it very carefully again to see if you can determine which mathematical operation you should use. Sometimes there are key words or phrases that will help you to decide.

3. *Estimate* what your answer should be. You just need to be close. What kind of an answer would be "sensible"?

4. *Perform* the mathematical operation(s) carefully, making sure you have copied the numbers correctly.

5. *Answer* the question that has been asked, and be sure to *label* your answer. Because word problems have a context, their answers are not just numbers, but numbers that mean something—for example, 12 feet, 57 people, $42.15, and so on.

6. *Reread* the problem, putting your answer in to see if it does, in fact, make sense. Is the exact answer reasonably close to the estimate?

To help determine which mathematical operation to use, study the following list of key words. These (and many other words) can tell you when to add, subtract, multiply, or divide.

Addition	Subtraction	Multiplication	Division
sum	difference	product	quotient
total	less	times	per
altogether	fewer	part of	split
increase	decrease	twice	each
more	remain	double	shared
in all	larger	area	average
combined	-er words	total	ratio

Sometimes getting the basic idea of what is happening in a problem will help you decide what to do. Sometimes a key word or idea will stand out and get you started. Remember, there are only four basic operations. If you are really stumped, estimate a reasonable answer and try all four operations. See which answer comes closest to your estimate and makes the most sense.

The best way to become successful with word problems is to do lots of them! The more you practice and follow the patterns, the more successful you become. Practice really helps. The first time you went into the deep end of a swimming pool, the result was probably not too graceful. Once you had gone swimming a few more times, you became more comfortable in the water, and your strokes became smoother and more effective. This can happen with word problems too, but it takes effort and persistence on your part. If you have been working a problem, and it just doesn't make sense to you, skip it and try another one. Spending a lot of time stewing over one or two problems does not help you learn. It only makes you frustrated. And sometimes if you leave a confusing problem and return to it later, you find you are suddenly able to solve it.

In each of the following examples, the six steps listed at the beginning of this section are given in detail. You should try to follow these steps each time you do a word problem. You do not always have to write everything down, but the pattern should be there in your thinking.

E X A M P L E   1   Margot weighed 153 pounds on March 15 and weighed 129 pounds on September 15. How much weight had she lost in those 6 months?

Steps 1 and 2: Read to see what is happening. Though there is no key word, you are asked to find the difference between two quantities. Subtract.

Solution   For estimating purposes, use 150 and 130. $150 - 130 = 20$

Step 3: You could use 155 and 130 just as well to get close to the answer.

$153 - 129 = 24$

Step 4: Subtract and borrow carefully.

24 pounds

Step 5: Be sure to label the answer as pounds.

Margot lost 24 pounds in that 6-month period of time.

Step 6: 24 pounds seems reasonable and it is close to the estimate of 20 pounds.   ◆

E X A M P L E   2   How far can you drive on your first day of vacation if you plan to drive for 8 hours and can average 55 miles per hour?

Steps 1 & 2: You will need to know the number of hours and the rate per hour to find the distance by multiplying.

Solution   8 hours $\times$ 60 mph = 480 miles

Step 3: It is easier to multiply by a number ending in zero.

$8 \times 55 = 440$

Step 4: Multiply carefully.

440 miles

Step 5: Distance will be measured in miles because the speed is in miles per hour.

You can plan to travel 440 miles during the first 8-hour day.

Step 6: This seems possible, and it is close to the estimate of 480 miles.   ◆

E X A M P L E 3

How many pounds of grass seed can you buy if the seed costs $2 per pound and you have only $8 in your wallet?

Steps 1 & 2: You only have $8 in your wallet, and you are going to split it up into $2 units. To find the number of parts, you will divide.

Solution

$8 divided by $2 = 4 pounds

Step 3: Both quantities are easy to work with, so you can use them as they are.

$$2\overline{)8}^{\,4}$$

Step 4: Perform the division.

4 pounds

Step 5: Be sure to label.

You can buy 4 pounds of seed for $8 if it costs $2 per pound.

Step 6: This answer makes sense and agrees with the estimate. ◆

E X A M P L E 4

Jose and Maxine went to the store to buy the supplies that they needed for their family picnic. They bought 2 packages of hot dogs for 99 cents each, 2 packages of buns for 59 cents each, a bag of potato chips for 79 cents, and a box of brownie mix for 85 cents. How much did they spend?

Steps 1 & 2: You need to find the total amount spent at the store, so you will add. Because they bought more than one package of the hot dogs and buns, you multiply to find the cost of these items.

Solution

2 times 100 cents = 200 cents
2 times 60 cents = 120 cents
1 times 80 cents = 80 cents
1 times 90 cents = 90 cents
Approximate cost = 490 cents, or $4.90.

Step 3: *Approximate* the item costs to the nearest dime, or even the nearest dollar or half dollar if you like, just to get an idea of the answer.

2 × 99 cents = 198 cents
2 × 59 cents = 118 cents
1 at 79 cents = 79 cents
1 at 85 cents = 85 cents
Total = 480 cents, or $4.80

Step 4: Perform the multiplication with the actual costs, and then add to find the total.

They spent $4.80 for the picnic items at the store.

Step 5: Give the answer to the problem, labeling it as money.

Considering the amount of each item and the fact that they bought only 6 items, the total seems to make sense.

Step 6: Reading the problem again, $4.80 seems like a reasonable answer and it is close to the estimate, $4.90. ◆

You certainly do not need to write out this many steps to do a simple word problem. That would be ridiculous. But you should be going through the steps mentally each time, particularly the estimating and checking steps that enable you to tell if your answer is reasonable. If it does not come out the way you think it should (or if it does not agree with the answer key), check your figuring first and then decide if perhaps you chose to do the wrong operation.

1.6 ◆ Practice Problems

Find the answer to each of the following problems. Estimate an answer first. You might use your calculator to check your work. Be sure to label your answers.

1. Max and Yvette decided to open a restaurant. They figured that it would cost them $2,250 for rent, $890 for taxes, $1,200 for utilities, $1,950 for supplies, and $2,400 for salaries for the first 2 months of operation. How much money would they need to cover expenses for those first 2 months?

2. Bentley had some unexpected car expenses. His faithful old car developed some serious problems. He was told that the repair bill would be $126 for new brakes, $82 for new shocks, $26 for an alignment, $87 for a rebuilt carburetor, and $17 for a new headlight. The garage would charge $220 for labor for all of these repairs. What will his total bill be?

3. Don planted 67 acres of spring wheat and 39 acres of corn. How many more acres did he plant with wheat?

4. Suzanna and Luis figured that it would cost them $418 a month to continue to rent their apartment. They could own their own home for $527 a month. How much more would they have to pay each month if they bought a home?

5. Aunt Millie died and left her estate to her 4 nieces and nephews. The total amount left to her 4 heirs was $81,452. How much was each heir to receive?

6. The Sun Valley Travel Association is planning a trip for the Senior Citizens Club from Marshall City. Hotel expenses will be $285 per person, and airfare will cost each traveler $487. How much will the trip cost if 18 members decide to go?

7. At holiday time, the Hill family's home is decorated with many strands of colored lights. The Hills' normal electric rate is $2 per day, but it jumps to $4 per day with the extra lights. How much extra will the Hills pay for the 3-week holiday season (21 days) when they use more electricity?

8. John, the manager of Noah's Restaurant, orders 36 12-ounce packages of cream cheese to make cheesecakes for the week. If each cheesecake uses 8 ounces of cream cheese, how many cheesecakes can he make?

9. A pair of glasses with a scratch-resistant surface costs $23 more than a pair without that surface. What will a pair of $78 glasses cost if you add the scratch resistance?

10. Farmer McDonald has a herd of 93 cattle. If each animal eats 2 pounds of grain at each of 3 feedings per day, how many pounds of feed will he need per day for his herd?

11. Traveling along U.S. Highway 36, the distance from Hannibal, Missouri, to St. Joe, Missouri, is 189 miles. Continuing on, the distance from St. Joe to Norton, Kansas, is 275 miles. How far is it along U.S. 36 from Hannibal to Norton?

12. In Indianapolis, Indiana, the average temperature in June is 75 degrees. In January the average temperature is 28 degrees. On the average, how much colder is Indianapolis in January than in June?

13. In Problem 11, how much farther is it from Norton to St. Joe than from Hannibal to St. Joe?

14. Cosmetic tint for glasses costs $19. If Marlene's new glasses cost $83 and she decides to add a tint, what will her total bill be?

15. Mr. Hutton's estate, valued at $13,486,835, is to be divided equally among his 5 heirs. How much money will each person inherit?

16. A Boeing 747 jet flew 2,348 miles on Monday, 1,456 miles on Tuesday, 1,089 miles on Wednesday, and 4,576 miles on Friday. How many total miles did it fly during those 4 days?

17. Your new car averages 19 miles per gallon. How far can you drive on 16 gallons of gas?

18. A flight from Winnipeg, Canada, to Honolulu is approximately 3,824 miles. If the flight takes 8 hours, how many miles does the plane fly in 1 hour?

19. Upholstery fabric for new sofa cushions will cost approximately $7 a yard. The trim for each cushion will be $6 a yard. Each cushion pattern calls for 3 yards of fabric and 2 yards of trim. What will it cost to make 4 of these large pillows?

20. The total bill for computer desks for the new lab is $4,296. If there are to be 24 work stations in the lab, how much does each desk cost?

1.7 Interpreting Data

Often in work or classroom situations, you will be expected to look at a collection of numeric facts (data) and see a pattern or relationship. Perhaps you will need to organize and display the data in an easy-to-interpret manner. This kind of work requires the use of graphs to display and organize the data and the use of statistics to interpret the data.

The basic statistical quantities that you will compute in this section are the mean, the median, the mode, and the range. By providing these four pieces of information, you are telling someone the average value (mean), middle value (median), the most often repeated value (mode), and the spread between the highest and lowest values (range). You have provided your audience with key information that should enable them to understand and make sound decisions.

The most common statistical term used with a set of values is the **mean**, or average. What if your grades on mathematics tests for this semester were 75, 84, 68, 93, 44, 86, 80, 70, and 84, and you would like to know what your average grade is in the class? To find the mean score for your tests, add all of the scores together and divide that sum by the number of grades.

$$\text{Mean} = \frac{\text{sum of the grades}}{\text{the number of grades}}$$

$$\text{Mean} = \frac{75 + 84 + 68 + 93 + 44 + 86 + 80 + 70 + 84}{9}$$

$$\text{Mean} = \frac{684}{9} = 76$$

The average score, or mean score, then, is 76.

The **median** of a set of scores is the score that is in the middle position when the scores are arranged in order of size.

93, 86, 84, 84, 80, 75, 70, 68, 44

The number in the middle of the nine scores is 80. Therefore, the median of these scores is 80. If the list of data had had an even number of entries, there is not a single middle value. The median for such data is the average of the two middle values.

The **mode** of a set of scores or data is the most often repeated value. Sets of data can have one mode, more than one mode, or no mode (if none of the values is repeated). For the grades on these mathematics tests, the only repeated score is 84, so that is the mode.

Another meaningful statistic is the **range**. The range shows how far apart the highest and lowest values are. The range for these test scores is 93 to 44. It is calculated as 93 − 44, or 49 points.

E X A M P L E 1 The daily low temperatures for Indianapolis last week were 58°, 49°, 46°, 38°, 35°, 38°, and 44°. Find the mean, median, mode, and range for these temperatures.

Solution
$$\text{Mean} = \frac{\text{sum of temperatures}}{\text{number of temperatures}}$$

$$= \frac{58 + 49 + 46 + 38 + 38 + 35 + 44}{7}$$

$$= \frac{308}{7} = 44°$$

Median: the middle value when the elements are arranged in order of size.
58°, 49°, 46°, 44°, 38°, 38°, 35°
The middle value, the median, is 44°.

Mode: most often repeated temperature
                  Because 38° appears twice, it is the mode.

Range: difference between the highest and lowest temperatures
                  $58° - 35° = 23°$
                  The range is 23°.   ◆

E X A M P L E **2**   For the past six months, Antonio's monthly gasoline bills have been $49, $30, $29, $32, $40, and $36. Find the mean, median, mode, and range for these data.

Solution   Mean $= \dfrac{49 + 30 + 29 + 32 + 40 + 36}{6}$

$= \dfrac{216}{6} = \$36$

The mean, or average, gasoline bill was $36.

Median: Arrange the values in order of size.
                  49, 40, 36, 32, 30, 29
                  There is an even number of entries, so the median is the average of the middle two values.
                  $\dfrac{36 + 32}{2} = \dfrac{68}{2} = 34$
                  Therefore, the median is $34.

Mode: No value is repeated, so there is no mode.

Range: $\$49 - \$20 = \$29$
                  The range of these bills is $29.   ◆

Although the mean, median, mode, and range provide a great deal of information about a set of numbers, they don't show the relationship among the values. Graphs are used to quickly show how values may be related or what trends or patterns exist.

A **bar graph** is used to compare several items. One of the scales is labeled to show the units represented by each interval. The other scale indicates the items being compared. The height or length of a bar indicates the value for that item.

In Figure 4, the vertical scale is labeled "Number of Students," and each increment in that scale represents 4,000 students. The horizontal scale is used to represent four different years in which the autumn quarter enrollments are being compared. Reading across the top of the 1988 bar to the vertical scale shows that the enrollment that year was 12,000 students. The 1990 enrollment must be estimated at about 14,000 students.

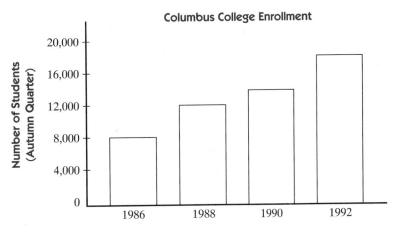

FIGURE 4

E X A M P L E 3 Use Figure 4 to answer the following questions.

a. What was the enrollment in 1986?

b. In what year was the enrollment 18,000?

c. How many more students enrolled in 1992 than in 1986?

d. Is the sum of the 1986 and 1988 enrollments greater than or less than the enrollment in 1992?

e. In what year did the enrollment show the least increase over the previous year studied?

Solutions **a.** 8,000 students

b. 1992

c. 18,000 − 8,000 = 10,000 more students

d. 8,000 + 12,000 = 20,000
     20,000 is greater than 18,000

e. The 1986–1988 increase was 4,000 students.
     The 1988–1990 increase was 2,000 students.
     The 1990–1992 increase was 4,000 students.
     The increase was the least in 1990. ◆

Bar graphs can also be arranged horizontally (Figure 5).

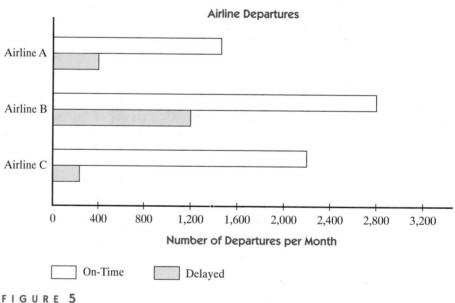

FIGURE 5

E X A M P L E 4 Use Figure 5 to answer the following questions.

a. How many delayed flights did Airline B have during the month?

b. Which airline flew the greatest number of flights during this period? How many?

c. How many more on-time departures did Airline B have than Airline C?

d. Is the sum of the delays for Airlines A and C greater than, less than, or equal to the number of delays for Airline B? By how much?

e. If all other factors are the same, which airline would you prefer to fly on? Explain your answer.

Solution a. Airline B had 1,200 delayed flights.

b. Airline B flew 4,000 flights.

c. Airline B had 2,800 and Airline C had 2,200. Therefore, Airline B had 600 more on-time departures.

d. Airlines A and C had a total of 600 delays. Airline B had 1,200. Therefore, Airline B had 600 more delays than Airlines A and C combined.

e. One possible response would be to fly Airline C because it has a "middle" number of on-time flights but experiences lots fewer delays. ◆

Some bar graphs use rows of pictures in place of the bars (Figure 6). The pictures suggest the items being represented. These bar graphs are called **pictographs** or **pictograms**. Each symbol represents a stated number of objects. And a part of a symbol represents a fractional part of the stated number.

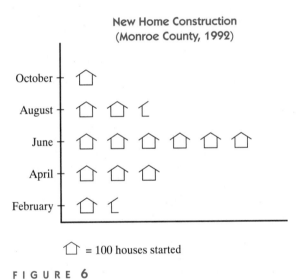

**New Home Construction
(Monroe County, 1992)**

⌂ = 100 houses started

F I G U R E **6**

E X A M P L E **5** Use Figure 6 to answer the following questions.

a. How many houses were started in April?

b. In what month were 250 houses started?

c. How many more houses were started in June than in February?

Solution a. 300 houses were started in April.

b. August had 250 housing starts.

c. 600 houses were started in June and only 150 in February. Therefore, there were 450 more houses started in June than in February. ◆

The data from Figure 4 might also be represented by a **line graph**. In such a graph, points rather than bars represent the values (Figure 7). A line connecting the points shows the change in enrollment over the period represented. The pattern, or trend, is evident in a line graph.

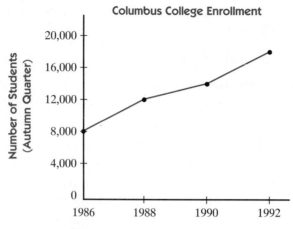

FIGURE 7

In Figure 8, two different quantities are represented in the same line graph. A solid line represents the average cost of a college algebra textbook, and the broken line represents the average cost of a psychology textbook.

E X A M P L E 6   Use Figure 8 to answer the following questions.

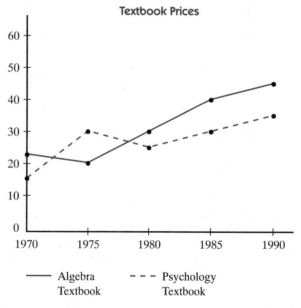

FIGURE 8

a. How much more did the psychology textbook cost than the algebra textbook in 1975?

b. In what 5-year interval did the psychology text increase the most?

c. Which book had a cost of $30 in 1980?

d. Which book price experienced a decrease in price of $5 in a 5-year interval?

Solutions   **a.** In 1975, the algebra text cost $20 and the psychology text cost $30.
$30 − $20 = $10 more

b. The increase in the price of the psychology text was the greatest between 1970 and 1975.

c. The algebra text cost $30 in 1980.

d. The price of the psychology text decreased by $5 between 1975 and 1980. ◆

1.7 ◆ Practice Problems

A. Find the answer to each of the following problems.

1. Find the requested statistical information for the given set of values.

      13, 19, 26, 19, 10, 21, and 18
   a. mean
   b. median
   c. mode
   d. range

2. Find the requested statistical information for the given set of values.

      12, 14, 18, 22, 12, 24, and 17
   a. mean
   b. median
   c. mode
   d. range

3. Find the requested statistical information for the given set of values.

      128, 218, 176, 204, 228, and 204
   a. mean
   b. median
   c. mode
   d. range

4. Find the requested statistical information for the given set of values.

      358, 313, 358, 332, 307, and 330
   a. mean
   b. median
   c. mode
   d. range

B. Using the graphs and tables that are given, find the answers to each of the following questions. For Problems 5 through 8, use Table 1.

TABLE 1

	Scores on French Tests			
	1	**2**	**3**	**4**
Maria	86	75	91	84
Igor	78	84	84	86
Chris	68	82	76	66

5. Find the mean, median, mode, and range of Maria's grades.

6. Find the mean, median, mode, and range of Chris's grades.

7. Which of the three students has the widest range of scores?

8. Which student has a mode score?

For Problems 9 and 10, use Figure 9.

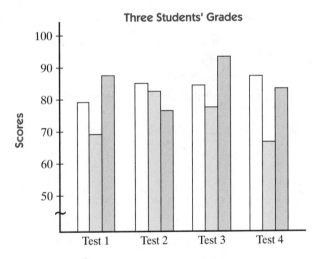

FIGURE 9

9. The data from Table 1 are displayed in Figure 9. However, the key that tells which student's scores are depicted by each bar has been left off. Whose scores do the white bars represent?

10. The data from Table 1 are displayed in Figure 9, but the key that tells which student's scores are depicted by each bar has been left off. Whose scores do the colored bars represent?

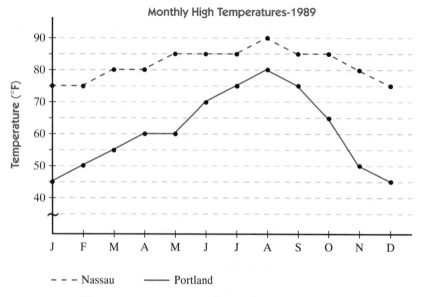

Monthly High Temperatures-1989

- - - Nassau   ———— Portland

F I G U R E **10**

For Problems 11 through 16, use Figure 10.

11. What was the monthly high temperature in Portland in April?

12. Which city had a high temperature of 85 in June?

13. How much warmer was Nassau than Portland in December?

14. How much cooler was Portland than Nassau in July?

15. During what month(s) was the high temperature in Portland closest to the high temperature in Nassau?

16. Between what months does the high temperature in Portland undergo the greatest change?

Use Figure 11 to answer Problems 17 through 24.

17. During which game did the Flyers score 15 points?

18. What was the Flyers' score in game 4?

19. How many more points were scored in the fourth game than in the third game?

20. Is the sum of the scores in game 2 and game 3 greater than, less than, or equal to the score in game 1?

21. What is the mean score for the five games?

22. What is the median of the scores?

23. What is the mode of the scores?

24. What is the range of the scores?

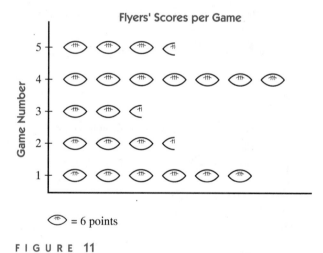

Flyers' Scores per Game

= 6 points

F I G U R E **11**

1.8 ▸ Chapter Review

The purpose of this introductory chapter is to give you many of the basics you will need to study mathematics. Because whole numbers are the most basic numbers, that is the place to begin. Most mathematical operations depend on an understanding of place value. The properties introduced here are used every day. The only one that requires some practice is the Distributive Principle. The practical side of basic calculations is the solving of word problems. You will be successful with them if you practice often.

When you have finished studying this chapter, you should be able to:

1. Explain the following terms: place value, place-holding zeros, hyphenated numbers, rounding, sum, addends, difference, product, factors, quotient, divisor, dividend, remainder, estimation, Associative and Commutative Properties, undefined answer, and Distributive Principle.

2. Read and write whole numbers using commas and hyphens correctly.

3. Do addition, subtraction, multiplication, and division problems without errors and by using the following hints.
   a. Remember to line up the ones digits in addition, subtraction, and multiplication problems.
   b. When borrowing in a subtraction problem, borrow 1 from the entire remaining quantity.
   c. In multiplication involving factors of more than one digit, remember to move over the partial product under the multiplying digit.
   d. Zeros appearing in either the dividend or the divisor in division problems require great care.

4. Recognize and identify the number properties.

5. Read and solve word problems, following the six-step process that was outlined in Section 1.6.

6. Interpret data from a graph.

7. Find the mean, median, mode, and range of given data.

Review Problems

A. Name the digit in the indicated place in each of the following problems.

1. 238,195
   a. tens place
   b. thousands place
   c. hundred-thousands place

2. 18,743,092
   a. thousands place
   b. hundreds place
   c. millions place

3. 5,264,381
   a. hundred-thousands place
   b. ones place
   c. millions place

4. 20,381,467
   a. ten-millions place
   b. ten-thousands place
   c. hundreds place

B. Write each of the following quantities in numeric form.

5. eighteen million, four thousand, seventeen

6. six billion, four hundred thousand, twelve

7. nine thousand, forty-two

8. twelve million, eighty-one thousand, six

9. seventeen billion, nine thousand, fifty-five

10. four hundred twenty-three thousand

11. eight hundred forty-six thousand

12. eight thousand, four hundred thirty-nine

C. Write each of the following numbers in words.

13. 32,189
14. 27,693
15. 891,407
16. 375,682
17. 1,424,856
18. 3,427,895
19. 77,923
20. 98,137
21. 492,108
22. 648,000
23. 23,011
24. 62,039

D. Round each number to the place indicated.

25. 184,563 to the nearest thousand
26. 22,761 to the nearest thousand
27. 358,901 to the nearest hundred
28. 26,511 to the nearest thousand
29. 117,241 to the nearest ten-thousand
30. 267,189 to the nearest ten
31. $92,341 to the nearest hundred
32. $1,457,832 to the nearest million
33. $389,726 to the nearest thousand
34. $4,223,971 to the nearest hundred

E. Perform the mathematical operations indicated.

35.
$$\begin{array}{r} 1,289 \\ \times\quad 35 \\ \hline \end{array}$$

36. $14\overline{)8,223}$

37.
$$\begin{array}{r} 247 \\ 1184 \\ +\ 936 \\ \hline \end{array}$$

38.
$$\begin{array}{r} 4,005 \\ -\ 428 \\ \hline \end{array}$$

39.
$$\frac{14,582}{46}$$

40.
$$\begin{array}{r} 16,236 \\ -9,277 \\ \hline \end{array}$$

41.
$$\begin{array}{r} 9,982 \\ \times\quad 48 \\ \hline \end{array}$$

42.
$$\begin{array}{r} 159 \\ 3,803 \\ +\ 547 \\ \hline \end{array}$$

43. Find the difference between 82,006 and 37,848.

44. Compute the quotient of 11,396 divided by 28.

45. Find the product of 4,509 and 38.

46. Find the sum of 53,472 and 3,906.

F. Match each of the statements in Column I with the appropriate property in Column II.

Column I

47. $19 \times 43 = 43 \times 19$

48. $173 + 0 = 173$

49. $17(85) = 17(80) + 17(5)$

50. $7 + (9 + 3) = (7 + 9) + 3$

51. $276 \times 1 = 276$

52. $92 + 79 = 79 + 92$

53. $(14 \times 3)5 = 14(3 \times 5)$

54. $0 = 46 \times 0$

Column II

a. Distributive Principle

b. Multiplication Property of One

c. Addition Property of Zero

d. Multiplication Property of Zero

e. Commutative Property of Addition

f. Associative Property of Addition

g. Commutative Property of Multiplication

h. Associative Property of Multiplication

G. Determine whether each of the following statements is true or false by using the Distributive Principle.

55. $9(14 - 6) = 9(14) - 9(6)$

56. $5(12) + 5(14) = 5(12 + 14)$

57. $11(6) - 11(4) = 11(6 + 4)$

58. $8(8 + 7) = 8(8) + 7(7)$

59. $7(14 + 6) = 7(14) + 7(6)$

60. $9(15 - 3) = 9(15) - 9(3)$

61. $6(8) + 6(9) = 6(8 + 6)$

62. $7(9) - 7(4) = 7(9 - 4)$

63. $22(5 + 8) = 22(5) + 22(8)$

64. $18(16 - 7) = 18(16) - 18(7)$

H. Find the answer to each of the following word problems. Estimate an answer first. Be sure to label your answers.

65. Martin swam 18 laps on Monday, 23 laps on Tuesday, 19 laps on Friday, and 32 laps on Saturday. How many laps did he swim on those 4 days?

66. How many feet of copper tubing will Eduardo need to purchase at the hardware store if he has to run three lines of 17 feet each?

67. Kelly and two of her friends split the cost of dinner at Rocco's Spaghetti House. If the total bill was $42, how much did each of them pay?

68. As a fund raiser, students at the Edgehill School are selling flats of bedding plants to raise money for a new freezer for the school cafeteria. The school makes $8 on each flat of plants sold. If the freezer will cost about $4,400, how many flats will the students need to sell to pay for the freezer?

69. The population of Columbus, Indiana, is approximately 30,800. The population of Columbus, Ohio, is about 566,100. How much larger is Columbus, Ohio?

70. The area of the state of Alaska is 586,412 square miles. The area of the state of Rhode Island is 1,214 square miles. How much larger is Alaska? Estimate your answer to the nearest hundred square miles.

71. Find the total cost of buying computers for the new lab if there are 24 work stations in the lab and each computer costs $1,078.

72. Textbooks for next term are being sold in the campus bookstore. Santiago needs an economics book ($38), a literature book ($17 used), and a mathematics book ($26). What will his textbook bill be for these three courses?

73. Gabrielle went on a 4-month diet to lose some extra pounds. She weighed 153 pounds when she started the diet and weighed 129 pounds at the end of the four months. How much weight did she lose?

74. Janet and Sam want to buy a new home. They need $3,600 more for a down payment. If they are saving $240 a month, how many more months will it be before they will have all the money saved?

75. In Problem 73, how many pounds was Gabrielle's average loss each month during her diet?

76. Gretchen needs to begin saving money for a down payment on a new car. She hopes that her old car will last another 18 months. She plans to try to save $15 a week for the next 18 months. If each month averages 4 weeks, how much money can she plan on saving during this period?

77. Marc, a salesman for Rayburn Electronics, travels an average of 372 miles each week. On the average, estimate how many business miles he puts on his car each year during the 49 weeks that he works. (You could use 370 miles and 50 weeks.)

78. The air distance from San Francisco to Tokyo is 5,410 miles. The distance from San Francisco to London is 5,980 miles. How much closer to San Francisco is Tokyo?

I.   Use Figure 12 to answer the following questions.

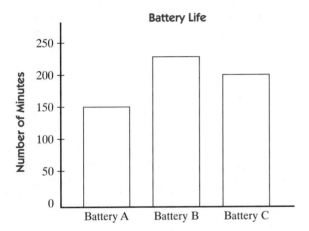

FIGURE 12

79. How many more minutes did Battery C last than Battery A?

80. Which battery lasted 150 minutes?

81. How long did Battery B last?

82. How many more minutes did Battery B last than Battery C?

J.   Answer the following questions by using the information in Figure 13.

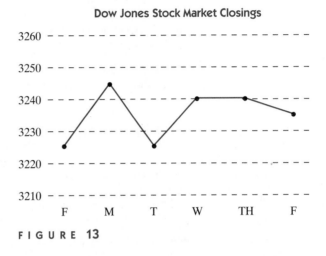

FIGURE 13

83. What was the mean for the stock market closings from Friday through Friday?

84. What was the median for the stock market closings from Friday through Friday?

85. What was the mode for this time period?

86. What was the range in closings for this period?

87. On which day was there the greatest decrease?

88. On which day was there the greatest increase?

89. What was the overall change from the first Friday to the next Friday?

90. Was this a good week for the stock market or not? Explain your answer.

Chapter Test

1. Write in digits forty-seven million, six thousand, eight hundred four.

2. Write 30,206,011 in words.

3. Round 176,852 to the nearest hundred.

4. Round 347,826 to the nearest hundred-thousand.

5. Find the product of 347 and 29.

6. Find the quotient of 149,036 divided by 37.

7. What is the difference between 40,082 and 27,995?

8. From the sum of 3,827 and 1,046 subtract 2,785.

9. Subtract 509 from 782.

10. Complete this statement using the Commutative Property of Addition: $29 + 83 = \underline{\hspace{1cm}}$

11. Complete this statement by applying the Associative Property of Multiplication:
$3 \times (7 \times 6) = \underline{\hspace{1cm}}$

12. $7 + 0 = 7$ is an example of the use of which property?

13. $0 \div 143 = \underline{\hspace{1cm}}$

14. $2,642 \div 0 = \underline{\hspace{1cm}}$

15. According to the Distributive Principle, is $7(8 - 5)$ equal to $7(8) - 7(3)$?

16. Tim and Julia went to the laundromat to wash several large blankets and area rugs. Each large washer required 8 quarters, and each oversized dryer used 5 quarters. How many quarters will they need to wash and dry five large loads?

17. In order to buy a new suit, Laura promised to save $5 for every 3 pounds that she lost on her diet. After six months, she had lost 42 pounds. How much money had she saved for the suit?

18. The United Campaign estimates that it will receive $17 from each employee working at American Limited. American Limited currently has 463 employees. How much money can the United Campaign expect to receive from this group?

19. Projected monthly expenses for Le Cafe are as follows: rent, $1,200; utilities, $486; insurance, $389; salaries, $7,385; taxes, $243; and telephone, $157. *Estimate*, to the nearest hundred dollars, the total amount needed to pay these expenses.

20. Malcolm's scores on his biology test for this semester were
    82, 76, 93, 84, 100, 76, and 84
    Find the following statistical information.
    a. mean
    b. median
    c. mode
    d. range

Thought-Provoking Problems

1. A friend is very uneasy about starting a new math class. Write down three or four hints about studying math that you find helpful and that might help your friend.

2. Explain in two or three sentences why students should study word problems.

3. Explain how to multiply 30 by 700 without using long multiplication.

4. The local income tax collected in Centerburg last year was $404,010. Estimate how much, on the average, each of Centerburg's 1,206 working citizens paid in local tax.
   a. Tell what values you could use to estimate this answer, and explain why you chose those values.
   b. Calculate the answer using the estimated values.
   c. Compare the estimated answer to the actual answer.

5. The Golden Trippers Club is considering a trip to Gatlinburg, Tennessee. The bus fare will be $137 per person, and the lodging costs for two nights will be $91 per person. Explain how you could estimate the approximate cost per person for transportation and lodging.

6. Write down an estimated answer (approximation) for each of the following problems. Then do each of the problems using your calculator. How good were your estimates? What could you have done to make them better?
   a. $11,206 + 5,437 + 2,861$
   b. $20,105 - 11,768$
   c. $32,638 \times 1,726$
   d. $485,616 \div 1,608$

7. Why is it just as important to estimate an answer when using a calculator as it is when calculating by hand?

8. Using your calculator, find the answer to each of the following problems. Write the display that the calculator shows each time. Explain each answer.
   a. $1,320 \div 24$ b. $1,208 \div 5$
   c. $13 \div 7$ d. $0 \div 18$ e. $27 \div 0$

9. Explain two ways in which you could apply the Distributive Principle to multiply 48 times 54.

10. Write a word problem that could be solved by other students in your class. Be sure it is clear and easy to understand but also interesting and real. (Now perhaps you can appreciate that this is sometimes not so easy for authors to do!)

Whole Number Concepts

2.1 Introduction

This chapter covers a number of interesting and useful topics. Some of them will follow each other very closely. Others won't seem to be related at all to the previous or following sections. For this reason, the chapter may seem confusing to you. Don't be discouraged. Take it one step at a time. Each section introduces a skill or an approach you will need in the chapters to come and in later courses.

Learning Objectives

When you have completed this chapter, you should be able to:

1. Simplify expressions involving exponents and roots.
2. Simplify problems using the Order of Operations Agreement.
3. Answer questions that require the use of a formula.
4. Recognize prime numbers.
5. List all the factors of a given whole number.
6. Use the divisibility tests for 2, 3, and 5.
7. Find the prime factorization of a given number.
8. Find the least common multiple of a group of numbers.
9. Find the greatest common factor of a group of numbers.

2.2 Exponents and Roots

Before going any further, you need to understand what exponents and roots are. The expression 7^2 is read "7 to the second power" or "7 squared." It means 7×7, or 49. The 7 is referred to as the **base**, and the 2 is referred to as the **exponent** or **power**. 4^3 means $4 \times 4 \times 4$, or 64. The 4 is the base, and the 3 is the exponent or power. The exponent tells how many times the base is used as a factor. The expression 3×4 means 4 will be used as an addend 3 times: $4 + 4 + 4$. The expression 4^3 means 4 will be used as a factor 3 times: $4 \times 4 \times 4$. An exponent can be any number, but the most commonly used exponents are 2 and 3. 4^2 means 4×4, or 16. 5^3 is read "5 to the third power" or "5 cubed"; it means $5 \times 5 \times 5$, or 125.

$\sqrt{49}$ is read "the square root of 49." The square root means the number that when squared gives 49. Therefore, $\sqrt{49} = 7$. $\sqrt{36}$ is read "the square root of 36," and it represents the number that when multiplied by itself gives 36. Therefore, $\sqrt{36} = 6$. The root symbol, $\sqrt{}$, is called a **radical symbol**, or just a **radical**, and the quantity inside the symbol is called the **radicand**. Problems involving exponents and radicals are easy to do once you understand what the symbols mean. Look at the pattern of numbers in Table 1, which gives the most commonly used squares, cubes, and square roots. In the third column, the radicands are all **perfect squares**—numbers produced when a digit is "squared," or multiplied by itself.

TABLE 1

Squares	Cubes	Square Roots
$1 \times 1 = 1^2 = 1$	$1 \times 1 \times 1 = 1^3 = 1$	$\sqrt{1} = 1$
$2 \times 2 = 2^2 = 4$	$2 \times 2 \times 2 = 2^3 = 8$	$\sqrt{4} = 2$
$3 \times 3 = 3^2 = 9$	$3 \times 3 \times 3 = 3^3 = 27$	$\sqrt{9} = 3$
$4 \times 4 = 4^2 = 16$	$4 \times 4 \times 4 = 4^3 = 64$	$\sqrt{16} = 4$
$5 \times 5 = 5^2 = 25$	$5 \times 5 \times 5 = 5^3 = 125$	$\sqrt{25} = 5$
$6 \times 6 = 6^2 = 36$	$6 \times 6 \times 6 = 6^3 = 216$	$\sqrt{36} = 6$
$7 \times 7 = 7^2 = 49$	$7 \times 7 \times 7 = 7^3 = 343$	$\sqrt{49} = 7$
$8 \times 8 = 8^2 = 64$	$8 \times 8 \times 8 = 8^3 = 512$	$\sqrt{64} = 8$
$9 \times 9 = 9^2 = 81$	$9 \times 9 \times 9 = 9^3 = 729$	$\sqrt{81} = 9$
$10 \times 10 = 10^2 = 100$	$10 \times 10 \times 10 = 10^3 = 1,000$	$\sqrt{100} = 10$
$11 \times 11 = 11^2 = 121$	$11 \times 11 \times 11 = 11^3 = 1,331$	$\sqrt{121} = 11$
$12 \times 12 = 12^2 = 144$	$12 \times 12 \times 12 = 12^3 = 1,728$	$\sqrt{144} = 12$

Try these exercises and see if your answers agree with the ones that follow. If possible, don't look up the values given in the table. Do the calculations on your own!

Exercises Find the value of each of the following expressions.

1. 3^2 2. 12^2 3. $\sqrt{64}$ 4. 2^4 5. $\sqrt{121}$
6. $\sqrt{4}$ 7. 3^3

Answers:

1. 9 2. 144 3. 8 4. 16 5. 11 6. 2 7. 27

What happens when the number inside the square root symbol, $\sqrt{}$, is not a perfect square? For instance, what is the value of $\sqrt{20}$? To answer this question, you may want to use a square root table or a calculator. Table 2 is a portion of a square root table. The values, except for the perfect squares, are rounded to three decimal places.

T A B L E 2

Number	Square Root	Number	Square Root
14	3.743	21	4.583
15	3.873	22	4.690
16	4	23	4.796
17	4.123	24	4.899
18	4.243	25	5
19	4.359	26	5.099
20	4.472	27	5.196

The values given in Table 2, except 16 and 25, are decimal approximations. When squared, each gives a number very close to the number whose square root you wish to find. In order to find the value of $\sqrt{20}$, it helps to realize that 20 is between 16 and 25. Because $\sqrt{16} = 4$ and $\sqrt{25} = 5$, $\sqrt{20}$ should have a value between 4 and 5, somewhat closer to 4 than to 5. Looking up the square root of 20 in the table reveals that $\sqrt{20} \doteq 4.472$. (The symbol $\doteq$ means "is approximately equal to." Some texts use $\approx$ to mean the same thing.) To see if this answer is correct, multiply 4.472 by itself.

$$4.472 \times 4.472 = 19.998784$$

Remember that the values in the table are approximations. Because 20 is not a perfect square, it has no exact whole number square root. A more comprehensive square root table is located on the inside back cover of your text. Use the values from that table to find the following square roots.

Exercises Find an approximate value for each of the following expressions. Use the square root table on the inside back cover.

1. $\sqrt{18}$ **2.** $\sqrt{7}$ **3.** $\sqrt{39}$ **4.** $\sqrt{42}$ **5.** $\sqrt{11}$

Answers:

1. 4.243   **2.** 2.646   **3.** 6.245   **4.** 6.481   **5.** 3.317

Radical problems do not always involve square roots. $\sqrt[3]{8}$ is asking you to find the cube root, or the third root, of 8—that is, the number that when cubed (used as a factor 3 times) gives 8. That number is 2 because $2 \times 2 \times 2 = 8$. Therefore, $\sqrt[3]{8} = 2$. The raised 3 on the outside of the root symbol is called its **index**. A raised 4 on the outside of the radical is asking you to find the fourth root of the radicand. $\sqrt[4]{81} = 3$, because $3 \times 3 \times 3 \times 3 = 81$. Are you wondering where the raised 2 is in square root problems? $\sqrt{25}$ could indeed be written $\sqrt[2]{25}$, but in writing square roots there is no need to include the index. You may assume you are working with a square root unless there is an index to tell you differently.

Use your own calculation abilities or Table 1 to simplify the following problems.

Exercises Find the value of each of the following expressions.

1. $\sqrt[3]{125}$ **2.** $\sqrt[4]{16}$ **3.** $\sqrt{64}$ **4.** $\sqrt[3]{1000}$

Answers:

1. 5   **2.** 2   **3.** 8   **4.** 10

2.2 ◆ Practice Problems

A. Simplify each of the following expressions involving exponents. Do the problems without using the tables.

1. 3^3 2. 4^2 3. 5^2 4. 6^3

5. 1^5 6. 0^4 7. 2^8 8. 1^7

9. 0^3 10. 3^5

B. Simplify each of the following expressions involving square roots. When necessary, refer to the table on the inside back cover.

11. $\sqrt{35}$ 12. $\sqrt{49}$ 13. $\sqrt{17}$ 14. $\sqrt{27}$

15. $\sqrt{36}$ 16. $\sqrt{85}$ 17. $\sqrt{52}$ 18. $\sqrt{100}$

19. $\sqrt{16}$ 20. $\sqrt{96}$

C. Simplify each of the following expressions involving radicals. When necessary, refer to the table on the inside back cover.

21. $\sqrt[3]{64}$ 22. $\sqrt[4]{1}$ 23. $\sqrt[4]{16}$ 24. $\sqrt[3]{729}$

25. $\sqrt[5]{1}$ 26. $\sqrt[2]{100}$ 27. $\sqrt[3]{1000}$ 28. $\sqrt[5]{0}$

29. $\sqrt[2]{121}$ 30. $\sqrt[3]{216}$

D. Answer each of the following questions.

31. What is the square root of 196?

32. What is the third root of 125?

33. What is the third root of 512?

34. What is the square root of 144?

35. Is 289 a perfect square? Why or why not?

36. Is 169 a perfect square? Why or why not?

37. Is 343 a perfect cube (third power)? Why or why not?

38. Is 64 a perfect cube (third power)? Why or why not?

39. What number used as a factor four times is 256?

40. What number used as a factor five times is 243?

E. Use your calculator to find the simplified form of each of the following expressions. Carry out your answer to four decimal places for any root problem.

41. 8^4 42. 7^5 43. $\sqrt[5]{32}$

44. $\sqrt[5]{3,125}$ 45. $\sqrt[4]{6,561}$ 46. 12^4

47. 16^5 48. $\sqrt[3]{4,913}$ 49. $\sqrt[3]{1,000,000}$

50. $\sqrt[4]{50,625}$

2.3 Order of Operations Agreement

In an example in Chapter 1 the total needed to pay for an order at the grocery store was calculated. To figure that total, you added after doing some multiplication, because the order included more than one of some items. Often in mathematical situations, you need to do more than one operation within a problem. In application problems such as the one at the grocery store, the order in which the operations must be done is obvious. In some calculation problems, however, that is not the case. Consider this simple problem:

$$25 - 4 \times 3$$

It would seem perfectly normal to do the problem from left to right because that is the direction in which you read. In some other countries, however, people read from top to bottom because their alphabet and communication customs are different from our own. Because mathematics is a universal language, it became necessary, centuries ago, for mathematicians to reach an agreement about how certain problems would be done. The **Order of Operations Agreement** used today is a result of that effort. All people throughout the world do mathematics by following these same steps. As you will see, there are four steps in the Order of Operations Agreement, and you must complete each step, if applicable, before going on to the next.

Order of Operations Agreement

When simplifying a mathematical expression involving more than one operation, follow these steps:

1. Simplify within parentheses or other grouping symbols.
2. Simplify any expressions involving exponents or square roots.
3. Multiply or divide in order from left to right.
4. Add or subtract (combine) in order from left to right.

Let's look again at the problem $25 - 4 \times 3$. This problem involves subtraction and multiplication. The agreement says to multiply before subtracting. Underlining each step as it is to be done helps you to keep on track. It is much easier to simplify one step at a time and sift your way down through the problem than to work "sideways." Therefore,

$$25 - \underline{4 \times 3} \qquad \text{Multiply before subtracting.}$$
$$= \underline{25 - 12} \qquad \text{Subtract from left to right.}$$
$$= 13$$

Grouping symbols, such as parentheses, are used to indicate an order other than the normal operations agreement. When parentheses appear, we work within them before following the normal procedure. If the problem had been written $(25 - 4) \times 3$, we would have done the operation within the parentheses first and then the multiplication.

$$\underline{(25 - 4)} \times 3 \qquad \text{Work within parentheses first.}$$
$$= \underline{21 \times 3} \qquad \text{Multiply from left to right.}$$
$$= 63$$

Here is another example where parentheses make a difference.

$$\underline{28 \div 4} + 3 \qquad \text{Divide before you add.}$$
$$= \underline{7 + 3} \qquad \text{Add from left to right.}$$
$$= 10$$

Now with the parentheses included in the problem, the order of the operations changes.

$$28 \div \underline{(4 + 3)} \qquad \text{Work within the parentheses first.}$$
$$= \underline{28 \div 7} \qquad \text{Divide from left to right.}$$
$$= 4$$

Wherever you see a number next to a parenthesis, with no operation symbol between them, that indicates multiplication. Look at this problem:

$$6\underline{(7 - 4)} \qquad \text{Work within parentheses first.}$$
$$= \underline{6(3)} \qquad \text{Multiply by 6.}$$
$$= 18$$

Here's another:

$$20 - 4\underline{(8 - 6)} \qquad \text{Work within parentheses first.}$$
$$= 20 - \underline{4(2)} \qquad \text{Multiply before you subtract.}$$
$$= \underline{20 - 8} \qquad \text{Subtract from left to right.}$$
$$= 12$$

Study each of these examples.

E X A M P L E   **1**

$5 + \underline{(14 - 3)} \times 2$

$= 5 + \underline{11 \times 2}$

$= \underline{5 + 22}$

$= 27$ ◆

E X A M P L E   **2**

$\underline{(36 \div 3)} + \underline{(12 \times 2 - 6)}$

$= 12 + \underline{(24 - 6)}$

$= \underline{12 + 18}$

$= 30$ ◆

E X A M P L E   **3**

$14 + \underline{12 \times 4} - 3$

$= \underline{14 + 48} - 3$

$= \underline{62 - 3}$

$= 59$ ◆

E X A M P L E   **4**

$45 - \underline{18 \div 3} + 3$

$= \underline{45 - 6} + 3$

$= \underline{39 + 3}$

$= 42$ ◆

E X A M P L E   **5**

$\underline{5^2} - \underline{6 \times 3} + 4$

$= \underline{25 - 18} + 4$

$= \underline{7 + 4}$

$= 11$ ◆

E X A M P L E   **6**

$\underline{\sqrt{81}} + \underline{24 \div 3} - 2$

$= \underline{9 + 8} - 2$

$= \underline{17 - 2}$

$= 15$ ◆

E X A M P L E   **7**

$\underline{12 \div 3} \times 4 \div 2$

$= \underline{4 \times 4} \div 2$

$= \underline{16 \div 2}$

$= 8$ ◆

E X A M P L E **8** $\underline{25 - 17} + 3 - 5$

$= \underline{8 + 3} - 5$

$= \underline{11 - 5}$

$= 6$ ◆

In Examples 9 and 10, you will notice more than one pair of grouping symbols within each problem. This happens often in mathematics. The symbols [ ], called brackets, serve the same function as a set of parentheses. They indicate that whatever is to be done inside of them should be completed before moving on. They are like another set of parentheses but are shaped differently so that you don't get confused by a set of parentheses within a set of parentheses. Work within the innermost set of grouping symbols first. You may then use the quantity that you have found to complete the problem. In most problems involving more than one set of grouping symbols, the parentheses are inside the brackets. You just need to remember to begin with the innermost grouping symbols and work your way out.

E X A M P L E **9** $(\underline{220 \div 10})(\underline{18 + 5})$

$= (22)(23)$

$= 506$ ◆

E X A M P L E **10** $[2(\underline{5 \times 5}) - 7][6 - (\underline{12 \div 3})]$

$= [\underline{2(25)} - 7][6 - (4)]$

$= [\underline{50 - 7}][\underline{6 - 4}]$

$= [43][2]$

$= 86$ ◆

E X A M P L E **11** $8 + [13 - 5(\underline{6 - 4})]$

$= 8 + [13 - \underline{5(2)}]$

$= 8 + [\underline{13 - 10}]$

$= \underline{8 + 3}$

$= 11$ ◆

E X A M P L E **12** $7[21 - 8(\underline{5 - 3})]$

$= 7[21 - \underline{8(2)}]$

$= 7[\underline{21 - 16}]$

$= \underline{7[5]}$

$= 35$ ◆

If there are no exponents or square roots within the problem, you can skip over step 2 of the Order of Operations Agreement. If there are no parentheses, you can begin with step 2. Remember, though, that it is important for you to check for each step *in order* in each problem. Order of operations means the order in which things are to be done. Don't be discouraged if it takes a while before you can do these problems correctly all the time. Start working the problems with the Order of Operations Agreement right in front of you, and go through all four steps each time. When you learn the Agreement and can put it away, you will quickly become successful with these problems.

2.3 ◆ Practice Problems

In Sections A, B, and C, simplify the mathematical expressions using the Order of Operations Agreement. Be sure to check for each of the four steps in every problem.

A.

1. $4 + (6 \times 3)$

2. $12 - (2 \times 5)$

3. $(8 + 2) - (12 \div 3)$

4. $(10 \div 2) + (4 \times 3)$

5. $19 - (3 \times 4)$

6. $28 + (9 \div 3)$

7. $(12 \times 8) \div (4 \times 4)$

8. $(20 \times 9) \div (5 \times 3)$

9. $(5 + 2)(8 \div 2)$

10. $(3 \times 5)(4 \div 2)$

11. $30 - (2 + 1)(5 \times 2)$

12. $42 - (3 \times 2)(6 + 1)$

13. $4 + 5 \times 3$

14. $6 + 4 \times 4$

15. $24 - 3 \times 6$

16. $18 - 2 \times 7$

17. $12 + 4 \times 5 - 6$

18. $10 + 5 \times 3 - 7$

19. $24 - 6 + 7 - 2$

20. $18 - 4 + 9 - 7$

B.

21. $14 + 6(5)$

22. $12 + 7(3)$

23. $36 \div 4 \times 3$

24. $42 \div 6 \times 7$

25. $20 - 3(4 + 2)$

26. $24 - 4(2 + 1)$

27. $20 - (3 \cdot 4) + 2$

28. $24 - (4 \cdot 2) + 1$

29. $28 \div (4 + 3)$

30. $36 \div (9 + 3)$

31. $(28 \div 4) + 3$

32. $(36 \div 9) + 3$

33. $24 + 12 \div 6 + 3$

34. $21 + 14 \div 7 - 2$

35. $(24 + 12) \div 6 + 3$

36. $(21 + 14) \div 7 - 2$

C.

37. $3^2 + \sqrt{25}$

38. $5^2 - \sqrt{49}$

39. $2^4 - \sqrt{16}$

40. $4^2 + \sqrt{36}$

41. $3 \cdot 4^2 \div 2^3$

42. $8 \cdot 2^4 \div 4^2$

43. $3\sqrt{49} - 3^2$

44. $4\sqrt{25} - 4^2$

45. $\sqrt{36} + \sqrt{81} - \sqrt{9}$

46. $\sqrt{16} + \sqrt{121} - \sqrt{64}$

47. $38 - 2\sqrt{25}$

48. $42 - 3\sqrt{9}$

49. $4 + [3(9 - 5)]$

50. $6 + [4(8 - 3)]$

51. $24 - 3[5 + (2 + 1)]$

52. $37 - 4[3 + (8 - 2)]$

53. $3[12 - 4 \times 2]$

54. $5[16 - 3 \times 4]$

55. $5 + 4[20 - 2(7 - 2)]$

56. $4 + 2[18 - 3(5 - 2)]$

57. $[2 + 3(4)][5 + 2(3)]$

58. $[4 + 2(3)][7 - 2(2)]$

59. $58 - 3[4 + 2(6)]$

60. $47 - 5[3 + 2(3)]$

D. Use your calculator to simplify the following expressions using the Order of Operations Agreement. Give your answer to three decimal places if necessary.

61. $5.2(8.1 - \sqrt{20})$

62. $17.9(8.76 - \sqrt{12})$

63. $(14.3)(19.72) - 6.4(8.9)$

64. $28.2[16.4 - 2.9(4.7)]$

65. $(17.3)^3 - (16.2)^2$

66. $(21.8)(14.7) - 9.2(12.45)$

67. $37.4[26.8 - 7.3(2.5)]$

68. $(23.5)^2 - (7.3)^3$

69. $(6.5)^2 - \sqrt{169}$

70. $(14.5)^2 - \sqrt{225}$

<table>
<tr><td>**2.4**</td><td># Using Formulas</td></tr>
</table>

Formulas are general statements that are always true and that can be applied to particular situations. Formulas are general statements that express relationships. For instance, if you were planning a trip, you might need to know how far you can travel in your first 8-hour day if you can average 52 miles per hour. There is a formula that states that the distance (D) you can travel is equal to your rate (r), or speed, times the length of time (t) that you travel. Written as a formula, it is $D = rt$. Whenever two **variables**, or letters that stand for numbers, appear next to each other with no symbol in between, the operation called for is multiplication. A number next to a variable also means to multiply. This formula can be used to determine distance whether you are talking about a trip by car, by plane, or by space shuttle. The relationship among distance, rate, and time is always the same. The values for the letters change, but the formula remains the same. To determine how far you can travel in the first 8 hours of your trip, write the formula, substitute the given values, and perform the mathematical calculation that is called for.

E X A M P L E 1 How far could you travel in your first 8-hour day if you could average 52 miles per hour?

Solution

$D = rt$	Write the formula.
$D = (52)(8)$	Substitute the given values.
$D = 52 \times 8$	Two letters next to each other tell you to multiply the values.
$D = 416$	You could travel 416 miles at 52 miles per hour during an 8-hour day. ◆

Suppose you are putting a fence around your back yard, and the yard has an unusual shape. You are putting a fence around the outside of the yard, so you need to calculate the perimeter of the yard. (Whenever you are looking for the distance around something, you are looking for **perimeter**. If you want to know how much space is contained inside the fence, you are looking for **area**.) Suppose the yard has five different sides. You could write the formula for the perimeter of any five-sided figure as $P = a + b + c + d + e$, where a, b, c, d, and e represent the lengths of the sides.

E X A M P L E 2 The measurements of the sides of your yard are 50 feet, 35 feet, 42 feet, 12 feet, and 47 feet. Use the formula to find the amount of fencing that you would need to buy.

Solution

$P = a + b + c + d + e$	Write the formula.
$P = 50 + 35 + 42 + 12 + 47$	Substitute the values. Perform the operation called for.
$P = 186$	You would need 186 feet of fencing for the yard. ◆

When you are given a mathematical formula and the values to be substituted for the letters, you will often need to remember how to apply the Order of Operations Agreement to do the calculations. Here is an example of a nonsense formula, one that doesn't have a real application.

EXAMPLE 3 Using the formula $L = M + K(N - P)$, solve for L when $M = 12$, $K = 4$, $N = 9$, and $P = 3$.

Solution

$L = M + K(N - P)$	Write the formula.
$L = 12 + 4(9 - 3)$	Substitute the given values.
$L = 12 + 4(6)$	Work within the parentheses.
$L = 12 + 24$	Multiply before adding.
$L = 36$	Add from left to right. ◆

Geometry figures and terms are often used in books other than geometry texts because pictures make it easier to remember a concept. Geometry is, in part, the study of shapes. Geometric figures, such as rectangles, are appropriate to this study of formulas because the formulas apply no matter how large or small the rectangle. The formula for the area of a rectangle is $A = L \times W$, or $A = LW$. (Remember that two variables next to each other tell you to multiply their values.)

EXAMPLE 4 Find the area of a rectangle whose length is 27 yards and whose width is 14 yards (see Figure 1).

Solution

$A = LW$	Write the formula.
$A = (27)(14)$	Substitute the given values. Perform the indicated operation.
$A = 378$	The area of the rectangle is 378 square yards. (Area is always measured in *square* units.) ◆

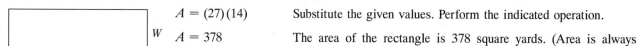

FIGURE 1

EXAMPLE 5 Find the perimeter of the rectangle in Example 4.

Solution

$P = 2L + 2W$	Because perimeter is equal to the sum of the sides, and because a rectangle has two long sides called **lengths** and two short sides called **widths**, this is the formula to use.
$P = 2(27) + 2(14)$	Substitute the given values.
$P = 54 + 28$	Perform the indicated operations.
$P = 82$	The perimeter of the rectangle is 82 yards. ◆

EXAMPLE 6 Find the perimeter of an **equilateral** (all sides equal) **triangle** if each of the three equal sides is 23 inches long.

Solution

$P = 3S$	The formula for perimeter of any triangle is $P = a + b + c$, where a, b, and c represent the lengths of the sides. In this case, however, all the sides are of equal length so $P = S + S + S$, or $P = 3S$ (see Figure 2).
$P = 3(23)$	Substitute the given values. Perform the indicated operation.
$P = 69$	The perimeter of this equilateral triangle is 69 inches. ◆

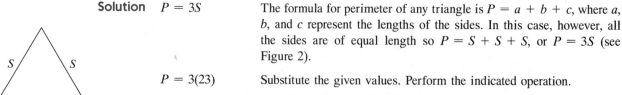

FIGURE 2

E X A M P L E 7 Find the area of the square in Figure 3. Each of its equal sides is 17 inches long.

Solution $A = S^2$ Because a square is a type of rectangle, use the formula for the area of a rectangle, $A = LW$. Because the figure is a square, L and W are the same; call them S. Therefore, $A = LW = S \times S = S^2$.

$A = 17^2$ Substitute.

$A = 289$ The area is 289 square inches.

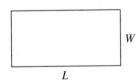

FIGURE 3

As you study further in this book, you will find a number of sections that deal with formulas. Some will be taken from geometry; others will involve interest, percents, or just nonsense! The process and steps remain the same in all problems involving a formula.

2.4 ◆ Practice Problems

A. Problems 1 through 4 use Figure 4 and these formulas:

$P = 2L + 2W$
$A = LW$

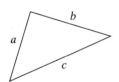

FIGURE 4

1. Find the perimeter for the rectangle if the length is 26 yards and the width is 5 yards.

2. Find the area of the rectangle in Problem 1.

3. Find the area of the rectangle if its width is 12 feet and its length is 40 feet.

4. Find the perimeter of the rectangle in Problem 3.

B. Problems 5 and 6 use Figure 5 and the formula
$P = a + b + c$

FIGURE 5

5. Find the perimeter of a triangle if its sides are 12 inches, 15 inches, and 19 inches.

6. Find the perimeter of a triangle whose sides are 8 miles, 15 miles, and 22 miles.

C. Problems 7 through 10 use the distance formula $D = rt$, which says that the distance you can travel is equal to the product of the rate, or speed, at which you move and the length of time that you travel.

7. Find the distance you can travel at 48 mph in 8 hours.

8. Find the distance you can travel at 52 mph in 7 hours.

9. Find the distance an object can travel at 1,200 feet per second in 15 seconds.

10. Find the distance that an object can travel at 1,500 feet per minute in 8 minutes.

D. Solve Problems 11 through 20 using the formulas given. (They have no specific application in the real world.)

11. Find P in $P = 2R + 3S$ when $R = 5$ and $S = 4$.

12. Find P in $P = 2R + 3S$ when $R = 4$ and $S = 5$.

13. Find K in $K = 3L - 2M$ when $L = 12$ and $M = 9$.

14. Find K in $K = 3L - 2M$ when $L = 10$ and $M = 12$.

15. Find Y in $Y = 3A + 2B - 4C$ when $A = 5$, $B = 4$, and $C = 3$.

16. Find Y in $Y = 3A + 2B - 4C$ when $A = 4$, $B = 5$, and $C = 2$.

17. Find R in $R = 4S - 3T + 2U$ when $S = 12$, $T = 8$, and $U = 5$.

18. Find R in $R = 4S - 3T + 2U$ when $S = 10$, $T = 4$, and $U = 8$.

19. Find A in $A = 3B^2 - 5C$ when $B = 4$ and $C = 7$.

20. Find A in $A = 3B^2 - 5C$ when $B = 6$ and $C = 12$.

E.   Solve the following problems using your calculator.

21. How far can a rocket travel in 126 seconds at 62 feet per second? (Use $D = rt$.)

22. How far does an airplane travel in 47 minutes if it averages 12 miles per minute? (Use $D = rt$.)

23. If the interest on America's national debt grows at the rate of \$23,132 per day, how much will it grow in September? (Growth = Rate × Days.)

24. How much will the interest in Problem 23 grow in a year? (Use $G = RD$.)

25. Find the perimeter of an athletic practice field that is 125 yards long and 35 yards wide. (Use $P = 2L + 2W$.)

26. Find the area of the practice field in Problem 25. (Use $A = LW$.)

27. If $A = 3B - 4D + 2C$, find A when $B = 124$, $C = 92$, and $D = 76$.

28. If $K = 4M - 2N + 5P$, find K when $M = 426$, $N = 195$, and $P = 28$.

29. Find R in $R = 5S^2 - 4T$ when $S = 27$ and $T = 32$.

30. Find W in $W = 7X^2 - 9Y$ when $X = 42$ and $Y = 26$.

2.5   Factors

In the last chapter, addends were mentioned when the topic was addition, and factors were mentioned in the discussion of whole number multiplication. Addends are numbers that are added. Factors are whole numbers that are multiplied. You will encounter factors throughout your study of math. *Divisors* are whole numbers that divide into a given number. *Factors* are divisors whose remainders are zero. They can also be whole numbers that multiply together to give a certain number. This sounds pretty complicated, but it is not confusing once you get the idea. Consider this example.

E X A M P L E   1   Find all the factors of 30. (This problem is asking for all the whole numbers that divide into 30 without a remainder).

Solution   You will need some system for doing this, because it asks for *all* the factors and you don't want to miss any of them.

What is the smallest whole number that is a factor of 30? 1, of course. Therefore, because $1 \times 30 = 30$, both 1 and 30 are factors of 30. The next number to try is 2. $2 \times 15 = 30$, so both 2 and 15 are factors of 30. And $3 \times 10 = 30$, so both 3 and 10 are factors of 30. There is no whole number to multiply by 4 to get 30, so 4 is not a factor of 30. $5 \times 6 = 30$, so both 5 and 6 are factors of 30. $6 \times 5 = 30$, but you already have 6 and 5, so you are finished. Let's look at this systematically.

$1 \times 30 = 30$
$2 \times 15 = 30$
$3 \times 10 = 30$
$4 \times *******$
$5 \times 6 = 30$
6×5, which you already have

Therefore, 1, 2, 3, 5, 6, 10, 15, and 30 are all the factors of 30.   ◆

E X A M P L E   2   List all the factors of 28.

Solution $1 \times 28 = 28$
$2 \times 14 = 28$
$3 \times *******$
$4 \times 7 = 28$
$5 \times *******$
$6 \times *******$
$7 \times 4 = 28$, which you already have

Therefore, 1, 2, 4, 7, 14, and 28 are the only whole numbers that divide evenly into 28. They are the only factors of 28. It is not necessary for you to write out the 3 × *****, the 5 × *****, and the 6 × ***** in your list as long as you remember to check all the numbers in numerical order until you come to a pair you already have. ◆

E X A M P L E   3   List all the factors of 29.

Solution $1 \times 29 = 29$
That's it. Those are the only two whole numbers that are factors of 29. A number that only has two factors, itself and 1, is called a **prime number**. Therefore, 29 is a prime number. ◆

It would be helpful if you could tell very quickly whether a number will be divisible by, say, 2. If you think about it, you already know how to tell. Is 2 a factor of 35? No. Is 2 a factor of 66? Yes. What is the difference between 35 and 66 that lets you know that? 35 is an *odd number*—a number whose last digit is 1, 3, 5, 7, or 9. Therefore, 2 will not divide into it evenly and is not a factor of 35. 66 is an *even number*—a number whose last digit is 0, 2, 4, 6, or 8. Therefore, 2 is a factor of 66. This simple test is actually a *divisibility test*. There are several of these tests. You will need to learn three of them.

Divisibility Tests

1. A number is divisible *by 2 if it ends in 0, 2, 4, 6, or 8*. 2 is then a factor of the number.
2. A number is divisible *by 3 if the sum of its digits is divisible by 3*. 3 is then a factor of the number.
3. A number is divisible *by 5 if it ends in 0 or 5*. 5 is then a factor of the number.

The divisibility tests for 2 and for 5 make good sense, and you probably would have been able to come up with the test for 5 on your own, just like you did the test for 2. The test for 3 is a strange one, though. It needs some explanation, and using it takes some practice. You don't need to know why it works. But solving problems will be much easier if you remember how it works.

E X A M P L E   4   Is 3 a factor of 114?

Solution   You could divide 114 by 3 and see that it divides evenly without a remainder. That would take some time. If you use the test for 3, you just add $1 + 1 + 4 = 6$. Because 3 is a factor of 6, you know that it is also a factor of 114. ◆

E X A M P L E   5   Is 3 a factor of 57?

Solution $5 + 7 = 12$. 3 is a factor of 12. Therefore, 3 is a factor of 57. ◆

E X A M P L E   6   Is 2 a factor of 58?

Solution   58 ends in 8, an even number, so 2 is a factor of 58. ◆

E X A M P L E   7   Is 5 a factor of 86?

Solution   86 does not end in 0 or 5, so 5 is not a factor of 86. ◆

E X A M P L E   **8**   Is 3 a factor of 46?

Solution $4 + 6 = 10$. Because 3 does not divide evenly into 10, 3 is not a factor of 46.   ◆

E X A M P L E   **9**   Decide if 2, 3, and 5 are factors of 180.

Solution   180 ends in 0, an even number, so 2 is a factor of 180.
$1 + 8 + 0 = 9$. 3 is a factor of 9, so 3 is a factor of 180.
180 ends in 0. Therefore, 5 is also a factor of 180.
You know that these answers are correct because $2 \times 90 = 180$, $3 \times 60 = 180$, and $5 \times 36 = 180$.   ◆

The reason why the test for 3 is so convenient is that the sum of the digits is usually a fairly small number, and most people recognize the early multiples of 3 from having learned the multiplication facts. There are divisibility tests for some of the other numbers, but you won't need them very often. You will be using the tests for 2, 3, and 5 many times.

Earlier we looked at the number 29 and found that it had only two factors, itself and 1. 29 is a **prime number** because that is the definition of a prime number.

Prime Number

> A prime number is a whole number that has exactly two factors, itself and 1.

Let's look at the first few numbers and see which of them are prime numbers. To prove that a number is not a prime, you need only find one factor of that number other than the number itself and 1.

$0 \times 4 = 0$, $0 \times 12 = 0$. $0 \times$ any number is 0. Therefore, 0 *is not* a prime number.

$1 \times 1 = 1$. 1 only has one factor. 1 *is not* a prime number.

$2 \times 1 = 2$. Therefore, 2 *is* a prime number.

$3 \times 1 = 3$. Therefore, 3 *is* a prime number.

$2 \times 2 = 4$. Therefore, 4 *is not* a prime number.

$5 \times 1 = 5$. Therefore, 5 *is* a prime number.

$2 \times 3 = 6$. Therefore, 6 *is not* a prime number.

You should continue the list in this manner up through the number 30 and see how many prime numbers there are between 0 and 30. There is no pattern to the prime numbers. You will just need to find and remember which numbers they are. The list of prime numbers goes on indefinitely, but the ones you will use the most are the ones less than 30. See if your list includes 2, 3, 5, 7, 11, 13, 17, 19, 23, and 29. These are the prime numbers less than 30. Remember them!

The next step in this trip through new topics is to put together the concepts of prime numbers and factors and discuss a concept called prime factorization. As the name implies, **prime factorization** involves *writing a given number as a product of its prime factors*. Once again, if you follow a pattern, it's easy. Two methods will be shown. You will have an opportunity to try both, and then you can decide which one you prefer. The first example uses a technique called a **factor tree**.

E X A M P L E **10**   Use a factor tree to find the prime factorization of 36.

Solution

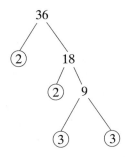

One way to make this tree is to begin with the smallest prime number that is a factor of 36. That number is 2. Write 36 as 2×18. Then take the smallest prime number that is a factor of 18. Next do the same for the 9 this step yields. That number is 3. Write 9 as 3×3. Because the remaining factor, 3, is already a prime number, you are finished. Therefore, $36 = 2 \times 2 \times 3 \times 3$.

Check   This type of problem is easy to check. 2 and 3 are indeed prime numbers, and when you multiply all the factors you do get 36.

You do not have to begin to factor 36 by using the smallest prime. You could have begun with 6 times 6 or with 4 times 9. Then each factor would need to be factored again until each branch ended in a prime. Look at the following two trees used to factor 36. Be sure to circle each prime as you find it so that you don't lose any of them along the way. Note that the prime factorization is the same no matter which two factors you use first.

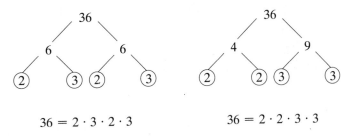

$$36 = 2 \cdot 3 \cdot 2 \cdot 3 \qquad\qquad 36 = 2 \cdot 2 \cdot 3 \cdot 3$$

Each answer could also have been written using exponent notation instead of expanded notation. In exponent notation, the prime factorization for 36 is written

$$36 = 2^2 \times 3^2 \quad \text{or} \quad 36 = 2^2 \cdot 3^2$$

because 2 and 3 are both used as a factor twice.   ◆

E X A M P L E **11**   Write 90 in its prime factored form. Use a factor tree.

Solution

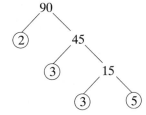

Because 3 and 5 are primes, you are finished. $90 = 2 \times 3 \times 3 \times 5$, or $90 = 2 \times 3^2 \times 5$.

Check   This is correct because 2, 3, and 5 are prime numbers and when you multiply, you get 90.   ◆

E X A M P L E **12**   Write the prime factorization of 80.

Solution

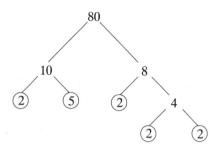

This time, factor 80 as 10×8 and then factor each branch until only primes remain. Remember to circle each prime factor as it appears.

Therefore, the prime factorization for 80 is $2 \cdot 2 \cdot 2 \cdot 2 \cdot 5$, or $2^4 \cdot 5$.

Check   2 and 5 are prime numbers, and when you multiply these factors together, you get 80.   ◆

E X A M P L E **13**   Write 40 in its prime factored form. Use the method of **repeated division**.

Solution   This time you will actually use the division process, but *each of your divisors must be a prime number.* You might begin with 2, the smallest factor that will divide without a remainder. Divide 2 into 40 and write the answer, 20, on the next line. Continue using 2 as the divisor until it no longer works. (The divisibility test will tell you this.) Then try 3, then 5, and so on through the list of prime numbers until the answer in the division is 1.

$2 \overline{)40}$ $40 \div 2 = 20$

$2 \overline{)20}$ $20 \div 2 = 10$

$2 \overline{)10}$ $10 \div 2 = 5$

$5 \overline{)\ 5}$ 2 is not a factor of 5 and neither is 3, but 5 is. $5 \div 5 = 1$.

$\quad 1$

Therefore, the prime factorization of 40 is $2 \times 2 \times 2 \times 5$. It could also be written using exponent notation: $2^3 \times 5$.

Check   All the factors are prime, and when you multiply, you get 40.   ◆

E X A M P L E **14**   Write 120 in its prime factored form. Use the method of repeated division. It may be easier to begin with the smallest prime that is a factor of 120, which is 2.

Solution $2 \overline{)120}$ $120 \div 2 = 60$

$2 \overline{)\ 60}$ $60 \div 2 = 30$

$2 \overline{)\ 30}$ $30 \div 2 = 15$

$3 \overline{)\ 15}$ $15 \div 3 = 5$

$5 \overline{)\ \ 5}$ 3 does not work, but $5 \div 5 = 1$.

$\quad 1$

Therefore, the prime factorization of 120 is $2 \times 2 \times 2 \times 3 \times 5$, or $2^3 \cdot 3 \cdot 5$.

Check   All the factors are prime numbers, and when you multiply, you get 120.   ◆

E X A M P L E   **15**   Find the prime factorization for 196. Use either method.

Solution

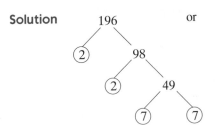

or

2)196
2) 98
7) 49
7) 7
   1

Therefore, the prime factorization of 196 is
$2 \times 2 \times 7 \times 7$, or $2^2 \cdot 7^2$.

Check   All the factors are prime numbers, and when you multiply, you get 196.   ◆

These examples show why it is so helpful to know the divisibility tests, recognize the early prime numbers, know what a factor is, and be confident of your multiplication facts. You use all of those skills in the prime factorization process. We will use prime factorization to reduce fractions in the next chapter. It will make some of the toughest-looking fractions reduce easily.

2.5 ◆ Practice Problems

A. List all the factors of the given number.

1. 8	**2.** 12	**3.** 21	**4.** 15
5. 7	**6.** 13	**7.** 32	**8.** 36
9. 18	**10.** 28	**11.** 45	**12.** 42
13. 70	**14.** 85	**15.** 120	**16.** 90
17. 150	**18.** 140	**19.** 200	**20.** 300

B. Answer the following questions without actually doing the division. Then explain your answer; that is, tell why or why not.

21. Is 14 divisible by 3?

22. Is 15 divisible by 2?

23. Is 216 divisible by 2?

24. Is 216 divisible by 3?

25. Is 216 divisible by 5?

26. Is 13,170 divisible by 2?

27. Is 13,170 divisible by 3?

28. Is 13,170 divisible by 5?

29. Is there a remainder when 525 is divided by 2?

30. Is there a remainder when 525 is divided by 3?

31. Is there a remainder when 525 is divided by 5?

32. Is 176 divisible by 3?

33. Is 2,142 divisible by 2?

34. Is 2,142 divisible by 3?

35. Is 2,142 divisible by 5?

36. Is 3 a factor of 153?

37. Is 2 a factor of 178?

38. Is 5 a factor of 270?

39. Is 3 a factor of 270?

40. Is 2 a factor of 270?

C. Find the prime factorization for each of the following numbers. You may use either a factor tree or the method of repeated division. You may write your answer in expanded form or use exponent notation.

41. 12	**42.** 18	**43.** 20	**44.** 30
45. 36	**46.** 50	**47.** 54	**48.** 42
49. 84	**50.** 98	**51.** 85	**52.** 100
53. 120	**54.** 66	**55.** 154	**56.** 132
57. 144	**58.** 225	**59.** 242	**60.** 250

D. Use your calculator to help you to find the prime factorization for each of the following numbers.

61. 2,805	**62.** 3,325	**63.** 3,927
64. 644	**65.** 14,553	**66.** 6,006
67. 4,095	**68.** 13,013	**69.** 16,445
70. 34,307	**71.** 10,780	**72.** 10,626

2.6 Least Common Multiple

Least Common Multiple

> The least common multiple (LCM) for two or more numbers is the smallest number into which each of those numbers divides evenly.

Try to find the least common multiple (LCM) for 8 and 12. The list of multiples of 12 begins

12, 24, 36, 48, . . .

The list of multiples of 8 begins

8, 16, 24, 32, 40, 48, . . .

You can see that 24 is the first common multiple for 8 and 12. There are many other numbers, such as 48, 72, and 96, that are also common multiples of 8 and 12, but 24 is the *least*. Therefore, 24 is the **least common multiple** (LCM) for 8 and 12.

You may remember that when you need to add or subtract fractions, you need a common denominator. Sometimes the process of finding a common denominator can be very difficult or require a lot of trial and error. In this section you will learn two methods for finding the least common multiple of several numbers. When those numbers are the denominators of fractions, these same methods will enable you to find the **least common denominator** for fractions. The method that looks so much like the prime factorization method is often the one that students prefer.

The approach of listing multiples works, but it is not very efficient. The other two methods you can use are less time-consuming. You need to remember that the name *least common multiple* explains exactly what you are trying to find. The examples that follow illustrate these two methods.

E X A M P L E 1 Find the least common multiple for 24 and 15 by using the method of repeated division by primes.

Solution This method looks a lot like the method for prime factorization described in the previous section. In fact, it is identical except that you divide into more than one number each time and multiply the factors together at the end. It is only necessary that the prime number you choose to divide by divides evenly into one of the numbers, although it may go into more than one of them. If it does not divide into a number, just bring that number down to the next line.

2)	24	15	Smallest prime into any is 2.
2)	12	15	Smallest prime into any is 2.
2)	6	15	Smallest prime into any is 2.
3)	3	15	Smallest prime into any is 3.
5)	1	5	Smallest prime into any is 5.
	1	1	All 1s across the bottom. Stop.

The least common multiple for 24 and 15 is $2 \times 2 \times 2 \times 3 \times 5 = 120$. ◆

E X A M P L E **2**   Find the least common multiple for 4, 6, and 20.

Solution   2) $\underline{\quad 4 \qquad 6 \qquad 20 \quad}$ Smallest prime into any is 2.

2) $\underline{\quad 2 \qquad 3 \qquad 10 \quad}$ Smallest prime into any is 2.

3) $\underline{\quad 1 \qquad 3 \qquad 5 \quad}$ Smallest prime into any is 3.

5) $\underline{\quad 1 \qquad 1 \qquad 5 \quad}$ Smallest prime into any is 5.

$\quad 1 \qquad 1 \qquad 1$ All 1s across bottom. Stop.

The LCM for 4, 6, and 20 is $2 \times 2 \times 3 \times 5 = 60$.   ◆

E X A M P L E **3**   Find the LCM for 14, 35, and 10.

Solution   2) $\underline{\quad 14 \qquad 35 \qquad 10 \quad}$ Smallest prime into any is 2.

5) $\underline{\quad 7 \qquad 35 \qquad 5 \quad}$ Smallest prime into any is 5.

7) $\underline{\quad 7 \qquad 7 \qquad 1 \quad}$ Smallest prime into any is 7.

$\quad 1 \qquad 1 \qquad 1$ All 1s across bottom. Stop.

The LCM for 14, 35, and 10 is $2 \times 5 \times 7 = 70$.   ◆

There is another way to find the least common multiple (LCM). This method parallels the one you will use later in algebra. Find the prime factorization for each of the given numbers. List the primes under each other. Then bring down each prime number and multiply those primes together. That product is the LCM of the given numbers. Study Example 4, which is done using this method. It is exactly the same problem as Example 3, so the answer should be the same.

E X A M P L E **4**   Find the LCM for 14, 35, and 10.

Solution $14 = 2 \times 7$

$35 = \qquad 7 \times 5$

$\underline{10 = 2 \times \qquad 5 \quad}$

$LCM = 2 \times 7 \times 5 = 70$

Compare this answer with that in Example 3.   ◆

E X A M P L E **5**   Find the LCM for 24 and 15.

Solution $24 = 2 \times 2 \times 2 \times 3 \qquad = 2^3 \times 3$

$\underline{15 = \qquad\qquad\quad 3 \times 5 \qquad = \qquad 3 \times 5 \quad}$

$LCM = 2 \times 2 \times 2 \times 3 \times 5 \qquad = 2^3 \times 3 \times 5 = 120$

Compare this answer with that in Example 1.
Note that when you use exponent notation for the prime factorization, each factor is used to its highest power in the least common multiple.   ◆

E X A M P L E   **6**   Find the LCM for 12, 20, and 18.

Solution
$$
\begin{array}{ll}
12 = 2 \times 2 \times 3 & = 2^2 \times 3 \\
20 = 2 \times 2 \qquad \times 5 & = 2^2 \qquad \times 5 \\
\underline{18 = 2 \qquad \times 3 \times 3} & \underline{= 2 \times 3^2} \\
\text{LCM} = 2 \times 2 \times 3 \times 3 \times 5 & = 2^2 \times 3^2 \times 5 = 180
\end{array}
$$

Again, each factor in the prime factorization appears in the LCM raised to its highest power.   ◆

2.6 ◆ Practice Problems

A. Find the least common multiple (LCM) for each of the following groups of numbers. Use either method.

1. 6 and 8
2. 4 and 10
3. 7 and 12
4. 5 and 9
5. 4, 5, and 6
6. 7, 8, and 10
7. 2, 5, and 8
8. 3, 6, and 10
9. 10 and 12
10. 12 and 20
11. 18 and 20
12. 15 and 12
13. 10, 15, and 18
14. 9, 12, and 20
15. 3, 4, 5, and 6
16. 3, 5, 8, and 10

17. 15, 20, and 24
18. 12, 18, and 20
19. 3, 7, 12, and 14
20. 5, 7, 10, and 21

B. Find the least common multiple for each of the following groups of numbers. You might find your calculator helpful.

21. 10, 12, 18, and 45
22. 16, 12, 20, and 15
23. 24, 30, 54, and 42
24. 60, 48, 108, and 36
25. 63, 45, and 25
26. 27, 45, and 12
27. 100, 60, and 75
28. 48, 30, and 72
29. 72, 180, and 108
30. 96, 72, and 120

2.7   Greatest Common Factor

Another use for prime factorization is to find the greatest common factor for a group of numbers. The **greatest common factor** (GCF) is the largest factor (divisor) common to all the given numbers. To find the greatest common factor, you will find the prime factorization for each number and then list those factors in an orderly fashion. You can then identify the factors that need to be multiplied together to produce the greatest common factor. When there are no common prime factors, the GCF is 1. Numbers whose greatest common factor is 1 are said to be **relatively prime**.

E X A M P L E   **1**   Find the greatest common factor for 20 and 30.

Solution   First, find the prime factorization for each number.

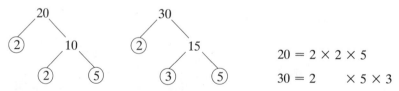

$$20 = 2 \times 2 \times 5$$
$$30 = 2 \qquad \times 5 \times 3$$

List the factors, with like factors under each other, leaving spaces for those that are missing. The greatest common factor (GCF) is the product of the factors that are *common to both 20 and 30.*

$$20 = 2 \times 2 \times 5$$
$$30 = 2 \times 3 \times 5$$
$$GCF = 2 \times 5 = 10$$

Therefore, 10 is the largest number that is a divisor (factor) of both 20 and 30. ◆

E X A M P L E 2 Find the greatest common factor for 12, 24, and 30.

Solution

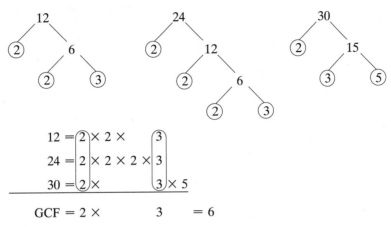

$$12 = 2 \times 2 \times 3$$
$$24 = 2 \times 2 \times 2 \times 3$$
$$30 = 2 \times 3 \times 5$$
$$GCF = 2 \times 3 = 6$$

The GCF for 12, 24, and 30 is 6. ◆

E X A M P L E 3 Find the greatest common factor for 8, 12, and 20.

Solution

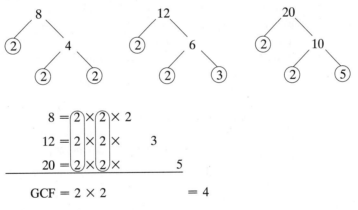

$$8 = 2 \times 2 \times 2$$
$$12 = 2 \times 2 \times 3$$
$$20 = 2 \times 2 \times 5$$
$$GCF = 2 \times 2 = 4$$

The GCF for 8, 12, and 20 is 4. ◆

E X A M P L E 4 Find the greatest common factor for 14 and 27.

Solution

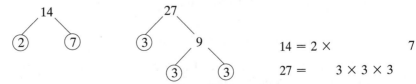

$$14 = 2 \times 7$$
$$27 = 3 \times 3 \times 3$$

Because there are no common prime factors for 14 and 27, their greatest common factor (GCF) is 1. These numbers are said to be *relatively prime* because their GCF is 1. ◆

As you practice finding the greatest common factors, your skills in using prime factorization and the divisibility tests will also improve.

2.7 ◆ Practice Problems

A. Find the greatest common factor (GCF) for each of the following sets of numbers. If the numbers are *relatively prime*, please include that in your answer.

1. 10 and 12

2. 15 and 20

3. 8 and 12

4. 8 and 20

5. 24 and 40

6. 24 and 60

7. 9 and 10

8. 12 and 35

9. 15 and 30

10. 12 and 48

11. 16, 18, and 20

12. 8, 12, and 20

13. 21 and 35

14. 28 and 42

15. 14 and 15

16. 15 and 28

17. 15, 20, and 30

18. 12, 16, and 20

19. 18, 30, and 42

20. 20, 30, and 40

 B. Find the greatest common factor (GCF) for each of the following groups of numbers. You might find it helpful to use your calculator.

21. 72, 180, and 108

22. 42, 105, and 63

23. 210, 336, and 294

24. 700, 420, and 1,120

25. 216, 360, and 432

26. 288, 180, and 108

27. 252, 504, and 315

28. 160, 192, and 256

29. 120, 216, and 144

30. 270, 360, and 315

2.8 Chapter Review

This chapter has been a real mixture of topics and ideas. The difficult thing is not mastering each one separately but knowing what to do when all of them are used together. Be sure you have studied each topic thoroughly and are comfortable with each one. Study related topics together, and you will see the progression. Use the Order of Operations Agreement to simplify some formulas. Factors and prime numbers are both involved in prime factorization. And the use of prime factorization ties together least common multiples and greatest common factors.

When you have finished studying this chapter, you should be able to:

1. Recognize and discuss the following terms and expressions: exponent, base, square root, radical, radicand, perfect square, root, formula, perimeter, area, factors, prime numbers, LCM, GCF, and relatively prime.

2. Simplify and find approximate values for exponent and radical expressions by using tables.

3. Use the Order of Operations Agreement to simplify expressions that include multiple operations and grouping symbols.

4. Substitute values into a formula, and evaluate to find the answer to a question.

5. Beginning with the least and greatest factors of a number, find *all* the factors in between.

6. Use the divisibility tests for 2, 3, and 5.

7. Find the prime factorization of any number by using a factor tree or repeated division by primes.

8. Find the least common multiple for any group of numbers by using repeated division by primes or by writing each factor to its highest power.

9. Find the greatest common factor for any group of numbers by writing each factor that is common to all the given values.

Review Problems

A. Simplify each of the following exponent or radical expressions.

1. 5^3

2. $\sqrt{36}$

3. $\sqrt[3]{64}$

4. 3^5

5. $\sqrt{144}$

6. $\sqrt[3]{1}$

7. $\sqrt[4]{16}$

8. $\sqrt[4]{81}$

9. 1^4

10. 0^6

B. Use the Order of Operations Agreement to simplify each of the following expressions.

11. $4 + 9 - 3 + 2$

12. $8 \div 2 \times 4 \div 2$

13. $24 \div 3 \times 2 \div 4$

14. $14 + 6 - 5 + 3$

15. $32 - 4 \times 3 \div 2$

16. $45 - 6 \times 3 \div 9$

17. $18 - 2 + 4 \times 3$

18. $22 - 5 + 3 \times 6$

19. $(8 - 5) + (12 \div 3)$

20. $(9 - 2) + (16 \div 2)$

21. $7(9 + 12)$

22. $9(6 + 11)$

23. $2^3 + 9 \div 3 - 1$

24. $3^2 - 8 \div 2 + 1$

25. $\sqrt{16} - 3 + 1$

26. $\sqrt{36} - 4 + 2$

27. $5^2 - \sqrt{49} + 8 \times 2$

28. $7^2 - \sqrt{81} + 7 \times 4$

29. $34 - 3(8 - 4)$

30. $26 - 4(7 - 3)$

31. $(18 \div 3)(12 - 6)$

32. $(30 \div 6)(9 + 2)$

33. $40 + 5[8 - 2(6 - 3)]$

34. $36 + 4[9 - 2(5 - 3)]$

C. Answer each of the following questions. Be sure to write a general formula first if one is not given.

35. Using $D = rt$, find out how far you can travel in 22 hours at 56 mph.

36. Find the perimeter of the form in Figure 6.

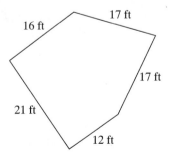

FIGURE 6

37. Find the area for the rectangle in Figure 7.
   $(A = LW)$

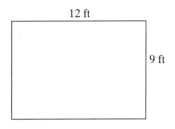

FIGURE 7

38. Using $D = rt$, see how far you can fly in 3 hours at 420 mph.

39. Find the perimeter of the form in Figure 8.

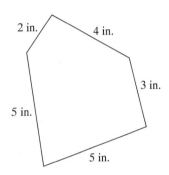

FIGURE 8

40. Find the area of the rectangle in Figure 9. $(A = LW)$

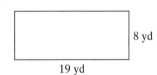

FIGURE 9

41. Using the table of square roots on the inside back cover, find the value of $\sqrt{29}$.

42. Using the table of square roots on the inside back cover, find the value of $\sqrt{43}$.

43. Find the value for A in $A = P + R(T - V)$ when $P = 8$, $R = 5$, $T = 7$, and $V = 3$.

44. Find the value for A in $A = P + R(T - V)$ when $P = 6$, $R = 8$, $T = 9$, and $V = 4$.

45. Find the value of K in $K = 2L + 3M$ when $L = 8$ and $M = 4$.

46. Find the value of R in $R = 5W - 3Y$ when $W = 8$ and $Y = 4$.

47. Find the value of H in $H = K^2 - 3L$ when $K = 9$ and $L = 5$.

48. Find the value of D in $D = E^2 - 5F$ when $E = 6$ and $F = 4$.

D. Respond to each of the following.

49. List all the factors of 18.

50. List all the factors of 24.

51. List all the factors of 30.

52. List all the factors of 72.

53. List all the factors of 120.

54. List all the factors of 105.

55. Is 126 divisible by 2? Why or why not?

56. Is 147 divisible by 3? Why or why not?

57. Is 280 divisible by 5? Why or why not?

58. Is 117 divisible by 5? Why or why not?

59. Is 118 divisible by 3? Why or why not?

60. Is 215 divisible by 3? Why or why not?

61. Is 318 divisible by 3? Why or why not?

62. Is 127 divisible by 2? Why or why not?

E. Find the prime factorization for each of the following numbers.

63. 80	64. 84	65. 210	66. 90
67. 114	68. 150	69. 75	70. 126
71. 66	72. 85	73. 132	74. 242

F. Find the least common multiple (LCM) for each of the following groups of numbers. Use any method.

75. 12 and 20	76. 18 and 30
77. 8, 12, and 15	78. 45 and 60
79. 3, 14, and 12	80. 4, 15, and 18
81. 36 and 60	82. 42 and 70
83. 70 and 105	84. 12, 20, and 15

G. Find the greatest common factor (GCF) for each of the following groups of numbers.

85. 42 and 70	86. 20 and 21
87. 30 and 77	88. 105 and 70
89. 12, 18, and 30	90. 16, 40, and 24

Chapter Test

1. List the prime numbers less than 25.

2. Is 3 a factor of 276? Why or why not?

3. Find the prime factorization for 630.

4. Find the prime factorization for 175.

5. Find the LCM for 14 and 15.

6. Find the LCM for 6, 20, and 8.

7. Find the LCM for 9, 10, and 12.

8. Find the GCF for 18 and 42.

9. Find the GCF for 12 and 35.

10. Simplify $38 - 4 \times 5 + 2$.

11. Simplify $12 - 4 + 3 - 2$.

12. Simplify $7 + 4[18 - 2(6 - 2)]$.

13. Simplify $4^3 - 3^2$.

14. List all the factors of 50.

15. Find the value of R in $R = 3S - 4T$ when $S = 24$ and $T = 13$.

16. If $K = 3A - 5B + 4C$, find the value of K when $A = 6$, $B = 2$, and $C = 3$.

17. If $M = 3X^2 - 4X + 2$, find the value of M when $X = 4$.

18. Using the formula $A = LW$, find the area of a rectangle that is 18 feet long and 7 feet wide.

19. Find the perimeter of the rectangle in Problem 18 by using the formula $P = 2L + 2W$.

20. Find the distance you can travel in 7 hours if you average 43 miles per hour. Use $D = rt$.

Thought-Provoking Problems

1. Now that you have finished studying Chapter 2, what advice would you give to someone just beginning to study it? (Be encouraging, but realistic.)

2. Comment on whether you find a fragmented chapter, like Chapter 2, easier or harder to study than a chapter like Chapter 1 that deals with just one topic.

3. Describe the relationship between squared numbers and square roots.

4. Write an application problem in which you would have to use two different mathematical operations. Is there more than one way to solve your problem? Does it matter which math operation you do first?

5. Use your calculator to find the value of each of the following.

   a. $4 \times 4 \times 4$ b. 4^3 c. $\sqrt{25}$
   d. $\sqrt{32}$ e. $\sqrt{36}$ f. $\sqrt[3]{64}$

6. a. Do the following problem on paper using the Order of Operations Agreement: $26 - 4 \times 5$. Now do the problem using your calculator. Are both answers the same? Did you get 6 both times? If your calculator display showed 6, it has built-in order of operations. This feature is called **hierarchy**. If your calculator display showed 110, it is not programmed to follow the Order of Operations Agreement, and you must be careful when doing this type of problem.

    b. Do the following problem on your paper.

    $$33 - 6 \times 4 \div 3 + 5 \times 2$$

    If your calculator has hierarchy (order of operations), key in the same problem and check the answer that you got when you worked the problem by hand.

7. Explain the difference between the LCM and the GCF. (*Hint*: You might use 12 and 30 as an example.)

8. Write a divisibility test for 6. Show that it works for at least three numbers. Explain why it works.

Multiplication and Division of Fractions

3.1 Fractions Again!

Here we are ready to start fractions and many of you are saying, "I can't do fractions" or "I hate fractions!" or "Oh no, not again!" We are going to try to turn that thinking around! Successful operations with fractions require only that you learn a few rules and that you use them carefully.

The material on fractions has been divided into two parts (Chapters 3 and 4) so that you can learn and digest some of it before going on to the rest. This should make fractions more manageable.

Learning Objectives

When you have completed this chapter, you should be able to:

1. Understand the terms numerator, denominator, proper fraction, improper fraction, mixed number, equivalent fraction, reduced to lowest terms, and reciprocal.
2. Change improper fractions to mixed numbers, and vice versa.
3. Write fractions equivalent to other fractions.
4. Reduce fractions to lowest terms by prime factorization and by repeated division.
5. Multiply fractions and mixed numbers.
6. Divide fractions and mixed numbers.
7. Solve word problems that require multiplication and division of fractions and mixed numbers.

<table>
<tr><td>

3.2 The Language of Fractions
</td></tr>
</table>

You need to begin by learning the vocabulary of fractions. Then you will understand what is being explained when terms such as *numerator* and *improper fractions* are used by your instructor or appear in the text.

Every fraction indicates division. When you say 3/4, you mean "3 divided by 4," or "3 parts out of 4." The rectangle in Figure 1 is divided into 4 equal parts. Of these 4 parts, 3 are shaded; that is, 3/4 of the rectangle is shaded. In this example, the top number represents the number of parts you are interested in, and the bottom number represents the total number of parts that the rectangle contains.

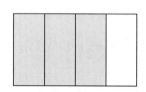

FIGURE 1

$$\frac{3}{4} \quad \frac{\text{number of parts that are shaded}}{\text{total number of parts in the rectangle}}$$

The top number is called the **numerator** and the bottom number is the **denominator**.

$$\frac{3}{4} \quad \frac{\text{numerator}}{\text{denominator}}$$

EXAMPLE 1 Write a fraction to represent each of the following statements.

a. Three students, in this math class of 25, are absent today. What fraction of the class is absent?

Solution $\dfrac{3 \text{ students absent}}{25 \text{ students enrolled in class}}$ or $\dfrac{3}{25}$ are absent.

b. This class of 25 people has 17 male students. What fraction of the class is male?

Solution $\dfrac{17 \text{ male students}}{25 \text{ students enrolled in class}}$ or $\dfrac{17}{25}$ are male.

c. What fraction of the class is female?

Solution 25 total − 17 male = 8 females

$\dfrac{8 \text{ female students}}{25 \text{ students enrolled in class}}$ or $\dfrac{8}{25}$ are female.

d. Another class contains 13 males and 19 females. What fraction of this class is female?

Solution You need the total number of students.

13 male + 19 female = 32 total

$\dfrac{19 \text{ female students}}{32 \text{ total students}}$ or $\dfrac{19}{32}$ are female.

e. Write the fraction that represents 17 minutes of an hour.

Solution The quantities must have the same units, and an hour contains 60 minutes.

$\dfrac{17 \text{ minutes}}{60 \text{ minutes}}$ or $\dfrac{17}{60}$ of an hour ◆

Any fraction whose numerator is less than its denominator is called a **proper fraction**.

$\dfrac{3}{4}, \dfrac{5}{18}, \dfrac{8}{9}, \dfrac{1}{2}, \dfrac{31}{64}$, and $\dfrac{10}{15}$ are all proper fractions.

All proper fractions have a value less than 1 because they involve only some of the total number of parts.

Any fraction whose numerator is *greater than or equal to* its denominator is called an **improper fraction**.

$\dfrac{5}{4}, \dfrac{19}{7}, \dfrac{5}{5}, \dfrac{12}{11},$ and $\dfrac{10}{10}$ are all improper fractions.

All improper fractions have a value of 1 or more. 5/4 means that all of one rectangle and 1/4 of another are shaded, as shown in Figure 2.

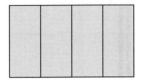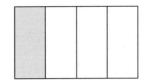

FIGURE 2

A number that is made up of a whole number *and* a fraction part is called a **mixed number**.

$1\dfrac{1}{2}, 2\dfrac{4}{5},$ and $17\dfrac{1}{8}$ are mixed numbers.

Figure 2 not only illustrates 5/4 but also shows the mixed number 1 1/4, which is read "one and one-fourth." One whole rectangle (4 parts out of 4) and one-fourth (1 part out of 4) more were shaded.

You will often need to change from one form to the other: improper fraction $\rightarrow$ mixed number or mixed number $\rightarrow$ improper fraction. To change from an improper fraction to a mixed number, divide the denominator into the numerator. This answer becomes the whole number. Then write the remainder over the divisor (denominator). In 5/3, 3 divides into 5 one time (1) with 2 left over.

$$\dfrac{5}{3} \rightarrow \text{mixed number} \qquad \text{divisor } 3\overline{)5} \begin{array}{l} 1 \text{ whole number} \\ \underline{-3} \\ 2 \text{ remainder} \end{array} \qquad = 1\dfrac{2}{3}$$

EXAMPLE 2   Change 17/5 to a mixed number.

Solution $\dfrac{17}{5} \rightarrow \text{mixed number} \qquad 5\overline{)17} \begin{array}{l} 3 \\ \underline{-15} \\ 2 \end{array} \qquad = 3\dfrac{2}{5}$

It isn't necessary to write out these steps if you can do them in your head.   ◆

Exercises   Change these improper fractions to mixed numbers.

1. $\dfrac{5}{3}$ **2.** $\dfrac{22}{7}$ **3.** $\dfrac{65}{4}$ **4.** $\dfrac{13}{9}$

Answers

1. $1\dfrac{2}{3}$ **2.** $3\dfrac{1}{7}$ **3.** $16\dfrac{1}{4}$ **4.** $1\dfrac{4}{9}$

To change from a mixed number to an improper fraction, you do the opposite operation from division: you multiply. Multiply the whole number times the denominator, and then add the numerator to this product and write the answer over the original denominator. This is not as complicated as it sounds. (See Figure 3.)

$3\dfrac{1}{4} \rightarrow$ improper fraction

$\dfrac{\text{whole number} \times \text{denominator} + \text{numerator}}{\text{original denominator}}$

$= \dfrac{3 \times 4 + 1}{4} = \dfrac{13}{4}$

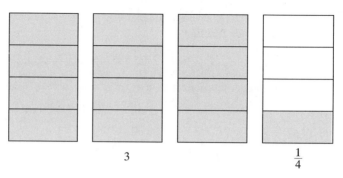

3    $\frac{1}{4}$

F I G U R E 3

E X A M P L E **3**   Change these mixed numbers to improper fractions. (Again, you don't have to write these steps out if you can do them in your head.)

a. $6\frac{4}{5}$ b. $2\frac{5}{9}$ c. $1\frac{2}{3}$

Solution   a. $6\frac{4}{5} = \frac{6 \times 5 + 4}{5} = \frac{34}{5}$

b. $2\frac{5}{9} = \frac{2 \times 9 + 5}{9} = \frac{23}{9}$

c. $1\frac{2}{3} = \frac{1 \times 3 + 2}{3} = \frac{5}{3}$ ◆

The definitions in this section can be stated in general form. That means that they are written with letters in place of the numbers to show that the statements are true for *all* numbers.

Defining Fractions

A **fraction** is a number in the form *a/b*, where *a* is any number and *b* is any number except zero (division by zero is undefined).

$$\frac{2}{3}, \frac{14}{9}$$

The **numerator** of the fraction *a/b* is *a*, and the **denominator** is *b*.
    In 2/3, 2 is the numerator and 3 is the denominator.
If *a* is less than *b* in *a/b*, then the fraction is **proper**.
    2/3 is proper because 2 is less than 3.
If *a* is greater than or equal to *b* in *a/b*, then the fraction is **improper**.

    9/8 and 16/16 are improper.

If *a/b* is an improper fraction, then it can be changed to a **mixed number** by dividing *b* into *a* and writing the remainder over *b*.

$$\frac{23}{7} = 3\frac{2}{7}$$

If *C a/b* is a mixed number, with *C* a whole number, then it can be changed to an improper fraction by following the formula:

$$C\frac{a}{b} = \frac{C \times b + a}{b}$$

$$5\frac{3}{8} = \frac{43}{8}$$

3.2 ◆ Practice Problems

A. Write a fraction to represent each statement.

1. An algebra class has 35 students enrolled. The number of female students is 13. Write
   a. The fraction of the total number of students that are female
   b. The fraction of the total number of students that are male

2. A button box contains 5 white buttons and 3 black buttons. Write
   a. The fraction of the total number of buttons that are white
   b. The fraction of the total number of buttons that are black

3. The Learning Center is open 13 hours each weekday. Write
   a. The fraction of the total hours in a day that the center is open
   b. The fraction of the total hours in a day that the Center is closed

4. The Ice Kreme Shoppe has 7 flavors that contain nuts and 15 flavors that do not. Write
   a. The fraction of the total number of flavors that do not contain nuts
   b. The fraction of the total number of flavors that do contain nuts

5. The office supply drawer contains 9 yellow legal pads and 11 white pads. Write
   a. The fraction of the total number of pads that are white
   b. The fraction of the total number of pads that are yellow

6. An hour of television time for a sporting event includes 11 minutes of commercials. Write
   a. The fraction of the hour that is sports coverage
   b. The fraction of the hour that is commercials

B. Change these improper fractions to mixed numbers.

7. $\dfrac{21}{5}$ 8. $\dfrac{13}{8}$ 9. $\dfrac{7}{2}$ 10. $\dfrac{3}{2}$

11. $\dfrac{5}{3}$ 12. $\dfrac{11}{9}$ 13. $\dfrac{29}{11}$ 14. $\dfrac{6}{5}$

15. $\dfrac{67}{32}$ 16. $\dfrac{101}{25}$ 17. $\dfrac{9}{9}$ 18. $\dfrac{24}{6}$

C. Change these mixed numbers to improper fractions.

19. $6\dfrac{2}{7}$ 20. $2\dfrac{2}{3}$ 21. $11\dfrac{3}{5}$

22. $14\dfrac{2}{3}$ 23. $1\dfrac{3}{8}$ 24. $9\dfrac{4}{9}$

25. $7\dfrac{5}{9}$ 26. $3\dfrac{7}{12}$ 27. $2\dfrac{4}{7}$

D. Consider the following set of numbers.

$$\left\{1\dfrac{1}{2}, \dfrac{5}{9}, \dfrac{19}{4}, \dfrac{8}{8}, 25, \dfrac{6}{5}, 4\dfrac{4}{9}, 3, \dfrac{1}{2}\right\}$$

List all the numbers among these that

28. Are mixed numbers in the given form
29. Are improper fractions in the given form
30. Are proper fractions in the given form
31. Have a value greater than 1
32. Have a value less than 1
33. Have a value equal to 1
34. Are improper fractions or could be expressed as improper fractions
35. Are mixed numbers or could be expressed as mixed numbers

3.3 Equivalent Fractions

Write the fraction that is represented by the shaded portion of each rectangle in Figure 4.

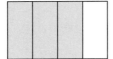   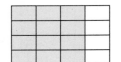

FIGURE **4**

Did you get 3/4, 6/8, 18/24, and 12/16? Even though the fractions look different, is the shaded portion of the rectangle different? No. In fact, if you were able to place the rectangles on top of each other, all the shaded portions would exactly match.

The fractions 3/4, 6/8, 18/24, and 12/16 all represent the same amount of shading in these identical figures. The fractions are all equal to each other and are **equivalent fractions**.

$$\frac{3}{4} = \frac{6}{8} = \frac{18}{24} = \frac{12}{16}$$

If you took 3/4 and multiplied *both* the numerator and the denominator by the *same nonzero number*, 2, you would get 6/8, an equivalent fraction.

$$\frac{3}{4} \times 1 = \frac{3}{4} \times \frac{2}{2} = \frac{3 \times 2}{4 \times 2} = \frac{6}{8}$$

If you multiply both parts of the fraction 3/4 by 6 or by 4, you get other equivalent fractions.

$$\frac{3 \times 6}{4 \times 6} = \frac{18}{24} \qquad \frac{3 \times 4}{4 \times 4} = \frac{12}{16}$$

In each of these problems, the original fraction is being multiplied by a form of 1, because 2/2, 6/6, and 4/4 all equal 1. The new fraction looks different from the original, but its value is the same because when you multiply a quantity by 1, the product is the same as the original number.

If you took 18/24 and divided *both* the numerator and denominator by the *same number*, you would also get an equivalent fraction.

$$\frac{18}{24} \div 1 = \frac{18 \div 6}{24 \div 6} = \frac{3}{4} \qquad \frac{18 \div 3}{24 \div 3} = \frac{6}{8}$$

You are using forms of 1 (6/6 and 3/3) to divide, so the value does not change even though the appearance changes.

E X A M P L E 1   Are $\dfrac{3}{4}$ and $\dfrac{9}{12}$ equivalent fractions?

Solution   They are if you can multiply 3 and 4 by the same factor and get 9 and 12.

Isn't $\dfrac{3 \times 3}{4 \times 3} = \dfrac{9}{12}$?   Then $\dfrac{3}{4}$ is equivalent to $\dfrac{9}{12}$, or $\dfrac{3}{4} = \dfrac{9}{12}$.   ◆

E X A M P L E 2   Are $\dfrac{18}{24}$ and $\dfrac{9}{12}$ equivalent?

Solution   They are if you can divide 18 and 24 by the same factor to get 9 and 12.

Isn't $\dfrac{18 \div 2}{24 \div 2} = \dfrac{9}{12}$?   Then $\dfrac{18}{24}$ is equivalent to $\dfrac{9}{12}$, or $\dfrac{18}{24} = \dfrac{9}{12}$.   ◆

If you multiply or divide both terms of a fraction by the same factor, you get an *equivalent fraction*. Let's state this in general terms:

Equivalent Fractions

If a, b, and c are any number (with b, $c \neq 0$), then

$$\frac{a}{b} = \frac{a \times c}{b \times c} \quad \text{and} \quad \frac{a}{b} = \frac{a \div c}{b \div c}$$

E X A M P L E   **3**   Are $\frac{8}{9}$ and $\frac{72}{81}$ equivalent?

Solution   Yes, because $\frac{8 \times 9}{9 \times 9} = \frac{72}{81}$. ◆

E X A M P L E   **4**   Are $\frac{24}{36}$ and $\frac{2}{3}$ equivalent?

Solution   Yes, because $\frac{24 \div 12}{36 \div 12} = \frac{2}{3}$. ◆

E X A M P L E   **5**   Are $\frac{21}{24}$ and $\frac{3}{8}$ equivalent?

Solution   No, because the only number that divides into both 21 and 24 is 3. But $\frac{21 \div 3}{24 \div 3} = \frac{7}{8}$

and $\frac{3}{8} \neq \frac{7}{8}$. ◆

The process of *multiplying both terms* of the fraction by the same factor *gives an equivalent fraction with larger numbers* than the original fraction had. You will apply this procedure extensively when you add and subtract fractions in the next chapter. The process of *dividing both terms* of the fraction by the same factor *produces an equivalent fraction with smaller numbers* than the original fraction. This very important concept is called **reducing**. You will apply it to every fraction answer so that the answers are always *reduced to lowest terms*.

When you solve a problem and get an answer of 3/4 and other students in the class get 12/16 or 9/12, how do you recognize that you all have the same answer? That is not a problem if all the answers are reduced to lowest terms. All those answers are 3/4 when reduced.

Simplifying all fraction answers to lowest terms is a universally accepted practice. Reducing requires that you divide out common factors that are in both the numerator and denominator.

When reducing 24/28, write the numerator and denominator as the product of prime factors.

$$\frac{24}{28} = \frac{2 \cdot 2 \cdot 2 \cdot 3}{2 \cdot 2 \cdot 7}$$

Divide out the two pairs of factors that are in both the numerator and the denominator.

$$= \frac{\overset{1}{\cancel{2}} \cdot \overset{1}{\cancel{2}} \cdot 2 \cdot 3}{\underset{1}{\cancel{2}} \cdot \underset{1}{\cancel{2}} \cdot 7}$$

Then multiply together the remaining factors.

$$= \frac{2 \cdot 3}{7} = \frac{6}{7}$$

The fraction is in lowest terms because there are no common factors other than 1.

E X A M P L E **6**   Reduce these fractions to lowest terms.

$$\text{a.} \quad \frac{14}{21} \qquad \text{b.} \quad \frac{18}{45} \qquad \text{c.} \quad \frac{21}{32} \qquad \text{d.} \quad \frac{44}{48}.$$

Solution   a. $\dfrac{14}{21} = \dfrac{2 \cdot \overset{1}{\cancel{7}}}{3 \cdot \underset{1}{\cancel{7}}} = \dfrac{2}{3}$ b. $\dfrac{18}{45} = \dfrac{2 \cdot \overset{1}{\cancel{3}} \cdot \overset{1}{\cancel{3}}}{\underset{1}{\cancel{3}} \cdot \underset{1}{\cancel{3}} \cdot 5} = \dfrac{2}{5}$

c. $\dfrac{21}{32} = \dfrac{3 \times 7}{2 \times 2 \times 2 \times 2 \times 2}$

$\dfrac{21}{32}$ is in lowest terms because there are no common factors.

d. $\dfrac{44}{48} = \dfrac{\overset{1}{\cancel{2}} \cdot \overset{1}{\cancel{2}} \cdot 11}{\underset{1}{\cancel{2}} \cdot \underset{1}{\cancel{2}} \cdot 2 \cdot 2 \cdot 3} = \dfrac{11}{12}$ ◆

You may not recognize this procedure as the method you've used to reduce fractions, but it is a similar process described in a more mathematical format. If you reduce fractions by dividing a number into both the numerator and denominator, then you are dividing out common factors, even though you didn't list the prime factors to do this. Part d of the previous problem can be done this way:

$$\frac{44}{48} = \frac{44 \div 2}{48 \div 2} = \frac{22 \div 2}{24 \div 2} = \frac{11}{12}$$

E X A M P L E **7**   Reduce $\dfrac{48}{60}$ to lowest terms.

Solution   The prime factors of 48 and 60 can be found by either the factor tree method or the repeated division method (see Section 2.5).

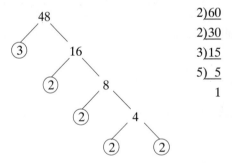

$$\frac{48}{60} = \frac{2 \times 2 \times 2 \times 2 \times 3}{2 \times 2 \times 3 \times 5}$$

$$= \frac{2}{2} \times \frac{2}{2} \times \frac{3}{3} \times \frac{2 \times 2}{5}$$

$$= 1 \times 1 \times 1 \times \frac{4}{5}$$

$$= \frac{4}{5}$$ ◆

E X A M P L E **8** Simplify $\dfrac{312}{384}$ by reducing to lowest terms.

Solution Find the prime factors of 312 and 384.

$$
\begin{array}{ll}
2)\overline{312} & 2)\overline{384} \\
2)\overline{156} & 2)\overline{192} \\
2)\overline{78} & 2)\overline{96} \\
3)\overline{39} & 2)\overline{48} \\
13)\overline{13} & 2)\overline{24} \\
1 & 2)\overline{12} \\
 & 2)\overline{6} \\
 & 3)\overline{3} \\
 & 1
\end{array}
$$

List the primes in the numerator and denominator.

$$
\frac{312}{384} = \frac{2 \times 2 \times 2 \qquad\qquad\quad \times 3 \times 13}{2 \times 2 \times 2 \times 2 \times 2 \times 2 \times 2 \times 3}
$$

Each factor of 2 in the numerator that matches a factor of 2 in the denominator can be written as a fraction with a value of 1. The pair of 3s also has a value of 1.

$$
\frac{312}{384} = 1 \times 1 \times 1 \times \frac{1}{2 \times 2 \times 2 \times 2} \times 1 \times \frac{13}{}
$$

Multiply together what remains in each part of the fraction.

$$
\frac{312}{384} = \frac{13}{16}
$$

Instead of rearranging and rewriting the factors several times, just cross out the pair of factors and write a 1 above each to indicate that the division has a value of 1. Then multiply together the factors that remain.

$$
\frac{312}{384} = \frac{\overset{1}{\cancel{2}} \times \overset{1}{\cancel{2}} \times \overset{1}{\cancel{2}} \times \overset{1}{\cancel{3}} \times 13}{\underset{1}{\cancel{2}} \times \underset{1}{\cancel{2}} \times \underset{1}{\cancel{2}} \times 2 \times 2 \times 2 \times 2 \times \underset{1}{\cancel{3}}} = \frac{13}{16} \qquad\qquad \blacklozenge
$$

E X A M P L E **9** Using prime factorization, reduce $\dfrac{315}{1,350}$ to lowest terms.

Solution

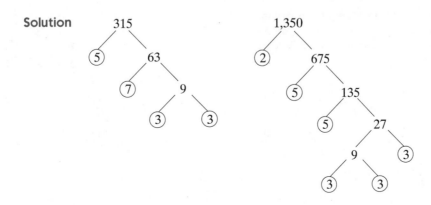

$$\frac{315}{1,350} = \frac{3 \times 3 \times 5 \times 7}{2 \times 3 \times 3 \times 3 \times 5 \times 5}$$

$$\frac{315}{1,350} = \frac{3 \times \cancel{3} \times \cancel{5} \times 7}{2 \times \cancel{3} \times \cancel{3} \times 3 \times \cancel{5} \times 5}$$

Cross out pairs of matching factors, and then multiply together the factors that remain.

$$= \frac{7}{30}$$

◆

E X A M P L E   10   Using prime factorization, reduce 1,617/1,372.

Solution

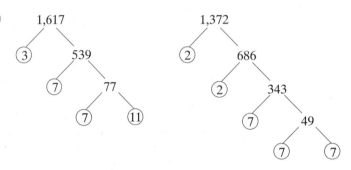

$$\frac{1,617}{1,372} = \frac{3 \times 7 \times 7 \times 11}{2 \times 2 \times 7 \times 7 \times 7} = \frac{3 \times \cancel{7} \times \cancel{7} \times 11}{2 \times 2 \times \cancel{7} \times \cancel{7} \times 7} = \frac{3 \times 11}{2 \times 2 \times 7} = \frac{33}{28}, \text{ or } 1\frac{5}{28}$$

It would have been correct to leave the improper fraction, 33/28, as an answer because it was in lowest terms. But if you are uncomfortable with that, change the improper fraction to a mixed number.   ◆

3.3 ◆ Practice Problems

A. Are these pairs of fractions equivalent fractions? Tell why. See Example 3, page 71.

1. 5/8 and 25/40
2. 18/21 and 6/7
3. 24/30 and 4/5
4. 3/5 and 21/35
5. 2/3 and 12/24
6. 10/12 and 20/24
7. 12/20 and 24/40
8. 45/54 and 5/9
9. 5/8 and 15/32
10. 12/16 and 6/8

B. Write an equivalent fraction with the denominator or numerator that is given by multiplying or dividing both the numerator and denominator by the same value.

11. $\dfrac{3}{4} = \dfrac{}{12}$
12. $\dfrac{1}{2} = \dfrac{}{10}$
13. $\dfrac{5}{8} = \dfrac{}{24}$

14. $\dfrac{3}{5} = \dfrac{}{15}$
15. $\dfrac{1}{2} = \dfrac{8}{}$
16. $\dfrac{2}{3} = \dfrac{10}{}$

17. $\dfrac{10}{12} = \dfrac{}{6}$
18. $\dfrac{18}{24} = \dfrac{9}{}$
19. $\dfrac{15}{18} = \dfrac{5}{}$

20. $\dfrac{16}{18} = \dfrac{}{9}$
21. $\dfrac{7}{1} = \dfrac{}{4}$
22. $2\dfrac{3}{4} = \dfrac{}{8}$

23. $3\dfrac{7}{8} = \dfrac{}{16}$
24. $\dfrac{4}{1} = \dfrac{}{9}$

C. Reduce to lowest terms.

25. $\dfrac{15}{18}$
26. $\dfrac{24}{36}$
27. $\dfrac{9}{15}$
28. $\dfrac{8}{16}$

29. $\dfrac{4}{16}$
30. $\dfrac{12}{14}$
31. $\dfrac{6}{8}$
32. $\dfrac{6}{9}$

33. $\dfrac{14}{21}$
34. $\dfrac{3}{6}$
35. $\dfrac{10}{16}$
36. $\dfrac{2}{4}$

37. $\dfrac{8}{15}$
38. $\dfrac{15}{22}$
39. $\dfrac{39}{21}$
40. $\dfrac{84}{36}$

41. $\dfrac{36}{80}$
42. $\dfrac{120}{200}$
43. $\dfrac{98}{144}$
44. $\dfrac{124}{236}$

45. $\dfrac{96}{124}$
46. $\dfrac{320}{480}$
47. $\dfrac{64}{84}$
48. $\dfrac{192}{336}$

49. $\dfrac{225}{300}$
50. $\dfrac{196}{320}$

3.4 Multiplying Fractions

If you packed one-half a box of books each day this week, how many boxes did you pack?

F I G U R E 5

The diagram in Figure 5 indicates that 7 half-boxes were packed. 1/2 + 1/2 + 1/2 + 1/2 + 1/2 + 1/2 + 1/2 is equal to "seven halves" and is expressed as 7/2. You could find this value by multiplying 7 times 1/2, or 7 × 1/2, because multiplication is just a short way of doing repeated addition. The same amount is packed each day for 7 days. Then 7/1 × 1/2 will equal 7/2 if you multiply the numerators together and multiply the denominators together.

$$\frac{7}{1} \times \frac{1}{2} = \frac{7 \times 1}{1 \times 2} = \frac{7}{2}$$ This is the process used to multiply any fractions together.

Multiplication of Fractions

> Given the fractions a/b and c/d (where b and d are not equal to zero), then
> $$\frac{a}{b} \times \frac{c}{d} = \frac{a \times c}{b \times d}$$

E X A M P L E 1   Multiply $\frac{3}{4}$ by $\frac{5}{8}$.

Solution $\frac{3}{4} \times \frac{5}{8} = \frac{3 \times 5}{4 \times 8} = \frac{15}{32}$

You don't have to write out the middle step.   ◆

E X A M P L E 2   Find the product of $\frac{2}{15}$ and 4.

Solution   *Product* means to multiply. Any whole number can be written in fraction form by writing it over 1. (Isn't 4/1 the same as 4 ÷ 1, which has a value of 4?)

$$\frac{2}{15} \times 4 = \frac{2}{15} \times \frac{4}{1} = \frac{8}{15}$$ ◆

In each of the previous examples the answer was already in lowest terms, but you need to check every answer to be sure that it is reduced.

E X A M P L E   **3**   Multiply $\frac{8}{15}$ by $\frac{25}{28}$.

Solution $\frac{8}{15} \times \frac{25}{28} = \frac{8 \times 25}{15 \times 28} = \frac{200}{420}$ This must be reduced to lowest terms.

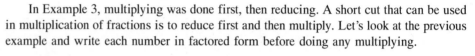

$$\frac{200}{420} = \frac{\cancel{2} \times \cancel{2} \times 2 \times \cancel{5} \times 5}{\cancel{2} \times \cancel{2} \times 3 \times \cancel{5} \times 7} = \frac{10}{21}$$ ◆

In Example 3, multiplying was done first, then reducing. A short cut that can be used in multiplication of fractions is to reduce first and then multiply. Let's look at the previous example and write each number in factored form before doing any multiplying.

$$\frac{8}{15} \times \frac{25}{28}$$

$$= \frac{2 \times 2 \times 2}{3 \times 5} \times \frac{5 \times 5}{2 \times 2 \times 7}$$ Cross out pairs of matching factors (their value is 1).

$$= \frac{\cancel{2} \times \cancel{2} \times 2 \times \cancel{5} \times 5}{3 \times \cancel{5} \times \cancel{2} \times \cancel{2} \times 7}$$ Multiply the factors that remain.

$$= \frac{10}{21}$$

Problems involving the multiplication of fractions can be done by reducing first then multiplying, or by multiplying first and then reducing.

E X A M P L E   **4**   Find the product of $\frac{15}{16}$ and $\frac{24}{39}$.

Solution   Multiplying First:

$$\frac{15}{16} \times \frac{24}{39}$$ Multiply 15 × 24 and 16 × 39.

$$= \frac{360}{624}$$ Find the prime factors of 360 and 624.

$$= \frac{2 \cdot 2 \cdot 2 \cdot 3 \cdot 3 \cdot 5}{2 \cdot 2 \cdot 2 \cdot 2 \cdot 3 \cdot 13}$$ Divide pairs of like factors.

$$= \frac{\cancel{2} \cdot \cancel{2} \cdot \cancel{2} \cdot 3 \cdot \cancel{3} \cdot 5}{\cancel{2} \cdot \cancel{2} \cdot \cancel{2} \cdot 2 \cdot \cancel{3} \cdot 13}$$ Multiply the remaining factors.

$$= \frac{15}{26}$$

Reducing First:

$$\frac{15}{16} \times \frac{24}{39}$$ Write each number in prime factored form.

$$= \frac{3 \times 5}{2 \times 2 \times 2 \times 2} \times \frac{2 \times 2 \times 2 \times 3}{3 \times 13}$$

$$= \frac{3 \times 5 \times 2 \times 2 \times 2 \times 3}{2 \times 2 \times 2 \times 2 \times 3 \times 13}$$ Cross out pairs of factors.

$$= \frac{3 \times 5 \times \overset{1}{\cancel{2}} \times \overset{1}{\cancel{2}} \times \overset{1}{\cancel{2}} \times \overset{1}{\cancel{3}}}{2 \times \underset{1}{\cancel{2}} \times \underset{1}{\cancel{2}} \times \underset{1}{\cancel{2}} \times \underset{1}{\cancel{3}} \times 13}$$

Multiply the factors that remain.

$$= \frac{15}{26}$$

◆

A popular technique for doing the reducing first lets you bypass the prime factoring step. You look mentally for pairs of factors (not necessarily primes). In this problem you might "see" the pair of 3s that could be divided out of 15 and 39. There is also a pair of 8s—one in 16 and one in 24.

$$\frac{15}{16} \times \frac{24}{39} = \frac{\overset{5}{\cancel{15}}}{\underset{2}{\cancel{16}}} \times \frac{\overset{3}{\cancel{24}}}{\underset{13}{\cancel{39}}}$$

Because there are no more pairs of factors, multiply the remaining factors together.

$$= \frac{15}{26}$$

The reason why this works is the same as in the earlier explanation about reducing (Section 3.2). The pairs divided have a value of 1.

$$\frac{15}{16} \times \frac{24}{39} = \frac{5 \times 3}{2 \times 8} \times \frac{8 \times 3}{3 \times 13}$$

$$= \frac{5 \times 3 \times 8 \times 3}{2 \times 3 \times 8 \times 13}$$

$$= \frac{5}{2} \times 1 \times 1 \times \frac{3}{13}$$

$$= \frac{15}{26}$$

This shortened version is called *cancellation*. Although the technique is not a "pure" mathematical process, it is all right to use as long as you see the connection to equivalent fractions. In fact, it is a very practical tool.

Cancellation can be done whenever you can divide any *one factor* in the numerator by any *one factor* in the denominator. You can keep doing this until there are no more common factors.

As we have said, cancellation is a very useful short cut, but remember that it can only be done while *multiplying* fractions. *Do not try to cancel if the fractions are linked by* $+, -, \div,$ *or* $=$ *signs.*

E X A M P L E 5   Find the product: $\dfrac{3}{4} \times \dfrac{5}{9} \times \dfrac{12}{20}$

Solution   By cancellation:

$$\frac{3}{4} \times \frac{5}{9} \times \frac{12}{20}$$

$$\frac{\overset{1}{\cancel{3}}}{\underset{1}{\cancel{4}}} \times \frac{\overset{1}{\cancel{5}}}{\underset{3}{\cancel{9}}} \times \frac{\overset{\overset{1}{\cancel{3}}}{\cancel{12}}}{\underset{4}{\cancel{20}}}$$

$$= \frac{1}{4}$$

The cancellation can be done in many different ways and still produce the same answer. In this example, 3 and 9 were divided by 3, 4 and 12 by 4, and 5 and 20 by 5. Then 3 and 3 were divided by 3.

By prime factorization:

$$\frac{3}{4} \times \frac{5}{9} \times \frac{12}{20}$$

$$= \frac{\overset{1}{\cancel{3}}}{2 \times 2} \times \frac{\overset{1}{\cancel{5}}}{\underset{1}{\cancel{3}} \times \underset{1}{\cancel{3}}} \times \frac{\overset{1}{\cancel{3}} \times \overset{1}{\cancel{2}} \times \overset{1}{\cancel{2}}}{\underset{1}{\cancel{5}} \times \underset{1}{\cancel{2}} \times \underset{1}{\cancel{2}}} \qquad$$ Divide two pairs of 2s, two pairs of 3s, and one pair of 5s. Multiply what remains.

$$= \frac{1}{4}$$ ◆

When asked to multiply mixed numbers such as 2 and 3 1/5, you must realize that 2 is not multiplying just 3, but 1/5 as well. From the discussion of the Distributive Principle in Chapter 1, you know you could write this as

$$2 \times 3 \ 1/5 = 2(3 + 1/5) = 2 \times 3 + 2 \times 1/5$$
$$= 6 + 2/5$$
$$= 6 \ 2/5$$

The problem is more commonly calculated by converting the mixed number and the whole number to improper fractions before doing the multiplication.

$$2 \times 3\frac{1}{5}$$

$$= \frac{2}{1} \times \frac{16}{5}$$

$$= \frac{32}{5} \quad \text{or} \quad 6\frac{2}{5}$$

The following multiplication examples that involve mixed numbers all illustrate this second procedure, changing mixed numbers to improper fractions before multiplying.

E X A M P L E **6** Multiply the mixed numbers: $2\frac{3}{4} \times 1\frac{2}{3}$

Solution $\qquad 2\frac{3}{4} \times 1\frac{2}{3}$ Change to improper fractions.

$$= \frac{11}{4} \times \frac{5}{3}$$ Because cancellation cannot be done, multiply the numerators and the denominators.

$$= \frac{55}{12} \quad \text{or} \quad 4\frac{7}{12}$$ ◆

Is this a reasonable answer? Quickly calculate the product of the next larger whole numbers: $3 \times 2 = 6$. Then use the next smaller whole numbers: $2 \times 1 = 2$. Is the actual answer between 2 and 6? It is, so you know you are at least within the reasonable range.

Because 55/12 is reduced to lowest terms, it would have been all right to leave the answer as an improper fraction. But, if you prefer, change it to a mixed number.

E X A M P L E **7** Find the product: $7\frac{1}{7} \times 8 \times 6\frac{3}{10}$

Solution First estimate the answer.

$8 \times 8 \times 7 = 448$ and $7 \times 8 \times 6 = 336$. The answer should be between 336 and 448.

$$7\frac{1}{7} \times 8 \times 6\frac{3}{10}$$

$$= \frac{50}{7} \times \frac{8}{1} \times \frac{63}{10}$$

$$= \frac{\overset{5}{\cancel{50}}}{\underset{1}{\cancel{7}}} \times \frac{8}{1} \times \frac{\overset{9}{\cancel{63}}}{\underset{1}{\cancel{10}}}$$

$$= \frac{360}{1} = 360 \qquad \text{This answer is within the estimated range.} \qquad \blacklozenge$$

E X A M P L E **8**   Find 3/4 of 90.

Solution   The instruction to find "part of" something indicates multiplication.

$\dfrac{3}{4}$ "of" 90 means $\dfrac{3}{4} \times 90$.

$$\frac{3}{4} \times \frac{90}{1} = \frac{3}{\underset{2}{\cancel{4}}} \times \frac{\overset{45}{\cancel{90}}}{1} = \frac{135}{2} \text{ or } 67\frac{1}{2} \qquad \blacklozenge$$

E X A M P L E **9**   Two-thirds of the students in this class are doing "B" or better work. If there are 27 students in the class, how many are doing "B" or better work?

Solution   The problem could be restated by saying that 2/3 of the 27 students are doing "B" or better work. It is a "part of" problem, so multiply. Estimate your answer by reasoning that 2/3 of the class will be over half, but not all, of the 27 students.

$$\frac{2}{3} \text{ of } 27 = \frac{2}{3} \times \frac{27}{1}$$

$$= \frac{2}{\underset{1}{\cancel{3}}} \times \frac{\overset{9}{\cancel{27}}}{1} = 18$$

Therefore, 18 students are doing "B" or better work. This answer is reasonable because it is over half of 27 but not the whole class. $\qquad \blacklozenge$

E X A M P L E **10**   On the Stay Trim diet program, Carolyn averaged a weight loss of 2 1/2 pounds per month. If she stayed on the program for 6 months, how many pounds did she lose?

Solution   First estimate the answer.

2 pounds per month for 6 months = 12 pounds

3 pounds per month for 6 months = 18 pounds

2 1/2 pounds every month for 6 months means that the same loss occurred six times.

$$2\frac{1}{2} \times 6 = \frac{5}{\underset{1}{\cancel{2}}} \times \frac{\overset{3}{\cancel{6}}}{1} = 15 \text{ pounds}$$

This is within the estimated range. $\qquad \blacklozenge$

E X A M P L E **11** Ray is ordering fence materials for his back yard. He knows that the length of the yard is 68 1/4 feet and that the width is 2/3 of the length. What is the width of the yard?

Solution The width is $\frac{2}{3}$ "of" the length $= \frac{2}{3} \times 68\frac{1}{4}$

$$\frac{\overset{1}{\cancel{2}}}{\underset{1}{\cancel{3}}} \times \frac{\overset{91}{\cancel{273}}}{\underset{2}{\cancel{4}}} = \frac{91}{2} = 45\frac{1}{2} \text{ feet}$$ ◆

Leaving the answer to a word problem as an improper fraction is not meaningful, so change it to a mixed number. Remember to label the answer with the correct units.

3.4 ◆ Practice Problems

A. Find the product and reduce all answers to lowest terms.

1. $\frac{2}{3} \times \frac{5}{9}$ 2. $\frac{4}{5} \times \frac{7}{9}$

3. $\frac{6}{7} \times \frac{2}{5}$ 4. $\frac{1}{3} \times \frac{4}{9}$

5. $\frac{3}{4} \times \frac{5}{18}$ 6. $\frac{4}{5} \times \frac{15}{16}$

7. $\frac{12}{18} \times \frac{27}{36}$ 8. $\frac{8}{25} \times \frac{10}{24}$

9. $\frac{3}{8} \times \frac{12}{15} \times \frac{2}{3}$ 10. $1\frac{2}{3} \times 3\frac{3}{5}$

11. $5\frac{2}{5} \times 2\frac{7}{9}$ 12. $\frac{1}{2} \times \frac{5}{6} \times \frac{8}{10}$

13. $12\frac{2}{3} \times 7\frac{1}{2}$ 14. $2\frac{7}{8} \times \frac{4}{5}$

15. $4 \times 5\frac{1}{2}$ 16. $12\frac{2}{3} \times 9$

17. $1\frac{5}{9} \times 2\frac{1}{3}$ 18. $\frac{13}{24} \times \frac{12}{39}$

19. $6 \times 4\frac{1}{3}$ 20. $5\frac{5}{6} \times \frac{3}{7}$

21. $11\frac{5}{6} \times 3\frac{1}{2}$ 22. $19\frac{1}{2} \times \frac{6}{7}$

23. $15 \times 19\frac{1}{3}$ 24. $9\frac{1}{5} \times 12$

B. Find the solution. Be sure that all answers are reduced to lowest terms.

25. 3/4 of 18   26. 5/6 of 24

27. 3/8 of 12   28. 4/9 of 90

29. 1/2 of 15   30. 8/5 of 16

31. 1 3/7 of 200   32. 2 4/9 of 36

33. 12/5 of 30   34. 1/5 of 20

35. 7/8 of 90   36. 5/16 of 84

C. Find a solution to each problem after estimating the answer. Then be sure that the answers are reasonable and that fractions are reduced to lowest terms.

37. Two-thirds of a class of 27 are going on a field trip. How many students are going?

38. Becky has delivered 3/4 of her 160 newspapers. How many have been delivered?

39. Niko has eaten 5/6 of the carton of ice cream bars. If the box had contained 12 bars, how many has he eaten?

40. Five-sixths of the students in this class passed the last exam. If there are 42 students, how many passed?

41. Three-fifths of the textbooks sold in the bookstore were paperbacks. If 48,395 books were sold, how many were paperbacks?

42. Victor's raise in salary means that he will receive one and one-eighth times his previous pay. If he had been earning $120 per week, what is his new salary?

43. Dr. Garfield has computed the averages of two-thirds of his students. How many averages has he done if he has 165 students this quarter?

44. A national diet program advertises that customers can expect a weight loss of 2 3/4 pounds per month. What is the total loss expected for six months?

45. A stock drops 3/8 point every day for 4 2/3 days. What is the total drop for that period?

46. Sarah wants to make 1 1/2 times her recipe for chocolate chip cookies. Please help her figure the amounts of flour, sugar, and vanilla to use if the original amounts were 2 1/2 cups of flour, 3/4 cup of sugar, and 1 1/2 teaspoons of vanilla.

47. In the Cavaliers' last basketball game, two-ninths of their 108 points were scored with three-point shots. How many points were scored this way?

48. Sue's new exercise plan calls for her to walk 3/4 mile for 8 days and then 1 1/4 miles for 8 days. What is the total distance she will walk during this period?

49. Mary addresses three-eighths of a box of envelopes each day. At this rate, how many boxes will she complete in 4 1/2 days?

50. If a candy bar contains 1 3/8 ounces of almonds, how many ounces of almonds are in 4 1/2 bars?

3.5 Dividing Fractions

Before learning how to divide fractions, you need to understand the concept of reciprocals. Look at these:

$$\frac{5}{8} \times \frac{8}{5} = \frac{\overset{1}{\cancel{5}}}{\cancel{8}} \times \frac{\overset{1}{\cancel{8}}}{\cancel{5}} = \frac{1}{1} = 1 \quad \text{or} \quad \frac{5}{8} \times \frac{8}{5} = \frac{40}{40} = 1$$

$$\frac{1}{6} \times 6 = \frac{1}{\cancel{6}} \times \frac{\overset{1}{\cancel{6}}}{1} = 1 \qquad \frac{11}{3} \times \frac{3}{11} = \frac{33}{33} = 1$$

In all of these examples, the product is 1. *If you multiply two fractions together and the product equals 1,* then the fractions are reciprocals of each other.

$\dfrac{5}{8}$ is the reciprocal of $\dfrac{8}{5}$, and $\dfrac{8}{5}$ is the reciprocal of $\dfrac{5}{8}$.

$\dfrac{1}{6}$ is the reciprocal of 6, and 6 is the reciprocal of $\dfrac{1}{6}$.

The reciprocal of a fraction is written by inverting the fraction.

E X A M P L E 1 Write the reciprocals of $\dfrac{1}{4}, \dfrac{3}{5}, \dfrac{7}{6}$, and $2\dfrac{5}{8}$.

Solution The reciprocals are $\dfrac{4}{1}$ (or 4), $\dfrac{5}{3}, \dfrac{6}{7}$, and $\dfrac{8}{21}$. The reciprocal of $2\dfrac{5}{8}$ was written after changing the mixed number to an improper fraction, $\dfrac{21}{8}$. ◆

Reciprocals are necessary for dividing fractions. Let us look at an illustration of fraction division before learning rules for the operation. One-third of the rectangle in Figure 6 is shaded. If the shaded portion is divided into two equal parts, what portion of the whole figure does one of these parts represent? The drawings show that the answer is 1/6 of the whole rectangle. Therefore,

$$\frac{1}{3} \div 2 = \frac{1}{6}$$

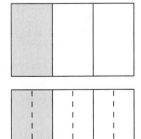

FIGURE 6

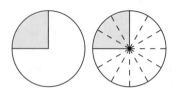

FIGURE 7

Here is another illustration. In Figure 7, the remaining pizza, one-fourth of the whole pizza, must be split into 3 equal portions. What part of the whole pizza does one of these portions represent? Do you see that it would be 1/12 of the whole pizza? Therefore, you could say that

$$\frac{1}{4} \div 3 = \frac{1}{12}$$

Look again at Figure 6. The only way that $\frac{1}{3} \div 2$ can equal $\frac{1}{6}$ is if $\frac{1}{3}$ is multiplied by $\frac{1}{2}$. Note that $\frac{1}{3} \div 2$ and $\frac{1}{3} \times \frac{1}{2}$ must be equivalent expressions, because they give the same answer, $\frac{1}{6}$. Thus $\frac{1}{3} \div 2$ can be solved by multiplying 1/3 by the reciprocal of the second fraction (divisor).

$$\frac{1}{3} \div 2 = \frac{1}{3} \div \frac{2}{1} = \frac{1}{3} \times \frac{1}{2} = \frac{1}{6}$$

The same pattern is true for Figure 7. The only way that $\frac{1}{4} \div 3$ can equal $\frac{1}{12}$ is if $\frac{1}{4}$ is multiplied by $\frac{1}{3}$. The correct answer to 1/4 ÷ 3 can be found by multiplying the first fraction (dividend) by the reciprocal of the second fraction (divisor).

$$\frac{1}{4} \div 3 = \frac{1}{4} \div \frac{3}{1} = \frac{1}{4} \times \frac{1}{3} = \frac{1}{12}$$

Division of Fractions

To divide fractions, multiply the dividend by the reciprocal of the divisor. In other words, multiply the first fraction by the reciprocal of the second fraction. For the fractions a/b and c/d (b, c, and $d \neq 0$),

$$\frac{a}{b} \div \frac{c}{d} = \frac{a}{b} \times \frac{d}{c} = \frac{ad}{bc}$$

E X A M P L E 2 Divide the fraction $\frac{2}{3}$ by $\frac{5}{6}$.

Solution $\frac{2}{3} \div \frac{5}{6} = \frac{2}{3} \times \frac{6}{5}$ Rewrite as multiplication by the reciprocal of the second fraction.

$\dfrac{2}{\underset{1}{\cancel{3}}} \times \dfrac{\overset{2}{\cancel{6}}}{5}$ Because it is now multiplication, cancellation can be done.

$= \dfrac{4}{5}$ ◆

E X A M P L E 3 Divide 3/5 by 9.

Solution $\dfrac{3}{5} \div 9 = \dfrac{3}{5} \div \dfrac{9}{1} = \dfrac{3}{5} \times \dfrac{1}{9}$

$= \dfrac{\overset{1}{\cancel{3}}}{5} \times \dfrac{1}{\underset{3}{\cancel{9}}} = \dfrac{1}{15}$ ◆

E X A M P L E **4**   Find the quotient of 2 1/2 divided by 8.

Solution $2\dfrac{1}{2} \div 8$ First change both numbers to improper fractions.

$= \dfrac{5}{2} \div \dfrac{8}{1}$

$= \dfrac{5}{2} \times \dfrac{1}{8}$ Now rewrite as multiplication by the reciprocal.

$= \dfrac{5}{16}$ ◆

E X A M P L E **5**   Divide: $12 \div 4\dfrac{7}{8}$

Solution $12 \div 4\dfrac{7}{8}$ Change the numbers to improper fractions.

$= \dfrac{12}{1} \div \dfrac{39}{8}$ Rewrite as multiplication by the reciprocal of the second fraction.

$= \dfrac{12}{1} \times \dfrac{8}{39}$

$= \dfrac{2 \cdot 2 \cdot \overset{1}{\cancel{3}}}{1} \times \dfrac{2 \cdot 2 \cdot 2}{\underset{1}{\cancel{3}} \cdot 13}$

$= \dfrac{32}{13}$ or $2\dfrac{6}{13}$ ◆

E X A M P L E **6**   Tony can drive 320 miles on 12 4/5 gallons of gas. How many miles does his car average on 1 gallon?

Solution   To find miles per gallon, divide the total miles by the number of gallons.

$320 \div 12\dfrac{4}{5}$

$= \dfrac{320}{1} \div \dfrac{64}{5}$

$= \dfrac{\overset{5}{\cancel{320}}}{1} \times \dfrac{5}{\underset{1}{\cancel{64}}}$

$= 25$ miles per gallon   ◆

E X A M P L E **7**   A 7 7/8-acre parcel of land is to be divided into nine equal-sized lots. What will be the size of each lot?

Solution   First estimate the answer. The numbers are very close in size, so dividing nine into a number that is nearly 8 should give an answer that is a little less than 1.

7 7/8 is to be *divided* into 9 equal parts.

$$7\dfrac{7}{8} \div 9 = \dfrac{63}{8} \div \dfrac{9}{1} = \dfrac{\overset{7}{\cancel{63}}}{8} \times \dfrac{1}{\underset{1}{\cancel{9}}} = \dfrac{7}{8} \text{ acre}$$

This agrees with the estimate.   ◆

E X A M P L E   **8**   How many gasoline cans, each with a capacity of 2 1/2 gallons, can be filled from a tank holding 23 3/4 gallons?

Solution   The total amount is 23 3/4 gallons. It is being divided into parts that are each 2 1/2 gallons in size.

Total amount ÷ size of each part = number of parts

$$23\frac{3}{4} \div 2\frac{1}{2} = \frac{95}{4} \div \frac{5}{2} = \frac{\overset{19}{\cancel{95}}}{\underset{2}{\cancel{4}}} \times \frac{\overset{1}{\cancel{2}}}{\cancel{5}} = \frac{19}{2} \text{ or } 9\frac{1}{2} \text{ cans}$$ ◆

A fraction whose numerator and/or denominator is a fraction is called a **complex fraction**. It's an appropriate name, because expressions like the ones that follow can be very confusing!

$$\frac{\frac{2}{3}}{\frac{3}{4}} \qquad \frac{8}{\frac{6}{5}} \qquad \frac{4\frac{2}{5}}{11}$$

Actually, they look much worse than they are. If you keep in mind that a fraction is an indicated division and use that concept, then you can rewrite complex fractions as the division of two fractions.

$$\frac{\frac{2}{3}}{\frac{3}{4}} \qquad \text{becomes} \qquad \frac{2}{3} \div \frac{3}{4} = \frac{2}{3} \times \frac{4}{3} = \frac{8}{9}$$

E X A M P L E   **9**   Simplify: $\dfrac{8}{\frac{6}{5}}$

Solution $\dfrac{8}{\frac{6}{5}}$ becomes $8 \div \dfrac{6}{5} = \dfrac{8}{1} \times \dfrac{5}{\underset{3}{\cancel{6}}} \overset{4}{} = \dfrac{20}{3} \text{ or } 6\dfrac{2}{3}$ ◆

E X A M P L E   **10**   Simplify: $\dfrac{4\frac{2}{5}}{11}$

Solution $\dfrac{4\frac{2}{5}}{11}$ rewritten with the division symbol becomes

$$4\frac{2}{5} \div 11 = \frac{22}{5} \div \frac{11}{1} = \frac{\overset{2}{\cancel{22}}}{5} \times \frac{1}{\underset{1}{\cancel{11}}} = \frac{2}{5}$$ ◆

E X A M P L E   **11**   Simplify: $\dfrac{8\frac{1}{6}}{11\frac{2}{3}}$

Solution $8\dfrac{1}{6} \div 11\dfrac{2}{3} = \dfrac{49}{6} \div \dfrac{35}{3} = \dfrac{49}{\underset{2}{\cancel{6}}} \times \dfrac{\overset{1}{\cancel{3}}}{\underset{5}{\cancel{35}}} \overset{7}{} = \dfrac{7}{10}$ ◆

3.5 ◆ Practice Problems

A. Write the reciprocal of each of the following.

1. $\dfrac{2}{3}$ 2. $\dfrac{5}{8}$ 3. 4 4. $\dfrac{4}{5}$

5. 12 6. $\dfrac{1}{6}$ 7. $\dfrac{9}{7}$ 8. 15

9. $\dfrac{13}{3}$ 10. 2

B. Divide and reduce answers to lowest terms.

11. $\dfrac{2}{3} \div \dfrac{3}{4}$ 12. $\dfrac{5}{8} \div \dfrac{2}{3}$ 13. $\dfrac{5}{6} \div \dfrac{1}{3}$

14. $\dfrac{4}{5} \div \dfrac{2}{15}$ 15. $8 \div \dfrac{5}{6}$ 16. $1\dfrac{2}{3} \div 3\dfrac{1}{8}$

17. $5\dfrac{5}{6} \div 2\dfrac{1}{3}$ 18. $4 \div \dfrac{2}{3}$ 19. $\dfrac{7}{15} \div \dfrac{4}{5}$

20. $63 \div \dfrac{3}{8}$ 21. $26 \div \dfrac{2}{9}$ 22. $\dfrac{5}{6} \div \dfrac{2}{18}$

23. $\dfrac{4}{5} \div \dfrac{7}{8}$ 24. $21\dfrac{3}{10} \div 7\dfrac{2}{5}$ 25. $2\dfrac{1}{3} \div 3$

26. $2\dfrac{1}{3} \div 1\dfrac{1}{9}$ 27. $\dfrac{2}{15} \div 32\dfrac{1}{5}$ 28. $11\dfrac{3}{8} \div 6$

C. Simplify these complex fractions.

29. $\dfrac{\frac{2}{3}}{\frac{3}{4}}$ 30. $\dfrac{\frac{8}{3}}{\frac{3}{4}}$ 31. $\dfrac{1\frac{1}{3}}{\frac{3}{8}}$

32. $\dfrac{\frac{3}{5}}{\frac{5}{6}}$ 33. $\dfrac{\frac{4}{9}}{\frac{2}{3}}$ 34. $\dfrac{1\frac{7}{8}}{1\frac{1}{4}}$

35. $\dfrac{7\frac{7}{8}}{9}$ 36. $\dfrac{5\frac{1}{3}}{8}$ 37. $\dfrac{4\frac{1}{3}}{\frac{8}{2\frac{1}{9}}}$

D. Solve these word problems after estimating an answer. Be sure to reduce fractions to lowest terms.

38. Divide 26 2/3 into eight equal parts.

39. If Lori evenly divides a 3 1/8-pound boneless roast among her nine dinner guests and herself, how much meat will each receive?

40. Each can of ready-to-eat soup contains 10 3/4 ounces. If the soup is divided into three equal servings, how many ounces will each person receive?

41. When at full production, a car assembly plant has a new car roll off the line every 2/3 hour. If the plant is operating around the clock, how many cars are produced in 24 hours?

42. A tanning booth is used for 12 hours a day. If each session is 3/4 hour long, how many sessions can be scheduled in a day?

43. If a plane flies 65 2/3 miles in 7 1/2 minutes, how far can it travel in one minute?

44. How many pieces of cloth, each 5 3/8 yards long, can be cut from a bolt containing 64 1/2 yards?

45. Laura earned 38 1/2 dollars for working 7 hours. How much does she earn per hour? Express the answer in fraction form.

46. A 3 1/8-pound roast costs $10. How much per pound does it cost? Give the answer in fraction form.

47. Marilyn bought 7 1/2 yards of material to make scarves. How many can she make if each requires 3/4 of a yard?

48. If 297 bushels of corn were produced by a 6 3/4-acre farm, how many bushels were produced per acre?

49. At the Colonial Cheese Shoppe, a clerk used a 20 1/4-pound block of cheese to make small packages, each 3/8 pound in size. How many packages did she make?

50. In five hours, a car was driven 240 1/2 miles. How many miles per hour did the car average?

51. A land developer plans to divide 70 acres of land into 1 1/4-acre plots. How many plots will be available?

52. A spool of wire 28 1/2 feet long is cut into 19 equal pieces. How long is each piece?

3.6   Chapter Review

Your success on the test for this chapter will depend on your understanding the vocabulary of fractions and on your being able to apply a few rules to perform multiplication and division. By now you should know that:

1. The *numerator* is the top number in a fraction, and the *denominator* is the bottom number.

2. An *improper fraction* has a numerator that is greater than or equal to the denominator. A *proper fraction* has a numerator that is smaller than the denominator.

3. A *mixed number* has a whole number part and a fraction part.

4. *To change from a mixed number to an improper fraction, multiply the whole number times the denominator and add the numerator.* Write this answer over the denominator.

5. *To change from an improper fraction to a mixed number, divide the denominator into the numerator.* This answer is written as the whole number, and the remainder is written over the denominator for the fraction part.

6. *Equivalent fractions* are fractions that have the same value but different forms. An equivalent fraction can be obtained by multiplying or dividing both numerator and denominator by the same factor.

7. A fraction has been *reduced to lowest terms* when there is no factor other than 1 common to both the numerator and the denominator.

8. *To multiply fractions*, multiply the numerators together, multiply the denominators together, and then reduce the answer to lowest terms.

9. *To divide fractions*, multiply the first fraction by the reciprocal of the second fraction.

10. *To multiply or divide mixed numbers*, first change them to improper fractions.

11. A fraction is a *reciprocal* of another fraction if their product is 1.

12. To simplify a *complex fraction*, rewrite it as one fraction divided by another.

13. All fractional answers *must be reduced to lowest terms.*

Review Problems

A. Change these improper fractions to mixed numbers, reduced to lowest terms.

1. $\dfrac{44}{15}$ 2. $\dfrac{12}{7}$ 3. $\dfrac{19}{3}$ 4. $\dfrac{5}{2}$

5. $\dfrac{22}{12}$ 6. $\dfrac{19}{2}$ 7. $\dfrac{61}{5}$ 8. $\dfrac{126}{10}$

B. Change these mixed numbers to improper fractions.

9. $3\dfrac{3}{8}$ 10. $5\dfrac{1}{2}$ 11. $12\dfrac{2}{3}$ 12. $1\dfrac{5}{8}$

13. $54\dfrac{2}{5}$ 14. $32\dfrac{3}{4}$ 15. $10\dfrac{1}{8}$ 16. $2\dfrac{2}{3}$

C. Reduce these fractions to lowest terms.

17. $\dfrac{27}{36}$ 18. $\dfrac{15}{21}$ 19. $\dfrac{8}{20}$ 20. $\dfrac{10}{18}$

21. $\dfrac{3}{9}$ 22. $\dfrac{7}{14}$ 23. $\dfrac{32}{80}$ 24. $\dfrac{24}{64}$

25. $\dfrac{24}{39}$ 26. $\dfrac{18}{48}$ 27. $\dfrac{54}{81}$ 28. $\dfrac{56}{96}$

29. $\dfrac{96}{126}$ 30. $\dfrac{140}{252}$ 31. $\dfrac{224}{320}$ 32. $\dfrac{156}{234}$

D. Find the correct numerator or denominator to make these fractions equivalent.

33. $\dfrac{2}{3} = \dfrac{}{12}$

34. $\dfrac{1}{4} = \dfrac{}{16}$

35. $\dfrac{3}{5} = \dfrac{}{15}$

36. $\dfrac{5}{8} = \dfrac{15}{}$

37. $\dfrac{24}{36} = \dfrac{8}{}$

38. $\dfrac{15}{18} = \dfrac{}{6}$

39. $\dfrac{3}{4} = \dfrac{}{24}$

40. $\dfrac{12}{16} = \dfrac{6}{}$

41. $\dfrac{8}{24} = \dfrac{}{3}$

42. $\dfrac{16}{1} = \dfrac{}{2}$

43. $\dfrac{11}{1} = \dfrac{44}{}$

44. $\dfrac{63}{35} = \dfrac{}{5}$

E. Multiply or divide as indicated and reduce answers to lowest terms.

45. $\dfrac{2}{3} \times \dfrac{5}{6}$

46. $\dfrac{4}{9} \times \dfrac{3}{8}$

47. $\dfrac{7}{8} \div \dfrac{1}{4}$

48. $\dfrac{3}{4} \div \dfrac{1}{8}$

49. $1\dfrac{2}{3} \times 4$

50. $8 \times 5\dfrac{1}{3}$

51. $6\dfrac{1}{4} \div \dfrac{5}{8}$

52. $4 \div 1\dfrac{3}{5}$

53. $\dfrac{3}{4}$ of 6

54. $2\dfrac{4}{9} \div \dfrac{3}{11}$

55. $\dfrac{2}{3} \div \dfrac{3}{4}$

56. $1\dfrac{5}{8}$ of 39

57. $\dfrac{3}{8} \times 1\dfrac{3}{5} \times 2\dfrac{2}{3}$

58. $1\dfrac{1}{2} \times 4\dfrac{2}{3} \times 1\dfrac{5}{6}$

59. $\dfrac{5}{6} \times 1\dfrac{2}{3} \times 3\dfrac{3}{8}$

60. $4\dfrac{1}{3} \times \dfrac{10}{39} \times 1\dfrac{1}{5}$

61. $\dfrac{2\dfrac{2}{3}}{1\dfrac{2}{4}}$

62. $\dfrac{\dfrac{3}{8}}{\dfrac{5}{6}}$

63. $\dfrac{\dfrac{5}{9}}{\dfrac{2}{3}}$

64. $\dfrac{1\dfrac{4}{5}}{3\dfrac{1}{3}}$

F. Find the solution to each problem. Be sure to reduce all answers to lowest terms.

65. McArches Corporation processes beef for its hamburger restaurants. It forms 3/8-pound patties from a package of ground beef weighing 24 pounds. How many of these patties can be formed from each package?

66. Five-sixths of the students doing a problem will get it correct. If 48 students are doing the problem, how many will get the correct answer?

67. Sara hates to get up in the morning, but she has decided to try harder. Three-fourths of the last 24 school days she has gotten up on time. How many days did she get up on time?

68. A farmer died and left his land to his children. The 74 2/3 acres are to be divided evenly among his four children. How much land will each inherit?

69. Five-eighths of the compact disks sold by Hi-Notes music store are by rock artists. If 1,280 disks were sold last month, how many were by rock artists?

70. Connie has read two-thirds of the book assigned for psychology class. How many of the 396 pages in the book has she read?

71. If a train travels 162 3/4 miles in 3 1/2 hours, what is its average speed in miles per hour?

72. At the Java Jungle Coffee Shop, 28 pounds of coffee beans are made into 7/8-pound packages. How many packages can be made?

Chapter Test

1. Change $\dfrac{19}{7}$ to a mixed number.

2. Change $4\dfrac{5}{8}$ to an improper fraction.

3. Reduce $\dfrac{36}{80}$ to lowest terms.

4. Reduce $\dfrac{160}{252}$ to lowest terms.

5. Find the correct numerator to make these fractions equivalent: $\dfrac{7}{8} = \dfrac{}{32}$.

6. Write the reciprocal of $\dfrac{3}{5}$.

Multiply or divide as indicated in the following problems. Reduce all fractions to lowest terms.

7. $\dfrac{2}{3} \times \dfrac{5}{7}$

8. $\dfrac{5}{9} \div \dfrac{1}{5}$

9. $2\dfrac{3}{4} \times \dfrac{8}{9}$

10. $3 \times 5\dfrac{1}{9}$

11. $12\dfrac{4}{5} \div \dfrac{3}{10}$

12. Find $\dfrac{5}{6}$ of 42.

13. $\dfrac{3}{8} \times 2\dfrac{3}{5} \times 2\dfrac{2}{3}$

14. $3\dfrac{2}{3} \div 2\dfrac{4}{9}$

15. $\dfrac{\dfrac{5}{6}}{\dfrac{4}{9}}$

16. $\dfrac{\dfrac{4}{3}}{8}$

Solve the following word problems.

17. Danny has repaid two-thirds of the money that he borrowed from his mother to buy a car. If he spent $2,970 for the car, how much has he repaid?

18. A 28-pound block of cheese is being cut into smaller pieces to be wrapped and sold. How many 4/5-pound packages can be made?

19. If Jan can drive 238 miles on 8 1/2 gallons of gas, what is her car's average number of miles per gallon?

20. A bow to decorate a wreath of dried flowers requires 4 1/2 feet of ribbon. If Marilyn has 9 wreaths to decorate, how much ribbon will she need?

Thought-Provoking Problems

1. In the rule for multiplication of fractions on page 75, why is the statement "where *b* and *d* are not equal to zero" included?

2. If you were helping someone learn the vocabulary of fractions, what "tricks" might you suggest she or he use to remember that the numerator is on top and the denominator is on the bottom?

3. Does zero have a reciprocal? Explain your answer.

4. Some students have difficulty understanding that a fraction division problem can produce an answer that is a larger number than the original numbers in the problem. Using the following example, describe this "phenomenon" to such a puzzled student.

    A 5 1/2-foot length of wire is to be cut into pieces that are each 5/6 of a foot in length. How many pieces will be produced?

5. The Bengals scored 42 points in last week's game. Four-sevenths of those points came on passing plays. Six points were scored on kicking plays. All other points came from running plays. How many points were scored by passing? What fraction of the total number of points were scored by kicking plays? How many points were scored by running?

6. According to the Order of Operations Agreement, in problems containing both multiplication and division, we do those operations in order from left to right. Simplify each of the following problems.

    a. $1\dfrac{3}{4} \div \dfrac{5}{8} \times \dfrac{15}{16}$

    b. $2\dfrac{2}{3} \times 1\dfrac{5}{8} \div \dfrac{7}{9}$

    c. $6 \div 5\dfrac{5}{8} \times \dfrac{2}{3} \times 1\dfrac{1}{4}$

7. Some new calculators have keys that enable you to input fractions. Does yours? The key is usually labeled $\boxed{\text{a b/c}}$. If you enter a fraction as 3 $\boxed{\text{a b/c}}$ 8, the display shows 3/8. If your calculator does not have this function key, you will have to enter the fraction as 3 $\boxed{\div}$ 8. What does your display show when you enter this? Can you describe the relationship between division and fractions?

8. Work the following problem two ways. First find the answer using pencil and paper, and then solve it using your calculator. Write the two answers side by side and describe how they represent the same answer. Are any of them mixed numbers? Are any improper fractions? What must be the decimal equivalent of two-fifths?

    $$\dfrac{4}{5} \times 5\dfrac{1}{2}$$

9. Section 3.4 showed multiplication of 2 × 3 1/5 using the Distributive Principle. Use a similar process to show how to multiply 5 1/8 × 3 2/3 as mixed numbers, not improper fractions.

10. Divide 3/4 by 1/10. Then divide 3/4 by 1/100 and then by 1/1000. What happened to the quotient as the divisor got smaller and smaller by powers of 10?

Addition and Subtraction of Fractions

4.1 ◆ Introduction

As we continue with fractions, keep in mind that you have already mastered many of the concepts of fractions, and only a few remain. In this chapter, you will study the operations of addition and subtraction.

◆ **Learning Objectives**

When you have completed this chapter, you should be able to:

1. Find the least common denominator of several fractions.
2. Add or subtract fractions.
3. Add or subtract mixed numbers.
4. Apply the Order of Operations Agreement to fraction problems.
5. Compare the sizes of several fractions.
6. Solve word problems involving all operations with fractions and mixed numbers.

Adding and Subtracting Fractions

It would be difficult to combine the fractions represented by the following shaded amounts, because each figure is divided into parts of a different size.

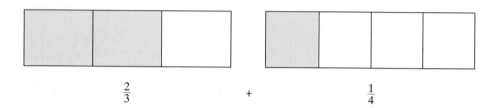

$$\frac{2}{3} \qquad + \qquad \frac{1}{4}$$

If equivalent figures were drawn—figures that had divisions of the same size—the problem would be easy to solve.

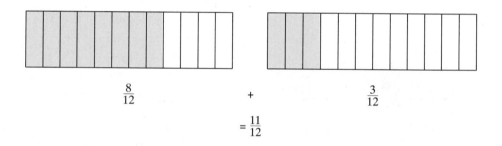

$$\frac{8}{12} \qquad + \qquad \frac{3}{12}$$

$$= \frac{11}{12}$$

If the same drawings were used to show subtraction, the same difficulty would arise.

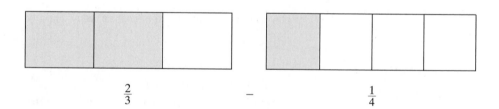

$$\frac{2}{3} \qquad - \qquad \frac{1}{4}$$

The difference can be seen only when the figures are divided into parts of the same size.

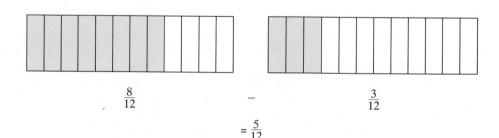

$$\frac{8}{12} \qquad - \qquad \frac{3}{12}$$

$$= \frac{5}{12}$$

This is the principle that must be applied to every addition or subtraction problem involving fractions. The quantities can be combined only if they have the same denominators. The following fractions can be added because they have the same denominator (they are divided into parts of the same size). Add the numerators together and write the answer over the common denominator.

$$\frac{4}{8} + \frac{1}{8} = \frac{5}{8}$$

The next two fractions can be subtracted because they have the same, or a common, denominator. Subtract the numerators and write the answer over the common denominator.

$$\frac{11}{16} - \frac{4}{16} = \frac{7}{16}$$

If the fractions that you are to add or subtract do not have the same denominator, then you must find a value that all denominators can divide into evenly. Equivalent fractions, with that new denominator, must be written before the addition or subtraction can be done. Remember, to change a fraction to an equivalent fraction, multiply it by a fraction whose value is 1 (4/4, 9/9, 2/2, and so forth; see Section 3.3).

Let's take, for example, the following problem:

$$\frac{2}{3} + \frac{1}{4}$$

12 is a number into which 3 and 4 can divide evenly. Use it as the common denominator and write equivalent fractions.

You may find it easier to make the change to equivalent fractions if the problem is written vertically.

$$\frac{2}{3} = \frac{8}{12} \qquad \text{Multiply: } \frac{2}{3} \times \frac{4}{4} = \frac{8}{12}$$

$$+\frac{1}{4} = +\frac{3}{12} \qquad \text{Multiply: } \frac{1}{4} \times \frac{3}{3} = \frac{3}{12}$$

$$\frac{11}{12}$$

What would have happened if you had used 24 or 36 as the common denominator? Let's see.

$$\frac{2}{3} \times \frac{8}{8} = \frac{16}{24} \qquad\qquad \frac{2}{3} \times \frac{16}{16} = \frac{32}{48}$$

$$+\frac{1}{4} \times \frac{6}{6} = +\frac{6}{24} \qquad\qquad +\frac{1}{4} \times \frac{12}{12} = +\frac{12}{48}$$

$$\frac{22}{24} = \frac{11}{12} \qquad\qquad \frac{44}{48} = \frac{11}{12}$$

All three gave the same answer, but you had to work with larger numbers in the last two, and you had to reduce the answer to lowest terms. Using the smallest common denominator, the **least common denominator**, gives you a problem that is a little easier to solve. Most of the time this value can be found by inspection. However, in problems with more than two denominators or with large denominators, the **least common multiple** method (Section 2.6) can be used. A least common multiple (LCM) and a least common denominator (LCD) are the same thing.

E X A M P L E   **1**   Find the sum of 4/7 and 3/4.

Solution

$$\frac{4}{7} = \frac{16}{28}$$

Because 28 (the product of the denominators 4 and 7) is a number that both 4 and 7 can divide into evenly, use it as the common denominator.

$$+\frac{3}{4} = +\frac{21}{28}$$

$$\frac{37}{28}$$

Write equivalent fractions, and then add the numerators and write that answer over 28.

or $1\frac{9}{28}$

Because 37/28 cannot be reduced, it can be left as the answer or changed to the mixed number.   ◆

E X A M P L E   **2**   Find the difference between 19/24 and 1/2.

Solution

$$\frac{19}{24} = \frac{19}{24}$$

The common denominator is the larger denominator (24), because 2 also divides into 24.

$$-\frac{1}{2} = -\frac{12}{24}$$

$$\frac{7}{24}$$

Write equivalent fractions, and then subtract. Be sure the answer is in lowest terms, which it is.   ◆

E X A M P L E   **3**   Add: 3/4 + 5/12 + 2/3

Solution

$$\frac{3}{4} = \frac{9}{12}$$

It is still easy to find the common denominator by inspection because 3, 4, and 12 all divide into 12.

$$\frac{5}{12} = \frac{5}{12}$$

Write equivalent fractions, and then add the numerators.

$$+\frac{2}{3} = +\frac{8}{12}$$

$$\frac{22}{12}$$

$$= \frac{11}{6} \text{ or } 1\frac{5}{6}$$

Reduce the answer to lowest terms.   ◆

E X A M P L E   **4**   Combine: $\frac{8}{15} + \frac{5}{18} - \frac{4}{9}$

Solution   Find the LCD for 15, 18, and 9.
Here we will use a method that was demonstrated in Section 2.6.

$$
\begin{array}{rl}
15 = & 3 \quad\;\; \times 5 \\
18 = & 2 \times 3 \times 3 \\
9 = & \quad\; 3 \times 3 \\
\hline
\text{LCD} = & 2 \times 3 \times 3 \times 5 = 90
\end{array}
$$

$$\frac{8}{15} + \frac{5}{18} - \frac{4}{9}$$

Write equivalent fractions:

$$\frac{8}{15} \times \frac{6}{6} = \frac{48}{90} \qquad \frac{5}{18} \times \frac{5}{5} = \frac{25}{90} \qquad \frac{4}{9} \times \frac{10}{10} = \frac{40}{90}$$

$$= \frac{48}{90} + \frac{25}{90} - \frac{40}{90} \qquad\qquad \text{Combine from left to right.}$$

$$= \frac{73}{90} - \frac{40}{90}$$

$$= \frac{33}{90} = \frac{11}{30}$$ Reduce to lowest terms.   ◆

EXAMPLE 5   Combine: $\frac{5}{9} + \frac{11}{15} + \frac{7}{24}$

Solution   The LCD is more difficult to find with these denominators. The methods for finding least common multiples (Section 2.6) can be used, because the LCM and LCD are the same thing. In one of those LCM methods, the numbers were arranged horizontally and then divided by a prime number that divided into at least one of them (it didn't have to divide into all of them). This process was continued until there were all 1s across the bottom. Apply that method to the denominators in this problem. List the denominators side by side with some space between them.

2)9	15	24
2)9	15	12
2)9	15	6
3)9	15	3
3)3	5	1
5)1	5	1
1	1	1

Begin dividing by a prime, and write the answers on the line below. If a number is not divisible by the prime, just bring the number down. Once you get a 1 in a row, continue to bring it down.

LCD $= 2 \times 2 \times 2 \times 3 \times 3 \times 5 = 360$ Now multiply the primes together to get the LCD.

This means that 360 is the least common denominator for this problem. Write the equivalent fractions and add.

$$\frac{5}{9} = \frac{200}{360} \qquad\qquad \frac{5}{9} \times \frac{40}{40} = \frac{200}{360}$$

$$\frac{11}{15} = \frac{264}{360} \qquad\qquad \frac{11}{15} \times \frac{24}{24} = \frac{264}{360}$$

$$+\frac{7}{24} = +\frac{105}{360} \qquad\qquad \frac{7}{24} \times \frac{15}{15} = \frac{105}{360}$$

$$\frac{569}{360} \text{ or } 1\frac{209}{360}$$ ◆

EXAMPLE 6   Simplify: $\frac{17}{21} - \frac{9}{15}$

Solution   Find the LCD of 21 and 15.

3)21	15
5) 7	5
7) 7	1
1	1

LCD $= 3 \times 5 \times 7 = 105$

$$\frac{17}{21} \times \frac{5}{5} = \frac{85}{105}$$

$$-\frac{9}{15} \times \frac{7}{7} = -\frac{63}{105}$$

$$\frac{22}{105}$$

Is 22/105 in lowest terms? Yes, because $\frac{22}{105} = \frac{2 \times 11}{3 \times 5 \times 7}$ and there are no factors common to both the numerator and the denominator.   ◆

EXAMPLE 7   Myra swam 7/8 of a mile on Monday, 3/4 mile on Wednesday, and 1/2 mile on Saturday. What was her total swimming distance for the week?

Solution   You are looking for a total of several different quantities, so addition is indicated.

$$\frac{7}{8} + \frac{3}{4} + \frac{1}{2}$$

$$= \frac{7}{8} + \frac{6}{8} + \frac{4}{8}$$

$$= \frac{17}{8} = 2\frac{1}{8} \text{ miles}$$

For word problems, the improper fraction answer is not meaningful. It should be changed to a mixed number.   ◆

EXAMPLE 8   Using the information in Example 7, find how much farther Myra swam on Monday than on Saturday.

Solution   When asked "how much farther (or shorter or whatever)," you need to subtract to find the difference. Be sure that the larger fraction is on top or to the left of the subtraction sign. (You may be able to tell which fraction is larger only when they have the same denominator.)

$$\frac{7}{8} = \frac{7}{8}$$

$$-\frac{1}{2} = \frac{4}{8}$$

$$\frac{3}{8} \text{ mile farther}$$ ◆

4.2   ◆   Practice Problems

A. If the given numbers were denominators in a fraction addition or subtraction problem, what would be the least common denominator?

1. 4 and 3
2. 7 and 14
3. 8 and 2
4. 6 and 9
5. 6 and 4
6. 5 and 3
7. 2, 3, and 4
8. 15, 3, and 60
9. 30, 40, and 60
10. 3, 4, and 5
11. 42 and 15
12. 39 and 15
13. 32 and 18
14. 48 and 30
15. 72, 15, and 48
16. 8, 12, 15, and 20
17. 5, 18, 15, and 20
18. 4, 24, 15, and 12
19. 18, 24, 6, and 3
20. 4, 6, 32, and 48
21. 12, 32, 8, and 18

B. Add or subtract as indicated. Reduce all answers to lowest terms. Your work from Section A will be of help.

22. $\frac{1}{8} + \frac{3}{8}$

23. $\frac{2}{9} + \frac{5}{9}$

24. $\frac{8}{11} - \frac{5}{11}$

25. $\frac{6}{5} - \frac{3}{5}$

26. $\frac{7}{12} + \frac{5}{6}$

27. $\frac{7}{15} + \frac{3}{5}$

28. $\frac{3}{4} + \frac{2}{3}$

29. $\frac{5}{7} - \frac{3}{14}$

30. $\frac{7}{8} - \frac{1}{2}$

31. $\frac{5}{6} + \frac{7}{9}$

32. $\dfrac{5}{6} - \dfrac{3}{4}$

33. $\dfrac{3}{5} + \dfrac{2}{3}$

34. $\dfrac{27}{32} - \dfrac{5}{6}$

35. $\dfrac{15}{16} - \dfrac{4}{5}$

36. $\dfrac{11}{18} + \dfrac{3}{4}$

37. $\dfrac{1}{2} + \dfrac{2}{3} + \dfrac{3}{4}$

38. $\dfrac{2}{5} + \dfrac{2}{3} + \dfrac{5}{6}$

39. $\dfrac{1}{3} + \dfrac{3}{4} + \dfrac{1}{6}$

40. $\dfrac{2}{3} + \dfrac{3}{4} - \dfrac{2}{5}$

41. $\dfrac{11}{12} - \dfrac{7}{15}$

42. $\dfrac{7}{9} - \dfrac{4}{15}$

43. $\dfrac{7}{12} - \dfrac{5}{18}$

44. $\dfrac{11}{18} + \dfrac{19}{20}$

45. $\dfrac{7}{12} + \dfrac{11}{15} - \dfrac{7}{18}$

46. $\dfrac{3}{8} + \dfrac{7}{12} + \dfrac{11}{15} + \dfrac{13}{20}$

47. $\dfrac{3}{5} + \dfrac{5}{18} + \dfrac{11}{15} + \dfrac{7}{20}$

48. $\dfrac{3}{4} + \dfrac{11}{24} + \dfrac{8}{15} + \dfrac{5}{12}$

49. $\dfrac{5}{18} + \dfrac{11}{24} + \dfrac{5}{6} + \dfrac{1}{3}$

50. $\dfrac{4}{9} + \dfrac{3}{4} + \dfrac{5}{12} + \dfrac{4}{5}$

C. Solve the following word problems. Be sure the answers are reduced to lowest terms.

51. Carol is making her own mixed-nut combinations to give as holiday gifts. She has blended 7/8 pound of cashews, 3/4 pound of pecans, 1/2 pound of almonds, and 3/8 pound of pistachio nuts. How much "deluxe mix" does she have?

52. In replacing the floor in Jo's kitchen, the contractor used plywood 3/8 inch thick and insulation material 1/4 inch thick. What was the total thickness of these materials?

53. In Problem 51, how many more pounds of pecans than pistachios were used?

54. In Problem 52, how much thicker is the plywood than the insulation material?

55. Raoul spent 1/2 hour on Monday, 7/8 hour on Tuesday, and 3/4 hour on Wednesday doing math homework. What is his total math study time? (If he can't solve this problem, then he didn't spend enough time!)

56. Scott is working out to get in condition for baseball season. He exercised 1/2 hour on Monday, 3/4 hour on Wednesday, and 7/8 hour on Friday. What was his total number of hours of exercise?

57. In Problem 55, how much more time did Raoul study on Tuesday than on Wednesday?

58. In Problem 56, how much more time did Scott exercise on Friday than on Monday?

59. Dominique studied her Spanish vocabulary words for 1/2 hour on Monday, 2/3 hour on Tuesday, and 3/4 hour on Wednesday. Altogether, how much time did she spend studying the vocabulary words on those days?

60. In Problem 59, how much longer did Dominique study on Wednesday than on Tuesday?

4.3 Adding and Subtracting Mixed Numbers

When adding or subtracting mixed numbers, you can either work with those numbers or change them to improper fractions. You will continue to follow the rules of addition or subtraction of fractions and write equivalent fractions with common denominators. Both methods will be demonstrated in the problems that follow. After studying these solutions, you will be able to choose the method that works best for you.

In the following problem, the proper fractions were written as equivalent fractions with the LCD of 10. Then the fraction parts and whole number parts were added.

$$1\dfrac{4}{5} = 1\dfrac{8}{10}$$

$$+3\dfrac{7}{10} = +3\dfrac{7}{10}$$

$$4\dfrac{15}{10} = 4\dfrac{3}{2} = 4 + 1\dfrac{1}{2} = 5\dfrac{1}{2}$$

Note: You cannot leave an improper fraction in a mixed number; 4 15/10 and 4 3/2 are not acceptable. Someone looking quickly at the answer might think that the answer is a little more than 4 because they see 4 and a fraction. In fact, because 3/2 is 1 1/2, you can (and must) add $4 + 1\ 1/2$ and get 5 1/2 as the answer.

Many mixed number subtraction problems proceed the same way as the previous problem. 14 7/8 − 5 2/3 can be done in this way.

$$14\frac{7}{8} = \quad 14\frac{21}{24}$$
$$-\ 5\frac{2}{3} = -\ 5\frac{16}{24}$$
$$\overline{\qquad\qquad 9\frac{5}{24}}$$

Mixed number addition and subtraction problems can be done by changing to improper fractions, though the numbers may become very large and awkward to work with. Doing the previous problem with improper fractions looks like this:

$$14\frac{7}{8} = \quad \frac{119}{8} = \quad \frac{357}{24} \qquad\qquad \frac{119}{8} \times \frac{3}{3} = \frac{357}{24}$$
$$-\ 5\frac{2}{3} = -\ \frac{17}{3} = -\ \frac{136}{24} \qquad\qquad \frac{17}{3} \times \frac{8}{8} = \frac{136}{24}$$
$$\overline{\qquad\qquad\qquad \frac{221}{24} \text{ or } 9\frac{5}{24}}$$

E X A M P L E 1   Solve: $9\dfrac{1}{2} + 11\dfrac{2}{3} - 6\dfrac{2}{15}$

Solution

$$9\frac{1}{2} = \quad 9\frac{15}{30} \qquad\qquad\qquad 21\frac{1}{6} = \quad 21\frac{5}{30}$$
$$+11\frac{2}{3} = +11\frac{20}{30} \qquad\qquad\qquad -\ 6\frac{2}{15} = -\ 6\frac{4}{30}$$
$$\overline{\qquad\quad 20\frac{35}{30}} \qquad\qquad\qquad\qquad \overline{\qquad\quad 15\frac{1}{30}}$$

$$= 20 + 1\frac{5}{30}$$

$$= 21\frac{5}{30}$$

$$= 21\frac{1}{6}$$

E X A M P L E 2   Solve: $5 + 6\dfrac{3}{4}$

Solution

$$5$$
$$+\ 6\frac{3}{4}$$
$$\overline{\quad 11\frac{3}{4}}$$

E X A M P L E   **3**   Solve: 5 1/3 − 2 3/4

Solution

$$5\frac{1}{3} = 5\frac{4}{12}$$

$$-2\frac{3}{4} = -2\frac{9}{12}$$

Once the equivalent fractions are written, a problem occurs in this mixed number subtraction because 9/12 cannot be subtracted from 4/12. When this happens, you must borrow. Borrow 1 from the whole number 5 and write it in the form most useful in this problem. Because 12 is the common denominator, write the 1 as 12/12 (12/12 = 1) and add that to the fraction already in the top number.

$$5\frac{4}{12} = \overset{4}{\cancel{5}}\frac{4}{12} + 1 = \overset{4}{\cancel{5}}\frac{4}{12} + \frac{12}{12} = 4\frac{16}{12}$$

$$-2\frac{9}{12} = -2\frac{9}{12} \qquad = -2\frac{9}{12} \qquad = -2\frac{9}{12}$$

$$2\frac{7}{12}$$ ◆

Borrowing is also necessary when a mixed number or fraction is subtracted from a whole number. In the following problem, 3/8 needs something from which to be subtracted. Borrow 1 from 7. Since 8 is the needed denominator, write the borrowed 1 as 8/8. Then the subtraction can be done.

$$7$$

$$-5\frac{3}{8}$$

$$7 = 6 + 1 = 6\frac{8}{8}$$

$$-5\frac{3}{8} = -5\frac{3}{8} \quad = -5\frac{3}{8}$$

$$1\frac{5}{8}$$

The previous two problems could have been worked with improper fractions. Example 3 asked us to subtract 5 1/3 − 2 3/4. The solution with improper fractions is

$$5\frac{1}{3} = \frac{16}{3} \times \frac{4}{4} = \frac{64}{12}$$

$$-2\frac{3}{4} = -\frac{11}{4} \times \frac{3}{3} = -\frac{33}{12}$$

$$\frac{31}{12} \text{ or } 2\frac{7}{12}$$

And improper fractions can be used as follows to solve $7 - 5\frac{3}{8}$.

$$7 - 5\frac{3}{8} = \frac{7}{1} - \frac{43}{8} \qquad \frac{7}{1} \times \frac{8}{8} = \frac{56}{8}$$

$$= \frac{56}{8} - \frac{43}{8} = \frac{13}{8} \text{ or } 1\frac{5}{8}$$

E X A M P L E   **4**   Subtract $121\frac{7}{8}$ from $146\frac{9}{16}$.

Solution   Using mixed numbers, we have

$$146\frac{9}{16} = \quad 146\frac{9}{16} = \quad \overset{145}{\cancel{146}}\frac{9}{16} + \frac{16}{16} = \quad 145\frac{25}{16}$$
$$-121\frac{7}{8} = -121\frac{14}{16} = -121\frac{14}{16} \qquad = -121\frac{14}{16}$$
$$24\frac{11}{16}$$

As in the previous examples, after the equivalent fractions are written with common denominators, the need to borrow is evident. The numerator 14 could not be subtracted from 9. We borrow 1 from the whole number 146 and write it as 16/16. This fraction is added to the 6/16 already there. Then the subtraction can be done.

Solution   Using improper fractions, we write

$$146\frac{9}{16} = \frac{2345}{16} = \frac{2345}{16}$$
$$-121\frac{7}{8} = -\frac{975}{8} \times \frac{2}{2} = -\frac{1950}{16}$$
$$\frac{395}{16} \text{ or } 24\frac{11}{16}$$ ◆

You will need to decide which method is better for you to use. In some problems it may be easier to use the mixed numbers and borrow, if necessary, in order to avoid large numbers in the improper fractions. You can make your choice of method after you see the problem.

E X A M P L E   **5**   Solve: $5 - 2\frac{5}{9}$

Solution   Using mixed numbers, we proceed as follows:

$$5 = \overset{4}{\cancel{5}}\frac{9}{9}$$

There has to be a fraction on top from which 5/9 can be subtracted. Borrow 1 from the 5 and write it as 9/9. Then subtract.

$$-2\frac{5}{9} = -2\frac{5}{9}$$
$$2\frac{4}{9}$$

Solution   Using improper fractions, we write

$$5 = \frac{5}{1} \times \frac{9}{9} = \frac{45}{9}$$
$$-2\frac{5}{9} = -\frac{23}{9} \qquad = -\frac{23}{9}$$
$$\frac{22}{9} \text{ or } 2\frac{4}{9}$$ ◆

E X A M P L E **6**   Find the difference between 6 3/8 and 5 1/2.

Solution
$$6\frac{3}{8} = 6\frac{3}{8}$$
$$-5\frac{1}{2} = -5\frac{4}{8}$$

Until the fractions are written with the common denominator, you do not know if you have to borrow. You will borrow here because 4 cannot be taken from 3.

$$6\frac{3}{8} = \overset{5}{\cancel{6}}\frac{3}{8} + \frac{8}{8} = 5\frac{11}{8}$$
$$-5\frac{4}{8} = -5\frac{4}{8} = -5\frac{4}{8}$$
$$\frac{7}{8}$$

Borrow 1 from 6 and write it as 8/8. Add 3/8 and 8/8. Then subtract.

This problem could also be worked by changing to improper fractions.

$$6\frac{3}{8} = \frac{51}{8} = \frac{51}{8}$$
$$-5\frac{1}{2} = -\frac{11}{2} \times \frac{4}{4} = -\frac{44}{8}$$
$$\frac{7}{8}$$

◆

E X A M P L E **7**   A cookie recipe combines 3 1/2 cups of flour, 1 2/3 cups of sugar, and 3/4 cup of oatmeal. What is the total amount of these dry ingredients?

Solution   You are looking for the *total amount* of several different quantities, so addition is indicated. First estimate, using convenient whole numbers: $4 + 2 + 1 = 7$. The answer should be a little less than 7.

$$3\frac{1}{2} = 3\frac{6}{12}$$
$$1\frac{2}{3} = 1\frac{8}{12}$$
$$+\frac{3}{4} = +\frac{9}{12}$$
$$4\frac{23}{12} = 4 + 1\frac{11}{12} = 5\frac{11}{12} \text{ cups}$$

◆

E X A M P L E **8**   Some computer disks are 5 1/4 inches in diameter, and others are 3 1/2 inches. How much difference in diameter is there?

Solution   The word *difference* means to subtract.

$$5\frac{1}{4} = 5\frac{1}{4} = \overset{4}{\cancel{5}}\frac{1}{4} + \frac{4}{4} = 4\frac{5}{4}$$
$$-3\frac{1}{2} = -3\frac{2}{4} = -3\frac{2}{4} = -3\frac{2}{4}$$
$$1\frac{3}{4} \text{ inches}$$

◆

4.3 ◆ Practice Problems

A. Add or subtract as indicated. Be sure to reduce all answers to lowest terms.

1. $\begin{array}{r} 4\frac{5}{8} \\ +\,3\frac{1}{2} \\ \hline \end{array}$
2. $\begin{array}{r} 7\frac{7}{8} \\ +\,1\frac{5}{6} \\ \hline \end{array}$
3. $\begin{array}{r} 6\frac{2}{3} \\ -\,4\frac{1}{2} \\ \hline \end{array}$

4. $\begin{array}{r} 17\frac{5}{8} \\ -\,12\frac{1}{3} \\ \hline \end{array}$
5. $\begin{array}{r} 2\frac{1}{4} \\ +\,7\frac{5}{6} \\ \hline \end{array}$
6. $\begin{array}{r} 10\frac{1}{3} \\ -\,4\frac{5}{16} \\ \hline \end{array}$

7. $\begin{array}{r} 22\frac{17}{18} \\ -\,10\frac{3}{4} \\ \hline \end{array}$
8. $\begin{array}{r} 13\frac{1}{12} \\ +\,4\frac{7}{10} \\ \hline \end{array}$
9. $\begin{array}{r} 16 \\ -\,7\frac{5}{6} \\ \hline \end{array}$

10. $\begin{array}{r} 12 \\ -\,5\frac{5}{7} \\ \hline \end{array}$
11. $\begin{array}{r} 15\frac{14}{15} \\ -\,10\frac{2}{3} \\ \hline \end{array}$
12. $\begin{array}{r} 30 \\ -\,17\frac{2}{3} \\ \hline \end{array}$

13. $\begin{array}{r} 11\frac{1}{4} \\ +\,4\frac{3}{7} \\ \hline \end{array}$
14. $\begin{array}{r} 18\frac{5}{9} \\ -\,17\frac{2}{30} \\ \hline \end{array}$
15. $\begin{array}{r} 21\frac{2}{5} \\ +\,10\frac{5}{8} \\ \hline \end{array}$

16. $\begin{array}{r} 4\frac{5}{8} \\ -\,3 \\ \hline \end{array}$
17. $\begin{array}{r} 11\frac{5}{9} \\ -\,5 \\ \hline \end{array}$
18. $\begin{array}{r} 6 \\ -\,2\frac{3}{8} \\ \hline \end{array}$

19. $\begin{array}{r} 8\frac{2}{3} \\ -\,7\frac{5}{6} \\ \hline \end{array}$
20. $\begin{array}{r} 126\frac{3}{4} \\ -\,125\frac{2}{3} \\ \hline \end{array}$
21. $\begin{array}{r} 296\frac{11}{16} \\ -\,152\frac{7}{8} \\ \hline \end{array}$

22. 2 5/8 − 1 15/16

23. 26 2/3 − 21 3/4

24. 13 4/5 − 11 13/15

25. 7 4/9 + 3 5/12 + 5/6

26. 2 3/4 + 5 1/8 + 2 2/3 + 3 1/2

27. 12 3/5 + 9 1/2 + 13 2/3 + 6 5/6

28. 2 7/16 + 5 3/8 + 7 3/24 + 2 1/2

29. 6 3/5 + 11 + 4 3/4 + 7 5/12

30. 1 9/16 + 3 3/5 + 9 + 2 1/2

B. Find the solution to each word problem. Reduce all answers to lowest terms.

31. How much more is 138 5/8 than 117 3/4?

32. Find the difference between 10 5/6 and 8 2/9.

33. Find the total of 14 1/3, 11 3/4, and 12 5/8.

34. Find the sum of 162 1/2, 131 4/9, and 120 3/5.

35. The Nut Shoppe blends its own deluxe mixed nuts. If 5 3/4 pounds of cashews, 2 1/2 pounds of almonds, 7 1/4 pounds of pecans, and 1 1/3 pounds of hazelnuts are combined, how much deluxe mix will there be?

36. One package of ground beef weighs 3 7/10 pounds and another weighs 2 1/2 pounds. How much ground beef is there?

37. Sue weighed 137 1/2 pounds when she started her diet. If she has lost 5 3/4 pounds, what is her new weight?

38. On three consecutive days of rain, Columbus, Ohio, had 1 1/2, 5/8, and 3/4 inches of rain. What was the total rainfall for that period?

39. After six weeks of dieting, Sue weighs 126 1/4 pounds. What is her total weight loss if she started at 137 1/2 pounds?

40. In Problem 38, how much more rain fell on the first day than on the second day?

41. A cake recipe calls for 2 1/2 cups of flour and 1 2/3 cups of sugar. How much more flour than sugar is used?

42. Jim bought 10 gallons of paint. After painting his house, he had 1 3/4 gallons left. How much paint did he use to paint the house?

43. Jan works part-time in the library. If she worked 4 1/2 hours on Monday, 2 3/4 hours on Thursday, and 5 1/3 hours on Saturday, what were her total hours for the week? Try estimating the answer before solving this problem.

44. On the stock market report, a certain stock opened at 42 1/8 points and closed at 43 3/8. How much did the stock increase?

45. In Problem 43, how many more hours did Jan work on Saturday than on Monday?

46. Mike took a long-distance bike trip. He rode 56 3/4 miles on the first day, rode 48 1/3 miles on the second day, and biked home by the same route. Altogether, how many miles did he ride? Try estimating the answer before solving this problem.

4.4   Order of Operations in Fraction Problems

If asked to solve a fraction problem that has more than one operation, apply the **Order of Operations Agreement** that was introduced in Section 2.3.

Order of Operations Agreement

When simplifying a mathematical expression involving more than one operation, follow these steps:

1. Simplify within parentheses or other grouping symbols.
2. Simplify any expressions involving exponents or square roots.
3. Multiply or divide in order from left to right.
4. Add or subtract (combine) in order from left to right.

In regard to Step 2, when a fraction is raised to the power indicated by an exponent, both the numerator and the denominator are raised to that power. When you are taking the square root of a fraction, you must take the square root of both parts of the fraction.

$$\left(\frac{2}{5}\right)^2 = \frac{(2)^2}{(5)^2} = \frac{4}{25} \qquad \left(\frac{2}{3}\right)^3 = \frac{(2)^3}{(3)^3} = \frac{8}{27}$$

$$\sqrt{\frac{4}{9}} = \frac{\sqrt{4}}{\sqrt{9}} = \frac{2}{3} \qquad \sqrt{\frac{49}{25}} = \frac{\sqrt{49}}{\sqrt{25}} = \frac{7}{5}$$

E X A M P L E 1   Simplify: $\dfrac{3}{4}\left(\dfrac{2}{3} + \dfrac{1}{2}\right)$

Solution $\dfrac{3}{4}\left(\dfrac{2}{3} + \dfrac{1}{2}\right)$ Work inside the parentheses first.

$= \dfrac{3}{4}\left(\dfrac{4}{6} + \dfrac{3}{6}\right)$ Find the common denominator and add.

$= \dfrac{3}{4}\left(\dfrac{7}{6}\right)$ 3/4 outside the parentheses says to multiply. Use the cancellation tool.

$= \dfrac{\overset{1}{3}}{4} \cdot \dfrac{7}{\underset{2}{6}} = \dfrac{7}{8}$ ◆

E X A M P L E 2   Simplify: $\dfrac{3}{8} \times \dfrac{5}{9} + \dfrac{3}{4}$

Solution $\dfrac{3}{8} \times \dfrac{5}{9} + \dfrac{3}{4}$ Do multiplication before addition.

$= \dfrac{\overset{1}{3}}{8} \times \dfrac{5}{\underset{3}{9}} + \dfrac{3}{4}$

$= \dfrac{5}{24} + \dfrac{3}{4}$ Then find a common denominator and add.

$= \dfrac{5}{24} + \dfrac{18}{24} = \dfrac{23}{24}$ ◆

EXAMPLE 3   Simplify: $\left(\dfrac{5}{6}\right)^2 + 1\dfrac{1}{3} \times \dfrac{7}{8}$

Solution $\left(\dfrac{5}{6}\right)^2 + 1\dfrac{1}{3} \times \dfrac{7}{8}$

$= \dfrac{25}{36} + 1\dfrac{1}{3} \times \dfrac{7}{8}$ Do exponents first, then multiplication, then addition.

$= \dfrac{25}{36} + \dfrac{\overset{1}{4}}{3} \times \dfrac{7}{\underset{2}{8}}$

$= \dfrac{25}{36} + \dfrac{7}{6}$

$= \dfrac{25}{36} + \dfrac{42}{36}$

$= \dfrac{67}{36}$ or $1\dfrac{31}{36}$

EXAMPLE 4   Simplify: $\left(\dfrac{5}{3}\right)^2 - \sqrt{\dfrac{16}{9}} \times 2$

Solution $\left(\dfrac{5}{3}\right)^2 - \sqrt{\dfrac{16}{9}} \times 2$ Do exponents and roots.

$= \dfrac{25}{9} - \dfrac{4}{3} \times 2$ Then do multiplication before subtraction.

$= \dfrac{25}{9} - \dfrac{4}{3} \times \dfrac{2}{1}$

$= \dfrac{25}{9} - \dfrac{8}{3}$

$= \dfrac{25}{9} - \dfrac{24}{9}$

$= \dfrac{1}{9}$

4.4 ◆ Practice Problems

A. Simplify each problem using the Order of Operations Agreement, and reduce answers to lowest terms.

1. $\dfrac{2}{3} + \dfrac{5}{6} \times \dfrac{1}{2}$

2. $\dfrac{3}{4} - \dfrac{5}{8} \times \dfrac{1}{2}$

3. $\dfrac{5}{9} \times \dfrac{3}{25} + \dfrac{7}{10}$

4. $\dfrac{9}{16} + \dfrac{3}{4} \times \dfrac{1}{2}$

5. $\dfrac{7}{8} - \dfrac{2}{3} \times \dfrac{3}{4}$

6. $\dfrac{6}{7} \times \dfrac{21}{24} + \dfrac{5}{8}$

7. $\sqrt{\dfrac{25}{36}} - \dfrac{2}{3}$

8. $\left(\dfrac{5}{8} + \dfrac{3}{4}\right)\left(\dfrac{5}{6} - \dfrac{3}{8}\right)$

9. $\dfrac{7}{8} \div \left(\dfrac{3}{4}\right)^2$

10. $\sqrt{\dfrac{16}{49}} - \dfrac{1}{3}$

11. $\left(\dfrac{3}{4} - \dfrac{2}{3}\right)\left(\dfrac{1}{2} - \dfrac{3}{8}\right)$

12. $\dfrac{5}{9} \div \left(\dfrac{2}{3}\right)^2$

13. $\left(2\dfrac{1}{4} - 1\dfrac{1}{2}\right) \times \dfrac{3}{4}$

14. $\left(3\dfrac{1}{2} + 1\dfrac{1}{4}\right) \times \dfrac{3}{5}$

15. $2\dfrac{1}{4} - 1\dfrac{1}{2} \times \dfrac{3}{4}$

16. $3\dfrac{1}{2} + 1\dfrac{1}{4} \times \dfrac{3}{5}$

17. $\left(\dfrac{3}{4}\right)^2 + \dfrac{1}{2} \times \dfrac{5}{6}$

18. $\sqrt{\dfrac{1}{9}} + \left(\dfrac{2}{3}\right)^2$

19. $\left(\dfrac{3}{4}\right)^2 + \sqrt{\dfrac{4}{9}}$

20. $\left(\dfrac{5}{6}\right)^2 + \dfrac{3}{4} \times 1\dfrac{5}{9}$

21. $3\dfrac{5}{6} - 1\dfrac{2}{3} + \dfrac{3}{4}$

22. $4\dfrac{4}{9} - 2\dfrac{3}{5} + \dfrac{2}{3}$

23. $3\frac{5}{6} - \left(1\frac{2}{3} + \frac{3}{4}\right)$ **24.** $4\frac{4}{9} - \left(2\frac{3}{5} + \frac{2}{3}\right)$ **27.** $1\frac{7}{15} \div 3\frac{2}{3} \times 1\frac{7}{16}$ **28.** $3\frac{5}{9} \div 2\frac{2}{3} \times 1\frac{3}{5}$

25. $\sqrt{\frac{16}{25}\left[\left(\frac{2}{3}\right)^2 - \frac{1}{2} \cdot \frac{3}{4}\right]}$ **29.** $\left(\frac{4}{9} \times \frac{5}{6}\right)\left(3\frac{1}{2} - 1\frac{3}{4}\right)$

26. $\left(1\frac{2}{3} + 3\frac{1}{2}\right)\left(\frac{5}{6} \times \frac{2}{3}\right)$ **30.** $\sqrt{\frac{1}{9}\left[\left(\frac{3}{4}\right)^2 + \frac{2}{3} \cdot \frac{3}{8}\right]}$

4.5 Comparing Fractions

In order to determine whether 5/16 is larger than 13/32, we must write the fractions with the same denominator. Then the fraction with the larger numerator is the larger fraction. The symbol > means "is greater than," and < means "is less than." Suppose you were asked, "Which fraction is larger, 5/16 or 13/32?" You would need to change 5/16 to 32nds so that both fractions had the same denominator. You would reason that 5/16 = 10/32, and clearly 13/32 > 10/32. Therefore, 13/32 > 5/16.

E X A M P L E 1   Determine which symbol, >, <, or =, should be between 2/5 and 3/5.

Solution   The fractions have the same denominator. Thus 2/5 is smaller than 3/5, because 2/5 contains one less "piece" than 3/5, and the pieces are the same size. Therefore, 2/5 < 3/5. ◆

E X A M P L E 2   Which is larger, 2/3 or 3/4?

Solution   The fractions must be changed to equivalent fractions with the denominator of 12 before they can be compared.

$$\frac{2}{3} = \frac{8}{12} \quad \text{and} \quad \frac{3}{4} = \frac{9}{12}, \quad \text{and} \quad \frac{9}{12} > \frac{8}{12}$$

Therefore, $\frac{3}{4} > \frac{2}{3}$. ◆

E X A M P L E 3   Determine whether 5/12 or 7/16 is the larger fraction, and then subtract the smaller number from the larger one.

Solution
```
2)12  16
2) 6   8
2) 3   4
2) 3   2
3) 3   1
   1   1
```
Find the LCD for 12 and 16 and write equivalent fractions.

LCD $= 2 \times 2 \times 2 \times 2 \times 3 = 48$

$$\frac{5}{12} = \frac{20}{48} \qquad \frac{7}{16} = \frac{21}{48}$$

Because $\frac{21}{48} > \frac{20}{48}$, subtract 20/48 from 21/48.

$$\frac{21}{48} - \frac{20}{48} = \frac{1}{48}$$ ◆

E X A M P L E **4**   Arrange these fractions in order from smallest to largest: 3/8, 1/3, 5/16, 5/12.

Solution

$$3 = 3$$
$$8 = 2 \times 2 \times 2$$
$$12 = 3 \times 2 \times 2$$
$$\underline{16 = 2 \times 2 \times 2 \times 2}$$
$$\text{LCD} = 3 \times 2 \times 2 \times 2 \times 2 = 48$$

Find the LCD for 8, 3, 16, and 12.

$$\frac{3}{8} = \frac{18}{48} \qquad \frac{1}{3} = \frac{16}{48}$$

$$\frac{5}{16} = \frac{15}{48} \qquad \frac{5}{12} = \frac{20}{48}$$

Change the fractions to equivalent fractions with a denominator of 48.

$$\frac{15}{48}, \frac{16}{48}, \frac{18}{48}, \frac{20}{48}$$

Arrange the new fractions from smallest to largest.

$$\frac{5}{16}, \frac{1}{3}, \frac{3}{8}, \frac{5}{12}$$

Replace these with their original form for the answer.   ◆

4.5 ◆ Practice Problems

A. **Determine which symbol should be between the fractions: >, <, or = (greater than, less than, or equal to).**

1. $\dfrac{3}{4}, \dfrac{2}{3}$ 2. $\dfrac{5}{9}, \dfrac{7}{12}$ 3. $\dfrac{5}{6}, \dfrac{3}{4}$

4. $\dfrac{7}{8}, \dfrac{49}{56}$ 5. $\dfrac{1}{6}, \dfrac{2}{12}$ 6. $\dfrac{10}{21}, \dfrac{3}{7}$

7. $\dfrac{4}{5}, \dfrac{5}{9}$ 8. $\dfrac{11}{12}, \dfrac{2}{3}$ 9. $\dfrac{4}{7}, \dfrac{5}{6}$

10. $\dfrac{11}{32}, \dfrac{2}{3}$ 11. $\dfrac{1}{2}, \dfrac{4}{5}$ 12. $\dfrac{15}{16}, \dfrac{11}{12}$

13. $\dfrac{48}{64}, \dfrac{3}{4}$ 14. $\dfrac{6}{36}, \dfrac{24}{144}$ 15. $\dfrac{19}{32}, \dfrac{7}{12}$

B. **Arrange these fractions in order from smallest to largest.**

16. $\dfrac{1}{2}, \dfrac{2}{5}, \dfrac{3}{4}$ 17. $\dfrac{3}{8}, \dfrac{5}{6}, \dfrac{2}{3}$

18. $\dfrac{4}{5}, \dfrac{5}{6}, \dfrac{7}{8}$ 19. $\dfrac{15}{16}, \dfrac{3}{4}, \dfrac{7}{8}$

20. $\dfrac{1}{4}, \dfrac{1}{2}, \dfrac{1}{6}$ 21. $\dfrac{2}{9}, \dfrac{2}{3}, \dfrac{4}{5}$

22. $\dfrac{4}{5}, \dfrac{2}{3}, \dfrac{5}{8}$ 23. $\dfrac{15}{16}, \dfrac{2}{3}, \dfrac{3}{4}$

24. $\dfrac{10}{12}, \dfrac{13}{16}, \dfrac{5}{8}, \dfrac{3}{4}$ 25. $\dfrac{2}{3}, \dfrac{5}{8}, \dfrac{3}{4}, \dfrac{1}{2}$

26. $\dfrac{15}{16}, \dfrac{5}{6}, \dfrac{7}{8}, \dfrac{11}{12}$ 27. $\dfrac{4}{7}, \dfrac{3}{8}, \dfrac{2}{3}, \dfrac{1}{2}$

28. $\dfrac{7}{12}, \dfrac{8}{15}, \dfrac{5}{9}, \dfrac{3}{4}$ 29. $\dfrac{11}{24}, \dfrac{5}{8}, \dfrac{7}{16}, \dfrac{17}{32}$

30. $\dfrac{3}{32}, \dfrac{5}{64}, \dfrac{1}{8}, \dfrac{3}{16}, \dfrac{1}{4}$

C. **Determine which fraction is larger and subtract the smaller fraction from it.**

31. $\dfrac{2}{3}, \dfrac{3}{4}$ 32. $\dfrac{5}{8}, \dfrac{5}{6}$ 33. $\dfrac{11}{12}, \dfrac{15}{16}$

34. $\dfrac{4}{5}, \dfrac{3}{4}$ 35. $\dfrac{5}{8}, \dfrac{9}{16}$ 36. $\dfrac{5}{12}, \dfrac{1}{3}$

37. $\dfrac{17}{16}, \dfrac{13}{12}$ 38. $\dfrac{19}{24}, \dfrac{5}{8}$ 39. $\dfrac{29}{32}, \dfrac{7}{8}$

40. $\dfrac{11}{18}, \dfrac{17}{27}$ 41. $\dfrac{54}{49}, \dfrac{31}{28}$ 42. $\dfrac{11}{16}, \dfrac{17}{32}$

4.6   Fraction Word Problems

Review the procedures and suggestions for solving word problems in Section 1.6. They apply to all types of word problems, including the fraction word problems in this section. Remember to review, as well, the rules for fractions that follow.

Basic Rules for Fraction Operations

1. Fractions must have common denominators to be added or subtracted, but not to be multiplied or divided.

2. Mixed numbers must be changed to improper fractions to be multiplied or divided, but they do not need to be changed to improper fractions when added or subtracted.

3. The cancellation short cut can be performed only when the fractions are being multiplied.

4. All fraction answers should be reduced to lowest terms. Improper fraction answers to word problems should be changed to mixed numbers.

The problems in this section involve all four operations with fractions, not just addition and subtraction. More than one step may be required to reach the solution.

E X A M P L E   1   Cans of soup normally have a net weight of 10 3/4 ounces. How many cans can be filled from a kettle containing 150 1/2 ounces?

Solution   Because the total amount is given (150 1/2 ounces) and the size of each part is given (10 3/4 ounces), division needs to be done. Remember to place the total amount before the division symbol. Estimate an answer by rounding to convenient whole numbers: $150 \div 10 = 15$ cans.

$$150\frac{1}{2} \div 10\frac{3}{4}$$

$$= \frac{301}{2} \div \frac{43}{4}$$

$$= \frac{301}{\overset{}{\underset{1}{2}}} \times \frac{\overset{2}{4}}{43}$$

$$= \frac{602}{43} = 14 \text{ cans} \qquad \text{This is close to the estimated answer.} \qquad \blacklozenge$$

E X A M P L E   2   Two-thirds of the students in a certain college drive their own car to campus each day. If there are 9,600 students attending, how many drive their own car?

Solution   2/3 of the total students drive. The total number of students is 9,600. 2/3 of the 9,600 drive. "Of" connecting two numbers usually indicates multiplication.

$$\frac{2}{\overset{}{\underset{1}{3}}} \times \frac{\overset{3200}{9600}}{1} = 6,400 \text{ students drive} \qquad \blacklozenge$$

E X A M P L E **3** A bolt of fabric contained 22 yards when it arrived at the store. After sales of 4 3/8 yards, 1 1/2 yards, and 3 3/4 yards, how much fabric remains on the bolt?

Solution The total amount of fabric that has been sold can be found by addition. Then "how much remains" suggests subtraction. Estimate the answer by using convenient whole numbers: $4 + 2 + 4 = 10$ and $22 - 10 = 12$ yards.

$$4\frac{3}{8} = \quad 4\frac{3}{8}$$

$$1\frac{1}{2} = \quad 1\frac{4}{8}$$

$$+ \ 3\frac{3}{4} = + \ 3\frac{6}{8}$$

$$8\frac{13}{8}$$

$$= \quad 8 + 1\frac{5}{8}$$

$$= \quad 9\frac{5}{8} \text{ yards have been sold}$$

$$22 = \quad 21\frac{8}{8}$$

$$- \ 9\frac{5}{8} = - \ 9\frac{5}{8}$$

$$12\frac{3}{8} \text{ yards remain on the bolt}$$

4.6 ◆ Practice Problems

Solve the following word problems, reducing fraction answers to lowest terms. Remember that the problem can require any operation, and in some problems you will need more than one step to get the solution. You may find that estimating an answer first will help you decide which operation to use.

1. Sarah's cookie recipe requires 2 2/3 cups of flour. How much flour will she need if she triples the recipe?

2. Five-eighths of an economics class are female students. If there are 32 students enrolled in that class, how many are female?

3. In Problem 1, how much flour will Sarah need if she makes only half of the recipe?

4. In Problem 2, what fraction of the class is male? How many students are male?

5. Sandy is trying to build up her endurance as a swimmer. She swam 5 1/2 laps on Monday, 6 3/4 laps on Wednesday, and 7 1/2 laps on Friday. How many laps did she swim altogether?

6. A box containing 15 pounds of hard candy is to be packaged into small baskets, each holding 3/4 of a pound. How many baskets can be filled?

7. In Problem 5, how much farther did Sandy swim on Friday than on Wednesday?

8. Sue has started another weight-loss program. If she lost 2 1/2 pounds the first week, 1 1/3 pounds the second week, and 3/4 pound the third week, how much has she lost so far?

9. In Problem 5, if each lap is one-third of a mile in length, how many miles did Sandy swim on Monday?

10. In Problem 8, if Sue's weight-loss goal is 10 pounds, how many more pounds does she need to lose to reach her goal?

11. If a car travels 95 1/2 miles in 2 3/4 hours, how far can it travel in 1 hour?

12. Marilyn is making a matching skirt and blouse. The blouse requires 1 7/8 yards of material, and the skirt requires 2 1/4 yards. How much fabric must she purchase to make both?

13. A tank contains 26 3/4 gallons of maple syrup. It will be used to fill bottles that hold 1/8 of a gallon. How many bottles can be filled?

14. In Problem 12, how much fabric would Marilyn need if she decided to make three of the blouses and two of the skirts?

15. Mark has formed his own lawn service business in order to earn money for college. He has contracted to mow the Engels' yard each week. The first time he mowed the yard, it took him 1 1/4 hours, but later in the summer he was able to do it in 5/8 hour. How much less time did it take late in the summer?

16. In a small town in Pennsylvania, one-third of the voters are registered as Democrats and nine-sixteenths are registered as Republicans. What fractional part of the registered voters are neither Democrats nor Republicans?

17. One-eighth of the players on the college's football team are maintaining an "A" average in their class work, and three-sixteenths have a "B" average. What fractional part of the team is doing less than "B" work?

18. If the town in Problem 16 has a population of 14,688 registered voters, how many are registered as Republican?

19. If there are 32 players on the football team in Problem 17, how many are doing "B" work?

20. A piece of wire that is 5 1/4 feet long is divided into three equal pieces. How long is each piece?

21. Maria is making matching skirts and blouses for her ethnic dance class. Each skirt requires 2 1/4 yards of fabric, and each blouse requires 1 3/8 yards. How much more fabric does a skirt need than a blouse?

22. Three-eighths of the children in a preschool class live in single-parent households. If there are 16 children in the class, how many live in single-parent households?

23. If Maria is making blouses and skirts for five members of the dance class (see Problem 21), how much fabric will she need altogether?

24. Forest rangers and volunteers have replanted trees on 16 3/4 acres of land that had been destroyed by a forest fire in Wilson State Park. If the area destroyed by the fire consists of 35 1/2 acres, how many acres remain to be replanted?

25. A pork roast is to be cooked 1/3 hour for each pound. If the roast weighs 5 1/6 pounds, how many hours should it be cooked?

26. In Problem 24, if an average of 2,420 trees were planted per acre, what is the total number of trees that have been planted so far in this recovery project?

27. How many volumes of an encyclopedia set will fit on a library shelf that is 34 inches long if each book is 2 1/4 inches thick?

28. A piece of pipe that is 7 1/3 feet long is to be cut into five equal pieces. Assuming there is no waste, find how long each piece will be.

4.7 Chapter Review

We hope you are feeling more confident about working with fractions. As you prepare for the test for this chapter, remember:

1. Fractions can be added or subtracted only if they have the same denominator.

2. Find the least common denominator by inspection or by the least common multiple method (Section 2.6), and then write equivalent fractions.

3. Mixed numbers do not have to be changed to improper fractions when you are adding or subtracting.

4. If borrowing is required in subtraction, borrow 1 from the whole number, write the 1 as the common denominator over itself, add together the fraction parts of the numerator, and then subtract.

5. Apply the Order of Operations Agreement to fraction problems containing more than one operation.

6. To square a fraction or to take the square root of a fraction, you must apply the operation to both the numerator and the denominator.

7. To compare fractions, write them as equivalent fractions (fractions that have the same denominator), and then compare the numerators.

8. To solve fraction word problems, use the same steps described for word problems with whole numbers.

9. Reduce all fraction answers to lowest terms.

Review Problems

A. Solve each problem, reducing answers to lowest terms.

1. $\dfrac{3}{4} + \dfrac{5}{6}$

2. $\dfrac{9}{16} - \dfrac{3}{8}$

3. $\dfrac{5}{24} - \dfrac{1}{8}$

4. $\dfrac{5}{9} + \dfrac{1}{2}$

5. $2\dfrac{3}{4} + 1\dfrac{5}{9}$

6. $3\dfrac{4}{5} - 1\dfrac{1}{3}$

7. $7\dfrac{6}{7} - 3\dfrac{3}{8}$

8. $4\dfrac{2}{3} + 2\dfrac{5}{8}$

9. $3\dfrac{1}{2} - 2\dfrac{3}{4}$

10. $1\dfrac{3}{4} + 5\dfrac{2}{3} + 7\dfrac{1}{6}$

11. $4\dfrac{1}{8} + 2\dfrac{2}{5} + 7\dfrac{3}{10}$

12. $\dfrac{4}{9} + \dfrac{11}{12} - \dfrac{3}{4}$

13. $\dfrac{5}{7} + \dfrac{3}{4} - \dfrac{5}{14}$

14. $9 - 2\dfrac{5}{8}$

15. $2\dfrac{1}{4} - \dfrac{3}{8}$

16. $8\dfrac{11}{16} + 5\dfrac{3}{4}$

17. $20 - 11\dfrac{2}{3}$

18. $4\dfrac{1}{8} - \dfrac{7}{16}$

19. $24\dfrac{5}{8} + 10\dfrac{1}{3}$

20. $5\dfrac{5}{9} - 2\dfrac{3}{4}$

21. $116 - 84\dfrac{5}{9}$

22. $2\dfrac{7}{12} + 3\dfrac{1}{3}$

23. $17\dfrac{5}{8} - 15$

24. $129 - 112\dfrac{17}{32}$

25. $11\dfrac{3}{5} + 9\dfrac{2}{9}$

26. $45\dfrac{13}{32} - 37$

27. $\dfrac{4}{15} + \dfrac{19}{32} + \dfrac{7}{12}$

28. $\dfrac{5}{16} + \dfrac{17}{18} + \dfrac{11}{24}$

29. $2\dfrac{2}{3} - 1\dfrac{5}{8} + \dfrac{5}{6}$

B. Solve each problem, applying the Order of Operations Agreement. Reduce answers to lowest terms.

30. $\dfrac{2}{3} - \dfrac{1}{2} + \dfrac{3}{4}$

31. $\left(\dfrac{3}{4}\right)^2 + \dfrac{1}{2} \times \dfrac{5}{8}$

32. $\dfrac{2}{3} + \dfrac{3}{8} - \sqrt{\dfrac{4}{9}}$

33. $\dfrac{7}{10} + \sqrt{\dfrac{9}{25}} - \dfrac{4}{5}$

34. $\left(\dfrac{5}{6}\right)^2 + \dfrac{3}{4} \times \dfrac{7}{9}$

35. $\dfrac{2}{3} - \dfrac{1}{2} \div \dfrac{3}{4}$

36. $1\dfrac{1}{2} + 2\dfrac{1}{3} \times \dfrac{3}{4}$

37. $5\dfrac{1}{4} \div \dfrac{7}{12} \times \dfrac{1}{2}$

38. $2\dfrac{2}{3} \div 1\dfrac{3}{5} \times \dfrac{3}{10}$

39. $2\dfrac{5}{8} + 3\dfrac{1}{2} \times 1\dfrac{3}{4}$

C. Arrange the fractions in order from *largest to smallest*.

40. $\dfrac{3}{4}, \dfrac{3}{8}, \dfrac{5}{16}$

41. $\dfrac{4}{9}, \dfrac{4}{5}, \dfrac{4}{15}$

42. $\dfrac{5}{8}, \dfrac{2}{3}, \dfrac{7}{9}$

43. $\dfrac{3}{16}, \dfrac{1}{4}, \dfrac{5}{12}$

44. $\dfrac{3}{4}, \dfrac{7}{12}, \dfrac{5}{6}$

45. $\dfrac{11}{18}, \dfrac{2}{3}, \dfrac{5}{9}$

46. $\dfrac{1}{5}, \dfrac{2}{7}, \dfrac{3}{10}$

47. $\dfrac{3}{8}, \dfrac{5}{12}, \dfrac{5}{16}$

48. $\dfrac{5}{6}, \dfrac{3}{5}, \dfrac{2}{3}$

49. $\dfrac{11}{12}, \dfrac{9}{10}, \dfrac{5}{6}$

D. Solve each problem and reduce answers to lowest terms. Remember that any operation may be required, not just addition and subtraction.

50. If a man drives 264 miles in 5 1/2 hours, how far does he drive in one hour?

51. Carol promised herself that she would study algebra and do homework for 1 1/4 hours every night of the week (7 days). What is the total number of hours that she will spend on algebra each week?

52. A stock opens at 16 3/8 points and closes at 20. How much of an increase did it experience?

53. Greg worked 5 3/4 hours on Tuesday, 4 hours on Wednesday, and 5 1/8 hours on Saturday. What was his total number of hours worked for the week?

54. Ann Marie will pay her baby sitter at the end of the week. How much is due if the sitter earns $3 per hour and worked 7 3/4 hours Tuesday, 6 1/2 hours Wednesday, and 8 5/8 hours Saturday?

55. An airplane requires 22 gallons of gasoline to travel 232 1/2 miles. How far does it travel on one gallon?

56. To make a batch of toffee, three-fourths of a pound of butter is needed. How much butter would be needed to make 5 batches?

57. In Problem 53, how many hours less did Greg work on Saturday than on Tuesday?

58. Two-thirds of the students at a certain college drive their own car or ride in a car to campus, rather than using public transportation. If 10,200 students are enrolled, how many drive or ride in a car?

59. A 12 1/2-acre plot of land is to be divided into building lots, each 5/8 acre in size. How many lots will there be?

60. Five-sixteenths of the voters in the last election cast ballots choosing candidates of only one political party. If 32,496 people voted in the election, how many cast one-party ballots?

Chapter Test

1. If 3, 15, and 8 are denominators of fractions in an addition or subtraction problem, what is the least common denominator?

Simplify as indicated. Reduce all answers to lowest terms.

2. $\dfrac{5}{6} + \dfrac{2}{3}$ **3.** $\dfrac{4}{5} - \dfrac{1}{2}$

4. $\dfrac{3}{8} + \dfrac{15}{16}$ **5.** $1\dfrac{2}{3} + 4\dfrac{3}{4}$

6. $3\dfrac{3}{5} - 2\dfrac{1}{3}$ **7.** $8 - 4\dfrac{3}{7}$

8. $\dfrac{7}{8} + \dfrac{2}{3} + \dfrac{5}{12}$ **9.** $\dfrac{7}{9} - \dfrac{7}{15}$

10. $4\dfrac{1}{4} - 2\dfrac{2}{3}$ **11.** $7\dfrac{1}{3} + 3\dfrac{4}{5} + 2\dfrac{4}{15}$

12. $9 - 2\dfrac{3}{5}$ **13.** $14\dfrac{2}{3} - 11$

14. $\left(\dfrac{2}{3}\right)^2 + 2\dfrac{1}{3} \times \dfrac{1}{2}$ **15.** $\dfrac{3}{10} + \sqrt{\dfrac{9}{25}} - \dfrac{2}{5}$

16. Arrange from largest to smallest: $\dfrac{3}{16}, \dfrac{3}{8},$ and $\dfrac{1}{4}.$

Solve the following word problems.

17. Laura worked 6 1/3 hours on Monday, 4 3/4 hours on Tuesday, and 7 5/8 hours on Wednesday. What was the total number of hours that she worked?

18. Using the information in Problem 17, find how many more hours Laura worked on Monday than on Tuesday.

19. Marc has mowed two-thirds of a 1 3/4-acre field. How many acres *remain* to be mowed?

20. If Tom can drive 132 miles on 5 1/2 gallons of gas, how far can he travel on one gallon?

Thought-Provoking Problems

1. Describe the entire process of writing equivalent fractions with larger numbers. You might want to describe an actual problem, such as $3/4 = 15/20$.

2. What part of fraction problems do you find to be the most difficult? Do you feel any better about fractions now?

3. Why aren't calculators very useful when we are solving fraction problems?

4. Solve the two problems that follow using paper and pencil. Then solve them using a calculator. What did you observe about the answers? Why did this occur?

$$\frac{3}{4} + \frac{5}{8} \qquad \frac{1}{3} + \frac{7}{12}$$

5. Is $\sqrt{\frac{49}{4}} = \frac{7}{4}$? Explain why or why not.

6. Is $\frac{5}{6} - \frac{1}{3} > \frac{1}{9} + \frac{8}{27}$?

7. Find the value of the following expressions, and then arrange the answers in order from largest to smallest.

$$\frac{5}{8} - \frac{3}{16} \qquad \frac{1}{4} + \frac{3}{5} \qquad \left(\frac{2}{3}\right)^2 \div \sqrt{\frac{4}{9}}$$

8. Find the value of $\left(\sqrt{\frac{25}{16}}\right)^3$.

9. Why must fractions have the same denominator before being added or subtracted?

10. Describe the process of borrowing in a mixed number subtraction problem.

11. How do the processes used to solve these two problems differ?

$$\begin{array}{c} 12 \\ - 7\frac{5}{8} \\ \hline \end{array} \qquad \begin{array}{c} 12\frac{5}{8} \\ - 7 \\ \hline \end{array}$$

12. Write several pairs of equivalent fractions. In each pair, compare the products when the numerator of each fraction is multiplied by the denominator of the other. Can you make an assumption about equivalent fractions?

13. Is $\frac{1}{2} \times \frac{1}{2} \div \frac{1}{2} = \frac{1}{2} \div \frac{1}{2} \times \frac{1}{2}$?

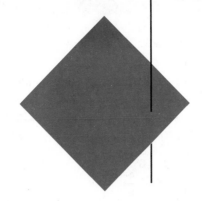

Cumulative Review, Chapters 1–4

Solve each of the following problems. Show all your work. Reduce all fraction answers to lowest terms.

A. In Problems 1 through 5, add.

1. $26 + 114 + 236$
2. $46,551 + 28,779$
3. $\dfrac{5}{8} + \dfrac{7}{8}$
4. $\dfrac{16}{25} + \dfrac{4}{5}$
5. $7\ 5/9 + 8\ 1/2$

B. In Problems 6 through 10, subtract.

6. $729 - 455$
7. $30,004 - 2,965$
8. $\dfrac{8}{15} - \dfrac{2}{15}$
9. $2\ 1/2 - 1\ 5/6$
10. $124 - 52\ 3/8$

C. In Problems 11 through 15, multiply.

11. 539×78
12. 45×100
13. $\dfrac{3}{4} \times \dfrac{9}{16}$
14. $1\ 2/3 \times 5/9 \times 12/25$
15. Find $7/8$ of 90.

D. In Problems 16 through 20, divide.

16. $429 \div 3$
17. $42\overline{)7,594}$
18. $\dfrac{5}{6} \div \dfrac{1}{3}$
19. $4\ 3/8 \div 2/5$
20. $\dfrac{20}{\dfrac{2}{3}}$

E. Find the answer to each of the following problems.

21. Is 1,593 divisible by 2? by 3? by 5?
22. Find the greatest common factor for 28 and 56.
23. Find the least common multiple for 7, 28, and 56.
24. Find the prime factorization for 63.
25. Find the prime factorization for 240.
26. $2^5 =$
27. $\sqrt{81} =$
28. List *all* the factors of 48.
29. Find the value of K in $K = L + 2M$ when $L = 6$ and $M = 5$.
30. Find the value of K in $K = L + 2M$ when $L = 2\ 1/2$ and $M = 1\ 1/3$.

F. Write the letter of the statement that exemplifies each of the properties listed in Problems 31 through 35.

31. Commutative Property of Multiplication
32. Multiplication Property of One
33. Associative Property of Multiplication
34. Distributive Property
35. Zero Property of Multiplication

a. $6 \times 1 = 6$
b. $6 \times 0 = 0$
c. $6 \times 1 = 1 \times 6$
d. $2(6 \times 1) = (2 \times 6) \times 1$
e. $2(6 + 1) = 2(6) + 2(1)$

G. **In Problems 36 through 43, simplify each of the expressions using the Order of Operations Agreement.**

36. $7 + 3 - 5 + 2$ **37.** $8 \div 2 \times 4 + 2$

38. $2^4 - 9 \div 3$ **39.** $\sqrt{81} + 2^3 \div 4$

40. $5/8 \times 2/3 + 1/2$ **41.** $\sqrt{\dfrac{4}{9}} \div 6 + \dfrac{4}{5}$

42. $7 + 3[8 - 2(6 - 3)]$ **43.** $2[4(18 \div 9 + 2)]$

H. **Find the answer to each of the following word problems.**

44. Three-fourths of the women students in this college are part-time students. If there are 6,400 women enrolled in the college, how many are part-time students?

45. In his job, Raoul drives an average of 115 miles a day. How many miles does he drive during a 5-day work week?

46. Anne bought a toy for each of her three children while she was shopping. She purchased a doll for $13, a game for $9, and a teddy bear for $10. How much did she spend for the three toys?

47. If Raoul can drive 228 miles on 12 gallons of gas, how far can he drive on one gallon?

48. An 8-pound block of cheese is to be cut into pieces that each weigh two-thirds of a pound. How many pieces can be cut from the block?

49. Find the difference between 26 2/3 and 14 3/8.

50. Find the mean of 76, 54, 83, and 63.

Decimals

Decimal numbers are usually made up of two parts. The whole numbers are written to the left of the decimal point, and a part of a whole number is written to the right of the decimal point. They are similar to mixed fractions. For instance, 45 7/100 is written 45.07 in decimal form. Our experience in dealing with money, which uses a decimal point to separate the dollars from the cents, makes this easy to remember.

Learning Objectives

When you have completed this chapter, you should be able to:

1. Read, write, round, and rank decimal numbers.
2. Add, subtract, multiply, and divide decimals.
3. Change decimal numbers to their fractional equivalents and fractions to their decimal values.
4. Estimate answers to word problems.
5. Solve word problems involving one or more operations with whole numbers, fractions, and decimal numbers.

5.2 Reading, Writing, Rounding, and Ranking Decimals

Almost all numbers can be written in decimal form using numerals and a decimal point. The whole number 8 can be written 8.0. The fraction 3/4 is equivalent to 0.75. 5/8 is 0.625. Because our monetary system is based on the decimal system, people are usually more comfortable and confident with decimals than with their fractional equivalents.

Decimal numbers have a place value system that is very similar to that of the whole number system. In fact, because the numbers to the left of the decimal point are whole numbers, the only new thing you need to learn is the place values to the right of the decimal point.

ten-millions	millions	,	hundred-thousands	ten-thousands	thousands	,	hundreds	tens	ones	• and	tenths	hundredths	thousandths	ten-thousandths	hundred-thousandths	millionths

FIGURE 1

In Figure 1, you can see that the progression of place names is the same for the "part" side (to the right of the decimal point) as it is for the whole number side, except that there are no "oneths." Because "tenths" means a number divided into "10" parts, "oneths" would mean a number divided into "1" part. A number divided by "1" is a whole number and belongs on the left of the decimal point. Therefore, the progression of place values on the right of the decimal point begins with tenths. Note also that the "ths" ending signifies place value on the right of the decimal point. Three hundred is written 300. Three-hundredths is written 0.03. If you think in terms of money, $300 is very different from $0.03. That "ths" ending is extremely important.

Exercises To practice, identify the digit in the given place values in 24,578.9061.

1. tenths place ________
2. hundreds place ________
3. ones place ________
4. thousandths place ________
5. thousands place ________

Answers:

1. 9 **2.** 5 **3.** 8 **4.** 6 **5.** 4

ten-millions	millions	,	hundred-thousands	ten-thousands	thousands	,	hundreds	tens	ones	and	tenths	hundredths	thousandths	ten-thousandths	hundred-thousandths	millionths
				2	4		5	7	8	•	9	0	6	1		

FIGURE 2

Place value plays an important role in the reading of decimal numbers. As an example, 1,624.038 would be read "one thousand six hundred twenty-four and thirty-eight thousandths." The whole number is read as discussed in Chapter 1; the decimal point is read "and." The part of the number to the right of the decimal point is read like a whole number that ends with its place value name, showing how many decimal places were used. Consider these examples:

14.23	fourteen and twenty-three hundredths
14.023	fourteen and twenty-three thousandths
14.0023	fourteen and twenty-three ten-thousandths

Exercises   Now write these numbers in words.

1.   26.025   _____
2.   3.145   _____
3.   76.421   _____
4.   147.05   _____
5.   0.0046   _____

Answers:

1.   twenty-six and twenty-five thousandths
2.   three and one hundred forty-five thousandths
3.   seventy-six and four hundred twenty-one thousandths
4.   one hundred forty-seven and five hundredths
5.   forty-six ten-thousandths

To translate from words to decimal numbers is just as easy. Look for the "and" that separates the whole number from the part. Notice the last word in the number; it lets you know how many decimal places to use. Then play "fill in the blanks." As an example, sixty-two and four thousandths looks like 62.__ __ __ (thousandths uses three spaces), so you write it 62.004, filling in the extra spaces with place-holding zeros. If you wrote 62.040, that would be read "sixty-two and forty thousandths." 62.400 would be read "sixty-two and four hundred thousandths." Sixty-two and four thousandths must be 62.004. Four thousand and sixteen hundredths would be 4,000.16. Forty-seven thousandths has no whole number part, and thousandths uses three decimal places, so it would be written 0.047. The leading zero in a number that has no whole number part is not absolutely necessary. It does help, however, to point out the fact that there is a decimal point in the number. 0.047 is easier to read than .047. Including the leading zero in the whole number ones place is recommended. Now write these numbers in digit form.

Exercises
1.   six and five hundredths   _____
2.   forty-three and fifteen thousandths   _____
3.   seventy-seven ten-thousandths   _____
4.   four hundred and four hundredths   _____
5.   six thousand two and two thousandths   _____

Answers:

1.   6.05   **2.**   43.015   **3.**   0.0077   **4.**   400.04   **5.**   6,002.002

Another non-computational process involving decimals is rounding. The rule for rounding decimal numbers is almost identical to the rule used for whole numbers. The main difference is that the places to the right of the rounded place are dropped and not replaced by zeros.

Rule for Rounding Decimal Numbers

1. Mark the place value being considered.
2. Look at the number to its right.
    a. If that digit is less than 5, the place marked remains as it is, and all following places are dropped.
    b. If that digit is 5 or more, the place marked increases by 1, and all following places are dropped.

A good hint to remember is that when you round decimals, the final answer must have *exactly* the number of digits named.

For example, round 921.6387 to the nearest hundredth.

1. Mark the hundredths place: 921.6387
       ∧

2. Look at the number to the right. Because it is 8, which is larger than 5, the 3 increases to 4, and the 87 is dropped.

3. 921.6387 becomes 921.64, nine hundred twenty-one and sixty-four *hundredths*. Hundredths is the second decimal place, and the answer must have exactly two decimal places.

Consider the following completed examples.

E X A M P L E **1**   Round 8.049 to the nearest hundredth.

Solution   8.049   9 is greater than 5, so the 4 increases by 1.
               ∧
            8.05   Drop all digits past hundredths.
            8.05   Eight and five hundredths.   ◆

E X A M P L E **2**   Round 16,047.346 to the nearest tenth.

Solution   16,047.346   4 is less than 5.
                  ∧
            16,047.3   Drop all digits past tenths.
            16,047.3   Sixteen thousand forty-seven and three tenths.   ◆

E X A M P L E **3**   Round 3,426.795 to the nearest hundredth.

Solution   3,426.795   5 is equal to 5, and therefore the 79 increases to 80.
                 ∧
            3,426.80   Keep the last zero so that the answer ends in hundredths.
            3,426.80   Three thousand, four hundred twenty-six and eighty hundredths. ◆

E X A M P L E **4**   Round $83.925 to the nearest cent.

Solution $83.925   Because the number to the right of the digit in the cents place is 5, the 2
               ∧   increases by 1.
            $83.93   Cents is two decimal places.
            $83.93   Eighty-three dollars and ninety-three cents.   ◆

E X A M P L E **5**   Round $47.884 to the nearest cent.

Solution $47.884   4 is less than 5, so the 8 remains the same.
            $47.88   Cents is always two decimal places.
            $47.88   Forty-seven dollars and eighty-eight cents.   ◆

While you are studying rounding, it is good to learn about another common mathematical practice. You do not write the answer to a problem as 5, for instance, unless the calculation comes out to be exactly the whole number 5. The number 5.0 is not the same as 5, because 5.0 indicates that the answer is approximately equal to ($\doteq$) 5.0. It could be 5.036, for example, rounded to the nearest tenth.

Suppose you were asked to decide which of two decimal numbers is the larger or to arrange three or four decimal numbers in order from the smallest to the largest. You would need some basis for comparing the sizes of the numbers, such as writing the numbers with the same number of decimal places. In order for us to compare two or more quantities, they must be described in the same way. For example,

"Which is larger, 0.26 or 0.264?"

If you annexed a zero to 0.26 (twenty-six hundredths), it would be written 0.260 (two hundred sixty thousandths). Adding a zero does not change the value of the decimal. You can see this by comparing the numbers in their fraction forms:

$$0.26 = \frac{26}{100} \qquad 0.260 = \frac{260}{1000} = \frac{260 \div 10}{1000 \div 10} = \frac{26}{100}$$

You can see that 0.26 and 0.260 are equivalent numbers. Annexing zeros to the end of a decimal number does not change its value. You can now compare 0.264 and 0.260. Because two hundred sixty-four thousandths is larger than two hundred sixty thousandths, 0.264 is larger than 0.260, which is equivalent to 0.26.

The same process would work if you were ranking three decimal numbers from the smallest to the largest. Consider this example.

E X A M P L E **6**   Rank 0.302, 0.31, and 0.306 in order from the *smallest to the largest*.

Solution   Write each decimal number with three decimal places and then order the numbers.

0.302 already has three decimal places.

0.31 becomes 0.310.

0.306 has three decimal places.

Therefore, the order would be

    0.302, 0.306, 0.310

Going back to the original form, the order is

    0.302, 0.306, 0.31   ◆

E X A M P L E **7**   Rank 1.21, 1.1, and 1.11 from the *largest to the smallest*.

Solution   Add zeros where necessary to write each number with two decimal places.

1.21 stays 1.21.

1.1 becomes 1.10.

1.11 stays as it is.

The order from largest to smallest is

    1.21, 1.11, 1.10 or 1.21, 1.11, 1.1   ◆

E X A M P L E   **8**   Insert > or < between 3.26 and 3.267.

Solution   The symbol > means "greater than" and < means "less than." In order to compare these two decimal numbers, you will need to add a zero to make both of them have three decimal places.

3.26 becomes 3.260.

3.267 stays as it is.

Because 3.260 is less than 3.267, the correct way to write the expression is 3.26 < 3.267. ◆

E X A M P L E   **9**   Write 6.32, 6.305, 6.23, and 6.31 in order using the symbol >.

Solution   First, write each number with three decimal places.

6.32 becomes 6.320.

6.305 stays as it is.

6.23 becomes 6.230.

6.31 becomes 6.310.

Because > means "greater than," begin with the largest number and rank the numbers down to the smallest.

6.320 > 6.310 > 6.305 > 6.230 or 6.32 > 6.31 > 6.305 > 6.23. ◆

Try the following problems using the reading, writing, rounding, and ranking skills demonstrated in this section.

5.2 ◆ Practice Problems

A. Look at the given number, and name the digit that is in the place requested.

1. 2,635.41
   a. tenths   b. tens
   c. hundreds   d. hundredths
2. 31.0569
   a. tenths   b. tens
   c. thousandths d. ten-thousandths
3. 0.43762
   a. thousandths b. tenths
   c. ones   d. hundredths
4. 143.567
   a. hundreds   b. tenths
   c. hundredths d. ones

B. Write the following numbers in words.

5. 160.6   6. 8.008   7. 182.034
8. 200.2   9. 0.0456   10. 143.143
11. 3.017   12. 0.68   13. 45.45
14. 12.1507

C. Write the following numbers in digits.

15. six and seventy-four thousandths
16. fourteen and four tenths
17. one hundred seven thousandths
18. eighty-three ten-thousandths
19. one hundred sixteen thousand
20. twenty-seven hundred

D. Round each of the following numbers to the place requested.

21. 16.047; hundredths   22. 43.186; tenths
23. 167.352; tens   24. 346.008; hundreds
25. 0.0753; hundredths   26. 6.01352; thousandths
27. 14.2635; thousandths 28. 26.396; hundredths
29. 357.985; tenths   30. 0.0063; hundredths
31. $286.452; dollar   32. $149.685; dollar
33. $24.8556; cent   34. $38.8947; cent
35. 267.521; whole number
36. 39.499; whole number

E. **Rank the following numbers in order from the smallest to the largest.**

37. 2.03, 2.2, and 2.12   **38.** 0.006, 0.01, and 0.02

39. 0.7, 0.635, and 0.67   **40.** 3.1, 3.095, and 3.09

41. 4.03, 4.003, and 4.304   **42.** 6.12, 6.001, and 6.102

F. **Rank the numbers using > or < as requested.**

43. 0.04, 0.4, 0.392, and 0.401; >

44. 6.07, 6.6, 6.76, and 6.067; <

45. 2.15, 2.123, 2.146, and 2.015; <

46. 0.28, 0.208, 0.21, and 0.218; >

47. 2.24, 2.004, 2.402, and 2.204; >

48. 7.18, 7.118, 7.081, and 7.108; <

49. 0.305, 0.31, 0.351, and 0.053; <

50. 0.451, 0.541, 0.405, and 0.514; >

5.3   Adding and Subtracting Decimals

Consider the following mixed number problem:

12 3/10 + 36 6/10

$$
\begin{array}{r}
12\ 3/10 \\
+36\ 6/10 \\
\hline
48\ 9/10
\end{array}
$$

First, the whole numbers are added; then the parts (with like denominators) are combined. The format of the numbers, larger numerals indicating whole numbers and the fraction bar indicating the parts, tells which quantities to combine.

This same thinking applies to decimal addition and subtraction. The whole numbers are aligned vertically, the decimal point separates the wholes from the parts, and the part values are on the right side of the decimal point. Below is the same problem, done in decimal form.

$$
\begin{array}{r}
12.3 \\
+36.6 \\
\hline
48.9
\end{array}
$$

Separating the whole numbers and the decimal quantities is the key.

In the next examples, you will notice that the practice of attaching or annexing zeros to the "part" (right) side of the decimal point is used. You can understand that this is the correct process to follow if you look at the fraction equivalents of some decimal numbers.

.7   = 7/10

.70 = 70/100, which reduces to 7/10

.700 = 700/1,000, which reduces to 7/10

Therefore, 0.7 = 0.70 = 0.700, which are all equal to 7/10.

.18   = 18/100, which reduces to 9/50

.180 = 180/1,000, which reduces to 18/100 = 9/50

.1800 = 1,800/10,000, which reduces to 9/50

Therefore, 0.18 = 0.180 = 0.1800, which are all equal to 9/50.

Now consider these problems in decimal addition and subtraction.

E X A M P L E   **1**   Add: 16.3 + 1.96 + 37.43

Solution   Annex a zero to three tenths, making it thirty hundredths (16.3 = 16.30), so you have like "denominations." Although this is not absolutely necessary, it really is a good idea.

$$
\begin{array}{r}
\overset{1\,1}{16.30} \\
1.96 \\
\underline{37.43} \\
55.69
\end{array}
$$

You could have estimated this answer first by adding 16, 2, and 37. The result, 55, would have been very close to the actual answer.   ◆

E X A M P L E   **2**   Add: 93.8 + 17 + 6.765

Solution
$$
\begin{array}{rcl}
93.8 & = & \overset{1\,1}{93.800} \\
17 & = & 17.000 \\
+\ 6.765 & = & +\ \ 6.765 \\
\hline
& & 117.565
\end{array}
$$
   Be sure to put 17 on the whole number side and to annex zeros as necessary.   ◆

E X A M P L E   **3**   Add: $2,063.80 + $149 + $13.65 + $0.40

Solution   You might first estimate the answer by adding $2,060 + $150 + $14 + $1. This will let you know that your actual answer should be close to $2,225.

$$
\begin{array}{rcl}
\$2,063.80 & = & \overset{1\,1\,1}{\$2,063.80} \\
149 & = & 149.00 \\
13.65 & = & 13.65 \\
+\ \ \ \ 0.40 & = & +\ \ \ \ 0.40 \\
\hline
& & \$2,226.85
\end{array}
$$
   ◆

You can see from these examples that annexing the zeros only serves to keep the number of decimal places the same in addition, subtraction, or comparison problems; it in no way affects the value of the number. One last example makes this principle easiest to see: $35 = $35.00.

The process of using decimals in subtraction is no different from using them in addition problems. Just be careful to keep the whole numbers with the whole numbers and the parts with the parts. And always remember to annex zeros when subtracting. You cannot subtract from thin air. This is demonstrated in Example 5.

E X A M P L E   **4**   Subtract 287.65 from 396.47.

Solution
$$
\begin{array}{r}
\overset{8\ ^{1}5\ 1}{39\cancel{6}.47} \\
-\ 287.65 \\
\hline
+\ 108.82
\end{array}
$$

Check   You can always check your subtraction answer by adding.

$$
\begin{array}{r}
287.65 \\
+\ 108.82 \\
\hline
396.47\ \checkmark
\end{array}
$$
   ◆

EXAMPLE 5   Find the difference between 83.42 and 17.864.

Solution   83.42   Look at this problem. There is no digit from which to subtract the 4.
            −17.864   Therefore, you must attach a zero to get 83.420.

You could estimate the difference between 83 and 18 and know that the actual answer should be close to 65.

$$\overset{7\,\,^{1}2\,\,^{1}3\,\,^{1}1}{83.4\cancel{2}0}$$
$$-17.864$$
$$+65.556$$
$$\overline{83.420}\ \sqrt{}$$ Always remember to check your answer.

EXAMPLE 6   Find the difference between 27.002 and 8.965.

Solution   Subtracting from zeros is always tricky. In this problem, you might think of borrowing 1 from 700, leaving 699, instead of borrowing from the zero, and then from another zero, and going all the way back to the 7. You learned how to do this with whole numbers in Section 1.3. It will help, too, to estimate an answer first by subtracting 9,000 from 27,000. You know then that the answer should be close to 18,000.

$$\overset{2\,6\,\,9\,9}{2\cancel{7}.0\cancel{0}2}$$
$$-\ \ 8.965$$
$$+18.037$$
$$\overline{27.002}\ \sqrt{}$$

5.3 ◆ Practice Problems

A. Perform the indicated operations. Check your answers by using your calculator, the answer key, or the procedures shown in this section.

1. 0.26 + 3.45 + 4.2
2. 17.5 + 2.689 + 0.81
3. 1.7 + 2.035 + 6 + 11.5
4. 2.4 + 3.086 + 4 + 15.1
5. 8.269 − 4.837
6. 9.578 − 3.684
7. 8.34 + 19 + 2.8
8. 16.3 + 142 + 9.82
9. 17.343 − 11.565
10. 21.822 − 14.954
11. 143.2 + 11.68 + 0.07
12. 43.86 + 0.09 + 173.4
13. 12.83 − 9.964
14. 11.76 − 5.893
15. 3.5 − 1.86
16. 7.4 − 4.83
17. $176 + $8.32 + $9.57
18. $18.53 + $43.96 + $147
19. $87 − $49.73
20. $27 − $19.97

B. Find the answer to the following problems without using your calculator.

21. Find the difference between 37 and 19.265.
22. What is the sum of 18.3, 43, and 1.79?
23. Find the sum of 47.5, 82, and 17.59.
24. What is the difference between 72 and 56.549?
25. From the sum of 17.6 and 38, take 22.94.
26. From the sum of 82.4 and 66, take 109.55.
27. How much change would you get back from a $50 bill if your purchases totaled $37.59?
28. How much change would you get back from a $20 bill if your purchases totaled $14.83?
29. How much larger is 85.1 than 27.42?
30. How much larger is 106.4 than 87.58?

C.   Use your calculator to find the answers to the following problems.

31. 117.365 + 829 + 16.8
32. 214.38 + 789 + 37.95
33. 21,000 − 17,952.43
34. 38,000 − 29,452.8
35. 16.8765 + 492.94
36. 837.64 + 58.2765
37. 1,273.9 − 287.854
38. 1,039.4 − 865.921
39. 209,075 − 43,869.57
40. 312,004 − 93,768.29
41. What is the difference between 867.4 and 12.982?

42. Find the sum of 1,002, 19.458, and 183.2.
43. What is the sum of 120,458.9 and 331.43?
44. What is the difference between 21,854 and 7,738.6?
45. From the sum of 1,876 and 339.45, take 994.534.
46. From the sum of 3,217.45 and 39,851.6, take 221.98.
47. From the sum of 0.00985 and 0.00874, take 0.00032.
48. From the sum of 0.0876 and 0.000321, take 0.00287.
49. Find the difference between 0.00875 and 0.000941.
50. Find the difference between 0.1024 and 0.00983.

5.4   Multiplying Decimals

Multiplying decimal numbers is perhaps the easiest of the operations. You multiply the quantities, ignoring the decimal points for the moment. In order to determine where to place the decimal point in the product (the result of the multiplication), count the total number of decimal places in the factors and move the decimal point that number of places from right to left in the answer. For example, consider

3.27×1.5

First perform the multiplication as if there were no decimal points in either number.

$327 \times 15 = 4905$

There are three decimal places altogether. Therefore, count three places from right to left, and position the decimal point between the 4 and the 9.

$3.27 \times 1.5 = 4.905$

Now let's look at 5.3×2.6:

$53 \times 26 = 1378$

There are two decimal places in the original factors, so your answer has two decimal places.

$5.3 \times 2.6 = 13.78$

Have you ever wondered where the multiplication rule for decimal placement comes from? If the rule says that 0.06×0.9 equals 0.054, try doing the problem in its fraction form.

$$0.06 = 6/100 \qquad 0.9 = 9/10$$

$$\frac{6}{100} \quad \times \quad \frac{9}{10} \quad = \frac{54}{1,000} = 0.054$$

hundredths ×   tenths   =   thousandths
(two places) × (one place) = (three places)

E X A M P L E 1   Multiply: 14.9 × 16

Solution   You could first estimate the answer by multiplying 15 × 16. The actual answer should be close to 240.

$$
\begin{array}{r}
14.9 \\
\times\ \ 16 \\
\hline
894 \\
149\ \ \\
\hline
2384 \\
\end{array}
$$

14.9 × 16 = 238.4   There is no decimal point in the whole number 16, so you have a total of one decimal place in the answer.

The actual answer is close to the estimate.   ◆

E X A M P L E 2   Multiply: 7.35 × 0.2

Solution $\begin{array}{r} 7.35 \\ \times\ \ 0.2 \\ \hline 1.470 \end{array}$ There is a total of three decimal places in the factors, so there will be three decimal places in the product. It is standard practice to drop off ending zeros after the decimal point has been placed.

7.35 × 0.2 = 1.47   ◆

E X A M P L E 3   Multiply: 1.006 × 0.29

Solution $\begin{array}{r} 1.006 \\ \times\ \ 0.29 \\ \hline 9054 \\ 2012\ \ \\ \hline 0.29174 \end{array}$ Five decimal places in the factors require five decimal places in the answer (product).

It is common practice in this type of answer to place a zero to the left of the decimal point to call attention to the decimal point. Either way is certainly correct, but .29174 is not as clear as 0.29174.

1.006 × 0.29 = 0.29174   ◆

E X A M P L E 4   Multiply: 0.823 × 0.07

Solution   You could estimate the answer by multiplying 0.8 by 0.07. 8 times 7 is 56 and you need three decimal places, so your estimate would be 0.056.

$\begin{array}{r} 0.823 \\ \times\ \ .07 \\ \hline 0.05761 \end{array}$ In this example, you need to move the decimal point five spaces left in the product, but there are only four digits. Add a zero to the left of your answer, and then move the decimal point a total of five places. Note also the leading zero to the left of the decimal point. Again, it is optional.

0.823 × 0.07 = 0.05761

The answer, 0.05761, is reasonably close to the estimate, 0.056.   ◆

E X A M P L E 5   Multiply: $83.48 × 0.055

Solution $\begin{array}{r} \$83.48 \\ \times\ \ 0.055 \\ \hline 41740 \\ 41740\ \ \\ \hline \$4.59140 \end{array}$

Because your answer is money, you must round to the nearest cent, even though the problem says nothing about rounding.

$83.48 × 0.055 = $4.59 ◆

E X A M P L E   6   Multiply: $143.84 × 1.7

Solution
$$
\begin{array}{r}
\$143.84 \\
\times \quad 1.7 \\
\hline
100688 \\
14384 \\
\hline
\$244.528
\end{array}
$$

This problem is money once again, so you need to round to the nearest cent. 2 is in the cents place and 8 is more than 5, so you should round up to 53 cents.

$143.84 × 1.7 = $244.53 ◆

E X A M P L E   7   Multiply: 246.2 × 100

Solution
$$
\begin{array}{r}
246.2 \\
\times \quad 100
\end{array}
$$
When you multiply by a number ending in zeros, you can simply annex that number of zeros to the right of the number, multiply by the nonzero number(s), and position the decimal point.

$$
\begin{array}{r}
246.2 \\
\times \quad 100 \\
\hline
24,620.0
\end{array}
$$
Because 100 has two zeros, add two zeros to the other factor and move the decimal point one place to the left. ◆

There is a short cut that you can use when multiplying a decimal number by a power of 10. In other words, you can use it when one of the factors in the problem is 10, 100, 1,000, or another power of 10. You will use this short cut often, especially when doing percent problems. Here it is:

Multiplication by Powers of Ten

To multiply by

10, move the decimal point one place to the right.

100, move the decimal point two places to the right.

1,000, move the decimal point three places to the right.

And so on.

In order to move the decimal point, you may need to add some place-holding zeros. The following examples illustrate this short cut.

E X A M P L E   8   26.4 × 100 = 26.40. = 2,640 ◆

E X A M P L E   9   0.056 × 10 = 0.0.56 = 0.56 ◆

E X A M P L E   10   17 × 100 = 17.00. = 1,700 ◆

Remember that in a whole number, the decimal point is understood to be to the right of the last digit.

5.4 ◆ Practice Problems

A. Multiply in each of the following problems. Check your answers by using your calculator or the answer key.

1. 17.42 × 2.8
2. 16.35 × 3.7
3. 4.6 × 0.009
4. 5.8 × 0.007
5. 26 × 3.9
6. 37 × 7.4
7. 0.07 × 0.5
8. 0.06 × 0.8
9. 3.48 × 10
10. 12.451 × 100
11. 248 × 11.6
12. 357 × 12.4
13. $18.39 × 0.055
14. $32.75 × 0.055
15. 23 × 100
16. 45 × 1,000
17. $47 × 8.6
18. $59 × 9.4
19. 632 × 14
20. 753 × 26
21. 3.82 × 0.035
22. 4.96 × 0.185
23. 84.9 × 16.7
24. 76.8 × 18.6
25. 0.59 × 100
26. 0.962 × 10
27. 0.0035 × 0.54
28. 0.00987 × 0.0032
29. 0.00176 × 0.0046
30. 0.0045 × 0.000825

31. Find the product of 28.97 and 16.3.
32. Find the product of 76.05 and 22.7.
33. Find the product of 21, 18.36, and 0.4.
34. Find the product of 0.9, 32, and 21.45.

B. Find the answer to each of the following problems by using your calculator.

35. 0.296 × 14.834
36. 72.35 × 0.849
37. 892.7 × 21.63
38. 678.4 × 37.42
39. 18.763 × 9.25
40. 34.638 × 7.45
41. 2,000.76 × 0.825
42. 7,012.4 × 1.97
43. 0.279 × 0.0038
44. 0.821 × 0.0762

45. From the product of 897.5 and 38.6, take 4,005.4.
46. To the product of 117.9 and 0.376, add 215.9.
47. To the product of 376.91 and 1.038, add 476.1.
48. From the product of 737.4 and 93.5, take 29,879.42.
49. From the product of 0.35 and 0.086, take 0.0009.
50. To the product of 0.0095 and 0.765, add 0.0045.

5.5 ◆ Dividing Decimals

If multiplication is one of the easiest decimal operations, division is the one that causes the most trouble. There is, however, a series of steps that, when followed in order, make the decimal division process much simpler.

First, some vocabulary is necessary. Study each of these forms:

$$\frac{\text{quotient}}{\text{divisor}\overline{)\text{dividend}}} \qquad 9\overline{)45}^{\,5}$$

$$\frac{\text{dividend}}{\text{divisor}} = \text{quotient} \qquad \frac{45}{9} = 5$$

$$\text{dividend} \div \text{divisor} = \text{quotient} \qquad 45 \div 9 = 5$$

Division with decimals, like division with whole numbers, often results in a remainder. This difficulty is usually handled by directions that state where the quotient should be rounded. These directions allow the division process to be halted before it runs off the page! For example, the quotient of 7 divided by 9 is

0.777777777777777777 . . . , or 0.$\overline{7}$

The three dots following the last 7 mean that the 7s continue indefinitely. If the quotient is rounded to the nearest thousandth, the answer is 0.778.

Now let's see how the series of steps for division are applied in a simple problem. They are listed in the order in which they should be considered.

Divide 14 by 1.5 and round your answer to the nearest hundredth.

1. *Is there a decimal point in the dividend?* If not, put it at the end (on the right side) of the whole number.

   No, there is no decimal point, so place one at the right of the whole number.

   $$1.5 \overline{)14.}$$

2. *Is the divisor a whole number?* If it is, then put the decimal point in the quotient right above the decimal point in the dividend. If it is not, swing the decimal point in the divisor to the right so that the divisor becomes a whole number. Then swing the decimal point in the dividend the same number of places to the right. Place the decimal point straight up in the quotient.

   You are dividing by a decimal number, so move the decimal point in both numbers one place to the right, writing 1.5 as 15 and 14 as 140. Place the decimal point in the quotient.

   $$1.5 \overline{)14.0.}$$

3. *Does the problem call for rounding?* If yes, annex enough zeros so that the quotient is one place longer than requested. If rounding is not called for, you may still need to annex zeros until the division ends with a zero remainder.

   Yes, it calls for hundredths, so you will add three zeros.

   $$1.5 \overline{)14.0.000}$$

4. Now, *divide.*

   $$1.5 \overline{)14.0.000} \quad 9.333$$

5. *Round* your quotient as requested.

   $$9.333 \doteq 9.33$$

6. You can *check* the answer by multiplication.

   $$9.33 \times 1.5 = 13.995 \doteq 14.0$$

   Since the quotient has been rounded, the check will be only approximately equal to the original dividend. You might find a calculator helpful when checking your answers.

Although the questions in these steps may seem cumbersome, they will become routine very quickly and will make dividing with decimals much easier. All of the decimal point and rounding questions are resolved before the dividing even begins.

Follow these examples. Each of the question steps is numbered for you.

E X A M P L E   1   Divide 1.83 by 2 and round to the nearest hundredth.

Solution   1.   Is there a decimal point in the dividend?

   Yes, 1.83 already has a decimal point.

   $$2 \overline{)1.83}$$

2. Are you dividing by a whole number?

Yes, 2 is a whole number. Therefore, place the decimal point straight up into the quotient.

$$2\overline{)1.83}$$

3. Does the problem call for rounding?

Yes, it calls for the nearest hundredth. You already have a digit in the hundredths place, so you need to add only one zero to have enough places for the rounding.

$$2\overline{)1.830}$$

4. Now, do the division.

$$\begin{array}{r} 0.915 \\ 2\overline{)1.830} \end{array}$$

5. Round to the nearest hundredth.

$$0.915 \doteq 0.92$$

6. Check. Because of the relationship between division and multiplication, you can check the solution by multiplying the quotient by the divisor. In the case of a rounded answer, however, the product may only be close to the original dividend.

$$0.92 \times 2 = 1.84 \doteq 1.83 \qquad \blacklozenge$$

E X A M P L E **2** What is the quotient when 0.367 is divided by 1.4? Round your answer to the nearest thousandth.

Solution 1. Is there a decimal point in the dividend?

Yes, there is.

$$1.4\overline{)0.367}$$

2. Is the divisor a whole number?

No, it is not. Therefore, move the decimal point one place to the right to make the divisor 14. Do the same thing to the dividend, and then place the decimal point in the quotient.

$$1.4\overline{)0.3.67}$$

3. Does the problem call for rounding?

Yes, you need to add two zeros to round to the thousandths place.

$$1.4\overline{)0.3.6700}$$

4. Do the division.

$$\begin{array}{r} 0.2621 \\ 1.4\overline{)0.3.6700} \end{array}$$

5. Round the quotient as requested.

$$0.2621 \doteq 0.262$$

6. Check by multiplication.

$$0.262 \times 1.4 = 0.3668 \doteq 0.367 \qquad \blacklozenge$$

E X A M P L E 3   Divide 18 by 0.36. Divide until the answer comes out even (has a remainder of zero).

Solution   1.   Is there a decimal point in the dividend?

No. Place it to the right of 18.

$$0.36\overline{)18.}$$

2.   Is the divisor a whole number?

No, so move both decimal points two places to the right.

$$0.36.\overline{)18.00.}$$

3.   Are there any rounding instructions?

The problem says to divide until the quotient comes out even and there is no remainder. You may have to add several zeros, or it may work out quickly. Add zeros one at a time until the remainder is zero.

$$0.36.\overline{)18.00.}$$

4.   Divide.

$$0.36.\overline{)18.00.}^{\,50.}$$

5.   Round as requested.

The answer is the whole number 50.

6.   Check the answer by using multiplication.

$50 \times 0.36 = 18.00$

The check is exact because the quotient was not rounded.   ◆

E X A M P L E 4   Simplify $\dfrac{2.007}{1.5}$ by dividing until the answer comes out even.

Solution   1.   Decimal point in dividend?

Yes.

$$1.5\overline{)2.007}$$

2.   Divisor a whole number?

No. Move the decimal point one place to the right.

$$1.5.\overline{)2.0.07}$$

3.   Divide until the answer comes out even.

Add two zeros and see what happens.

$$1.5.\overline{)2.0.0700}$$

4. Divide.

$$1.5.\overline{)2.0.0700} \quad \frac{1.338}{}$$

5. Round as requested.

The answer comes out even, so

$$\frac{2.007}{1.5} = 1.338$$

6. Check by multiplication.

$$1.338 \times 1.5 = 2.007 \qquad \blacklozenge$$

E X A M P L E 5   Divide 6 by 7. Round the answer to the nearest thousandth if it does not come out even before that time.

Solution   1. Decimal point in dividend? No.

$$7\overline{)6.}$$

2. Dividing by a whole number? Yes. Move decimal point up into the quotient.

$$7\overline{)6.} \quad \overset{.}{}$$

3. Rounding? Yes, to the nearest thousandth, so add four zeros—three for thousandths and the fourth to round.

$$7\overline{)6.0000} \quad \overset{.}{}$$

4. Divide.

$$7\overline{)6.0000} \quad \frac{.8571}{}$$

5. Round to the nearest thousandth.

$$0.8571 \doteq 0.857$$

6. Check by using multiplication.

$$0.857 \times 7 = 5.999 \doteq 6 \qquad \blacklozenge$$

Here is a summary list of the steps that can help you do long division problems successfully. You don't need to write the five questions out each time, but you should follow those steps, in order, when you are doing the problems.

Five Steps for Dividing Decimals

1. Is there a decimal point in the dividend?
2. Is the divisor a whole number?
3. Does the problem call for rounding?
4. Now, divide.
5. Round the quotient if required.

Try Exercises 1 through 5 on some scrap paper and see if you can follow the five steps demonstrated in Examples 1 through 5. Check to see if your answers are the same as those given. If not, review Examples 1 through 5 and try again. In case you need to ask your instructor or a friend for help, be sure to show all your work so that together you can find your errors.

Exercises
1. Divide 0.8 into 3.67 and round the answer to the nearest hundredth.
2. Find the quotient of 0.3744 divided by 2.4, and divide until your answer comes out even.
3. $36 \div 9.3$ Round to the nearest hundredth.
4. Simplify 4/26 and round to the nearest thousandth.
5. Divide 70.5 by 28.2 and round to the nearest thousandth if the answer does not come out even before that time.

Answers:

1. $4.587 \doteq 4.59$ **2.**   0.156   **3.**   3.87   **4.** $0.1538 \doteq 0.154$
5.   2.5

There is a division short cut that parallels the multiplication short cut given in Section 5.4. When a quantity is divided by a power of 10 (10, 100, 1,000, and so on), the easiest approach is to move the decimal point to the left. The number of places in the move is determined by the number of zeros in the divisor.

Division by Powers of Ten

To divide by

    10, move the decimal point one place to the left.

    100, move the decimal point two places to the left.

    1,000, move the decimal point three places to the left.

And so on.

Consider these examples.

E X A M P L E 6   Divide: $146.5 \div 100$

Solution   Move the decimal point two places to the left.

$$146.5 \div 100 = 1.46.5 = 1.465$$ ◆

E X A M P L E 7   Divide: $0.082 \div 100$

Solution   Move the decimal point two places to the left.

$$0.082 \div 100 = 0.00.082 = 0.00082$$ ◆

E X A M P L E 8   Divide: $963 \div 10$

Solution   Move the decimal point one place to the left.

$$963 \div 10 = 96.3. = 96.3$$ ◆

E X A M P L E **9**   Divide: 197 ÷ 1,000

Solution   Move the decimal point three places to the left.

$$197 ÷ 1,000 = 0.197. = 0.197$$

◆

5.5 ◆ Practice Problems

A. Divide in each of the following problems. Follow the five steps, positioning the decimal point first. Divide until the answer comes out even. (Each will eventually have a remainder of zero.) You should check your answers by multiplication, perhaps using a calculator.

1. 6 ÷ 0.3   2. 8 ÷ 0.04
3. 128 ÷ 0.4   4. 327 ÷ 0.3
5. 1.5 ÷ 6   6. 2.8 ÷ 5
7. 2.45 ÷ 100   8. 98.2 ÷ 10
9. 2,000 ÷ 5   10. 1,400 ÷ 8
11. 1.6)‾32‾   12. 2.4)‾120‾
13. 10)‾2.65‾   14. 1000)‾3.275‾
15. 0.9)‾0.081‾   16. 0.7)‾0.0035‾
17. 2.5)‾1.6‾   18. 3.2)‾2.4‾
19. 0.15)‾90‾   20. 0.18)‾9‾
21. $\dfrac{3.6}{8}$ 22. $\dfrac{4.2}{12}$
23. $\dfrac{18.34}{100}$ 24. $\dfrac{2.762}{100}$

B. Divide in each of the following problems. Round your answer to the nearest thousandth if the answer does not come out even before that place. You might check your answers by using your calculator, but remember that the checks may not be exact if the answers have been rounded.

25. 13 ÷ 0.6   26. 22 ÷ 0.7
27. 2.6 ÷ 3   28. 1.8 ÷ 8

29. 0.23 ÷ 0.7   30. 0.34 ÷ 0.6
31. 12 ÷ 2.3   32. 27 ÷ 2.6
33. 26 ÷ 1,000   34. 1,452 ÷ 100
35. 0.028 ÷ 4   36. 0.059 ÷ 9
37. 2 ÷ 3.5   38. 5 ÷ 5.4
39. 0.38 ÷ 1.4   40. 0.45 ÷ 2.3
41. 0.556 ÷ 2.6   42. 0.489 ÷ 3.2
43. 66 ÷ 2.3   44. 57 ÷ 1.8
45. 5.935 ÷ 100   46. 0.267 ÷ 1,000
47. 0.023 ÷ 35   48. 0.036 ÷ 63
49. 0.007 ÷ 10   50. 0.093 ÷ 100

C. Find the answers to the following problems by using your calculator. Round each to the place indicated.

51. 827.6 ÷ 0.0192; hundredths
52. 769.8 ÷ 0.925; thousandths
53. 18.276 ÷ 12.85; ten-thousandths
54. 34.598 ÷ 18.37; ten-thousandths
55. 1,000.5 ÷ 27.38; thousandths
56. 24,015.9 ÷ 387.95; thousandths
57. 356.2 ÷ 36.895; tenths
58. 185.4 ÷ 18.692; hundredths
59. 283.5 ÷ 0.824; tenths
60. 369.5 ÷ 0.377; hundredths

5.6 ◆ Decimals and Fractions

The ability to change from decimal form to fraction form and from fraction form to decimal equivalent is a very useful skill when you are solving problems that involve both fractions and decimals. For example, to figure the cost of 2 1/2 pounds of tomatoes·at $0.47 a pound, you need to multiply 2 1/2 by $0.47. You can change 2 1/2 to 2.5, following the process outlined below, and then multiply 2.5 by $0.47, which equals 1.175. Because you are talking about money, round your answer to the nearest cent. The cost is $1.18. The key first step was to change 1/2 to its decimal equivalent as follows:

Changing Fractions to Decimals

> To change a fraction to its decimal equivalent, divide the numerator by the denominator and follow the rounding instructions if any are given.

Change 1/2 to a decimal by dividing 1 by 2.

$$\begin{array}{r} 0.5 \\ 2\overline{)1.0} \end{array}$$

Let's change a few more fractions to their equivalent decimal form.
Change 3/4 to a decimal.

$$\begin{array}{r} 0.75 = 0.75 \\ 4\overline{)3.00} \end{array}$$

Change 5/7 to a decimal, rounded to the nearest thousandth.

$$\begin{array}{r} 0.7142 \doteq 0.714 \\ 7\overline{)5.0000} \end{array}$$

Change 2 5/12 to a decimal, rounded to the nearest hundredth.

$$2\ 5/12 = 29/12 \qquad \begin{array}{r} 2.416 \doteq 2.42 \\ 12\overline{)29.000} \end{array}$$

or you can leave the 2 alone and change 5/12 to 0.416. Then 2 5/12 = 2.416 ≐ 2.42.
The example with which we began, find the cost of 2 1/2 pounds of tomatoes at $0.47 a pound, could also be solved using fractions.

$0.47 = 47/100
so 2 1/2 × 47/100
   = 5/2 × 47/100
   = 235/200
   = $1 7/40

Although this answer is not very useful, it is also a correct solution. $0.47 was changed to its equivalent fraction form by the following steps:

Changing Decimals to Fractions

> To change a decimal number to its fraction equivalent,
>
> Read, Write, and Reduce.

0.08 is 8 hundredths, 8 over 100, 8/100 = 2/25

1.473 is 1 and four hundred seventy-three thousandths, 1 and 473 over 1,000, 1 473/1,000

24.6 is twenty-four and six tenths, 24 and 6 over 10, 24 6/10 = 24 3/5

It is useful to memorize a few of the most common fraction and decimal equivalencies to save time when doing problems. The following short list may be very helpful.

$1/2 = 0.5$	$1/10 = 0.1$	$1/8 = 0.125$
$1/4 = 0.25$	$1/100 = 0.01$	$3/8 = 0.375$
$3/4 = 0.75$	$1/1000 = 0.001$	$5/8 = 0.625$
$1/5 = 0.2$		$7/8 = 0.875$
$2/5 = 0.4$		
$3/5 = 0.6$		
$4/5 = 0.8$		

Look at the following examples involving decimals and fractions. Does it seem to matter which way the problems are done? Generally not, if the fraction is one that changes exactly to decimal form. The solutions could be found using the fractions or their decimal equivalents. With 1/2, 1/4, 1/10, and so on, it is usually easier to do the problem with decimals. Fractions such as 1/3, 5/6, and 7/9, however, do not change to exact decimals. Instead, $1/3 = 0.333\ldots$, $5/6 = 0.833\ldots$, and $7/9 = 0.777\ldots$. (The three dots following a repeating digit let you know that that digit continues to repeat.) Such quantities are easier to work with in fraction form. Remember: When a problem requires a money answer, the answer must be changed to decimal form even if the problem was worked using fractions.

E X A M P L E 1   Add: $1/2 + 0.8$

Solution

Fractions	Decimals
$1/2 + 8/10$	$0.5 + 0.8$
$= 5/10 + 8/10$	$= 1.3$
$= 13/10$ or $1\ 3/10$	

◆

E X A M P L E 2   Subtract: $6.4 - 2\ 3/5$

Solution

Fractions	Decimals
$6\ 4/10 - 2\ 6/10$	$6.4 - 2.6$
$= 5\ 14/10 - 2\ 6/10$	$= 3.8$
$= 3\ 8/10 = 3\ 4/5$	

◆

E X A M P L E 3   Multiply: $5/8 \times 1.7$

Solution

Fractions	Decimals
$5/8 \times 1\ 7/10$	0.625×1.7
$= 5/8 \times 17/10$	$= 1.0625$
$= 85/80 = 17/16$	
$= 1\ 1/16$	

◆

E X A M P L E 4   Multiply: $7/9 \times 1.1$

Solution

Fractions	Decimals
$7/9 \times 1\ 1/10$	Should not be done using decimals because
$= 7/9 \times 11/10$	$7/9 = 0.77777\ldots$
$= 77/90$	

◆

5.6 ◆ Practice Problems

A. Write the following decimals in their fraction or mixed number form. Be sure to reduce where possible.

1. 0.14
2. 0.86
3. 2.003
4. 9.017
5. 17.8
6. 23.96
7. 0.072
8. 0.825
9. 36.125
10. 11.008

B. Change each of the following fractions or mixed numbers to their decimal form. Round to the nearest thousandth where necessary.

11. 3/11
12. 7/12
13. 2 5/6
14. 7 5/9
15. 1/12
16. 4/7
17. 21/2
18. 23/8
19. 16 4/19
20. 14 8/13

C. Do each of the following problems. Change to the same type of numbers, either fractions or decimals.

21. 3/4 × 2.9
22. 5/8 × 3.4
23. 1.6 + 2 1/2
24. 3/10 + 2.8
25. 19.4 ÷ 1/2
26. 13.5 ÷ 1/3
27. 7 1/4 − 3.8
28. 9 3/4 − 6.9
29. 4 3/8 × 1.2
30. 8 3/4 × 2.4

 D. Use your calculator to change the following fractions and mixed numbers to their decimal equivalents. Round your answers to the places indicated.

31. 35/76; thousandths
32. 84/79; hundredths
33. 3 95/112; hundredths
34. 7 44/53; thousandths
35. 2356/8429; thousandths
36. 3568/4579; hundredths
37. 9 56/67; hundredths
38. 4 67/74; thousandths
39. 128/3655; ten-thousandths
40. 359/5678; ten-thousandths

 E. Perform the following calculations using your calculator. Round division answers to the nearest ten-thousandth.

41. 19 5/8 + 39.79
42. 17 3/4 + 84.92
43. 38 1/4 × 73.6
44. 126 − 78 9/10
45. 68.5 ÷ 18 7/8
46. 85.4 × 29 5/8
47. 284 − 97 3/5
48. 14 3/5 ÷ 7.8
49. 89.4 × 26 3/4
50. 37 5/8 − 19 3/4

5.7 Applications of Decimals

Before looking at application problems, it will be helpful for you to review the list of key words and expressions given in Section 1.6. Also, remember that you can use the following ideas to help you decide which type of problem you are being asked to do.

Identifying the Operation

When putting quantities together, you might expect to add or multiply. When separating quantities, you are likely to subtract or divide.

Read each of the following examples carefully and then, before you read the explanation that is given, see if you can decide what you would do. Estimating an answer first is usually a good idea so that you can tell if you have chosen the correct operation before actually doing the calculations.

E X A M P L E   **1**   You are taking your daughter's Brownie Troop to the skating rink. There are 11 little girls going. The tickets are $2.25 apiece. How much will it cost?

Solution   You could add $2.25 + $2.25 + $2.25 + $\cdots$ + $2.25, up 11 times, or you could simply multiply 11 $\times$ $2.25. The result would be $24.75 in either case. If you had estimated the amount first in your head, using $2+ (a little more than $2) per Brownie, you would have known that 11 Brownies would cost slightly more than $22.   ◆

E X A M P L E   **2**   You have $18 to spend at the grocery store on the hamburger for a cookout. Hamburger is on special for $1.20 a pound. How many pounds can you buy?

Solution   You could begin with the $18 and subtract $1.20 to get $16.80, and then subtract $1.20 again to get $15.60, and continue until you ran out of money. The simpler way, of course, is to divide $18 by $1.20.

$$1.20.\overline{)18.00.00} \quad \frac{15.00}{}$$

So you find out that you can buy 15 pounds of hamburger for $18.   ◆

E X A M P L E   **3**   Fabric for the new jacket you are making costs $6.98 a yard. The pattern calls for 4 3/4 yards of fabric. How much will the fabric cost?

Solution   Estimating, you reason that if the fabric cost $7 a yard and you needed 5 yards, you would multiply $7 by 5 and get $35, so your actual answer should be about $35. Now that you have decided which operation to use, you need to multiply $6.98 by 4 3/4 yards. You are looking for cost, so a decimal answer is appropriate.

1.   Change 4 3/4 to 4.75.

2.   Multiply: 6.98 $\times$ 4.75 = 33.155

3.   Round to the nearest cent, $33.16.

The fabric for the jacket will cost $33.16. The answer is reasonably close to the estimated cost of $35.   ◆

E X A M P L E   **4**   A travel agent booking a tour to Orlando, Florida, for the 17 members of the Seniors Club needs to collect $137.80 for airfare and $114.95 for hotels from each member. How much money must the agent collect?

Solution   There are two different ways to do this problem. One method is to multiply $137.80 by 17 to get the total amount due for airfare and to multiply $114.95 by 17 to get the total amount due for hotels. Then you would add $2,342.60 and $1,954.15 to get the total amount to be collected, $4,296.75.

 The second method is to figure each person's individual total by adding $137.80 and $114.95. Then multiply that sum, $252.75, by 17 to get $4,296.75, the total amount due. The answer should be the same using either method.   ◆

Example 4 was a two-step problem. Many times there is more than one way to arrive at an answer, or more than one calculation that must be done. Read each problem carefully and plan your approach before you begin to figure.

5.7 ◆ Practice Problems

A. **Decide which mathematical operation to use, and then solve each of the following problems. It will help to estimate your answers first. Be sure to label your answers.**

1. Mark has $138.14 in his checking account. He writes checks for $8.32, $62.20, and $19.85. How much money is left in his account?

2. Four friends in a card club decide to go to dinner. The total bill, including the tip, is $46.60. How much is each person's share?

3. An auditor travels 218 miles. Her mileage payment for using her own car is 22 1/2 cents per mile. How much money will she receive for mileage on this trip?

4. Your son's birthday party guests are going bowling. How much will it cost for your son and his five guests if the cost of renting shoes and bowling three games is $3.75 for each child?

5. If pork chops are selling for $2.89 per pound, how much will you pay for 3 1/2 pounds?

6. Bradbury Travel is planning a trip to Calgary for a group of 32 people. The airline tickets cost $289 per person, and each person's hotel bill will be $185.40. How much money will have to be collected from the group to pay for the trip?

7. Apples are selling for $0.37 per pound. How much will you pay for 3 3/4 pounds?

8. A young college student's monthly utility bills were $18.26 for electricity, $12.35 for gas, $7.84 for water, and $31.18 for the telephone. How much were this month's bills altogether?

9. Black fabric for Sara's Halloween costume costs $3.69 a yard. Her witch's outfit calls for 4 2/3 yards. How much will the fabric cost?

10. Your car averages 22 miles per gallon. How much gas will you need for a trip to Washington, D.C., a distance of 408 miles? (Round to the nearest tenth of a gallon.)

11. If super unleaded gas now costs $0.93 per gallon, how many gallons can you buy with a $10 bill? (Round to the nearest tenth of a gallon.)

12. Candace worked 6.5 hours on Monday, 8 hours on Tuesday, 11 1/4 hours on Wednesday, 7 1/2 hours on Thursday, and 9.3 hours on Friday. How many hours did she work altogether?

13. A druggist puts 10.5 ounces of medicine into separate vials, each containing 0.75 ounces. How many vials can be filled?

14. Becca makes $0.40 per month for each paper that she delivers. If her route covers 103 houses, how much does she earn each month for delivering the papers?

15. Gold is currently selling for $415 an ounce on the trader's market. How much will Mr. Hutton pay for a gold bar that weighs 3/4 ounce?

16. How many pounds of ground beef can you buy with $8 if it is selling for $1.19 per pound? (Round to the nearest tenth of a pound.)

17. If the current balance in your checking account is $406.74, how much will be in the account after you pay bills totaling $328.96?

18. How much will your tuition be for next quarter if the current fees are $41.50 per credit hour and you are taking 13 hours?

19. Santiago is planning a trip to his native country and wants to know how much he will have to save each month for the next six months to be able to pay his plane fare. The cost of the plane ticket is $715.56.

20. Phillippe and Janis are buying new living room furniture. The cost for the furniture is $1,126.95. How much will they need to pay each month for their account to be paid in full in 12 months?

B. **You may use your calculator to help you find the answers to the following problems. Be sure to label your answers where appropriate.**

21. From the product of 82.4 and 0.65, take the sum of 18.7 and 9.26.

22. From the quotient of 533.875 divided by 6.5, take 79.8.

23. Divide the product of 583.9 and 218.6 by 32,506. Round your answer to the nearest thousandth.

24. Decrease the quotient of 546,823 divided by 879.42 by 495.9. Round your answer to the nearest thousandth.

25. Find the average of 458.9, 376.4, and 112.67.

26. Find the average of 872.4, 915.76, and 622.

27. The population of Columbus, Ohio, is approximately 30.4 times the population of Columbus, Nebraska. If the population of Columbus, Ohio, is about 566,100 people, what is the approximate population of Columbus, Nebraska?

28. The area of the state of New Mexico is approximately 121,660 square miles. The area of the country of Mexico is approximately 6.2 times as large. What is the approximate area of Mexico?

29. If 87,396 fans paid $23.85 each to see a college football game, what was the total amount taken in gate receipts?

30. If the national debt grows at the rate of $17,234.28 per day, by how much will the debt increase in a year?

5.8 Chapter Review

This has been a very long chapter with many problems to be done. Normally, students master decimals readily because of their familiarity with prices and money. You should now be quite comfortable performing the standard operations and processes with decimals. In the following chapters you will be applying all these skills to whole numbers, fractions, and decimals.

When you have finished studying this chapter, you should be able to:

1. Discuss the following terms: place value, attaching zeros, rounding, sum, difference, product, quotient, dividend, divisor, powers of 10, fraction equivalent, and decimal equivalent.

2. Rank decimals using > and <.

3. Solve division problems using these five steps:
    a. Decimal point in the dividend?
    b. Divisor a whole number?
    c. Rounding? Attach zeros as needed.
    d. Divide.
    e. Round the quotient as requested.

4. Change decimal values to their equivalent fraction values by using "Read, Write, and Reduce."

5. Change fractions to their decimal equivalents by dividing the denominator into the numerator and rounding as requested.

6. Estimate the solution to decimal word problems.

7. Solve word problems involving decimals, fractions, and whole numbers.

Review Problems

A. Write each of the following decimal numbers in words.

1. 24.063   **2.** 3.0058
3. 6.047   **4.** 35.98
5. 203.023   **6.** 607.204
7. 1,413.003   **8.** 3,548.058

B. Write each of the following verbal phrases as a decimal number.

9. four and sixteen thousandths
10. seventeen and five thousandths
11. four hundred nine and nine hundredths
12. six hundred two and twelve thousandths
13. four hundred eight ten-thousandths
14. seventy-seven ten-thousandths
15. twelve thousand and twelve thousandths
16. sixty thousand and sixteen thousandths

C. Round each of the following numbers to the place indicated.

17. 4.658 to the nearest tenth
18. 7.609 to the nearest hundredth
19. 45.675 to the nearest ten
20. 716.654 to the nearest hundred
21. 92.452 to the nearest tenth
22. 0.006 to the nearest hundredth
23. 0.1009 to the nearest thousandth
24. 37.863 to the nearest unit (one)

D. Arrange each of these groups of numbers from the smallest to the largest.

25. 0.61, 0.602, 0.609

26. 0.73, 0.729, 0.7

27. 2.2, 2.02, 2.002

28. 3.005, 3.5, 3.05

29. 0.801, 0.81, 0.08

30. 1.54, 1.405, 1.504

E. Arrange each of the following sets of numbers using > or < as requested.

31. 17.3, 17.03, 17.32, 17.2; >

32. 0.205, 0.255, 0.25, 0.22; <

33. 0.06, 0.006, 0.606, 0.066; <

34. 4.51, 4.503, 4.505, 4.05; >

F. Perform the indicated operations.

35. 45.2 + 31.567 + 14

36. 26 × 9.42

37. 18.5 + 37 + 92.386

38. 0.458 × 100

39. 145.8 − 87.982

40. $85 − $29.81

41. 3,446.9 − 674.86

42. 35 × 0.021

43. 24.8 × 76.5

44. 2.63 × 10

45. 18.95 × 2.8

46. 27.32 × 1,000

47. 18 + 92.7 + 24.78

48. 1.49 + 38 + 16.5

49. 14.9 + 27 + 33.86

50. 16.8 × 1,000

51. $84 − $38.83

52. 140.08 − 87.596

53. 91.5 + 37 + 86.43

54. 205.4 − 97.65

G. Divide in each of the following problems, and round to the nearest hundredth if the answer does not come out even before that place.

55. 0.085 ÷ 1.2

56. 0.096 ÷ 2.3

57. 1.46 ÷ 0.7

58. 3.45 ÷ 0.8

59. 17 ÷ 0.19

60. 23 ÷ 0.15

61. 0.02 ÷ 10

62. 4.56 ÷ 100

63. 24 ÷ 0.3

64. 35 ÷ 0.9

65. 0.24 ÷ 7

66. 0.68 ÷ 9

67. 5.46 ÷ 100

68. 8.72 ÷ 1,000

69. 74.5 ÷ 15

70. 86.3 ÷ 23

71. 82 ÷ 17

72. 56 ÷ 24

73. 65.4 ÷ 1,000

74. 0.159 ÷ 10

H. Change each of the following decimals to fraction form. Be sure to reduce to lowest terms.

75. 0.28

76. 0.96

77. 22.45

78. 37.85

79. 0.008

80. 0.0045

81. 45.215

82. 93.148

I. Change each of the following fractions or mixed numbers to decimal forms. Round to the nearest thousandth, where necessary.

83. 5/8

84. 3/5

85. 7/9

86. 4/7

87. 3 5/6

88. 5 3/8

89. 4/1,000

90. 16/10,000

J. Solve each of the following word problems by estimating an answer, writing a mathematical problem, and solving that problem. Be sure to label your answers and round money answers to the nearest cent.

91. 12 1/2 pounds of birdseed costs $7.07. Find the cost per pound.

92. To train for the upcoming bicycle race, Mike rode 2.4 miles on Tuesday, 6.5 miles on Thursday, 9.8 miles on Saturday, and 5.9 miles on Sunday. How many miles did he ride that week?

93. Sara needs 11 yards of yarn for the garland she is making for holiday decorations. If the package that she bought has 7.2 yards of yarn in it, how many more yards does she need to buy?

94. Your car gets 19 miles to the gallon of gas. How many gallons of gas will you need to drive to Chicago, a distance of 376 miles? Round your answer to the nearest tenth of a gallon.

95. The paint mixture for a special color calls for 4 gallons of base to be mixed with 1.2 quarts of tint. If you have already put in 0.85 quart of tint, how much more must you add?

96. How many syringes of serum, each containing 2.3 cc (cubic centimeters), can be filled from a vial containing 19.55 cc of serum?

97. Mike's utility bills average $56.25 per month. How much must he budget per year to pay for his utilities?

98. Cinnamon is selling for $0.69 per ounce. How much will 2 1/4 ounces cost?

99. A pair of jeans that normally sells for $24.95 is on sale for $20.98. How much can you save by buying two pair during this sale?

100. The new sheet metal product you are helping to develop must be within 0.07 inch of the specifications required for a particular job. If the job calls for the metal to be 1.4 inches thick, what is the thinnest piece that could be used?

Chapter Test

1. Write 0.032 using words.

2. Write sixteen and four hundred six ten-thousandths in number form.

3. Round 315.286 to the nearest tenth.

4. Round 27.885 to the nearest hundredth.

5. Round $37.698 to the nearest cent.

6. Rank 0.23, 0.023, and 0.203 in order from the smallest to the largest.

7. Find the sum of 0.38, 4.195, 12, and 7.2.

8. Find the difference between 17.68 and 9.895.

9. Subtract 14.834 from 52.

10. Find the product of 0.03 and 0.17.

11. Multiply $16.27 by 0.065. (Because your answer will be money, it should be rounded to the nearest cent.)

12. Multiply 3.217 by 100.

13. Find the quotient of 216 divided by 0.03.

14. Find the quotient of 12.85 divided by 17. Round your answer to the nearest tenth.

15. Divide 0.215 by 1,000.

16. Change 0.184 to its simplest fraction form.

17. Change 11/12 to its decimal form rounded to the nearest thousandth.

18. How many gallons of gas can you buy with a $20 bill if the price per gallon is $1.26? Round your answer to the nearest tenth of a gallon.

19. If the current balance in your checking account is $400.27, how much will remain in the account after you pay bills $69.25 and $217.34?

20. What will a 12 3/4-pound turkey cost if turkey is selling for $0.65 a pound this week?

Thought-Provoking Problems

1. Explain the similarities between the ways decimal and fraction problems are solved. Include addition, subtraction, and multiplication.

2. Explain in your own words the five steps to be followed in a decimal division problem.

3. Using your calculator, change 2/3 to a decimal. Write down the exact answer that the calculator gives in the display and explain it.

4. Using your calculator, complete the following decimal equivalents and answer the questions.

   a. 1/9 =
       2/9 =
       3/9 =
       4/9 =
       5/9 =
       6/9 =
       7/9 =

   b. What do you suppose 8/9 will look like? Why?

   c. What do you suppose 11/9 will look like? Why?

   d. What do you suppose 18/9 will look like? Why?

5. Using your calculator, compute the answer to 22,672 × 34,185.

   a. What does the display look like? What do you think it means?

   b. You could get an approximate answer if you multiplied 23,000 × 34,000. Try it on the calculator. What happens?

   c. Explain how you could get an approximate answer by using your calculator and applying some other things you know about multiplying by numbers that end in zero.

6. What happens when you try to multiply 0.0003762 by 0.000258 on your calculator? Can you think of a way to find the answer to this problem by using your calculator and applying some other things you know about decimal multiplication? Explain your ideas.

7. Using your calculator, compute the answer to this problem and round your answer to the nearest hundredth.

   $$1.56\overline{)26.732}$$

   Which value needs to be keyed into the calculator first? Why?

8. For each of the following problems, estimate an answer using paper and pencil and following the Order of Operations Agreement. Then use your calculator to solve the problems. How does the actual answer compare to your paper-and-pencil estimate? Does your calculator have built-in Order of Operations (hierarchy), or do you have to compensate and set up the steps yourself?

   a. 386.42 − 15.1(43.09 − 37.926)

   b. $(27.328)^3$

   c. 82.9 × 17.34 − 76.812 × 16.7

   d. 21,735.4 − 14.16 × 10^3

Ratios and Proportions

6.1 ◆ Introduction

Ratios and proportions have many applications outside the math classroom. You would use them in science, business, and drafting classes, and they are very useful in your own kitchen or workshop. Proportions are used to adjust quantities in recipes, to find measurements in the course of enlarging or reducing a pattern, and to determine mileage between cities on a map.

Because of the practical nature of ratios and proportions, many of you are going to understand this material very quickly and be very successful with it.

◆ **Learning Objectives**

When you have completed this chapter, you should be able to:

1. Write a ratio of two quantities in lowest terms.
2. Write ratios of measurement quantities in the same units.
3. Find rates and unit prices.
4. Determine if proportions are true by using cross products.
5. Solve a proportion for the missing quantity.
6. Solve word problems that involve proportions.
7. Convert measurements using unit fractions.

6.2   Ratios

Ratios and rates are used to compare two quantities. If you have a bookcase that contains 23 westerns and 45 mysteries, you might express the comparison of westerns to mysteries as 23/45 or as 23 to 45 or as 23:45, but the fraction form is most commonly used. Each expresses the relationship between the number of westerns and the number of mysteries. This comparison is called a **ratio**.

When writing a ratio, be sure always to write the first quantity that is given as the numerator or the first value in the statement. The second quantity given will be the denominator or the value after the symbol. If asked to write the ratio of the number of female students (5,400) to male students (4,800) in a college, you would write: 5,400/4,800. Is the relationship easier to understand as 5,400/4,800 or as 9/8? Anyone reading the information will have a better "feel" for the relative size of the female group compared to the male group if the fraction is in lowest terms. Simplify (reduce) all ratios and rates to lowest terms. (It is easier to remember to do this when the fraction form is used.) Leave improper fractions as comparisons rather than changing them to their mixed number form.

If the ratio of male to female students was requested, then 4,800/5,400, or 8/9, would be used. Because the total number of students is 5,400 + 4,800, or 10,200, the ratio of females to all students would be 5,400/10,200, or 9/17. The ratio of male students to the total student body is 4,800/10,200, or 8/17.

Leaving fraction or decimal quantities in a ratio is unacceptable. By definition, a ratio is the comparison of *two whole numbers*. Fractions within a ratio are simplified by rewriting the expression as the division of two fractions and completing that operation.

The ratio 5/6 to 7/9 must be simplified so that both numerator and denominator are whole numbers. Rewrite the complex fraction as division of two fractions.

$$\frac{\frac{5}{6}}{\frac{7}{9}} = \frac{5}{6} \div \frac{7}{9}$$

$$= \frac{5}{6} \times \frac{9}{7}$$

$$= \frac{5 \times \overset{1}{\cancel{3}} \times 3}{2 \times \underset{1}{\cancel{3}} \times 7} \quad \text{or} \quad \frac{5}{\underset{2}{\cancel{6}}} \times \frac{\overset{3}{\cancel{9}}}{7}$$

$$= \frac{15}{14} \qquad \text{Leave the improper fraction as it is because}$$
a ratio compares two numbers.

Decimals are changed to whole numbers by multiplying by a value of 1 that is in the form of 10/10, 100/100, or the like. The fraction form of "1" to be used is determined by the maximum number of places that the decimal point needs to move to make all the decimals into whole numbers (see Section 5.4).

A ratio of 3.2 to 0.06 is simplified by multiplying by 100/100, because the maximum number of decimal places in either number is two, and multiplying by 100 will move the decimal point two places to the right.

$$\frac{3.2}{0.06} = \frac{3.2}{0.06} \times \frac{100}{100} = \frac{320}{6} = \frac{160}{3}$$

EXAMPLE 1　Write these ratios in fraction form. Remember to reduce all ratios to lowest terms. This means that in the simplified form, nothing but whole numbers should remain.

   a.　16 to 12

   b.　3/4 to 5/8

   c.　1.6 to 0.8

   d.　The Franklin Flying Disk factory can manufacture 12,600 disks each day. If 800 of those disks are defective, write the ratio of defective disks to the total number produced.

Solution　**a.** $\dfrac{16}{12} = \dfrac{4}{3}$

   b. $\dfrac{\dfrac{3}{4}}{\dfrac{5}{8}}$

     $= \dfrac{3}{4} \div \dfrac{5}{8}$　　　Rewrite the complex fraction as division.

     $= \dfrac{3}{4} \times \dfrac{8}{5}$　　　Invert and multiply.

     $= \dfrac{3}{\underset{1}{4}} \times \dfrac{\overset{2}{8}}{5}$　or　$\dfrac{3 \times \overset{1}{2} \times \overset{1}{2} \times 2}{\underset{1}{2} \times \underset{1}{2} \times 5}$

     $= \dfrac{6}{5}$　　　Because a ratio is the comparison of two quantities, leave the answer as an improper fraction.

   c. $\dfrac{1.6}{0.8}$　　　Multiply both parts of the fraction by a value (such as 10, 100, or 1000) that will make the decimal numbers whole numbers (Section 5.4).

     $= \dfrac{1.6 \times 10}{0.8 \times 10}$

     $= \dfrac{16}{8} = \dfrac{2}{1}$　　　Then reduce, but leave the answer as a fraction.

   d. $\dfrac{800}{12{,}600}$　　　Write the number of defective disks over the total number of disks, and reduce to lowest terms.

     $= \dfrac{4}{63}$　　　　　　　　　　　　　　　　　　　◆

When writing the ratio of measurement numbers (length, volume, time, money, and so on), we must write the quantities in the same units. Otherwise the ratio is meaningless. In comparing 10 minutes to 1 hour, we cannot write 10 min/1 hr. Rather, we change 1 hour to 60 minutes. Then

$$\frac{10 \text{ min}}{60 \text{ min}} = \frac{\overset{1}{10 \text{ min}}}{\underset{1}{60 \text{ min}}} = \frac{1}{6}$$

Note that units cancel just like numbers do.

Table 1 presents some common measurement conversion factors that you should already know. If you do not, now is a good time to memorize them. Ask your instructor whether you will be expected to know them for the test on this chapter.

TABLE 1   Units of Measure

Length	Volume
1 foot (ft) = 12 inches (in.)	1 pint (pt) = 2 cups (c)
1 yard (yd) = 3 feet (ft)	1 quart (qt) = 2 pints (pt)
1 mile (mi) = 5,280 feet (ft)	1 gallon (gal) = 4 quarts (qt)

Time	Liquid Volume
1 minute (min) = 60 seconds (sec)	1 cup (c) = 8 fluid ounces (oz)
1 hour (hr) = 60 minutes (min)	1 pint (pt) = 16 fluid ounces (oz)
1 day = 24 hours (hr)	**Weight**
1 week = 7 days	
1 year (yr) = 52 weeks	1 pound (lb) = 16 ounces (oz)
1 year (yr) = 12 months	1 ton (T) = 2,000 pounds (lb)

EXAMPLE 2   Write the ratio of 4 yards to 8 feet.

Solution   If 1 yard = 3 feet, then 4 yards = 4 × 3 or 12 feet.

$$\frac{4 \text{ yards}}{8 \text{ feet}} = \frac{12 \text{ feet}}{8 \text{ feet}} = \frac{3}{2}$$

◆

EXAMPLE 3   Write the ratio of 4 nickels to 2 dollars.

Solution   If 1 dollar = 20 nickels, then 2 dollars = 2 × 20 or 40 nickels.

$$\frac{4 \text{ nickels}}{2 \text{ dollars}} = \frac{4 \text{ nickels}}{40 \text{ nickels}} = \frac{1}{10}$$

If you had changed both measurements to cents, you would have had

4 nickels = 4 × 5 = 20 cents

2 dollars = 2 × 100 = 200 cents

$$\frac{4 \text{ nickels}}{2 \text{ dollars}} = \frac{20 \text{ cents}}{200 \text{ cents}} = \frac{1}{10}$$

Note that the ratios are the same whether you use nickels or cents.   ◆

A **rate** compares two different kinds of items. Comparisons such as miles per gallon, feet per second, and cost per ounce are common examples of rates. In miles per hour, the number of miles traveled in *one* hour is a **unit rate**.

EXAMPLE 4   On a recent vacation, Juan drove 350 miles in 6 hours.

a. Write his rate of travel.

b. Find his average *unit* rate of travel in miles per hour.

Solution   **a.** $\dfrac{350 \text{ miles}}{6 \text{ hours}} = \dfrac{175 \text{ miles}}{3 \text{ hours}} = $ rate of travel

b. To find the number of miles traveled in 1 hour (unit rate), divide the miles by the number of hours.

$$\frac{350 \text{ miles}}{6 \text{ hours}} \quad (350 \div 6)$$

$$= \frac{58 \ 1/3 \text{ miles}}{1 \text{ hour}} = 58 \ 1/3 \text{ or } 58.3 \text{ miles per hour (mph)} \qquad \blacklozenge$$

E X A M P L E 5   On that 350-mile trip, Juan used 12 gallons of gasoline. How many miles per gallon (unit rate) did he average?

Solution   To get the unit rate, the number of miles on 1 gallon, divide the miles by the gallons.

$$\frac{350 \text{ miles}}{12 \text{ gallons}} = \frac{29 \ 1/6 \text{ miles}}{1 \text{ gallon}} = 29 \ 1/6 \text{ or } 29.2 \text{ miles per gallon (mpg)} \qquad \blacklozenge$$

Another common use of unit rates is in unit pricing. A **unit price** is the cost of one item or one unit, such as the cost per ounce.

E X A M P L E 6   Oranges are priced at 6 for 99 cents. What is the cost of one orange (unit price)?

Solution $\dfrac{99 \text{ cents}}{6 \text{ oranges}}$ Divide 99 by 6.

$$= \frac{16.5 \text{ cents}}{1 \text{ orange}}$$

Unit price = 16.5 cents or $0.165 per orange $\blacklozenge$

E X A M P L E 7   A jar of peanut butter is selling for $1.29 for 18 ounces. The 24-ounce jar is priced at $1.89. Which is the better buy? Round these unit prices to the nearest thousandth of a dollar.

Solution   The way to compare the prices is to look at the unit price (price per ounce) of each jar.

$$\frac{\$1.29}{18 \text{ oz}} = \frac{\$0.0716}{1 \text{ oz}} \text{ or } \$0.072 \text{ per ounce}$$

$$\frac{\$1.89}{24 \text{ oz}} = \frac{\$0.0787}{1 \text{ oz}} \text{ or } \$0.079 \text{ per ounce}$$

The smaller jar is the better buy. $\blacklozenge$

Definitions of Ratios and Rates

> A *ratio* is the relationship of two whole numbers that represent *like items*.
>
> A *rate* is the relationship of two whole numbers that represent *unlike items*.
>
> A *unit rate* has a denominator of 1.
>
> If a and b are whole numbers ($b \neq 0$), the ratio or rate can be written a/b, $a{:}b$, or a to b.

As you work with ratios and rates, remember to:

1. Write ratios and rates in the order given.
2. Reduce ratios and rates to lowest terms.
3. Express unit rates with a denominator of 1.
4. Write ratios of measurement quantities with the same units of measure.

6.2 ◆ Practice Problems

A. Write these ratios in fraction form. Remember to reduce to lowest terms.

1. 3 to 9	**2.** 16 to 48	**3.** 5:9
4. 0.8 to 1.2	**5.** 3.2 to 0.24	**6.** 72:64
7. 6:3/4	**8.** 2/3 to 3/4	**9.** 1.2:4
10. 2/3 to 7	**11.** 1/2:5/8	**12.** 5:3.2
13. 1 5/9 to 6	**14.** 8.6 to 0.14	**15.** 11.6:5.04
16. 4/5:2	**17.** 9 to 3/8	**18.** 14 to 3 3/8

B. Write these ratios in lowest terms.

19. 10 feet to 80 inches	**20.** 3 weeks to 14 days
21. 15 seconds to 2 minutes	**22.** 11 nickels:6 dimes
23. 3 dimes to 21 nickels	**24.** 14 inches to 3 feet
25. 6 pints to 5 quarts	**26.** 18 cups to 2 pints
27. 2 yards to 2 feet	**28.** 7 yards to 17 feet
29. 8 cents to 2 nickels	**30.** 1 hour to 19 minutes
31. 12 hours to 2 days	**32.** 3 pounds to 14 ounces
33. 2 pounds to 30 ounces	**34.** 5 gallons to 5 quarts
35. 3 cups to 2 pints	**36.** 400 pounds to 2 tons
37. 15 minutes to 3 hours	**38.** 3 quarts to 7 pints
39. 3.5 quarts to 14 pints	**40.** 5.5 feet to 9 inches

C. Answer each question with a ratio in lowest terms.

41. A vacation trip consisted of 640 miles traveled by airplane, 300 miles by ship, and 60 miles by car.

a. What is the ratio of air travel to car travel?

b. What is the ratio of ship travel to air travel?

c. What is the ratio of car travel to air travel?

d. What is the ratio of ship travel to the total trip?

e. What is the ratio of air travel to the total trip?

42. A basket of fruit contains 3 bananas, 4 pears, 5 apples, and 10 plums.

a. What is the ratio of pears to bananas?

b. What is the ratio of apples to the total number of pieces of fruit?

c. What is the ratio of apples to plums?

d. What is the fraction relationship of bananas to plums?

e. What is the ratio of total pieces of fruit to the number of pears?

43. The total enrollment of Metro Community College is 7,600. Of this number 4,200 are full-time students, and the rest are part-time. There are 150 full-time faculty members at the college.

a. Write the ratio of full-time students to total students.

b. What is the ratio of full-time to part-time students?

c. What is the ratio of full-time faculty to full-time students?

d. Write the ratio of full-time faculty to total students.

44. Yesterday the Easton Tool Company had a daily production total of 1,220 rakes and 1,400 shovels. Sixty of the rakes and 100 shovels were defective.

a. What is the ratio of defective rakes to total rakes?

b. What fraction of the shovels were defective?

c. What is the ratio of the number of shovels to the total number of tools manufactured that day?

d. What is the ratio of the total tools produced to the number of rakes?

D. For each question write a ratio or rate in simplified form. If necessary, round answers to the nearest thousandth.

45. A 16-ounce box of cereal sells for $2.40. What is the price per ounce?

46. A recipe calls for 3 cups of flour and 1 1/2 cups of sugar. What is the ratio of sugar to flour?

47. A button manufacturer produces 24,000 buttons each day. If 600 each day are defective, what is the ratio of total buttons to defective ones?

48. A 22-ounce bottle of dish detergent costs $1.10. What is the cost per ounce?

49. A jogger ran 2 1/2 miles on Monday and 5 miles on Saturday. What is the ratio of Saturday's run to Monday's?

50. A grocery store receives 4,000 dozen eggs per week. If 200 dozen contain broken or cracked eggs, what fraction of the total number of dozens is damaged?

51. A motorist travels 240 miles on 10 gallons of gasoline. How many miles per gallon did he average?

52. An airplane can travel 750 miles in 2.5 hours. What is its average speed in miles per hour?

53. A race car can travel 165 miles in 1.5 hours. What is its average speed in miles per hour?

54. A speedboat can travel 50 miles on 4 gallons of gas. What is the boat's fuel consumption in miles per gallon?

55. An airplane can travel 560 miles on 16 gallons of fuel. What is its rate of fuel usage in miles per gallon?

56. A motorcyclist can travel 225 miles on 5 gallons of gas. Find the number of miles per gallon.

57. Socks are on sale this week at Kay's Mart. Six pairs cost $5.97. What is the price per pair?

58. Rory's Roast Beef has a special price on sandwiches. Four sandwiches cost $5.00. How much does each sandwich cost?

59. Ten bags of cypress mulch cost $25.00. What is the cost per bag?

60. Soup is priced at 6 cans for $4.00. What is the price per can?

E. **In Problems 61 through 70, determine whether item A or item B is less expensive, and give its unit cost. If necessary, round answers to the nearest thousandth.**

	Item A	Item B
61.	6 for $0.98	15 for $2.49
62.	$1.49 for 12 oz	$0.98 for 8 oz
63.	$2.80 per dozen	$2.10 for 8
64.	8 for $6.00	3 for $2.25
65.	12 for $8.88	5 for $4.00
66.	$2.49 for 32 oz	$1.49 for 22 oz
67.	$0.72 per pint	$0.35 for 8 oz
68.	$1.56 per pint	$3.00 per quart
69.	10 oz for $1.79	16 oz for $2.49
70.	3 lb for $4.47	5 lb for $6.90

 F. **Round answers to the nearest hundredth.**

71. A car traveled 213 miles on 13.2 gallons of gas. Find how many miles per gallon the car averaged. If gas costs $1.139 per gallon, how much was spent for gas for this trip?

72. How many gallons of gas were pumped into a car if gas costs $1.179 per gallon and the total charged was $13.69. Round the answer to the nearest hundredth.

73. A 10.75-ounce can of soup costs $0.89. What is the cost per ounce to the nearest thousandth of a dollar?

74. If 22.5 square yards of carpet cost $337.05, what is the cost per square yard?

6.3 ◆ Proportions

As we have said, **ratio** is the comparison of two numbers, usually written as a fraction. A **proportion** is two equal ratios.

$$\frac{3}{12} = \frac{1}{4} \quad \text{and} \quad \frac{6}{19} = \frac{24}{76} \text{ are proportions.}$$

There are other ways of writing these proportions. One is

$$3:12::1:4 \quad \text{and} \quad 6:19::24:76.$$

Another is to write "3 is to 12 as 1 is to 4" and "6 is to 19 as 24 is to 76."

Proportions

$\dfrac{a}{b} = \dfrac{c}{d}$ is a proportion because it expresses the equality of two ratios (where $b, d \neq 0$).

For all proportions, the cross products are equal.

$$\frac{a}{b} \bowtie \frac{c}{d} \qquad a \cdot d = b \cdot c$$

When the $a:b::c:d$ form is used, a and d are called the **extremes**, and b and c are the **means**. The cross-product rule is then stated as "The product of the means equals the product of the extremes."

You can test to see if a statement is a proportion by comparing cross products. This method is usually faster and easier than reducing both fractions to see if they reduce to identical fractions.

E X A M P L E **1**   Is $\dfrac{3}{12} = \dfrac{1}{4}$ a proportion? Prove your answer.

Solution   Cross-multiply $3 \times 4 = 12$ and $12 \times 1 = 12$.

$$\frac{3}{12} = \frac{1}{4}$$

It is a proportion because $12 = 12$. (Showing that the cross products are equal proves your answer.)   ◆

The process of computing cross products to show that a statement is a proportion is a convenient method, but it is not a true mathematical process. To show the equality of the two ratios, we should multiply both sides of the equation by a value that is a common denominator for the fractions. For the previous problem, the least common denominator would be 48. Multiply both ratios by 48 and see if that produces a true statement.

$$48 \cdot \frac{3}{12} = \frac{1}{4} \cdot 48$$

$$\frac{\overset{4}{\cancel{48}}}{1} \cdot \frac{3}{\underset{1}{\cancel{12}}} = \frac{1}{\underset{1}{\cancel{4}}} \cdot \frac{\overset{12}{\cancel{48}}}{1} \quad \text{or} \quad \frac{\overset{1}{\cancel{2}} \cdot \overset{1}{\cancel{2}} \cdot 2 \cdot 2 \cdot \overset{1}{\cancel{3}} \cdot 3}{\underset{1}{\cancel{2}} \cdot \underset{1}{\cancel{2}} \cdot \underset{1}{\cancel{3}}} = \frac{1 \cdot \overset{1}{\cancel{2}} \cdot \overset{1}{\cancel{2}} \cdot 2 \cdot 2 \cdot 3}{\underset{1}{\cancel{2}} \cdot \underset{1}{\cancel{2}}}$$

$$12 = 12 \qquad \text{True.}$$

This is the same answer you got when using the short cut of cross products. Either method can be used in the following problems.

E X A M P L E **2**   Is $\dfrac{4}{64} = \dfrac{16}{192}$ a proportion? Why?

Solution   Using cross products to find out yields
$4 \times 192 = 768 \qquad 64 \times 16 = 1{,}024$
No, it is not a proportion because $768 \neq 1{,}024$.   ◆

E X A M P L E **3**   Is $\dfrac{2\ 1/2}{3/4} = \dfrac{15}{4\ 1/2}$ a proportion? Why?

Solution   Do not let the fractions change your approach. You still want to find out if the cross products are equal.
Multiply $2\ 1/2 \times 4\ 1/2$ and $3/4 \times 15$.
$2\ 1/2 \times 4\ 1/2 = 5/2 \times 9/2 = 45/4$
$3/4 \times 15 = 45/4$
Yes, because $45/4 = 45/4$.   ◆

E X A M P L E **4**   Is $\dfrac{7.2}{0.96} = \dfrac{35}{4}$ a proportion? Why?

Solution $7.2 \times 4 = 28.8$ $0.96 \times 35 = 33.6$

No, because $28.8 \neq 33.6$.

The property of equal cross products is also used to find missing parts of a proportion. If three out of the four parts of a proportion are known, the fourth part can be found by computing the cross products and then dividing. The missing number is represented by a letter. Any letter can be used.

To solve for the value for N that makes

$$\frac{2}{3} = \frac{N}{9}$$

a proportion, find the cross products $2 \cdot 9$ and $3 \cdot N$. Set the products equal to each other.

$$3 \cdot N = 2 \cdot 9$$

Writing 3 and N side by side shows multiplication, so

$$3N = 18$$

To get N by itself, *undo the multiplication*, $3N$, by doing the *opposite* operation. *Divide both sides by 3, then simplify.*

$$\frac{3N}{3} = \frac{18}{3}$$

$$1 \cdot N = 6 \quad \text{or} \quad N = 6$$

Thus 6 is the number missing from the original proportion.

Check   Check your answer. Is $2/3 = 6/9$ a proportion?

$$2 \times 9 = 18 \quad \text{and} \quad 3 \times 6 = 18$$

$$18 = 18 \quad \text{Yes}$$

E X A M P L E **5**   Find the unknown value in $\dfrac{6}{p} = \dfrac{15}{8}$.

Solution $\dfrac{6}{p} = \dfrac{15}{8}$ Cross-multiply $15 \cdot p$ and $6 \cdot 8$. Set the products equal to each other.

$15p = 48$

$\dfrac{15p}{15} = \dfrac{48}{15}$ Divide both sides by 15 to isolate p.

$p = \dfrac{16}{5}$ or $3\dfrac{1}{5}$ or 3.2

Check   Is $\dfrac{6}{3.2} = \dfrac{15}{8}$ a proportion?

$$6 \times 8 = 48 \quad \text{and} \quad 3.2 \times 15 = 48$$

$$48 = 48.0 \quad \text{Yes.}$$

E X A M P L E 6 Solve $\dfrac{4/5}{1/8} = \dfrac{1/10}{B}$ for B and check your answer.

Solution $\qquad \dfrac{4/5}{1/8} = \dfrac{1/10}{B}$ $\qquad$ Cross-multiply $4/5 \cdot B$ and $1/8 \cdot 1/10$.

$$\frac{4}{5}B = \frac{1}{80}$$

$$\frac{4}{5}B \div \frac{4}{5} = \frac{1}{80} \div \frac{4}{5}$$ $\qquad$ Divide by 4/5. Be sure that the number you are dividing *by* is placed after the division sign.

$$\frac{4}{5}B \cdot \frac{5}{4} = \frac{1}{80} \cdot \frac{5}{4}$$ $\qquad$ Multiply by the reciprocal.

$$\overset{1}{\underset{1}{\frac{\cancel{4}}{\cancel{5}}}}B \cdot \overset{1}{\underset{1}{\frac{\cancel{5}}{\cancel{4}}}} = \frac{1}{\underset{16}{\cancel{80}}} \cdot \overset{1}{\frac{\cancel{5}}{4}} \quad \text{or} \quad \frac{\overset{1}{\cancel{2}} \cdot \overset{1}{\cancel{2}} \cdot B \cdot \overset{1}{\cancel{5}}}{\cancel{5} \cdot \underset{1}{\cancel{2}} \cdot \underset{1}{\cancel{2}}} = \frac{1 \cdot \overset{1}{\cancel{5}}}{2 \cdot 2 \cdot 2 \cdot 2 \cdot \underset{1}{\cancel{5}} \cdot 2 \cdot 2}$$

$$B = \frac{1}{64}$$

Check $\quad$ Is $\dfrac{4/5}{1/8} = \dfrac{1/10}{1/64}$ a proportion?

$4/5 \times 1/64 = 1/80 \quad$ and $\quad 1/8 \times 1/10 = 1/80$

$1/80 = 1/80$. Yes.

6.3 ◆ Practice Problems

A. Determine whether each of the following is a proportion. Prove your answer by showing the cross products.

1. $\dfrac{2}{3} = \dfrac{12}{18}$ $\qquad$ 2. $\dfrac{3}{4} = \dfrac{15}{20}$

3. $\dfrac{5}{9} = \dfrac{35}{45}$ $\qquad$ 4. $\dfrac{6}{7} = \dfrac{48}{56}$

5. $\dfrac{4}{49} = \dfrac{28}{343}$ $\qquad$ 6. $\dfrac{99}{55} = \dfrac{81}{45}$

7. $\dfrac{1.2}{3.6} = \dfrac{0.48}{1.44}$ $\qquad$ 8. $\dfrac{3.5}{0.49} = \dfrac{70}{9.8}$

9. $\dfrac{3.9}{5.85} = \dfrac{3}{4.5}$ $\qquad$ 10. $\dfrac{1/2}{3/4} = \dfrac{1/8}{3/16}$

11. $\dfrac{5/8}{15/2} = \dfrac{1/4}{3}$ $\qquad$ 12. $\dfrac{9/4}{8} = \dfrac{4/3}{2/3}$

B. Find the missing value in each proportion. If necessary, round your answers to the nearest hundredth.

13. $\dfrac{5}{9} = \dfrac{45}{N}$ $\qquad$ 14. $\dfrac{6}{W} = \dfrac{54}{36}$

15. $\dfrac{R}{7} = \dfrac{24}{56}$ $\qquad$ 16. $\dfrac{15}{16} = \dfrac{M}{80}$

17. $\dfrac{3}{8} = \dfrac{15}{A}$ $\qquad$ 18. $\dfrac{5}{6} = \dfrac{B}{36}$

19. $\dfrac{4}{c} = \dfrac{28}{35}$ $\qquad$ 20. $\dfrac{T}{9} = \dfrac{24}{36}$

21. $\dfrac{N}{3} = \dfrac{7}{5}$ $\qquad$ 22. $\dfrac{12}{9} = \dfrac{15}{F}$

23. $\dfrac{36}{5} = \dfrac{48}{R}$ $\qquad$ 24. $\dfrac{19}{6} = \dfrac{G}{4}$

25. $\dfrac{A}{19} = \dfrac{7}{8}$ $\qquad$ 26. $\dfrac{11}{M} = \dfrac{9}{14}$

27. $\dfrac{97}{B} = \dfrac{24}{47}$ $\qquad$ 28. $\dfrac{36}{49} = \dfrac{13}{W}$

29. $\dfrac{12}{21} = \dfrac{T}{28}$ $\qquad$ 30. $\dfrac{51}{81} = \dfrac{17}{M}$

31. $\dfrac{0.07}{40} = \dfrac{0.28}{A}$ $\qquad$ 32. $\dfrac{3\ 1/4}{2} = \dfrac{B}{2/3}$

33. $\dfrac{W}{1\ 1/8} = \dfrac{3\ 1/2}{1\ 1/2}$ $\qquad$ 34. $\dfrac{0.04}{0.006} = \dfrac{C}{5.4}$

35. $\dfrac{3/4}{5/8} = \dfrac{N}{1/2}$ $\qquad$ 36. $\dfrac{1.2}{P} = \dfrac{0.8}{3.9}$

37. $\dfrac{A}{0.05} = \dfrac{1.2}{0.006}$ $\qquad$ 38. $\dfrac{5\ 1/2}{B} = \dfrac{3\ 1/2}{4}$

39. $\dfrac{9.82}{4.1} = \dfrac{7.6}{R}$ $\qquad$ 40. $\dfrac{11.2}{M} = \dfrac{7.65}{18.8}$

C. Answer the following questions. You may use your calculator.

41. Is $\dfrac{7.59}{5.22} = \dfrac{25.3}{17.4}$ a proportion? Why?

42. Is $\dfrac{0.0009}{0.004} = \dfrac{36,000}{1,600,000}$ a proportion? Explain your answer.

Find the value of N in each of the following proportions. If necessary, round answers to the nearest hundredth.

43. $\dfrac{0.076}{1.34} = \dfrac{0.24}{N}$ 44. $\dfrac{253}{891} = \dfrac{N}{567}$

45. $\dfrac{N}{965} = \dfrac{13,490}{2,340}$ 46. $\dfrac{0.018}{N} = \dfrac{2.96}{6.45}$

6.4 Applications of Proportions

You will find proportions very useful in solving word problems that involve the relationship between two different types of quantities. Follow these steps to set up and solve this kind of word problem.

1. After reading the problem carefully, decide on a letter to represent the missing value.

2. Write a ratio of two of the known values.

3. Set this ratio equal to a ratio containing the missing value. Be sure that the ratios are written in the same order. Writing a word fraction on each side might help keep the order the same.

4. Solve the resulting proportion.

5. It is a good idea to check your answer by testing to see if it makes sense in the problem.

E X A M P L E 1   In a bag of "Melt In Your Mouth" candies, the ratio of yellow candies to red candies is 3 to 4. Find the number of yellow candies if there are 124 red ones.

Solution   If Y is used to represent the number of yellow candies, then:

$$\frac{\text{yellow}}{\text{red}} \quad \frac{3}{4} = \frac{Y}{124} \quad \frac{\text{yellow}}{\text{red}}$$

If yellow/red is the first ratio, be sure that the order is the same on the other side. Now solve the proportion.

$4Y = 372$ Set cross products equal.

$\dfrac{4Y}{4} = \dfrac{372}{4}$ Divide both sides by 4.

$Y = 93$ yellow candies

Check   Is 93 yellow candies reasonable? If the ratio of yellow to reds is 3 to 4, then there should be fewer yellows than reds. $93 : 124$ is reasonable.   ◆

If you check your answer by substituting your solution in the proportion and finding the cross products, you are checking only the accuracy of your calculations. In word problems involving proportions, merely substituting would not catch errors of incorrectly written proportions. Estimating answers and testing to see if your answer is reasonable are more foolproof checks.

E X A M P L E 2   If 18 apples cost $3.20, how much do 7 apples cost? Round your answer to the nearest cent.

Solution $$\frac{\text{apples}}{\text{cost}} \quad \frac{18}{3.20} = \frac{7}{c} \quad \frac{\text{apples}}{\text{cost}}$$

In this proportion, the number of apples was written over the cost. This is just one of four correct proportions that could have been written. The others are

$$\frac{18}{7} = \frac{3.20}{c} \qquad \frac{7}{18} = \frac{c}{3.20} \qquad \frac{3.20}{18} = \frac{c}{7}$$

All would give the same cross products. Continuing with the solution, we find that

$$18c = 7 \times 3.20$$
$$18c = 22.4$$
$$\frac{18c}{18} = \frac{22.4}{18}$$
$$c = \$1.244 \text{ or } \$1.24$$

Check   Is $1.24 reasonable? Since 7 is less than half of 18, then the cost should be less than half of $3.20.

$$\$3.20 \div 2 = \$1.60 \quad \text{and} \quad \$1.24 < \$1.60 \qquad \blacklozenge$$

E X A M P L E 3   On a map, the scale says that a distance of 2 inches represents 15 miles. What is the actual distance between two cities if the map shows 4 1/2 inches between them?

Solution   Let d = distance.

$$\frac{\text{inches}}{\text{miles}} \quad \frac{2}{15} = \frac{4\ 1/2}{d} \quad \frac{\text{inches}}{\text{miles}}$$
$$2d = 135/2$$
$$2d \div 2 = 135/2 \div 2$$
$$d = 135/4 \text{ or } 33\ 3/4 \text{ miles}$$

or $$\frac{2}{15} = \frac{4.5}{d}$$
$$2d = 67.5$$
$$\frac{2d}{2} = \frac{67.5}{2}$$
$$d = 33.75 \text{ miles}$$

Check   Is 33.75 miles a reasonable answer? Because 4 1/2 is a little more than 2 times 2, the distance should be a little more than 2 times 15 miles.

$$2(15) = 30 \quad \text{and} \quad 33.75 > 30 \qquad \blacklozenge$$

6.4 ◆ Practice Problems

A. Write a proportion for each word problem and solve it. If necessary, round answers to the nearest hundredth.

1. In a manufacturing plant, two VCRs are defective out of every 72 that are made. At this rate, how many would be defective out of 450 VCRs?

2. While evaluating its telephone system, a company finds that it receives 172 phone calls in 2 days. At that same rate, how many calls would it receive in 5 days?

3. If Carlos drove 159 miles in three hours, how far could he drive in five hours?

4. On a map the scale says that 1 1/2 inches represent 12 miles. How many miles between two cities that are 5 1/4 inches apart on the map?

5. A doll house is built to scale so that one inch is equivalent to 1.5 feet. If a window in a real house measures 2.4 feet wide, how wide would it be in the doll house?

6. If a 6.4-pound roast costs $15.48, what is the cost per pound of that piece of meat? (Cost per pound means cost for one pound, or *C*/1.)

7. If a construction crew can frame a house in 2 1/4 days, how many houses can that crew frame in six days if they work at the same rate?

8. If a plane flies 320 miles in 5 hours, how far can it fly in 3 hours?

9. How much per pound does chicken cost if 3 3/4 pounds cost $3.00?

10. If five out of every 250 parts manufactured by Camco Corporation are defective, how many are defective out of 1,200 parts?

11. If three pounds of onions cost 89 cents, how much would two pounds cost?

12. Dan received $42 interest on his savings account, which contained $700. How much interest would he receive on $1,750?

13. The hospitality committee plans to have five cans of soft drinks for every three guests attending the homecoming dance. How many cans will be needed if 150 guests attend?

14. A property tax rate is $1.09 for each $1,000 of property value. How much tax would be charged on a house valued at $58,500?

15. A concrete mixture calls for 1,000 pounds of gravel for every 6 bags of cement. How much gravel is needed for 33 bags of cement?

16. How long will an enlarged photograph be if its width will be 10 inches and the original picture is 2 1/2 inches wide and 3 1/4 inches long?

17. Monique can type nine pages of text in two hours. How long will it take her to type a 54-page report?

18. If two pounds of ground beef cost $2.96, how much would five pounds cost?

19. Ishmial earned $15 in interest on the $120 he had in his savings account. How much interest would the account have earned if he had $600 deposited in the account?

20. Don's cookbook says that 1 1/3 pounds of boneless roast should serve four people. How much meat will he need to serve 10 guests?

 B. Write a proportion, and then solve it using your calculator. When necessary, round answers to the nearest hundredth.

21. Anja earned $36.94 on the $528 she had in her savings account. How much interest would the account have earned if she had deposited $1,350?

22. If a 5.43-pound roast costs $27.10, what is the cost per pound?

23. Property tax on a house valued at $96,500 is $1247.43. What would the tax be on an $80,400 house?

24. Of every 100 widgets that are manufactured, nine are defective. How many defective widgets will be made during a 40-hour work week if 1,750 widgets are produced per hour?

25. Suburban Telephone has an average daily absentee rate of 3 employees per 100. What is the average daily number of absentee employees if 7,148 people work at the facility?

6.5 ◆ Ratios and Conversion of Units

A very important application of ratios in scientific and technical courses is converting measurements to other units. The process involves writing conversion factors as fractions.

Because 1 foot = 12 inches, that fact could be written as the ratio 1 ft/12 in. or as the ratio 12 in./1 ft. Both fractions have a value of 1 because the quantities are equal. For this reason, they are called **unit fractions**. Use this principle to change 6 feet to inches. Write the measurement in fraction form, and multiply it by the conversion factor, written as a unit fraction. The form of the fraction that you should choose is the one that allows you to cancel out the feet from the problem. This is easier than it sounds!

$$\frac{6 \text{ ft}}{1} \times ? = \text{inches}$$

We must use 1 ft/12 in. or 12 in./1 ft.

$$\frac{6 \ \cancel{\text{ft}}}{1} \times \frac{12 \text{ in.}}{1 \ \cancel{\text{ft}}} = \frac{6 \times 12 \text{ in.}}{1} = 72 \text{ inches}$$

12 in./1 ft was used instead of 1 ft/12 in. so that the feet would cancel.

E X A M P L E   **1**   Change 30 miles per hour to miles per minute.

Solution   The conversion factor needed is 1 hr = 60 min. Which unit fraction, $\frac{1 \text{ hr}}{60 \text{ min}}$ or $\frac{60 \text{ min}}{1 \text{ hr}}$, will cause the hours to cancel and leave miles to minutes?

$$\frac{30 \text{ mi}}{1 \text{ hr}} \times ? = \frac{\text{mi}}{\text{min}}$$

$$\frac{30 \text{ mi}}{1 \ \cancel{\text{hr}}} \times \frac{1 \ \cancel{\text{hr}}}{60 \text{ min}} = \frac{30 \text{ miles}}{60 \text{ minutes}}$$

Now divide 30 by 60 to get a rate for 1 minute.

$$\frac{30 \text{ miles}}{60 \text{ minutes}} = 0.5 \text{ mile per minute}$$

Check   Look at the answer to see that only the desired units remain there.   ◆

E X A M P L E   **2**   Change 45 miles per hour to miles per second.

Solution   Two conversion factors will be needed: hours to minutes and minutes to seconds. Write the unit fractions so that all units cancel except miles in the numerator and seconds in the denominator.

$$\frac{45 \text{ mi}}{1 \text{ hr}} \times \frac{1 \text{ hr}}{60 \text{ min}} \times \frac{1 \text{ min}}{60 \text{ sec}} \times \frac{45 \text{ mi}}{1 \ \cancel{\text{hr}}} \times \frac{1 \ \cancel{\text{hr}}}{60 \ \cancel{\text{min}}} \times \frac{1 \ \cancel{\text{min}}}{60 \text{ sec}} = \frac{45 \text{ mi}}{3600 \text{ sec}}$$

Divide 45 by 3600.

$= 0.0125 \text{ mile per second}$ ◆

E X A M P L E   **3**   Using the conversion factors 1 cup = 8 ounces and 1 pint = 2 cups, change 6 pints to ounces.

Solution   The conversion factors will take you from pints to cups to ounces. Note that you have to write 6 pints as 6 pints/1 so that it will be in fraction form when you multiply by the other unit fractions.

$$\frac{6 \text{ pt}}{1} \times \frac{2 \text{ c}}{1 \text{ pt}} \times \frac{8 \text{ oz}}{1 \text{ c}} = \frac{6 \ \cancel{\text{pt}}}{1} \times \frac{2 \ \cancel{\text{c}}}{1 \ \cancel{\text{pt}}} \times \frac{8 \text{ oz}}{1 \ \cancel{\text{c}}} = 96 \text{ ounces}$$ ◆

E X A M P L E   **4**   Using the conversion factors 1 inches = 2.54 centimeters and 1 meter = 100 centimeters, change 30 inches to meters.

Solution $$\frac{30 \text{ in.}}{1} \times \frac{2.54 \text{ cm}}{1 \text{ in.}} \times \frac{1 \text{ m}}{100 \text{ cm}}$$

$$= \frac{30 \ \cancel{\text{in.}}}{1} \times \frac{2.54 \ \cancel{\text{cm}}}{1 \ \cancel{\text{in.}}} \times \frac{1 \text{ m}}{100 \ \cancel{\text{cm}}} = \frac{30 \times 2.54 \text{ m}}{100} = 0.762 \text{ meters}$$ ◆

E X A M P L E **5** Change 88 feet per second to miles per hour.

Solution Feet must be changed to miles, and seconds must be changed to hours. These conversion factors will be needed: 1 mile = 5,280 feet, 1 hour = 60 minutes, and 1 minute = 60 seconds. Write the unit fractions so that all units cancel except miles in the numerator and hours in the denominator.

$$\frac{88 \text{ ft}}{1 \text{ sec}} \times \frac{1 \text{ mi}}{5{,}280 \text{ ft}} \times \frac{60 \text{ sec}}{1 \text{ min}} \times \frac{60 \text{ min}}{1 \text{ hr}}$$

$$= \frac{88 \text{ ft}}{1 \text{ sec}} \times \frac{1 \text{ mi}}{5{,}280 \text{ ft}} \times \frac{60 \text{ sec}}{1 \text{ min}} \times \frac{60 \text{ min}}{1 \text{ hr}}$$

$$= \frac{88 \times 60 \times 60 \text{ mi}}{5{,}280 \text{ hr}} = \frac{316800 \text{ mi}}{5{,}280 \text{ hr}} = 60 \text{ miles per hour}$$

6.5 ◆ Practice Problems

A. Use unit fractions to convert the following measurements. You may use the chart of conversion factors, Table 1 in Section 6.2. If necessary, round answers to the nearest tenth.

1. Change 18 cups to quarts.

2. Change 68 inches to yards.

3. Change 3 days to minutes.

4. Change 33 yards to inches.

5. Change 8 hours to seconds.

6. Change 14 gallons to pints.

7. Change 40 miles per hour to miles per minute.

8. Change 20 miles per gallon to feet per gallon.

9. Change 100 miles per hour to feet per second.

10. Change 50 miles per hour to feet per second.

B. Change the following measurements using unit fractions. The conversion factors in Table 2 may be needed, as well as those in Table 1. If necessary, round answers to the nearest tenth.

T A B L E 2   Conversion Factors for Length in the English and Metric Systems

Metric System

1 centimeter (cm) = 10 millimeters (mm)

1 decimeter (dm) = 10 centimeters (cm)

1 meter (m) = 100 centimeters (cm)

1 kilometer (km) = 1,000 meters (m)

English ↔ Metric Conversion Factors

1 inch (in.) = 2.54 centimeters (cm)

1 meter (m) = 39.37 inches (in.)

1 mile (mi) = 1.61 kilometers (km)

11. Change 3 meters to decimeters.

12. Change 0.4 kilometer to meters.

13. Change 150 millimeters to centimeters.

14. Change 44 decimeters to meters.

15. Change 28 inches to centimeters.

16. Change 3.2 miles to kilometers.

17. Change 4 miles per hour to kilometers per hour.

18. Change 14 yards to meters.

19. Change 65 centimeters to feet.

20. Change 101.6 centimeters to feet.

C. Use your calculator, unit fractions, and the conversion tables in this chapter to find the following measurements.

21. Change 0.165 meter to decimeters.

22. Change 421.5 kilometers to miles.

23. Change 42,000 millimeters to inches.

24. Change 5.6 meters to inches.

25. Change 67.2 yards to meters.

26. Change 49 miles per hour to kilometers per hour.

27. Change 32.9 feet per second to miles per hour.

28. Change 635 centimeters to feet.

29. Change 486 inches to centimeters.

30. Change 0.85 yard to meters.

6.6   **Using Data**

Information is often presented in chart or graph form. The ability to read these visual displays accurately is an important skill. You will be given practice in this section as you write ratios and proportions with data taken from a chart or graph.

E X A M P L E   1   Use the information in Figure 1 to answer the questions that follow.

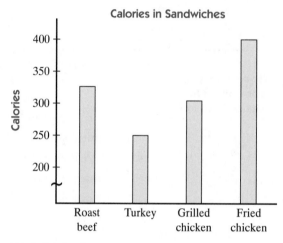

F I G U R E   1

a.   What is the ratio of the calories in the grilled chicken sandwich to the calories in the fried chicken sandwich?

b.   What is the ratio of the calories in the fried chicken sandwich to the calories in the turkey sandwich?

c.   What is the ratio of calories in the roast beef sandwich to the calories in the grilled chicken sandwich?

Solutions   a.   Grilled chicken sandwich = 300 calories

Fried chicken sandwich = 400 calories

$$\frac{300 \text{ calories}}{400 \text{ calories}} = \frac{3}{4}$$

b.   Fried chicken sandwich = 400 calories

Turkey sandwich = 250 calories

$$\frac{400 \text{ calories}}{250 \text{ calories}} = \frac{40}{25} = \frac{8}{5}$$

c.   The calories for the roast beef sandwich must be estimated. The top of the bar appears to be halfway between 300 and 350, so 325 calories is a reasonable estimate.

$$\frac{325 \text{ calories}}{300 \text{ calories}} = \frac{13}{12}$$

EXAMPLE 2 The number of male and female students in five sections of Introductory Economics are displayed in Chart 1. Use that information to answer the questions that follow.

CHART 1 **Number of Students by Sex**

	Section 111	Section 112	Section 113	Section 001	Section 002
Males	13	15	7	20	17
Females	18	12	13	11	13

a. What is the ratio of males to females in section 112?

b. What is the ratio of the female students to the total number of students in section 001?

c. What is the mean (average) number of students in the five sections?

d. What is the median number of male students in the five sections?

e. What is the mode for the female students in the five sections?

Solutions a. $\dfrac{15}{12} = \dfrac{5}{4}$

b. The total number of students in section 001 is $20 + 11 = 31$. The ratio is $\dfrac{11}{31}$.

c. For mean, median, and mode, review Section 1.7. To find the mean, add the total number of students in each section and divide by 5.

$$\text{Mean} = \frac{31 + 27 + 20 + 31 + 30}{5}$$

$$= \frac{139}{5} = 27.8 \text{ students per section}$$

d. The median is the middle number when the numbers are arranged in order of size: 20, 17, 15, 13, 7. The median number of male students is 15.

e. The mode for the female students is the most often repeated number of females. Mode = 13 females ◆

EXAMPLE 3 Use Chart 2 to answer the questions that follow.

CHART 2

Number of Shirts	1	2	4	6
Total Cost ($)	16.95	33.90	67.80	101.70

a. Write a proportion to find the cost of three shirts.

b. Use a proportion to find the cost of five shirts.

Solutions Any of the ratios of the number to the cost could be used as the first part of the proportion.

a. $\dfrac{\text{number}}{\text{cost}} \quad \dfrac{1}{16.95} = \dfrac{3}{c} \quad \dfrac{\text{number}}{\text{cost}}$

$1c = 50.85$

The cost of three shirts is $50.85.

b. Another ratio we could use for this part is

$\dfrac{2}{33.90} = \dfrac{5}{c} \quad 2c = 169.50 \quad \dfrac{2c}{2} = \dfrac{169.50}{2} \quad c = 84.75$

The cost of five shirts is $84.75. ◆

6.6 ◆ Practice Problems

All ratios should be written in simplest form. Proportion answers should be rounded to the nearest hundredth.

A. In Problems 1 through 6, use Figure 2.

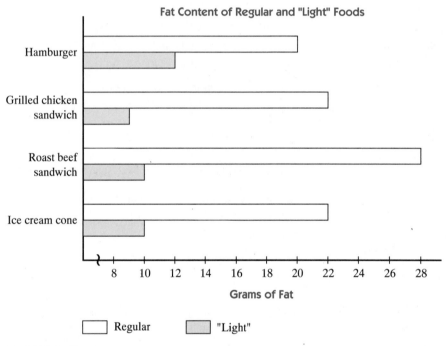

FIGURE 2

1. What is the ratio of grams of fat in "light" to grams of fat in regular hamburgers?

2. What is the ratio of grams of fat in regular to grams of fat in "light" ice cream cones?

3. What is the ratio of grams of fat in a regular hamburger to grams of fat in a regular roast beef sandwich?

4. What is the ratio of grams of fat in a "light" chicken sandwich to grams of fat in a regular roast beef sandwich?

5. What is the mean (average) number of grams of fat in the four regular food items displayed?

6. What is the mean (average) number of grams of fat in the four "light" food items displayed?

B. In Problems 7 through 12, use Chart 3.

CHART 3   Cardinal Season Records

	1988	1989	1990	1991
Won	8	10	16	12
Lost	10	6	1	6
Tied	0	2	1	1

7. What was the ratio of wins to losses in 1989?

8. What was the ratio of losses to ties in 1989?

9. What was the ratio of losses to wins in 1990?

10. What was the ratio of wins to losses in 1988?

11. What was the mean number of wins for the four seasons?

12. What was the mean number of losses for the period?

C. **In Problems 13 through 18, use Figure 3.**

FIGURE 3

13. Write the ratio of Maria's March sales to her May sales.

14. What is the ratio of Martin's May sales to Matt's May sales?

15. What is the mean of the three sales associates' March sales?

16. What is the mean of the three sales associates' May sales?

17. If these are the only salespersons in the company, what were the total sales for the company for March?

18. If these are the only salespersons in the company, what were the total sales for the company for May?

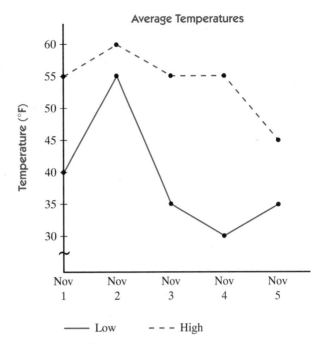

FIGURE 4

D. **In Problems 19 through 24, use Figure 4.**

19. What is the ratio of the average low temperature to the average high temperature for November 4?

20. What is the ratio of the average high temperature to the average low temperature for November 5?

21. What is the ratio of the average high temperature on November 3 to the average high temperature on November 2?

22. Write the ratio of the average low temperature on November 1 to the average low temperature on November 4.

23. What is the mean average low temperature for the five-day period?

24. What is the mean average high temperature for the five-day period?

E. **In Problems 25 and 26, use Chart 4.**

25. Write a proportion, and use it to find the number of miles traveled in three hours.

26. Write a proportion, and use it to find the number of miles traveled in seven hours.

CHART 4   Distance at 48 Miles per Hour

Hours	1	2	4	5	8
Miles	48	96	192	240	384

6.7 ◆ Chapter Review

Hopefully you found ratios and proportions to be interesting. Certainly, you have seen their usefulness. Upon completion of this chapter, you should understand all the vocabulary involved and should be able to use ratios and proportions successfully. Keep in mind that

1. A *ratio* is the comparison of two like numbers.
2. A *rate* is the comparison of two unlike numbers.
3. A *unit rate* has a denominator of 1.
4. A *proportion* is a statement that two ratios or rates are equal to each other.
5. The *cross products* of a proportion are always equal to each other.
6. To prove that a statement is a proportion, show that the cross products are equal.
7. To *solve a proportion* for an unknown quantity, cross-multiply and then divide by the number that is with the unknown.
8. To *check* the solution to a proportion, substitute the answer into the proportion and check the cross products.
9. For *word problems*, choose a letter for the unknown, then write two ratios or rates in the same order, and solve the proportion.
10. To *convert measurement quantities* to other units, multiply by the unit fraction that will enable you to cancel the unwanted units.

Review Problems

A. Write each ratio in lowest terms.

1. 14 to 63
2. 16 to 56
3. 28 to 8
4. 192:72
5. 6.4 to 0.12
6. 3/2:7/8
7. 2/5 to 9/15
8. 8 1/2:2/3
9. 0.8 to 7.2
10. 0.72 to 1.8
11. 6/5 to 3/8
12. 4 1/2 to 2
13. 6 minutes to 50 seconds
14. 11 feet to 29 inches
15. 20 cents to 5 dimes
16. 6 pints to 4 cups
17. 2 pounds to 14 ounces
18. 17 hours:90 minutes
19. On a business trip, a sales representative traveled 120 miles by car and 460 miles by plane. Write the following ratios:
    a. Car mileage to plane mileage
    b. Car mileage to total mileage
    c. Plane mileage to total mileage
20. An appliance factory manufactures 120 washers and 200 dryers each day. Write the following rates:
    a. Number of washers to number of dryers
    b. Number of washers to the total number of both
    c. Number of dryers to the total number of both

B. Find the unit price for each of the following. If necessary, round your answers to the nearest thousandth.

21. 6 apples for $0.78
22. $1.56 for 22 ounces
23. 5 cans for $4.90
24. 3 cans for 89 cents
25. 3 pounds for $1.29
26. $1.78 for 18 ounces
27. 3 shirts for $49.77
28. 4.2 pounds for $4.70
29. 6 donuts for $2.20
30. twelve for $2.28

C. Find out whether each of the following is a proportion. Prove each answer by showing your cross products.

31. $\dfrac{4}{5} = \dfrac{28}{35}$
32. $\dfrac{63}{72} = \dfrac{7}{8}$
33. $\dfrac{78}{16} = \dfrac{9}{2}$
34. $\dfrac{3}{2} = \dfrac{6.72}{4.48}$
35. $\dfrac{1.2}{0.36} = \dfrac{0.4}{0.12}$
36. $\dfrac{4/9}{3/2} = \dfrac{1/9}{3/8}$

37. $\dfrac{110}{212} = \dfrac{3}{8}$ 38. $\dfrac{39}{48} = \dfrac{15}{16}$

39. $\dfrac{3/4}{5/8} = \dfrac{36}{25}$ 40. $\dfrac{6.8}{9.5} = \dfrac{0.6}{1.2}$

D.   Solve for the unknown in each proportion. If necessary, round your answers to the nearest hundredth.

41. $\dfrac{15}{40} = \dfrac{C}{8}$ 42. $\dfrac{2}{3} = \dfrac{12}{A}$

43. $\dfrac{12}{27} = \dfrac{M}{6}$ 44. $\dfrac{15}{P} = \dfrac{8}{21}$

45. $\dfrac{C}{5} = \dfrac{36}{25}$ 46. $\dfrac{12}{39} = \dfrac{15}{R}$

47. $\dfrac{4}{F} = \dfrac{2.4}{0.28}$ 48. $\dfrac{19}{57} = \dfrac{N}{9}$

49. $\dfrac{14}{96} = \dfrac{7}{K}$ 50. $\dfrac{7.8}{5.6} = \dfrac{2.3}{M}$

51. $\dfrac{4/5}{3/8} = \dfrac{T}{1/2}$ 52. $\dfrac{1\,2/3}{2\,1/4} = \dfrac{3/4}{H}$

E.   Solve each word problem by using a proportion. If necessary, round to the nearest hundredth.

53. If six pens cost $1.98, how much would ten cost?

54. An airplane uses 20 gallons of fuel to fly 190 miles. How many miles could it fly on 25 gallons?

55. A map scale says that 1 1/2 inches represent 30 miles. How many miles do 7 inches represent?

56. If six dozen rolls cost $2.88, how much would 10 dozen cost?

57. The sales tax on a $360 television is $27. What would be the sales tax on the $98 table purchased at the same time?

58. A 6 1/2-pound roast costs $14.30. How much would an 8 1/4-pound roast cost?

59. At Mike's last dorm party, eight students ate six pizzas. Only 5 students will be present this time. How many pizzas will he need?

60. If a 12-ounce can of soda has 160 calories, how many calories will be in eight ounces of the beverage?

61. On a map, 3 inches represent 175 miles. How far apart on the map are two cities that are really 220 miles apart?

62. If a 6-foot man casts a 4-foot shadow, how tall is a tree that casts a 15-foot shadow?

F.   Convert these measurements as indicated. You may need to refer to Table 1 on page 144 and Table 2 on page 155. If necessary, round answers to the nearest hundredth.

63. 16.25 inches to feet

64. 8 centimeters to inches

65. 5 hours to seconds

66. 12 cups to pints

67. 4 gallons to cups

68. 15 miles per hour to feet per second

69. 22 feet per second to miles per hour

70. 120 inches to meters

71. 5 feet to centimeters

72. 55 miles per hour to feet per second

G.   Use the data represented in Figure 5 to do Problems 73 through 80.

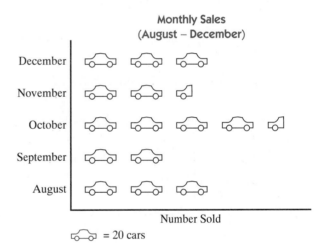

FIGURE 5

73. What is the ratio of August sales to October sales?

74. What is the ratio of September sales to August sales?

75. What is the mean of monthly sales for the period?

76. What is the median number of cars sold?

77. What is the mode of the sales for the period?

78. What is the ratio of October sales to the total sales for the five-month period?

79. If the scale showed that each car pictured represented 10 cars sold, would the ratio you found in Problem 73 change? Explain your answer.

80. If the scale showed that each car pictured represented 30 cars sold, would the ratio in Problem 74 change? Explain your answer.

Chapter Test

A. **In Problems 1 through 5, write a ratio or rate in lowest terms.**

1. 28 to 63   **2.** 7.2 to 0.12

3. 2/3 : 5/9   **4.** 2 dimes to 3 quarters

5. 5 feet to 17 inches

6. Find the unit price if 12 lemons cost $2.46.

7. Which is the better buy, a 12-ounce jar of coffee for $3.29 or an 8-ounce jar for $2.20? Tell which is the better buy, and state its unit price.

8. Change 4,680 seconds to hours using the conversion factors 1 hour = 60 minutes and 1 minute = 60 seconds.

9. Change 10 meters to inches using the conversion factor 1 meter = 39.37 inches.

10. Is $\dfrac{21}{35} = \dfrac{3}{5}$ a proportion? Why or why not?

11. Is $\dfrac{3.2}{0.12} = \dfrac{0.12}{0.04}$ a proportion? Why or why not?

B. **In Problems 12 through 16, solve for the unknown. If necessary, round answers to the nearest hundredth.**

12. $\dfrac{5}{9} = \dfrac{35}{W}$ 13. $\dfrac{12}{C} = \dfrac{15}{32}$

14. $\dfrac{63}{X} = \dfrac{7}{8}$ 15. $\dfrac{A}{14.5} = \dfrac{1.2}{0.5}$

16. $\dfrac{3/2}{3/4} = \dfrac{T}{5/8}$

17. If two pizzas cost $12.50, how much do 5 pizzas cost?

18. On the scale of a map, 2 inches represent 85 miles. How far apart are two cities that are 4 1/2 inches apart on the map?

19. If a 5-foot woman casts a 3-foot shadow, how tall is a flagpole that casts a 15-foot shadow?

20. If a 4-pound roast costs $9.36, what does a 2 1/2-pound roast cost?

Thought-Provoking Problems

1. Describe the similarities and the differences between a ratio and a rate.

2. Use your calculator to determine the unit price of each item. Round answers to the nearest thousandth.

   a. 39 ounces of coffee for $4.78

   b. 10 3/4-ounce can of soup for $0.89

   c. 12.9 gallons for $14.50

   d. Ten dozen rolls for $9.12 (Find cost per roll.)

3. The proportions in this chapter have been *direct proportions*. That is, as one quantity got larger, its corresponding value also got larger. For instance, if you pay $2.16 sales tax on an item costing $36, then you would expect to pay more than $2.16 on an item costing $50. Some items, however, are related inversely rather than directly. If 3 men can frame a house in 4 days, would 5 men do that job in more time or in less time than the 3-man crew? Write an *inverse* proportion for this problem, and solve it.

4. Inverse proportionality is described in Problem 3. Cite two or three pairs of quantities that are inversely related.

5. In the definition of a proportion (page 147), why is the statement "where $b, d \neq 0$" included?

6. The property of proportions that says cross products are equal actually comes from the use of a common denominator to multiply each ratio in the proportion.

$$\frac{a}{b} = \frac{c}{d} \quad \text{The LCD for the ratios is } bd.$$

Multiply both sides of the proportion by bd and see if you get the cross products ($ad = bc$). Show your work. *Hint*: Try the process with a number proportion, such as $3/4 = 9/12$, and then repeat the steps with the letters in place of the numbers.

7. Use your calculator to solve these proportions. You will first need to change the fractions to decimals. Round decimal answers to the nearest hundredth.

   a. $\dfrac{3.37}{14.6} = \dfrac{m}{21.56}$ b. $\dfrac{w}{0.124} = \dfrac{19.6}{5.44}$

   c. $\dfrac{1/6}{F} = \dfrac{7/8}{3/4}$ d. $\dfrac{3.2}{5/32} = \dfrac{3/8}{Y}$

Percent

We've all had experience with percents. Everyone who has taken a test understands that a 95 (95% correct) is a very good score. If your employer gives you an 8% raise, you are pleased that your good work has been recognized. These amounts, 95% and 8%, seem to be very different when you look at them as raw numbers. What makes them meaningful is the context in which they are discussed. Eight percent is a nice raise but a terrible grade on a test. Ninety-five percent is a grade to be proud of, but a 95% raise in salary is ridiculous—really nice, but unrealistic! In spite of the seemingly large difference between them, both 8% and 95% are very positive quantities in the situations we have mentioned. The point of all this is to help you realize that percents have no meaning all by themselves.

Percent problems involve practical situations, such as sales tax, commission, percent increase or decrease, discount, and interest. We will solve these problems using either the percent formula, $A = RB$, or the percent proportion, $A/B = P/100$. Both approaches are algebraic in nature, but don't let that scare you. Most students are very successful with these problems once they learn to use the methods shown.

When you have competed this chapter, you should be able to:

1. Understand the meaning of percent.
2. Interchange quantities among fraction form, decimal form, and percent form.
3. Identify A, R, and B in a statement given in standard form.
4. Use the percent formula, $A = RB$, to find the value of A, R, and B.
5. Solve a percent problem using the percent proportion, $A/B = P/100$.
6. Translate problems about practical situations into statements in standard form.
7. Find the solutions to word problems involving percents.
8. Solve problems using the simple interest formula.
9. Interpret data from graphs.

Learning Objectives

7.2 ◆ The Meaning of Percent

The word *percent* comes from two Latin words, *per* and *centum*, which mean "by (or through) 100." The word *cent* comes from that same Latin word: a penny is a cent, one-hundredth of a dollar. If you can remember that **percent** means "by 100," much of what you will do in this chapter will be very easy and will make good, logical sense.

If there were 100 students on the roster in your math class, and 87% of them were in class today, that would mean 87 students were present out of the 100 students on the roster. If the sales tax rate in your city is 6%, that means you pay 6 cents in tax on a purchase that costs $1.00, and $6 in tax on a purchase that costs $100.

Suppose there were 100 students taking economics this term, and 17% of those students were doing "A" work. How many students would be earning an A? How many students would be doing less than A work? In order to answer these questions, you need to remember that percent means "per 100." If 17% of the 100 students are earning an A for the course, then 17 out of the 100 students will receive an A. If you need to figure out how many students are earning less than an A, you can approach the problem two different ways.

Method 1. If 17% are getting an A, then the rest are not. 100% represents the whole class. 100% − 17% = 83%, which represents the part of the class *not* earning an A. 83% of 100 students is 83 students.

Method 2. If 17% of a class of 100 students is earning an A, then 17 students are earning a grade of A. If 17 students are earning an A, then the rest are not. In a class of 100 students, 100 − 17 = 83 students are not earning an A.

It doesn't matter whether you begin with 100% and subtract the 17%, or figure the number of students first and subtract that number from the total of 100 students. What is important is to realize that you need to subtract *17%* from *100%*, or *17 students* from *100 students*. (Recall from previous chapters that you can add or subtract only the same kind of quantities.)

Because not all surveys ask 100 people to respond, and not all tests have exactly 100 questions, it is necessary to develop methods to find the number of people asked, or the percent of problems correct, in situations where the total quantity is not 100. That will be coming up soon.

7.2 ◆ Practice Problems

Answer each of the following questions based on percents.

1. 100 people voted for their favorite flavor at Richard's Ice Cream Store. Of those 100, 62% voted for chocolate, 21% voted for strawberry, and 12% voted for peanut butter cup.

   a. How many people voted for chocolate?

   b. How many people voted for peanut butter cup?

   c. How many people voted for a flavor other than chocolate, strawberry, or peanut butter cup?

2. One hundred people were stopped on the street downtown and asked to name the mayor of the city. Of those 100, 43% gave the correct name, 29% gave the wrong name, and the rest admitted they didn't know.

   a. How many people knew the mayor's name?

   b. How many people gave the wrong name?

   c. How many people said they didn't know?

3. Loleetha was pleased to have earned an 89% on her vocabulary quiz. There were 100 questions on the quiz.

   a. How many questions did she answer correctly?

   b. How many did she answer incorrectly?

   c. What percent did she get wrong?

4. As part of a usage study, Sidney surveyed the first 100 students who walked into the library on a Wednesday morning. Eighteen of them were returning books, 36 were checking out books, and 38 were there to study. The rest of the students questioned were there for other reasons.

   a. What percent were returning books?

   b. What percent were there to study?

   c. What percent were there for activities other than returning a book, checking out a book, or studying?

5. If 72.6% of the people in a recent election voted in favor of the mass transportation levy, what percent voted against it?

6. When asked about a smoking ban for the school cafeteria, 63.8% of the students voted to have a "no smoking" policy in the cafeteria. What percent of the students thought smoking in the cafeteria should be allowed?

7. Marcus got a 92% on his last math exam. What percent of the questions did he miss?

8. Rose wants to make at least an 89% on her next French translation test to ensure that she maintains her "B" average. How many items can she afford to miss on this 100-question test?

9. When the state police set up a safety check on a well-traveled highway, they discovered that 18 of the first 100 cars stopped had one headlight that was not working. What percent of the cars stopped had a defective headlight?

10. The selling price of an item is a combination of the dealer's cost and the dealer's markup. (Markup covers profits, salaries, utilities, rent, insurance, and so on.) If the dealer's cost is 72% of the selling price, what percent represents the markup?

7.3 Converting Among Fractions, Decimals, and Percents

Your ability to do percent problems will be much greater if you first learn how to change quantities from fractions and decimals to percents, and from percents to fractions and decimals. This process seems at first to have no real significance. In a sense, it is just a matter of following the rules, playing the game, and not worrying about why you would want to do it. In the next section, however, you will do problems in which you find the interest earned on money saved in a bank account. In order to do that kind of problem, you need to use the form of the percent that makes the calculations possible.

In fact, this ability to change percents to their fraction or decimal equivalents is crucial in many situations. Although you can add or subtract percents, as you did in the previous section, you cannot multiply or divide by any percents except 100%, which is a form of 1. Because you need to be able to use a fraction or a decimal equivalent in most problems involving percents, you will divide by 100%, as a form of 1, to change a percent number to a usable fraction or decimal form. You will also multiply by 100%, as a form of 1, to change from fractions or decimal numbers to their percent form when you are asked for a percent.

You already have a basic understanding of some of the relationships involving percents and fractions. If you score 100% on a test, you got one whole test completely right (100% = 1). If you were to get only 50% on a test, you would have done only one-half of the problems correctly (50% = 1/2). Getting 75% on a test means that you did 3/4 of the problems correctly (75% = 3/4). Students usually recognize these relationships from their previous experiences. Some of the other relationships or equivalences are not so common. You need to learn methods for changing from one form to another.

Changing Decimals to Percents   Let's first consider changing from a decimal to a percent.

Decimals to Percents

> To change from a decimal to a percent, multiply the decimal quantity by 100%.

Consider the following examples to see how to change from decimal form to percent form. Remember that *percent* means "by 100" and that 100% is another name for 1. When you multiply by 100%, you are multiplying by 1. This operation does not change the value of the quantity you are working with; it only changes the way the quantity looks. You used the same kind of process when you changed 3/8 to sixteenths. You multiplied 3/8 by 2/2 to get 6/16. You multiplied by the form of 1 that would produce the equivalent fraction you needed (Section 3.3).

E X A M P L E   **1**   Change 0.45 to a percent.

Solution   When multiplying a decimal by 100, move the decimal point two places to the right (Section 5.4).

$$0.45 \times 100\% = 0.45.\% = 45\%$$ ◆

E X A M P L E   **2**   Change 0.384 to a percent.

Solution $$0.384 \times 100\% = 0.38.4\% = 38.4\%$$ ◆

E X A M P L E   **3**   Change 1.85 to a percent.

Solution $$1.85 \times 100\% = 1.85.\% = 185\%$$

This answer should not seem too large. Because 1.85 is more than 1, the equivalent percent should be more than 100%.   ◆

E X A M P L E   **4**   Change 0.006 to a percent.

Solution $$0.006 \times 100\% = 0.00.6\% = 0.6\%$$

0.006 is less than one-hundredth (0.01), so 0.006 in percent form should be less than 1%.   ◆

Changing Fractions to Percents

In order to change fractions to percents, you might change the fraction to its decimal form first and then change that decimal to a percent. In Section 5.6, you learned that to change a fraction to its decimal form, we divide the numerator by the denominator. Use that skill first, and then multiply the decimal answer by 100%.

Fractions to Percents

> To change from a fraction to a percent, first change the fraction to its decimal form and then multiply the decimal by 100%.

E X A M P L E   **5**   Change 3/4 to a percent.

Solution $3/4 = 3 \div 4 = 0.75$ Change 3/4 to a decimal.

$0.75 \times 100\% = 0.75.\% = 75\%$ Change the decimal to a percent.

Therefore, $3/4 = 75\%$.   ◆

E X A M P L E **6**   Change 5/8 to a percent.

Solution $5 \div 8 = 0.625$

$0.625 \times 100\% = 0.62.5\% = 62.5\%$

Therefore, $5/8 = 62.5\%$. ◆

E X A M P L E **7**   Change 7/9 to a percent, rounded to the nearest tenth of a percent.

Solution   As you well know, not all division comes out even, and some division answers need to be rounded. In the changing of *fractions to percents* in this book, the division answer is *carried out to four places, multiplied by 100%*, and then *rounded* to the nearest *tenth of a percent*. This is accurate enough for most situations.

$7 \div 9 \doteq 0.7777$

$0.7777 \times 100\% = 0.77.77\%$

$77.77\% = 77.8\%$, rounded to the nearest tenth of a percent

Therefore, $7/9 = 77.8\%$. (This is read "seventy-seven and eight-tenths percent.") ◆

E X A M P L E **8**   Change 1/12 to a percent, rounded to the nearest *tenth of a percent*.

Solution $1 \div 12 \doteq 0.0833$

$0.0833 \times 100\% = 0.08.33\% = 8.3\%$, rounded to the nearest tenth of a percent

Therefore, $1/12 = 8.3\%$. ◆

E X A M P L E **9**   Change 1 3/5 to a percent.

Solution   First change 1 3/5 to an improper fraction, 8/5. Then change 8/5 to a decimal, $8/5 = 1.6$.

$1.6 \times 100\% = 1.60.\% = 160\%$

Therefore, $1\ 3/5 = 160\%$. This answer makes sense; because 1 3/5 is more than 1, its percent form should be more than 100%. ◆

When doing Example 9, you could also change only the fraction form, 3/5, to its decimal equivalent, 0.6, and then rejoin it with the whole number, 1.

$1\ 3/5 = 1 + 3/5 = 1 + 0.6 = 1.6$

Then change the decimal form to its percent equivalent.

$1.6 \times 100\% = 160\%$

Try these exercises and see if your answers match the ones given below. If they don't, look back over the examples carefully and see if you can find your mistake. If you are still having trouble, be sure to ask your instructor for help.

Exercises   Change the following decimals and fractions to their percent form. When necessary, round your answer to the nearest tenth of a percent.

1. 0.26   **2.** 0.789   **3.** 3.28   **4.** 0.0376   **5.** 4/5   **6.** 1/9
7. 2 3/8   **8.** 1/15

Answers:

1. 26%   **2.** 78.9%   **3.** 328%   **4.** 3.8%   **5.** 80%
6. 11.1%   **7.** 237.5%   **8.** 6.7%

Changing from Percents to Decimals and Fractions

Changing from percent form to decimal or fraction form is the opposite of what you have just done. Because changing to a percent involves multiplying by 100%, it makes sense that *to change from a percent* involves *dividing by 100%*. Remember that to divide a number by 100, you move the decimal point two places to the left (Section 5.5).

Consider the following examples to see how this can be done.

E X A M P L E   **10**

Change 64% to its decimal form.

Solution

The decimal point in 64% currently follows the 4. Move it two places to the left and cross out the percent sign.

$$64\% \div 100\% = .64.\% \div 100\% = 0.64$$

Therefore, 64% = 0.64.   ◆

E X A M P L E   **11**

Change 0.7% to its decimal form.

Solution

$$0.7\% \div 100\% = 0.00.7\% \div 100\% = 0.007$$

Therefore, 0.7% = 0.007.   ◆

E X A M P L E   **12**

Change 258% to its decimal form.

Solution

$$258\% \div 100\% = 2.58.\% \div 100\% = 2.58$$

Therefore, 258% = 2.58. Because 258% is more than 100%, the decimal answer should be more than 1.   ◆

To change a percent to a fraction, you could go through the decimal form, because you already know how to go from a decimal to an equivalent fraction. However, that is not the only way to do the problem. Consider these examples.

E X A M P L E   **13**

Change 48% to a fraction.

Solution

$$48\% \div 100\% = 0.48.\% \div 100\% = 0.48$$

0.48 is read "forty-eight hundredths," so it is written 48/100 in fraction form. You need to reduce 48/100 (by fours) to 12/25.
Therefore, 48% = 48/100 = 12/25.   ◆

E X A M P L E   **14**

Change 0.8% to a fraction.

Solution

This problem has several steps. First change 0.8% from a decimal percent to a decimal number. Then change the decimal to its reduced fraction form.

$$0.8\% \div 100\% = 0.00.8\% \div 100\% = 0.008$$

0.008 is read "eight thousandths," so it is written 8/1,000. Reduce 8/1000 (by eights) to 1/125.
Therefore, 0.8% = 1/125.   ◆

E X A M P L E   **15**

Change 145% to a fraction.

Solution

$$145\% \div 100\% = 1.45.\% \div 100\% = 1.45$$

1.45 is read "one and forty-five hundredths," so it is written 1 45/100. The last step is to reduce 45/100 (by fives) to 9/20.

Therefore, 145% = 1 9/20.   ◆

E X A M P L E 16   Change 8 1/4% to a fraction.

Solution   This problem also has several steps. First you must change 8 1/4 to 8.25. Therefore, 8 1/4% = 8.25%.

8.25% ÷ 100% = 0.08.25% ÷ 100% = 0.0825

0.0825 is read "eight hundred twenty-five ten-thousandths," so it is written 825/10,000. Reduce 825/10,000 (by twenty-fives) to 33/400. Therefore, 8 1/4% = 33/400. This answer makes sense: 8 1/4% seems like a small quantity, and 33/400 seems equally small.
◆

Try these exercises and check your answers. Make sure that you are comfortable changing percents to their fraction or decimal forms.

Exercises   Change each percent to its decimal or fraction form as indicated. Round all decimal answers to the nearest thousandth, and be sure to reduce all fraction answers.

1.   Change 84% to a decimal.

2.   Change 84% to a fraction.

3.   Change 275% to a decimal.

4.   Change 275% to a fraction.

5.   Change 0.4% to a decimal.

6.   Change 12.5% to a fraction.

7.   Change 12 1/2% to a decimal.

8.   Change 8 3/4% to a fraction.

Answers:

1.   0.84   **2.**   21/25   **3.**   2.75   **4.**   2 3/4 or 11/4   **5.**   0.004

6.   1/8   **7.**   0.125   **8.**   7/80

Sometimes when you are doing problems like these, it is easy to get lost in the steps. Just keep these rules in mind:

Rules for Changing to and from Percents, Decimals, and Fractions

1.   To change *from decimal form to the percent form, multiply by 100%*.

2.   To change *from the percent form to decimal form, divide by 100%*.

3.   If starting with a *fraction, change to its decimal* form first, and then *multiply by 100%* to get the percent form.

4.   To change a *percent to a fraction, change to the decimal form first*, and then *read, write, and reduce*.

5.   Always check rounding instructions.

6.   Always reduce fraction answers to their lowest terms.

7.3 ◆ Practice Problems

A. Change each of the following fractions or decimals to its equivalent percent form. When necessary, round your answers to the nearest tenth of a percent.

1. 0.32	2. 0.43	3. 0.698
4. 0.821	5. 3.47	6. 1.56
7. 0.4639	8. 0.2155	9. 12
10. 6	11. 3/5	12. 4/8
13. 5/7	14. 6/11	15. 9/4
16. 12/5	17. 3 1/2	18. 1 4/5
19. 1/16	20. 3/17	

C. Change each of the following percents to its equivalent fraction form. Remember always to reduce fractions to their lowest terms.

33. 43%	34. 19%	35. 52%
36. 24%	37. 800%	38. 650%
39. 0.8%	40. 0.12%	

B. Change each of the following percents to its equivalent decimal form. When necessary, round your answer to the nearest thousandth.

21. 12%	22. 37%	23. 126%
24. 214%	25. 0.9%	26. 0.26%
27. 800%	28. 200%	29. 10 1/2%
30. 9 1/4%	31. 25 1/4%	32. 2 1/2%

D. Complete the following table. Reduce all fraction answers to lowest terms. Round each decimal answer to three places. Round each percent answer to the nearest tenth of a percent.

	Fraction	Decimal	Percent
41.		0.286	
42.			85%
43.	4/5		
44.	5/9		
45.			2 1/2%
46.		1.4562	
47.			3/4%
48.		0.375	
49.	7/12		
50.			17.8%

7.4 ◆ Percent Formula

When you say "16 is one-half of 32" or "16 is 50% of 32," you are making a statement in what is called standard form. The 16 is a part of the 32. The 32 is the whole, original quantity that you started with. "One-half," or "50%," tells just how large a part of 32 the quantity 16 is. When you say that the amount of sales tax you pay is $5.00 on an item that costs $100.00 in a city where the sales tax is 5%, you could also be making a statement in standard form. You could restate it as "$5.00 is 5% of $100.00." $5.00 is the part of the bill that you need to pay in sales tax based on the purchase price, or whole amount, of $100.00. The 5% is the *rate* of sales tax that is charged in your area.

When you talk about the relationship between two quantities in terms of "what percent one is of the other," you can restate that relationship in standard form. **Standard form** says that *the amount is a percent of the base.* You will use the percent formula to make this statement shorter and easier to work with. The **percent formula** says that *the amount (or part), A, is equal to the rate, R, times the base, B.* You can shorten all of that further and say $A = RB$. You need to remember that in formulas, two letters next to each other tell you to multiply those values. To work these formulas, you need to determine which quantities in a problem are the amount (A), the rate (R), and the base (B).

The rate, R, is always the easiest to identify. It is the quantity with the percent sign, or it signifies the relationship between the two quantities. In "16 is one-half of 32," one-half is the rate. Another name for one-half, 50%, is the rate in "16 is 50% of 32." The rate in the sales tax example was 5%.

The base, B, is the next easiest to identify. It very often follows the word "*of*" and stands for the original quantity. In the first two sample problems given in this section, 32 is the base; in the third example, $100.00 is the base because it is the amount before the tax was computed.

The amount or part, A, is the quantity that is yet to be identified. It represents the amount that is part of the base. In the first two examples, 16 is part of 32, so 16 is A. In the third example, $5.00 is the part that is paid in sales tax, based on the $100.00 purchase price. Therefore, $5.00 is A in the third example.

When you change problems to statements in standard form, pick out the value for R, for B, and then for A. To summarize from the earlier examples:

In "16 is one-half of 32," $R = 1/2$, $B = 32$, and $A = 16$.

In "16 is 50% of 32," $R = 50\%$, $B = 32$, and $A = 16$.

In "$5.00 is 5% of $100.00," $R = 5\%$, $B = \$100$, and $A = \$5$.

Consider each of the following examples. The statement is first put into standard form; next, R, B, and A are identified; and then the quantities are substituted into the percent formula.

E X A M P L E **1**   4 is 2% of 200.

Solution   Already in standard form.

2% is the rate, R.

200 follows "of" and is the base, B.

4 is the part or amount, A.

Therefore, substituting the values into $A = RB$ yields

$4 = 2\% \times 200$.

Check   Use 2% in its decimal form, 0.02, and multiply it by 200: $0.02 \times 200 = 4.00$, or 4.

4 *is* 2% of 200.   ◆

E X A M P L E **2**   70% of 20 is 14.

Solution   Stated in standard form: 14 is 70% of 20.

70% is the rate, R.

20 follows "of" and is the base, *B*.

14 is the amount or part, *A*.

Therefore, *A* = *RB*, or 14 = 70% × 20.

Check   Change 70% to its decimal form, 0.70, and multiply it by 20: 0.70 × 20 = 14.00, or 14. ◆

70% of 20 *is* 14.

Finding A   The question "What number is 3.5% of 63?" could be looked at as a question in standard form. Because "is" translates to =, and "of" to ×, you could rewrite the question as "What number = 3.5% × 63?" If you were to use the percent formula, *A* = *RB*, *R* would be 3.5%, *B* would be 63, and you would be looking for *A*. Consider how this problem could be worked out using the percent formula.

A = *RB*	Write the formula.
A = 3.5% × 63	Substitute 3.5% for *R* and 63 for *B*.
A = 0.035 × 63	Change 3.5% to decimal form, 0.035, and then multiply.
A = 2.205	

Therefore, 2.205 is 3.5% of 63.

Consider the following examples. You should first state the question in standard form and then identify the *R*, *B*, and *A* values. Substitute these into the formula, leaving the letter in place of the quantity you are trying to find. Finally, follow the steps for using a formula.

E X A M P L E **3**   What number is 42% of 87?

Solution   Already in standard form.

R is 42%. 87 follows "of" and is the value for *B*. *A* must be the value you are to find.

A = *RB*	Write the formula.
A = 42% × 87	Substitute the known values.
A = 0.42 × 87	Change 42% to its decimal form.
A = 36.54	Perform the indicated operation. In this case, multiply.

Therefore, 36.54 is 42% of 87. ◆

E X A M P L E **4**   8 1/2% of 56 is what number?

Solution   Written in standard form: What number is 8 1/2% of 56?

R is 8 1/2%, or 8.5% when you change 1/2 to its decimal form. 56 follows "of" and is the value for *B*. *A* must be the value you are looking for.

A = *RB*
A = 8.5% × 56
A = 0.085 × 56
A = 4.760

Therefore, 4.76 is 8 1/2% of 56. ◆

When you are using the percent formula, finding the value for *A* is the easiest procedure to do. Now let's use the percent formula to find the base, *B*, and the rate, *R*.

Finding *B* Suppose the question is "35 is 15% of what number?" In this case, you are given the rate, 15%, but the quantity following "of" is the one that is missing. In this type of problem, you are to find the value for *B*—the base—that will satisfy the question. You begin just as you did when finding the value for *A*. You follow the four steps to be used when solving a problem by applying a formula. To do that, remember to state the problem in standard form first and then pick out the values for *R*, *B*, and *A*. Study the following problem to see how this can be done.

30 is 15% of what number? This is already in standard form.

$R = 15\%$.

B is the quantity to be found, because there is no quantity that follows "of."

A must then be 30.

Write the formula: $A = RB$

Substitute the given values: $30 = 15\% \times B$

Change 15% to its decimal form: $30 = 0.15 \times B$

This is where you run into a problem. Before, to find the value for *A*, you had two quantities to multiply. This time the missing value, *B*, is not alone. It is multiplied by 0.15. To get the *B* value by itself, you need to divide by 0.15 to "undo" the multiplication. You used this same process in Chapter 6 when you solved proportions. When working with an equation (an equality statement), it is necessary to do the same operation to both sides of the equals sign. This keeps the equation in balance. Here, then, you need to divide 30, as well as $0.15 \times B$, by 0.15. This is how the steps look:

$$30 = 0.15 \times B$$

$$\frac{30}{0.15} = \frac{0.\overset{1}{\cancel{15}} \times B}{\underset{1}{\cancel{0.15}}} \qquad\qquad 0.15)\overline{30.00.}^{\;2\,00.}$$

$$200 = B$$

Therefore, 30 is 15% of 200. To check the answer, multiply 0.15 by 200, which is indeed equal to 30.00.

The following examples show how to find the value for *B* when you use the formula. Always follow the same steps and you won't get lost.

E X A M P L E 5 40% of what number is 25?

Solution Write the problem in standard form.

25 is 40% of what number?

R is 40%. *B* is the value you are to find, because no quantity follows "of." *A* is 25.

$A = RB$	Write the formula.
$25 = 40\% \times B$	Substitute the values.
$25 = 0.40 \times B$	Change percent to a decimal.

$$\frac{25}{0.40} = \frac{0.\overset{1}{\cancel{40}} \times B}{\underset{1}{\cancel{0.40}}} \qquad \text{Divide both sides by 0.40 to "undo" the multiplication and isolate } B.\ 25 \div 0.40 = 62.5$$

$$62.5 = B$$

Therefore, 25 is 40% of 62.5.

Check Multiply 0.40 by 62.5, which is equal to 25.00. ◆

E X A M P L E **6**   22.42 is 23.6% of what number?

Solution   The problem is already in standard form.

R is 23.6%. You must be looking for B, because no quantity follows the word "of." Therefore, 22.42 is A.

$$A = RB$$

$$22.42 = 23.6\% \times B$$

$$22.42 = 0.236 \times B$$

$$\frac{22.42}{0.236} = \frac{0.236 \times B}{0.236}$$

$$B = 95$$

Therefore, 22.42 is 23.6% of 95.

Check   Multiply 0.236×95, which is equal to 22.42.   ◆

Finding R   When the percent, or rate, is the quantity you are asked to find, you follow a process very similar to the one for finding B. After you have written the problem in standard form, follow the four steps for solving a problem by using a formula:

1. Write the formula.

2. Substitute the values you know.

3. Perform the indicated calculations.

4. Check the answer in the original problem.

In this type of problem, you will not have a percent value. This is the value you are looking for. Remember that once you have found the decimal answer, you *must* write it in percent form. Consider the following problems to see how this will be done.

"25 is what percent of 200?" The statement is already in standard form.

Because there is no percent given, you are looking for R. 200, which follows "of," is B. A must therefore be 25.

$$A = RB$$

$$25 = R \times 200$$

You are now at the same place where you were when looking for the base, B. The letter B was not alone. To isolate it, you divided both sides of the equality by the number next to B to "undo" the multiplication. This time you will divide both sides by 200 to isolate the R.

$$\frac{25}{200} = \frac{R \times 200}{200} \qquad \text{200 over 200 cancels. } 25 \div 200 = 0.125$$

$$0.125 = R$$

Note that you found a decimal answer, but the question asked for a percent. You know from Section 7.3 that you need to multiply a decimal form by 100% to change it to a percent.

$$R = 0.12.5 \times 100\% = 12.5\%$$

Therefore, 25 is 12.5% of 200.

Check the answer by multiplying 12.5% by 200:

$$0.125 \times 200 = 25.000 = 25$$

     All problems that involve finding the rate, R, are done in this same manner. They are manageable when you use the percent formula and always follow the same steps. A formula is a real help. It shows what operation to perform with the numbers you are given in a problem. Consider each of the following examples and see how these problems are done.

E X A M P L E 7 14 is what percent of 700?

Solution The problem is already in standard form. There is no percent given, so you are looking for R. 700, which follows "of," is B. 14 must then be A.

$A = RB$	Write the percent formula.
$14 = R \times 700$	Substitute the known values.
$\dfrac{14}{700} = \dfrac{R \times \overset{1}{\cancel{700}}}{\underset{1}{\cancel{700}}}$	To isolate R, you must divide by 700. $14 \div 700 = 0.02$
$0.02 = R$	
$R = 0.02 \times 100\%$	Once you have found the decimal answer, it is necessary
$R = 2\%$	to change it to a percent, so you multiply by 100%.

Therefore, 14 is 2% of 700.

Check Multiply 2% by 700: $0.02 \times 700 = 14.00 = 14$ ◆

E X A M P L E 8 What percent of 36 is 90?

Solution 90 is what percent of 36?   Write the question in standard form.

There is no percent given, so you are looking for R. $B = 36$ because 36 follows "of." Therefore, A must be equal to 90.

$A = RB$	Write the percent formula.
$90 = R \times 36$	Substitute the given values.
$\dfrac{90}{36} = \dfrac{R \times \overset{1}{\cancel{36}}}{\underset{1}{\cancel{36}}}$	Divide both sides of the equation by 36 to "undo" the multiplication. 36 over 36 cancels.
$2.5 = R$	$90 \div 36 = 2.5$
$2.5 \times 100\% = 250\%$	To change from decimal to percent form, multiply by 100%.

Therefore, 90 is 250% of 36.

Check Multiply 250% by 36: $2.5 \times 36 = 90.0$

Note that when the amount is greater than the base, the percent is greater than 100%. ◆

E X A M P L E 9 12 is what percent of 33? Round your answer to the nearest tenth of a percent.

Solution 12 is what percent of 33?   Already in standard form.

There is no percent value given, so you are looking for R. B is 33 because 33 follows "of." A must therefore be 12.

$A = RB$	Write the percent formula.
$12 = R \times 33$	Substitute the given values.
$\dfrac{12}{33} = \dfrac{R \times \overset{1}{\cancel{33}}}{\underset{1}{\cancel{33}}}$	Divide both sides of the equation by 33 to "undo" the multiplication. 33 over 33 cancels. $12 \div 33 = 0.3636$
$0.3636 = R$	

$$0.3636 \times 100\% = 36.36\%$$ Change from decimal form to percent form, and then round to the nearest tenth of a percent.

$$R = 36.4\%$$

Therefore, 12 is 36.4% of 33.

Check   Multiply 36.4% $\times$ 33: $0.364 \times 33 = 12.012$

The product will not be exactly 12, because 36.4% is a rounded answer.   ◆

Always remember to change the decimal answer to its percent form when asked to find the percent value. If percent answers are to be rounded to the nearest tenth of a percent, be sure to calculate your decimal answer to four places, multiply your answer by 100% to change it to a percent, and then round to the nearest tenth of a percent. If you were asked to round to the nearest hundredth of a percent, you would carry the division one place farther, multiply the decimal value by 100%, and then round to the nearest hundredth of a percent. This process could be carried out to the nearest thousandth of a percent (and beyond), but this is unlikely to be required in real-world situations. The nearest tenth of a percent is generally accurate enough.

7.4 ◆ Practice Problems

A. For each of the following problems, identify the values for *R*, *B*, and *A*. Then substitute the given values in the percent formula, and solve it to find the unknown value. Round all base and amount answers to the nearest hundredth and all percent answers to the nearest tenth of a percent. You may use your calculator to do the multiplication and division in this type of problem if your instructor agrees.

1. What is 23% of 80?

2. What is 143% of 60?

3. What is 7% of 95?

4. What is 0.5% of 93?

5. Find 124% of 70.

6. Find 2 1/2% of 80.

7. Find 12 1/2% of 100.

8. 20 is 40% of what number?

9. 35 is 7% of what number?

10. 30 is 125% of what number?

11. 15 is 0.5% of what number?

12. 2.8% of what number is 140?

13. 32% of what number is 8?

14. 74% of what number is 80?

15. 12% of what number is 43?

16. 40 is what percent of 50?

17. 12 is what percent of 10?

18. 45 is what percent of 50?

19. What percent of 40 is 4.8?

20. What percent of 120 is 50?

21. 16 is what percent of 30?

22. 125 is what percent of 200?

23. 7 is what percent of 8?

24. 24 is what percent of 30?

25. 16.5% of 42 is what number?

26. What percent of 18 is 11?

27. 24% of what number is 12?

28. What is 18.5% of 22?

29. 21 is what percent of 65?

30. 28 is 5 1/2% of what number?

B. Use your calculator to answer the following questions. Follow the rounding instructions given in Part A.

31. What is 77.3% of 92,476?

32. 17.5 is 21.8% of what number?

33. 91.4 is what percent of 146?

34. 11.9 is 43% of what number?

35. Find 29.4% of 37,645.

36. What percent of 8,372 is 43.5?

37. What percent of 8,276 is 189.4?

38. 81.2 is what percent of 375?

39. 127 is 38.4% of what number?

40. 83.6% of 42,756 is what number?

41. 67.4 is 82% of what number?

42. Find 17.9% of 46,384.

43. 35.4 is 46.5% of what number?

44. 238.5 is 21.6% of what number?

45. 1,276 is what percent of 8,375?

Another method for solving percent problems is to use the **percent proportion**. Because percent means "per 100," one side of the proportion can be expressed as a ratio of the percent number or rate (P), to 100. The other side of the proportion will be the ratio of the part or amount (A) to the total or base (B). The formula that expresses this relationship is

$$\frac{A}{B} = \frac{P}{100}$$

The most difficult part of solving percent problems by this method is identifying A, B, and P. Note that the percent quantity is denoted by P in the proportion. You will use the actual percent value, not its decimal equivalent as you did in the percent formula for R. Once you have identified the values and written the proportion, you will use the proportion solving skills that you learned in Chapter 6. Consider the following statement:

16 is 25% of 48.

Here 16 is A, 25 is P, 48 is B.

P and B are easy to identify, so do them first. P will be the number right before the percent symbol or the word *percent*. B always follows the word "of." Whatever is left will be A.

E X A M P L E 1 Identify P, B, and A in the following statements.

a. 65 is 50% of 130.

b. 12 is 75% of 16.

Solution **a.** 50% means that 50 is P. The words "of 130" mean B is 130. Therefore, 65 must be A.

b. 75% means that 75 is P. The words "of 16" mean that 16 is B. Therefore, 12 is A.

Once you have identified the parts of the problem, substitute the numbers for the letters in the percent proportion. Usually one of the parts is not known, and you will solve the proportion to get that value. But let's test the percent proportion with the two preceding examples. Substitute the parts you have identified, and see whether you get a true proportion when you compute the cross products. (See Section 6.3.)

$$\frac{65}{130} \diagup\hspace{-1em}\diagdown \frac{50}{100} \qquad\qquad \frac{12}{16} \diagup\hspace{-1em}\diagdown \frac{75}{100}$$

$$65 \cdot 100 = 50 \cdot 130 \qquad\qquad 120 \cdot 100 = 75 \cdot 16$$

$$6{,}500 = 6{,}500 \quad \text{True.} \qquad\qquad 1{,}200 = 1{,}200 \quad \text{True.}$$

E X A M P L E 2 Use the percent proportion to answer this question: What percent of 8 is 4?

Solution

$$\frac{A}{B} = \frac{P}{100}$$ Begin with the percent proportion. B is 8, you are looking for P, and A is 4.

$$\frac{4}{8} = \frac{P}{100}$$ Substitute the numbers for the letters.

$$8 \cdot P = 4 \cdot 100$$ Multiply $4 \cdot 100$ and $8 \cdot P$.

$$\frac{\overset{1}{\cancel{8}}P}{\underset{1}{\cancel{8}}} = \frac{400}{8}$$ Divide both sides by 8.

$$P = 50$$

$$P = 50\%$$ Because P is a percent number, use the % symbol. 50% of 8 is 4.

Check   50% of 8 is 4, so 4 is 50% of 8.

$$4 = 0.50 \times 8$$

$$4 = 4.00 \quad \text{True.}$$

Note that when you finish solving the proportion for P, the decimal point is in the correct place and does not have to be moved. In fact, you won't have to worry about moving decimal points at all, except in decimal multiplication or division.   ◆

E X A M P L E   3   5% of 90 is what number?

Solution $$\frac{A}{B} = \frac{P}{100}$$ P is 5, B is 90, and A is unknown.

$$\frac{A}{90} = \frac{5}{100}$$ Substitute.

$$100A = 450$$ Cross-multiply: $100 \cdot A$, $5 \cdot 90$

$$\frac{\overset{1}{\cancel{100}}A}{\underset{1}{\cancel{100}}} = \frac{450}{100}$$ Divide by 100.

$$A = 4.5$$ 4.5 is 5% of 90.

Check   4.5 is 5% of 90.

$$4.5 = 0.05 \times 90$$

$$4.5 = 4.50 \quad \text{True.}$$ ◆

E X A M P L E   4   12% of what number is 80?

Solution $$\frac{A}{B} = \frac{P}{100}$$ 12 is P, B is unknown, and A is 80.

$$\frac{80}{B} = \frac{12}{100}$$

$$12B = 8000$$

$$\frac{\overset{1}{\cancel{12}}B}{\underset{1}{\cancel{12}}} = \frac{8000}{12}$$

$$B = 666 \ 2/3, \text{ or } 666.7$$

80 is 12% of 666 2/3.

Check   80 is 12% of 666.7.

$$80 = 0.12 \times 666.7$$

$$80 = 80.004$$ The answer is a rounded answer, so the check statement is very close, but not exact.   ◆

7.5 ◆ Practice Problems

A. Solve these percent problems using the percent proportion, $A/B = P/100$. If necessary, round answers to the nearest hundredth, except for percents, which should be rounded to the nearest tenth of a percent. If you need more practice, do some of the practice problems from Section 7.4, using the percent proportion this time. Again, you may want to use your calculator to help you do the actual calculations.

1. 24 is 15% of what number?

2. 22% of what number is 38?

3. What percent of 40 is 12?

4. 18 is what percent of 96?

5. 94% of 30 is what number?

6. What is 16% of 128?

7. 6 1/2 is 18% of what number?

8. 4.8 is what percent of 2.6?

9. 30 is what percent of 40?

10. 24.5 is 8% of what number?

11. What is 18% of 40?

12. What number is 12.5% of 200?

13. 48 is 8% of what number?

14. What percent of 70 is 4.5?

15. What percent of 80 is 45?

16. Find 210% of 65.

17. 1.8% of what number is 54?

18. 6 is what percent of 9?

19. What is 14.8% of 300?

20. 46% of what number is 70?

21. 36 is 8 1/2% of what number?

22. What is 3.4% of 96?

23. 52 is 120% of what number?

24. What percent of 12 is 7?

25. What percent of 32 is 60?

25. What is 25% of 84?

27. 45 is 150% of what number?

28. 41 is what percent of 65?

29. 21.5 is what percent of 38?

30. 48 is 110% of what number?

 B. Use your calculator to solve the following problems by applying the percent proportion. Follow the rounding instructions in Part A.

31. What percent of 2,006 is 18.4?

32. Find 176.2% of 37,658.

33. 21 is 43.2% of what number?

34. 70.2 is what percent of 9,840?

35. What is 67.5% of 13,465?

36. 82.3% of what number is 17,260?

37. What percent of 18,540 is 623?

38. 27.5% of 11,864 is what number?

39. 46.8% of what amount is 1,765?

40. What percent of 33,674 is 18,265?

7.6 ◆ Applications Using Percent

One use for percent is to find interest. If you were to put some money into a savings account, this procedure would enable you to find the amount of interest that your deposit would earn in a year. To keep things uncomplicated, assume that this bank is paying only simple interest, not compound interest, and that you will earn interest only if the bank keeps your money for one year. Simple interest is figured once in a specific time period, and the interest is paid only on the principal. Compound interest is figured over and over again—compounded—so the principal increases each time the interest is added to it. With compound interest, an account earns interest on the principal and on the previously earned interest. The formula for compound interest is more complex than the simple interest formula, $I = PRT$. (Of course, earning only simple interest is not a very good deal financially, but it is good for our classroom purposes right now.)

E X A M P L E   **1**   Find the amount of interest that your account will earn in 1 year if you deposit $1,200 and the bank pays 9% interest per year.

Solution   You could reword the problem and ask,

"How much is 9% of $1,200?"

Change 9% to its decimal form, 0.09, and then multiply. 9% of $1,200 = 0.09 × 1,200 = 108.00, or $108.

Your account would earn $108 in interest if you invested $1,200 at 9% annual interest for 1 year.   ◆

Now expand the idea of finding interest by using the interest formula. It is a common formula, similar to $A = RB$, and is used when you need to compute simple interest for periods other than one year. The formula is an easy one:

$$I = PRT$$

where I is the amount of interest the account will earn.

P is the principal, the amount deposited or borrowed, the base quantity.

R is the rate or percent in its decimal form.

T represents the number of years involved.

E X A M P L E   **2**   Find the amount of interest you could earn on an account that pays 11% annually if you have invested $3,200 for 4 years.

Solution   Determine from the information given which letters in the formula you know the values of.

P = the money you invested = $3,200

R = the rate of interest paid = 11% annually (per year)

T = the length of time that the money was invested = 4 years

$I = PRT$	Write the formula.
$I = (\$3,200)(11\%)(4 \text{ years})$	Substitute given values.
$I = (3,200)(0.11)(4)$	Change the percent to a decimal.
$I = 1,408.00$ or $1,408	Multiply.

The account will earn $1,408 over a 4-year period if you have deposited $3,200 at 11% annual interest.   ◆

E X A M P L E   **3**   Find the amount of interest you would have to pay if you borrowed $1,450 for your tuition and you were charged 7% annual interest for 3 years.

Solution   Determine which letters you know the values of.

P = the money which you borrowed = $1,450

R = the rate of interest charged = 7% annual interest

T = the length of time for which the money was borrowed = 3 years

$I = PRT$	Write the formula.
$I = (\$1,450)(7\%)(3 \text{ years})$	Substitute given values.
$I = (1,450)(0.07)(3)$	Change the percent to a decimal.
$I = 304.50$ or $304.50	Multiply.

You would be charged $304.50 in interest for the 3-year loan. To find out how much you would have to pay back at the end of the 3 years, you need to add the interest to the *principal*, the amount you borrowed.

$1,450 + $304.50 = $1,754.50 to be paid back.   ◆

Not all percent problems involve the interest formula. Most of the time when a problem is not one that requires the use of the interest formula, you will want to solve it by using either the percent formula or the percent proportion. It does not matter which approach you take. One familiar type of problem that involves finding a percent of a quantity concerns **sales tax**. Almost every state now has a sales tax based on the price of the item bought. Suppose you live in a state where the sales tax is 6%. To find the amount of tax on any item purchased, you need to find 6% of its selling price. **Commission**, based on the dollar amount an employee has sold; **discount**, the amount in dollars saved by buying an item on sale; and **percent of increase** or **percent of decrease** are other situations where percents are used in everyday life.

When you are using the percent formula to solve application problems, the only difficulty is identifying the values that will be substituted in place of R, B, and A in the formula. Follow the same steps that you have been using in the preceding sections. Identify R first; it is the percent (%). B stands for the base, the original quantity, the whole or total. A stands for the part, or amount. Consider the following situation and see how to pick out the values.

> A sales tax is charged on the price of each item purchased in Franklin County. The sales tax rate is 5.5%. The tax is based on the purchase price of the item. The part that is charged in tax is added to the purchase price of the item to give the total bill. Find the amount of sales tax charged in Franklin County on an item costing $43.95.
>
> $A = RB$

R is the sales tax rate, 5.5%. Sales tax is based on the purchase price, so B is $43.95. You are to find the amount paid in tax, A.

$A = RB$

$A = 5.5\% \times \$43.95$

$A = 0.055 \times 43.95$

$A = 2.41725$

The amount of tax charged is $2.42, rounded to the nearest cent.

> The salesperson said that the tax on the new pair of slacks that Michael bought at the store would be $1.10. Can you figure out the price of the pants if he bought them at MAX's in Franklin County?
>
> $A = RB$

R is the sales tax rate, 5.5%. Sales tax is based on the purchase price, B, which is what you are trying to find. $1.10 is the amount paid in tax, A.

$$A = RB$$
$$1.10 = 5.5\% \times B$$
$$1.10 = 0.055 \times B$$

$$\frac{1.10}{0.055} = \frac{\overset{1}{0.0\cancel{55}} \times B}{\underset{1}{0.0\cancel{55}}}$$

$$20.00 = B$$

Therefore, the slacks cost $20.00.

> The same pair of slacks in Chicago would carry a tax of $1.40. What is the sales tax rate in Chicago?
>
> $A = RB$

You are looking for the sales tax rate, R. This tax is based on the purchase price of $20.00, which is B. The amount paid in tax is $1.40, A.

$$A = RB$$

$$\$1.40 = R \times 20$$

$$\frac{1.40}{20} = \frac{R \times \overset{1}{\cancel{20}}}{\underset{1}{\cancel{20}}}$$

$$0.07 = R$$

Remember to change the decimal value for R to its percent form by multiplying by 100%.

$$0.07 \times 100\% = 7\%$$

Therefore, the sales tax rate in Chicago is 7%.

$1.40 is 7% of $20.

These three examples illustrate simple uses of the percent formula or percent proportion. A calculator can help you do the actual calculations once you have identified the values and substituted them into the formula or proportion.

As you can see from these problems, you proceed by identifying first the value for R or P (the percent), next the value for B (the base), and then the value for A (the part or amount). The following descriptions will help you to pick out these values in various word problems.

A (Amount or Part)	R or P (Rate or Percent)	B (Base)
Amount of sales tax	Rate of sales tax	Original price
Amount of discount	Rate of discount	Original price
Amount of interest	Rate of interest	Original amount borrowed or saved
Amount of commission	Rate of commission	Amount of sales on which commission is based
Amount of decrease	Rate of decrease	Original quantity before the decrease
Amount of increase	Rate of increase	Original quantity before the increase

Study the following examples carefully to see if you can identify the values for R, B, and A. Once you have done that, you simply follow the steps outlined in the previous sections to solve for A, B, or R (when using the percent formula) or A, B, or P (when using the percent proportion). Even if your instructor wants you to practice only one method, you should at least read through all of the following examples to work on determining the values for A, B, P, or R. If you are working only on the percent formula, also study the method used in Examples 4, 5, 6, and 7. If you are concentrating on the percent proportion, study the method used in Examples 8, 9, 10, 11, and 12.

EXAMPLE 4   Find the sales tax on an item costing $42.85 if the sales tax rate is 6%.

Solution R is 6%. Sales tax is based on the price of an item, so $B = \$42.85$.

$$A = RB$$

$$A = 6\% \text{ of } \$42.85 = 0.06 \times 42.85$$

$$A = \$2.5710$$

Rounded to the nearest cent, a 6% tax on an item selling for $42.85 is $2.57.   ◆

E X A M P L E 5 Wendy wants to know how much she will save by buying a skirt at Sara's Shoppe while the store is having a "20% Off Sale." The skirt normally sells for $22.95.

Solution The rate of discount is 20%, and discount is based on the original price of the skirt, so B is $22.95.

$A = RB$

$A = 20\% \times \$22.95 = 0.20 \times 22.95$

$A = 4.59$

Therefore, Wendy can save $4.59 by buying the skirt on sale.   ◆

E X A M P L E 6 16 members on the board of directors voted to pass a no-smoking policy. This proposal passed with 80% of the board members voting for it. How many members are there on the board?

Solution $A = RB$

80% voted to pass the proposal, so R is 80%. You are looking for the total membership on the board, B. 16 represents A, the part of the board who voted to pass the policy.

$A = RB$	Write the percent formula.
$16 = 80\% \times B$	Substitute the given values.
$16 = 0.80 \times B$	Change 80% to its decimal form.
$\dfrac{16}{0.80} = \dfrac{0.\overset{1}{\cancel{80}} \times B}{\underset{1}{\cancel{0.80}}}$	Divide both sides by 0.80 to "undo" the multiplication and isolate B.
$20 = B$	$16 \div 0.80 = 20$
16 is 80% of 20.	There are 20 members on the board. ◆

E X A M P L E 7 Maria bought a new car a year ago and paid $12,500 for it. Today the value of the car is $8,500. What is the percent of decrease in the value of the car?

Solution You are looking for R, the rate of decrease. Decrease is based on the original value before the decrease, so $B = 12,500$. The amount of decrease, A, is the difference between the original value and the current value. Therefore, $A = 12,500 - 8,500$, or 4,000.

$$A = RB$$

$$4,000 = R \times 12,500$$

$$\frac{4,000}{12,500} = \frac{R \times 12,\overset{1}{\cancel{500}}}{\underset{1}{\cancel{12,500}}}$$

$$0.32 = R$$

$$0.32 \times 100\% = 32\%$$

Therefore, the percent of decrease was 32%.   ◆

The same type of approach is used when you are solving a word problem by applying the percent proportion, $A/B = P/100$. Begin by identifying the values for P, B, and A, and then substitute those values into the percent proportion. Solve this proportion as you would any other according to the steps you learned in Chapter 6.

E X A M P L E **8**   Ishmael earns a 15% commission on all of the furniture that he sells for Furniture Warehouse. Last week his sales totaled $3,627. How much commission did he earn?

Solution   The rate of commission, R, is 15%. Commission is based on the amount sold, so $B = \$3,627$.

$$\frac{A}{B} = \frac{P}{100}$$

$$\frac{A}{3,627} = \frac{15}{100}$$

$$100 \cdot A = 15 \cdot 3,627$$

$$100 \cdot A = 54,405$$

$$\frac{\overset{1}{\cancel{100}} \cdot A}{\underset{1}{\cancel{100}}} = \frac{54,405}{100}$$

$$A = 544.05$$

Therefore, Ishmael's commission last week was $544.05. ◆

E X A M P L E **9**   How many problems must you get right on a 40-question test to show 85% mastery?

Solution   Let's solve this problem using the percent proportion.

$$\frac{A}{B} = \frac{P}{100}$$

P is 85, the mastery level rate.

B is 40, the total number of problems on the test. You are looking for A, the amount necessary to show mastery of 85% of the test material.

$$\frac{A}{B} = \frac{P}{100} \qquad \text{Write the percent proportion.}$$

$$\frac{A}{40} = \frac{85}{100} \qquad \text{Substitute the given values.}$$

$$100 \cdot A = 40 \cdot 85 \qquad \text{Solve the proportion for } A.$$

$$100 \cdot A = 3,400$$

$$\frac{\overset{1}{\cancel{100}} \cdot A}{\underset{1}{\cancel{100}}} = \frac{3,400}{100}$$

$$A = 34$$

Therefore, you must get 34 problems right to show 85% mastery of the material on a 40-question test.

34 is 85% of 40. ◆

E X A M P L E **10**   A $16 discount is given on a dress that originally cost $80. What is the percent or rate of discount? Use the percent proportion.

Solution   What percent of 80 is 16? Restate in simple terms: 16 is what % of 80?

$$\frac{16}{80} = \frac{P}{100} \qquad P \text{ is unknown, } B \text{ is 80, and } A \text{ is 16.}$$

$$80P = 1,600$$

$$\frac{80P}{80} = \frac{1,600}{80}$$

$P = 20$ The rate of discount is 20%.   ◆

E X A M P L E 11 Enrollment for spring term is expected to be up 18% from last spring. If 8,250 students attended last spring, how many students can be expected on campus this spring term? Use the percent proportion.

Solution $\dfrac{A}{B} = \dfrac{P}{100}$

18% is the rate of increase, R. The increase is based on last year's figure, so B is 8,250. You need to find the amount of the increase, A.

$$\frac{A}{B} = \frac{P}{100}$$

$$\frac{A}{8,250} = \frac{18}{100}$$

$$100 \cdot A = 18 \cdot 8,250$$

$$100 \cdot A = 148,500$$

$$\frac{\overset{1}{\cancel{100}} \cdot A}{\underset{1}{\cancel{100}}} = \frac{148,500}{100}$$

$$A = 1485$$

Therefore, the amount of increase will be 1,485 students.

$8,250 + 1,485 = 9,735$ The total number of students expected on campus is the sum of last spring's enrollment, 8,250 students, and the amount of the increase, 1,485.

Therefore, 9,735 students are expected for spring term.   ◆

E X A M P L E 12 According to the 1980 census, the population of Glace Bay was 26,300. The new 1990 census reported a population of 28,141. What was the percent of increase over the 10-year period?

Solution You are not asked what percent one number is of the other; rather, you are asked to find the *percent increase*. The increase is based on the original population in the 1980 census. Therefore, B is 26,300. The amount of the increase, A, must be calculated.

$28,141 - 26,300 = 1,841$

Therefore, A is 1,841. You are looking for P, the percent of increase. Use the percent proportion.

$$\frac{A}{B} = \frac{P}{100}$$

$$\frac{1,841}{26,300} = \frac{P}{100}$$

$$26,300 \cdot P = 100 \cdot 1,841$$

$$26,300 \cdot P = 184,100$$

$$\frac{\overset{1}{\cancel{26,300}} \cdot P}{\underset{1}{\cancel{26,300}}} = \frac{184,100}{26,300}$$

$$P = 7$$

Therefore, the population increased by 7%.   ◆

7.6 ◆ Practice Problems

A. **Use the simple interest formula, *I = PRT*, to find the answers to these questions.**

1. How much interest will you be charged to borrow $800 for 3 years if the bank charges 12% annual interest?

2. How much interest will your account earn in 3 years if you deposit $1,450 and it earns 9% annual interest?

3. How much interest will you earn on a deposit of $600 if you leave it in the bank for 2 years and it earns 9% per year?

4. How much interest will you have to pay on a loan of $850 if you borrow the money for 1 year at 11% interest per year?

5. Martin borrowed $560 to pay his tuition for this term. He knows that he will pay it back in 1 year, and the bank is charging him 12% annual interest. How much interest will he have to pay when he pays back the loan?

6. Jacqueline received an inheritance from her grandfather. He left her $3,500 to help with her school expenses. She wants to leave the money in a savings account for 3 years. The bank will pay her 10 1/2% annual interest. How much interest will she earn in the 3 years?

7. How much interest will you pay on a $2,500 loan if you borrow the money for 2 years at 12 1/2% annual interest?

8. If you deposit $1,200 in a savings account that earns 9 1/4% annual interest for 2 years, how much interest will the account earn?

9. Amanda needs to apply for a car loan from her credit union. The credit union is charging 12 1/2% annual interest on used-car loans if the money is borrowed for 4 years. Amanda wants to borrow $3,250 for the car she is buying. How much interest will she have paid at the end of the 4 years?

10. Paula deposited her holiday bonus check of $750 in a special savings account. This account earns 10 1/2% interest per year. How much money will she have in her account at the end of 2 years?

B. **Find the answers to the following word problems. Use the percent formula or the percent proportion. Round percent answers to the nearest tenth of a percent. Be sure to label the answers, because you are finding the solution to a word problem.**

11. Blue Bird Airlines is offering a 35% discount on regular round-trip airfares to Florida. If the ticket usually costs $178, how much will you save by buying it during this sale?

12. National Car Company is offering a 12% discount on the sticker price of its new vans. If one of these vans normally sells for $17,895, how much would you save by buying it during this sale?

13. What will you pay for the discounted airplane ticket in Problem 11?

14. How much will you pay for the van in Problem 12 if you buy it during this sale?

15. A house originally purchased for $86,500 sold this month for $96,880. What was the percent of increase in the value of this house?

16. Allen County's sales tax rate is 6 1/2%. How much county sales tax will you pay for a new deluxe vacuum cleaner that costs $122.95?

17. In a recent survey, 44% of the people who were asked said that they preferred candies with nuts to candies without. If 176 people preferred the nut candies, how many people were asked their preference?

18. In a recent taste test at Dwight Marketing Research Center, 24% of the people who participated in the tests like cereal B better than cereal A. Cereal B was preferred by 120 people. How many people participated in the taste test?

19. How much will you save on a car priced at $8,285 if the dealer is advertising a "12% off sale"?

20. A new washer at Mayrun Appliance regularly sells for $425.50. In order to compete with the new store down the street, Mayrun has recently reduced the price of this washer to $374.44. What is the percent of decrease for this sale?

21. Marty's last credit statement showed an interest charge of $5.40. If the rate of interest on the card is 1.8%, what was his balance due on the account?

22. Rachel earned $176 in interest on her certificate of deposit at the Bank of Lewiston. This bank pays 11% simple interest per year. How much money did Rachel have invested in her CD last year?

23. Your part-time job pays $5.80 an hour. Your supervisor has just told you that because you have been doing such a good job, you have been given a raise and will be making $6.50. What is the percent of increase that you have earned? Round to the nearest whole percent.

24. The average price for a gallon of gasoline increased by 225% in the 30-year period from 1958 to 1988. In 1958, the average price for a gallon of gas was approximately $0.52. Calculate the average price for a gallon of gasoline in 1988.

25. The population in Mayes Beach, Georgia, increased from 22,650 in 1970 to 24,462 in 1980. What was the percent growth?

26. The average price of a new family car in 1978 was $8,250. The average price for the same type of car was $10,065 in 1987. What was the percent increase during that period?

27. Marshall City's sales tax rate is 6 1/2%. How much city sales tax will you be charged on the purchase of a new dining room set priced at $595?

28. A portable tool chest is on sale for $88.95. The sales tax rate in your area is 5 1/2%. How much sales tax will you be charged on the purchase of this deluxe tool chest?

29. José is selling furniture for Ortega and Sons. He earns a weekly salary of $225 plus a 6% commission on everything he sells. Last week his sales totaled $2,348. How much money did he earn last week?

30. Salespeople working for Goodman Insurance Company earn a commission of 0.5% on the value of the policies they sell. They are also paid a monthly salary of $830. In April, Martina sold policies with a total worth of $320,000. What was her income for the month of April?

7.7   Interpreting Data

A **circle graph**, often called a **pie graph**, is circular and is divided into portions, or slices. The pie represents one whole body of data, and the slices represent different parts of that data. The sizes of the slices are proportional to the sizes of the different components. Most often, percents are used to represent the portions that make up the whole (100%) pie graph.

Consider the following example, which uses the pie chart shown in Figure 1 to convey information about the grading procedures in Dr. Newton's American Literature course.

In Dr. Newton's American Literature course, a student's grade depends on three different components: test scores, oral presentations, and written projects. The three parts do not contribute equally to the final grade. The distribution is illustrated by the pie chart in Figure 1. It is quickly evident that the test grades have more impact on the final grade than do the other components. Use Figure 1 to answer the following questions.

a. What percent of the course grade comes from test scores?

b. What percent comes from written projects?

c. Is the sum of the oral and written components equal to, greater than, or less than the test score component? By what percent?

d. If the most it is possible to earn in the course is 800 points, how many points can come from test scores?

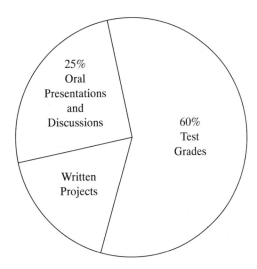

FIGURE 1

Answers:

a. The graph shows that 60% of the grade comes from test scores.

b. The percent that comes from written projects can be found by adding the other components together and subtracting that total from 100%.
60% + 25% = 85%
100% − 85% = 15%
Therefore, 15% of the final grade comes from written projects.

c. Test scores account for 60% of the course grade. 15% + 25% = 40%, and 40% is less than 60%. Therefore, the sum of the written and oral presentations is less than the test score component. 60% − 40% = 20%, so the other components account for 20% less than the test scores.

d. 60% of the final grade comes from test scores.
60% of 800 points = 0.60 × 800 = 480 points

E X A M P L E **1**   Use Figure 2 to answer the following questions.

a. What percent of the income is spent on utilities?

b. What percent is spent on miscellaneous items?

c. If the monthly income is $2,400, how much is spent each month for the mortgage payment?

d. From a monthly income of $2,400, how much is spent in paying taxes?

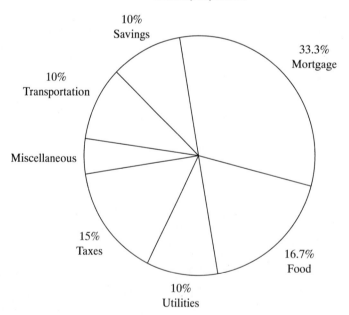

Monthly Expenses

F I G U R E **2**

Solution   From the graph, it can be determined that

a. 10% is spent on utilities.

b. 5% is spent on miscellaneous items.

c. You need to find 33.3% of $2,400.
33.3% of $2,400 = 0.333 × $2,400 = $799.20
Therefore, the mortgage payment is $799.20.

d. Taxes take 15% of the monthly income.
15% of $2,400 = 0.15 × $2,400 = $360
Therefore, $360 is withheld for taxes each month.   ◆

E X A M P L E 2 Use the following data to construct a pie graph. Label the components (or slices) in percents.

College Student's Monthly Expenses

Rent = $250   Transportation = $50

Food = $165   School materials = $35

Solution To calculate the percent parts, first find the monthly total.

$250 + $165 + $50 + $35 = $500

Find each component as a part of the whole, and then write it in percent form.

$$Rent = \frac{250}{500} = 0.5 = 50\%$$

$$Transportation = \frac{50}{500} = 0.1 = 10\%$$

$$Food = \frac{165}{500} = 0.33 = 33\%$$

$$School\ materials = \frac{35}{500} = 0.07 = 7\%$$

Draw a circle and mark off ten equally spaced sections. Each section represents 10%. See Figure 3.

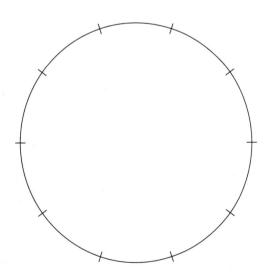

FIGURE 3

Mark off the expense percents, estimating where necessary and labeling each part. See Figure 4.

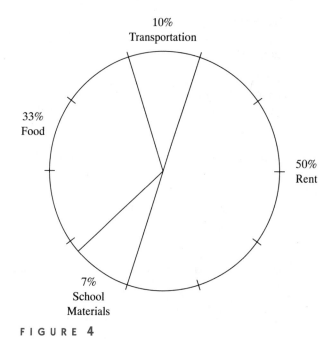

FIGURE 4

EXAMPLE 3   Use the bar graph in Figure 5 to answer these questions.

a. What was the computer operator's salary in 1988?

b. What was the annual salary in 1992?

c. What was the percent increase, to the nearest tenth of a percent, between 1986 and 1988?

d. What was the percent increase between 1988 and 1992?

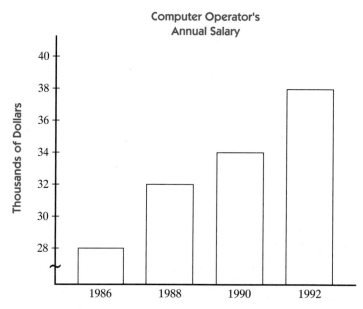

FIGURE 5

Solution   a. The 1988 salary was $32,000.

b. The 1992 salary was $38,000.

c. The change in salary was
$32,000 − $28,000 = $4,000
Percent increase is the amount of increase divided by the original quantity, so
4,000 ÷ 28,000 = 0.143 = 14.3%

d. The change in salary is $6,000. The percent increase is
6,000 ÷ 32,000 = 0.188 = 18.8%

You can see from these examples that it is sometimes very helpful to use a pie chart to give yourself and others a picture of what you are discussing. The procedures that you use to interpret the information are the same as those you have been using all along in this chapter.

7.7 ◆ Practice Problems

A. **Use Figure 6 for Problems 1 through 5.**

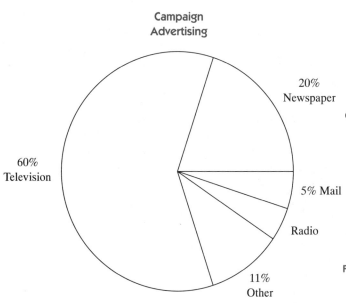

FIGURE 6

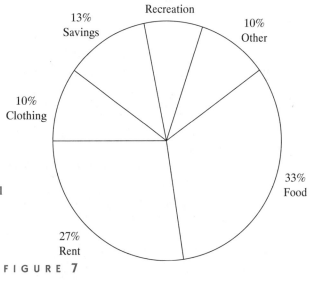

Monthly Allocation of Income

FIGURE 7

1. In Senator Joe Bucher's reelection campaign, what method of advertising cost the most?

2. What percent of Senator Joe Bucher's reelection expenses paid for newspaper advertising?

3. If the Senator spent $180,000 for the campaign, how much was spent on mailings?

4. How much of the $180,000 spent during the campaign was spent on "other" types of advertising?

5. What amount was spent on radio advertising?

B. **Use Figure 7 for Problems 6 through 12.**

6. What percent of this family's income goes into savings?

7. What percent is used for food?

8. What percent is spent on recreation?

9. If the monthly income is $3,200, how much is spent for food?

10. How much is spent on rent if the monthly income is $3,200?

11. If the monthly income increases to $3,500, but the budget is maintained, how much will now be spent for clothing?

12. How much will be spent for recreation if the monthly income increases to $3,500?

C. Use Figure 8 for Problems 13 through 18.

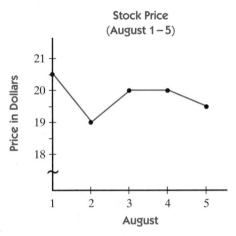

Stock Price
(August 1–5)

FIGURE 8

13. What was the stock's value on August 1?

14. What was the value of the stock on August 3?

15. What was the percent decrease in the value of the stock from August 4 to August 5?

16. What was the percent decrease from August 1 to August 2?

17. What was the average (mean) value of the stock for the 5-day period?

18. What were the median and the mode (see Section 1.7) for the stock in the 5-day period?

D. Use Table 1 for Problems 19 through 26.

TABLE **1**

Monthly Budget		
	Latoya	Juanita
Rent	$325	$400
Utilities	$100	$150
Food	$200	$175
Car	$175	$260
Miscellaneous	$200	$215

19. What is Latoya's total monthly amount budgeted?

20. How much does Juanita budget for the month?

21. What percent of Latoya's total budget goes for rent?

22. Food is what percent of Juanita's total budget?

23. Car expenses are what percent of Latoya's budget?

24. What percent of Juanita's total budget is spent for utilities?

25. Complete the pie graph shown in Figure 9 to display Latoya's monthly budget. Label each portion with a percent rounded to the nearest tenth of a percent.

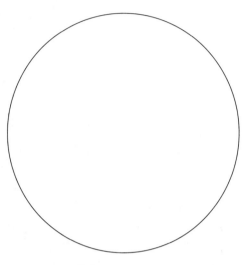

FIGURE **9**

26. Complete the pie graph shown in Figure 10 to display Juanita's monthly budget. Label each portion with a percent rounded to the nearest tenth of a percent.

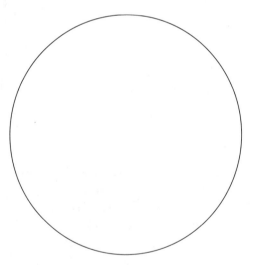

FIGURE **10**

7.8 Chapter Review

In this chapter, you have studied percents. Because you cannot multiply or divide by a percent (other than 100% as a form of 1), you had to learn some techniques for changing percents to their decimal or fraction forms. In order to do this, you needed to realize that 100% is another form of 1. Because multiplying or dividing by 1 does not change the value of the quantity being used, multiplying or dividing by 100% does not change the *value* of the fraction, decimal, or percent. It changes only the way that figure looks.

The principles for changing among forms are:

1. Percent means *per hundred.*
2. To change *from a decimal form to a percent form, multiply by 100%.*
3. To change *from a fraction form to a percent form*, first *change the fraction to its decimal form* by dividing, and then *multiply the decimal by 100%.*
4. To change *from a percent to a decimal or fraction, divide by 100%* and then *simplify* the answer.

It is important to remember that when you do problems using the percent formula, $A = RB$, you must first identify the values for R, B, and A. These values must then be substituted into the percent formula. If a percent value, R, is given, it must be changed to its decimal equivalent before any calculating can be done. If both number values are on one side of the equals sign, the formula tells you to multiply. However, if one of the numbers is on the same side of the equals sign as a letter, it is necessary to "undo" the indicated multiplication to isolate the letter. You undo multiplication by dividing. If you are asked to find the percent, R, it is necessary to change the final decimal answer to its percent equivalent.

When you are doing problems by using the percent proportion, $A/B = P/100$, follow the same steps as for solving any formula. Identify the values you are given, substitute them into the formula, and solve for the missing value. The one thing you need to remember when using the percent proportion is that the P value is the percent value in its percent form. You do not need to change it to a decimal first before substituting into the proportion.

In the case of word problems, it is necessary to identify from the statement of the problem the values for the rate, R or P; the base, B; and the part or amount, A. The rate, or percent, is the easiest to spot. It comes before the percent sign, %. The base, B, is usually the next easiest to pick out. It is the original value—the quantity before anything happens. The amount, A, is the part that you are finding. It might be the part paid in tax, the amount of decrease or discount, or the number of people who did a certain activity. It does not make any difference whether you choose to solve the problem using the percent formula or the percent proportion. The more word problems you do, the better you will become at picking out the appropriate values for R or P, B, and A.

When you have an interest problem to solve, you must use the simple interest formula, $I = PRT$. Identify the values you are given, and solve for the one you are asked to find.

To summarize, when you are solving a problem involving percent, follow these steps:

1. State the problem in standard form. (You don't always have to write it out, but at least think it through that way in your mind.)
2. Identify the values for R or P, B, and A in that order.
3. Write the percent formula, $A = RB$, or the percent proportion, $A/B = P/100$.
4. Substitute the values you know.
5. Decide from the formula whether you are to multiply or divide to find the answer, and then solve for the unknown value.

6. Give the answer in the form requested, such as a percent (%), or money, or rounded to a particular place.

7. Reread the problem to see if your answer makes sense.

8. When you are asked to interpret percent data from a chart, you need to use all the skills that you learned before when you studied graphs.

This chapter has been a long but important one. If you are going on to study algebra, the procedures you have learned here will be very helpful. If you go on to business math instead, or to applied math courses, you will often use these "formula skills."

Review Problems

A. Answer each of the following questions.

1. Sixty-two percent of the students who were asked thought that their school should sponsor a food drive to support a shelter for the homeless near their campus. What percent of the students who were asked disagreed with this idea?

2. Miguel has decided to try to save 28% of his paycheck for the next 3 months. He will use the rest to pay expenses. What percent will he use for paying his bills?

3. What percent of the problems could you afford to miss on your final exam and still keep your "B" average, 85%?

4. Eighteen percent of the people in Miss Blake's composition class failed to turn in their final drafts on time. What percent of the students met the deadline?

B. Fill in the following chart. Be sure to reduce fraction answers. Round decimal answers to three places, and round percent answers to the nearest tenth of a percent.

	Fraction	Decimal	Percent
5.		0.16	
6.			20%
7.	1/4		
8.		0.38	
9.	1/8		
10.			6.5%
11.		1.54	
12.	7/100		
13.			112%
14.		0.006	
15.	1/12		
16.			12 3/4%

C. Use the simple interest formula, *I = PRT*, to answer the following questions. You may use a calculator if you wish.

17. What is the interest earned in 3 years at 10% annual interest on a deposit of $1,200?

18. What is the amount of interest that you must pay on a 4-year loan if you borrow $800 at 12% annual interest?

19. How much interest will a $2,500 savings account earn in 2 years if it pays 9 1/4% annual interest?

20. How much will the same savings account earn if the bank pays 10 1/2% annual interest and the money is left on deposit for 3 years?

21. How much money must be paid back at the end of 5 years by a student who borrows $7,000 at 8.5% annual interest on a simple interest loan?

D.   Answer the following problems by using the percent formula, $A = RB$, or the percent proportion, $A/B = P/100$. Some of the statements may need to be written in standard form first. Be sure to identify the values for R or P, B, and A before you begin to work the problem. Round all percent answers to the nearest tenth of a percent if necessary. All other answers can be rounded to the nearest hundredth. You may use your calculator to help you with the calculations if you choose.

22. Find 42% of 80.

23. Find 19% of 77.

24. Find 24.5% of 420.

25. What is 11.2% of 90?

26. 30 is 50% of what number?

27. 45 is 60% of what number?

28. 80 is 15% of what number?

29. 42 is 25% of what number?

30. 2.8% of what number is 70?

31. 12 is what percent of 60?

32. 30 is what percent of 120?

33. 32 is what percent of 80?

34. 2.4 is what percent of 80?

35. What percent of 35 is 15?

36. What percent of 67 is 30?

37. 60 is 8.5% of what number?

38. 4.5% of what number is 135?

39. 1.5 is what percent of 60?

40. What percent of 45 is 12?

41. Find 12.8% of 120.

42. 9 1/2% of 126 is what number?

43. 12 1/4% of what number is 88?

44. 4 1/4% of what number is 80?

45. 5 1/2% of what number is 50?

46. Sixteen percent of what number is twelve?

47. Nine is what percent of sixty?

48. Twenty-four percent of 96 is what number?

49. Eighty-two percent of 16 is what number?

50. Seventy-two is what percent of 20?

E.   Solve the following word problems. Round percent answers to the nearest tenth of a percent, and round money answers to the nearest cent. Be sure to label your answers. You may use a calculator if you wish.

51. Sixteen people, stopped on a corner downtown, could not give the mayor's name. This represented 40% of the people who were asked. How many people were questioned?

52. A car purchased last year for $12,500 now has a market value of $9,500. What is the percent of decrease in the value of the car?

53. Sara has 110 math review problems to do. She has finished 95 of those problems. What percent of the problems has she completed? Round your answer to the nearest tenth of a percent.

54. Franklin County is considering raising its sales tax to 6 3/4%. If it does so, how much county sales tax will you have to pay on an item that costs $42.00?

55. How much money will you need to deposit in your savings account if you want to earn $450 in interest this year? Your account pays 9% simple interest per year.

56. Village Antiques is having a "12% off sale." How much will a quilt cost on sale if it is normally priced at $55?

57. Dalane is working on commission, selling service contracts on copiers. He makes 30% on what he sells. His most recent sale was for $235 for a year's service. How much commission will he make on that sale?

58. A college is experiencing an 18% growth rate from one year to the next. How many students can be expected on campus next fall term if there are 10,500 students on campus this fall?

59. Viola's savings account earned $120 in interest last year. If her bank pays 8% annual interest, how much money did she originally deposit in her account?

60. Of the students registered for biological sciences this term, 45 earned an "A" in the course. If there are 320 students taking that course, what percent earned an "A"? (Round to the nearest tenth of a percent.)

61. Of the 1,850 widgets produced by Carroll–Williams Manufacturing Company, 130 would not operate. What percent of the widgets were defective? Round to the nearest tenth of a percent.

62. The normal rate of error on tax returns is 8%. How many tax returns would have to be sampled if the audit staff wanted to study 100 faulty returns?

63. A dress is on sale for 30% off the original price. If a $24 discount is given, what was the original price of the dress?

64. Nineteen of 25 students are in class today. What percent of the class is present?

65. Kawai is paid a 3% commission on his total sales at Class Act Furniture. His sales total last week was $3,219.95. How much commission did he earn?

F. Use Figure 11 for Problems 66 through 70.

66. What percent of the salary goes to pay city taxes?

67. What percent of the pay is for federal taxes?

68. What percent was paid for insurance?

69. If the biweekly income is $1,750, how much was deducted for retirement?

70. What amount (in dollars) of the $1,750 salary is take-home pay?

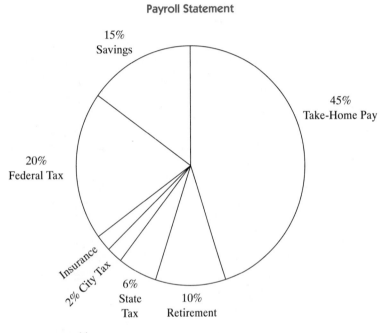

Payroll Statement

FIGURE 11

Chapter Test

1. Identify *R*, *B*, and *A* in this percent statement: 16% of 40 is 6.4.

2. In Shirley's division, 89% of the faculty were rated excellent by their students. What percent did not receive the excellent rating?

3. Change 7/1,000 to a percent.

4. Change 16 3/4% to a decimal.

5. Change 7 5/8 to a percent.

6. Change 14.5% to a fraction reduced to lowest terms.

7. Use *I = PRT* to find what amount of interest must be paid on a 4-year, simple interest loan if $900 is borrowed at an annual rate of 9%.

8. You must pay $17.60 in sales tax on a car stereo that sells for $320. What is the sales tax rate? Write the percent formula used to solve this problem, and then answer the question.

9. Write the percent proportion used to solve Problem 8, and then answer the question.

10. What is 8 3/4% of 400?

11. 16 is what percent of 800?

12. What percent of 82 is 16.646?

13. 25.2 is 36% of what number?

14. 1.5% of what number is 27?

15. How much commission will Alaine earn on sales of $2,740 if his commission rate is 5%?

16. A dress that regularly sells for $70 is on sale at 25% off. What is the sale price of the dress?

17. An airline ticket to Mexico City that usually costs $328 is on sale for $288.64. What is the percent of decrease in the cost?

18. What is the regular sticker price of a car that is selling for $14,025 at a "15% off sticker" sale?

19. How much money must you invest at 11% annual interest to earn $2,200 in 1 year?

20. In a pie chart used to show the breakdown of the college's annual budget, the money set aside for health insurance is 8.6%. How much is the cost of health insurance for the school if the annual budget is $13,560,000?

Thought-Provoking Problems

1. Explain the difference between the *R* in the percent formula and the *P* in the percent proportion.

2. Compare the values for *I*, *P*, and *R* in the interest formula with the values for *R*, *B*, and *A* in the percent formula. In what situations would you have to use the interest formula rather than the percent formula?

3. Write a word problem involving percent of increase. Are your values reasonable? Is your wording clear and concise?

4. Using your calculator, find the amount of money in an account at the end of 4 years if $2,000 was deposited originally at 8.5% and the interest is compounded annually. (*Hint:* In annual compounding, the interest is computed at the end of each year, and the principal is increased by that amount. Thus, there is a new base amount each year.)

5. Using your calculator, find the answer to each of the following problems. Round answers to the nearest tenth.

   a. 1,308 is what percent of 4,529?

   b. How much simple interest will an account of $13,525 earn in 3 years at 12 3/4%?

   c. A rare coin whose original value was $0.25 now has a value of $12,250. What is the percent of increase in the value of the coin?

   d. Alex earns a 5 3/8% commission on the computers that he sells. If his last commission check was $2,284.38, what was the value of his computer sales for that period?

6. How can you figure 10% of $36 in your head? How can you figure 15% of $36 in your head? Explain the process that you are using.

7. Explain how you can figure in your head the amount of a 20% tip on the bill for a restaurant dinner costing $17.95.

8. Arrange from the smallest to the largest:

   a. 16.5%, 1/8, 0.17

   b. 3 3/4, 280%, 3.4

9. Is there a difference between investing $10,000 at 10% for 1 year and investing $10,000 at 5% for 6 months (1/2 year) and then reinvesting the new amount for another 6 months at 5%? Explain your answer. Is 10% for 1 year the same as 5% twice a year?

10. How much would you need to repay at the end of 10 years on a simple interest loan if the agreement was for 8 3/4% and the amount of interest charged was $2,304.75?

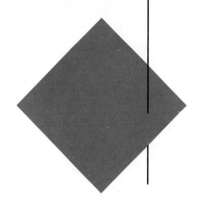

Cumulative Review, Chapters 1–7

Solve each of the following problems. Show all your work. Reduce all fraction answers to lowest terms. In Problems 1 through 5, add.

1. $6.3 + 18 + 24.6$

2. $\dfrac{1}{2} + \dfrac{3}{4}$

3. $1,214 + 2,765$

4. $2\,5/8 + 7\,2/3$

5. $0.065 + 1.2 + 25.62$

In Problems 6 through 10, subtract.

6. $14\,7/9 - 5\,2/3$

7. $900 - 371$

8. $64.36 - 8.97$

9. $156 - 8.3$

10. $49 - 18\,4/5$

In Problems 11 through 15, multiply.

11. Find $8/15$ of 90.

12. 0.08×13.75

13. $\$491 \times 23$

14. $5\,3/4 \times 8\,3/8$

15. 0.076×100

In Problems 16 through 20, divide. If necessary, round to the nearest hundredth.

16. $76.4 \div 0.06$

17. $1,488 \div 4$

18. $2\,3/8 \div 3/4$

19. $0.7\overline{)29}$

20. $6\,3/16 \div 3$

Find the answer to each of the following problems.

21. Change $11\,1/4$ to an improper fraction.

22. What is the missing numerator in $\dfrac{3}{5} = \dfrac{}{20}$?

23. Reduce $65/210$ to lowest terms.

24. Change 0.45 to a percent.

25. Change $12\,1/2\%$ to a fraction.

26. Round 65.187 to the nearest tenth.

27. Round 0.0053 to the nearest thousandth.

28. Change 2.375 to a fraction.

29. Change $5/12$ to a decimal rounded to the nearest hundredth.

30. Write the ratio of 8 feet to 6 inches in simplest form.

31. Write the ratio of 3.2 to 2.4 in simplest form.

32. Find 62% of 13.8.

33. 12 is what percent of 90? Round the answer to the nearest tenth of a percent.

34. 87 is 60% of what number?

Simplify Problems 35 through 39 using the Order of Operations Agreement.

35. $2 + 3[4 - (6 \div 3)]$

36. $\left(\dfrac{2}{3}\right)^2 \div \sqrt{\dfrac{25}{36}}$

37. $(11.8 - 2.9)(0.06 \times 1.4)$

38. $81.9 \div 0.9 \times 2$

39. $1/2 \times 2/3 + 5/6 \times 1/5$

In Problems 40 through 43, solve for N.

40. $\dfrac{7}{15} = \dfrac{56}{N}$

41. $\dfrac{11}{N} = \dfrac{27}{18}$

42. $\dfrac{5.6}{18.3} = \dfrac{N}{6.1}$

43. $\dfrac{N}{40} = \dfrac{36}{100}$

Find the answer to each of the following word problems.

44. If the sales tax rate is 6%, find the amount of tax on an item selling for $158.95.

45. Two-thirds of the pieces of candy in an assorted box of chocolates have soft centers. If there are 36 pieces of candy in the box, how many have soft centers?

46. Max's checking account balance was $189.64 before he wrote checks for $12.96, $56.37, and $24.20. How much money remains in his account?

47. If 21 of the 35 points in the football game were scored on passing plays, what percent of the points was scored on passes?

48. Using the information from Problem 47, find what fraction of the points was scored on passing plays.

49. If 15 bagels cost $3.75, what do 8 bagels cost?

50. The scale on a map shows that 2 inches equal 15 miles. How far apart are cities that are 5 1/2 inches apart on the map?

Practical Geometry

Introduction

You may be wondering what this chapter about geometry is doing in the middle of an arithmetic review and beginning algebra text. That's a good question and one that has several answers. But before we answer it, you need to understand what is usually included in a geometry course.

There are two main functions of a beginning geometry course. The first is to acquaint the student with the basic figures, definitions, and relationships that are part of plane geometry. The second and most important aspect of such a course is the development of formal proof. This is the backbone of any complete geometry course, and it is one of the means used to develop critical thinking in mathematics students.

The definitions and relationships in geometry are the practical nuts and bolts used in applications; they are also the concepts that appear on many standardized tests. Another use for these geometry topics is to help "visual learners" understand some of the algebraic concepts and problems via figures and hands-on applications. Although you will see a few formal geometric proofs in this chapter, you will be studying definitions and relationships and working with practical situations. The process of formal proof takes much longer to master and is entirely too important to try to begin here. We will leave that for a full geometry course.

Because there are a number of formulas in this chapter, you should ask your instructor which, if any, you need to memorize. You may also find it helpful to use a calculator to simplify the arithmetic involved in working some of the problems.

Learning Objectives

When you have completed this chapter, you should be able to:

1. Understand the concepts of point, line, and plane.
2. Define perpendicular lines, parallel lines, and intersecting lines.
3. Define and recognize acute, obtuse, right, straight, vertical, adjacent, complementary, and supplementary angles.
4. Define and recognize right, equilateral, isosceles, obtuse, scalene, congruent, and similar triangles.
5. Use the Pythagorean Theorem.
6. Solve problems involving similar triangles.
7. Find the area and perimeter of plane figures.
8. Find the area and circumference of a circle.
9. Calculate the volume and surface area of certain three-dimensional figures.

8.2   Lines and Angles

To begin the study of geometry, we must consider the most basic of all geometric quantities, a point. From this simple concept, we can build everything that we need in practical geometry. If you were to take a straight pin and poke it through a piece of paper, making a hole in the paper, you would have produced a representation of a point. There are other, more scientific ways to describe a point, but you get the idea. A **point** is represented by a single dot or period. If you were to put a number of points next to each other in a row, stretching out from that first point in exactly opposite directions, and if you continued that process indefinitely, you would have drawn a representation of a straight line. A **straight line** looks like a piece of string stretched out indefinitely with no bends or curves. If you were to place a number of straight lines right next to each other on a flat surface and if you extended that process indefinitely, you would have a representation of a surface called a plane. A **plane** looks like a flat surface made up of points, extending indefinitely in all directions, with no spaces in between.

This indefinite and infinite language makes these concepts difficult to understand. Point, line, and plane are in fact *undefined* terms, so you need to have an idea in your mind of what they look like. Lines, unless otherwise described, are straight, and planes are flat surfaces (see Figure 1).

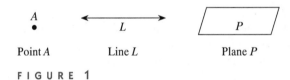

Point *A*   Line *L*   Plane *P*

FIGURE 1

Try thinking of the paper you write on as a plane; the pencil you use can present the line, and the pencil tip is a point. Remember, however, that lines and planes extend forever. They don't really have edges or ends.

Two lines in a plane **intersect** (meet at some point), or they are **parallel** (never meet and always are the same distance apart), or they coincide (one lies on top of the other). The rails in a straight stretch of railroad track are parallel. The distance between the rails is always the same, and they never intersect. Two streets that cross one another on a map intersect at that point. The notation for parallel lines is $L_1 \| L_2$, and the notation for intersecting lines is $L_1 \cap L_2$.

$L_1 \| L_2$ says that line 1 is parallel to line 2.

$L_1 \cap L_2$ says that lines 1 and 2 intersect at some point.

A section of line is called a **line segment** (Figure 2). It has two end points. A section of a line that has one end point and extends forever in the other direction is called a **ray**. Note that the end point of a ray is named first. Be sure to use the arrow on one end of the ray to show that it continues.

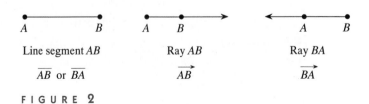

Line segment *AB*   Ray *AB*   Ray *BA*

$\overline{AB}$ or $\overline{BA}$ $\overrightarrow{AB}$ $\overrightarrow{BA}$

FIGURE 2

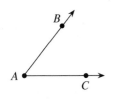

$\angle$ *BAC* or $\angle$ *CAB* or $\angle$ *A*

F I G U R E **3**

When you connect three or more line segments, you make geometric figures. We will examine some of those in Sections 8.3, 8.4, and 8.5. When you join two rays at their end points you have formed an **angle**. $\angle$ is the symbol used to name an angle. Figure 3 shows an angle. In naming an angle with three letters, be sure to indicate the **vertex**, or the corner of the angle, as the middle letter. You can also use the single vertex letter to name the angle when there is no possibility of its being confused with another angle. Remember that the sides of an angle are rays that extend forever.

Before we define different kinds of angles, you need to know how they are measured. The instrument used to measure angles is called a **protractor** (Figure 4). For an angle to be measured on a protractor, its sides must be extended out past the scale marks on the protractor. Make sure that the center of the protractor is on the vertex of the angle and that one side of the angle is on the 0° (zero degree) mark of the protractor. Figure 5 shows that $\angle$ *CDE* has a measurement of 62°. This could be written $m \angle CDE = 62°$. Figure 6 shows how to measure $\angle$ *XYZ*.

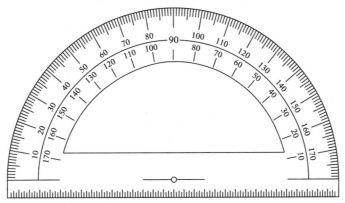

F I G U R E **4**

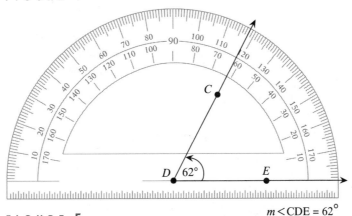

F I G U R E **5** $m < CDE = 62°$

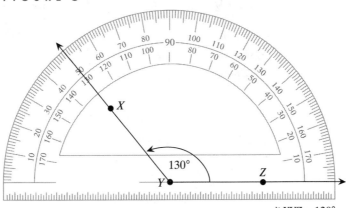

F I G U R E **6** $m \angle XYZ = 130°$

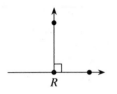

R is a right angle.

$m \angle R = 90°$

F I G U R E **7**

There are several kinds of angles that you need to learn about and be able to recognize. The most widely known angle is a **right angle**. Its sides form a square corner. They are said to be **perpendicular** (⊥), and the measurement of a right angle is 90°. (See Figure 7, where the symbol ⌐ inside the angle shows that it is a right angle.)

Angles that are greater than 0° but less than 90° are called **acute angles**. ∠*CDE* in Figure 5 is an acute angle. An angle whose measure is greater than 90° but less than 180° is called an **obtuse angle**. ∠*XYZ* in Figure 6 is an obtuse angle. ∠*PQR* in Figure 8 is a **straight angle**. Its measure is exactly 180°, and it forms a straight line.

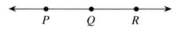

∠*PQR* measures 180° and is a straight angle.

F I G U R E **8**

Angles whose measures are equal are said to be **congruent**. ≅ is the symbol for congruence. If $m \angle DEF = 82°$ and $m \angle KLM = 82°$, then ∠*DEF* ≅ ∠*KLM*. Their sides could seem to be of different lengths because they are rays and extend indefinitely. However, if the measures of the angles between the rays are the same, the angles are said to be congruent.

There are a few other special relationships between angles that you should know. Two angles are said to be **complementary** if their sum is 90°. Two angles are said to be **supplementary** if their sum is 180°.

If $m \angle ABC = 43°$ and $m \angle RST = 47°$, ∠*ABC* and ∠*RST* are complementary angles because 43° + 47° = 90°.

If $m \angle ABC = 43°$ and $m \angle GHI = 137°$, ∠*ABC* and ∠*GHI* are supplementary angles because 43° + 137° = 180°.

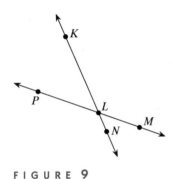

F I G U R E **9**

The opposite angles formed by two intersecting lines are called **vertical** angles. Two angles that share a common side and vertex are called **adjacent** angles. Figure 9 shows that ∠*KLM* and ∠*PLN* are vertical angles. (Can you pick out and name the other pair of vertical angles?) These intersecting lines form four pairs of adjacent angles. ∠*KLM* and ∠*MLN* are adjacent angles. ∠*KLM* is also adjacent to ∠*KLP*. What do you know about the sum of the measures of ∠*KLM* and ∠*MLN*?

E X A M P L E   **1**

If $m \angle KLM$ in Figure 9 is 120°, find the measure of

a. ∠*MLN*   **b.** ∠*NLP*   **c.** ∠*PLK*

Solution

a. ∠*KLM* and ∠*MLN* are adjacent angles, and ∠*KLN* is a straight angle because *KLN* is a straight line. Therefore,

$m \angle KLM + m \angle MLN = 180°$	Definition of straight angle.
$120° + m \angle MLN = 180°$	Substitute; $m \angle KLM = 120°$.
$m \angle MLN = 60°$	Subtract 120° from 180°.

b. ∠*MLN* and ∠*NLP* are adjacent angles, and $m \angle MLP = 180°$ because ∠*MLP* is a straight angle. Therefore,

$m \angle MLN + m \angle NLP = 180°$	Definition of straight angle.
$60° + m \angle NLP = 180°$	From part (a).
$m \angle NLP = 120°$	

c. ∠*NLP* and ∠*PLK* are adjacent angles, and ∠*NLK* is a straight angle with a measure of 180°. Therefore,

$m \angle NLP + m \angle PLK = 180°$	Definition of straight angle.
$120° + m \angle PLK = 180°$	From part (b).
$m \angle PLK = 60°$	

◆

From Example 1, you can see that the intersection of two lines produces four angle⌐. There are four pairs of adjacent, supplementary angles. There are two pairs of vertical angles, and those vertical angles are congruent because their measures are equal: $m \angle KLM = m \angle NLP$, and $m \angle MLN = m \angle KLP$. This is always true and can be proved using formal geometric proofs that are similar to the way these facts were proved in Example 1.

Geometric relationships that can be proved from known or accepted facts are called **theorems**. The process used in Example 1 is an example of a formal proof. The facts and information given were stated. And for each statement made, its source, verification, or explanation was listed in the column on the right. This continued until the desired result was reached. You will not be required to learn formal proof in this chapter, but you may want to study it further in another course. This logical, deductive process appeals to many students. For some people, laying out a deductive proof is similar to solving a mystery or working a puzzle.

8.2 ◆ Practice Problems

A. For each of the following statements, fill in the missing word, words, or symbols.

1. Two angles whose sum is 180° are _____ .

2. An angle less than 90° is a(n) _____ angle.

3. A line consists of _____ points.
   (how many?)

4. A(n) _____ is a part of a line that has one end point and extends indefinitely in the opposite direction.

5. Two angles whose sum is 90° are _____ .

6. A(n) _____ is a part of a line that has two end points.

7. An angle whose measure is between 90° and 180° is a(n) _____ angle.

8. Angles that share a common vertex and a common side are called _____ .

9. Opposite angles formed by two intersecting lines are called _____ angles.

10. An angle of exactly 180° is a(n) _____ angle.

11. The device used to measure angles is called a(n) _____ .

12. An angle of exactly 90° is a(n) _____ angle.

13. Angles are measured in units called _____ .

14. ∥ is the symbol that means that two lines are _____ .

15. ∩ is the symbol that means that two lines are _____ .

16. A plane has _____ edges.
    (how many?)

17. If $\angle A$ and $\angle B$ are complementary, and $m \angle A = 29°$, what is the measure of $\angle B$?

18. If $\angle P$ and $\angle R$ are supplementary angles, and the measure of $\angle R = 97°$, what is the measure of $\angle P$?

19. The measure of $\angle K = 107°$. $\angle K$ and $\angle L$ are supplementary angles. What is the measure of $\angle L$?

20. If the measure of $\angle C$ is 43°, and $\angle C$ and $\angle D$ are complementary, find the measure of $\angle D$.

B. Solve each of the following problems using Figure 10.

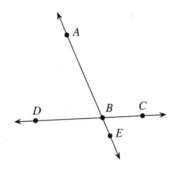

FIGURE 10

21. Find the measure of $\angle DBE$ if $m \angle ABC = 127°$.

22. Find the measure of $\angle ABD$ if $m \angle ABC = 127°$.

23. Find the measure of $\angle DBE$ if $m \angle CBE = 62°$.

24. Find the measure of $\angle EBC$ if $m \angle ABD = 43°$.

C. Draw two lines PQ and RS intersecting at Point T. Let the measure of $\angle PTR$ be 52°.

25. Find the measure of $\angle STQ$.

26. Find the measure of $\angle PTS$.

27. Find the measure of $\angle PTQ$.

28. Find the measure of $\angle RTQ$.

29. $\angle PTS$ and $\angle RTQ$ are _____ and _____ .

30. $\angle STQ$ and $\angle QTR$ are _____ and _____ .

8.3 Three-Sided Figures

When you use three line segments to join three points in a plane that are not all on the same straight line, you form a **triangle**. The symbol $\triangle$ is used to denote a triangle. A triangle, as its name implies, has three angles and, of course, three sides. Figure 11 shows samples of different kinds of triangles. The line segments are called sides and the "corners" are called **vertices**. We describe the sides by using line segment notation. That is, each of these triangles has three sides: $\overline{AB}$, $\overline{BC}$, and $\overline{AC}$. We denote the *lengths of these sides* by AB, BC, and AB without the line segment notation. Each triangle also has three vertices, A, B, and C. $\angle A$ is the **angle opposite** side $\overline{BC}$. $\overline{AC}$ is the **side opposite** $\angle B$. $\angle C$ is the angle opposite $\overline{AB}$.

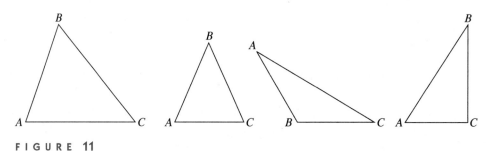

F I G U R E **11**

To show that two sides have the same length, you can say $AB = BC$ (the length of AB is the same as the length of BC), or you can say that $\overline{AB} \cong \overline{BC}$ (line segment $\overline{AB}$ is congruent to ($\cong$) line segment $\overline{BC}$). **Congruent to** means that one is an exact replica of the other. To show that two angles have the same measure, you could say

$$m \angle ABC = m \angle DEF$$

or you could say

$$\angle ABC \cong \angle DEF \text{ (Angle } ABC \text{ is congruent to angle } DEF.)$$

There are several different types of triangles that you need to be able to recognize. Triangles are described in two ways—by the type(s) of angle(s) that they have, and by their sides. We shall first consider angle classification by examining Chart 1.

C H A R T 1 **Classification by Angles**

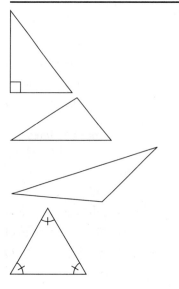

Right triangle: one right angle; the other angles are acute; the side opposite the right angle is called the *hypotenuse*.

Acute triangle: all angles are acute (less than 90°).

Obtuse triangle: one obtuse angle (greater than 90°); both other angles are less than 90°.

Equiangular triangle: all three angles are the same size—are congruent; the lengths of the sides are equal as well.

In this chart, angles with the same number of tick marks are congruent angles, or angles whose measures are equal.

Now let's use Chart 2 to consider classification by sides. The tick marks on the sides of the triangles indicate sides of the same length (same number of tick marks) or of different lengths (different numbers of tick marks).

C H A R T 2   Classification by Sides

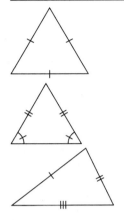

Equilateral triangle: three equal (or congruent) sides; the measures of the angles are equal as well.

Isosceles triangle: only two congruent sides; the angles opposite those sides are also congruent, and they are called **base angles**.

Scalene triangle: all three sides have different lengths; all three angles have different measures as well.

Triangles, therefore, can be described by using angle classification, or side classification, or both. You can talk about an obtuse isosceles triangle. This triangle has two congruent sides and one obtuse angle. A right isosceles triangle has two congruent sides and one right angle. An equilateral triangle is also equiangular. Figure 12 shows several different triangles and gives some information about their parts.

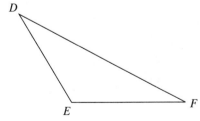

$\triangle ABC$ is a right triangle.
$\angle C$ is a right angle.

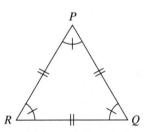

$\triangle PQR$ is an equilateral triangle.
$PQ = QR = PR$,
and $\angle P \cong \angle Q \cong \angle R$.

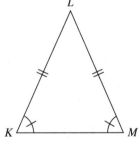

$\triangle KLM$ is an isosceles triangle.
$KM = LM$, $\angle K$ and $\angle M$
are base angles,
and $\angle K \cong \angle M$.

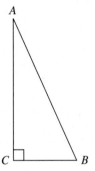

$\triangle DEF$ is an obtuse triangle.
$\angle E$ is an obtuse angle.

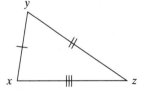

$\triangle xyz$ is a scalene triangle.
All sides and angles are different

F I G U R E 12

The perimeter of any geometric figure is the sum of the lengths of its sides. The formula for the perimeter of any triangle with side lengths a, b, and c is $P = a + b + c$.

E X A M P L E **1**   Find the perimeter of the triangle in Figure 13. *Note*: *BC* means the length of side $\overline{BC}$.

Solution

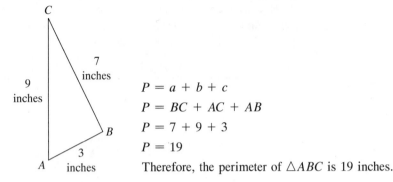

$$P = a + b + c$$
$$P = BC + AC + AB$$
$$P = 7 + 9 + 3$$
$$P = 19$$

Therefore, the perimeter of $\triangle ABC$ is 19 inches.   ◆

F I G U R E **13**

A very important fact that you need to remember is that in plane geometry, *the sum of the measures of the angles in a triangle is always equal to 180°.*

E X A M P L E **2**   Find the measure of $\angle A$ in $\triangle ABC$ in Figure 13 if $m \angle B = 105°$ and $m \angle C = 30°$.

Solution

$$m \angle A + m \angle B + m \angle C = 180°$$
$$m \angle A + 105° + 30° = 180°$$
$$m \angle A + 135° = 180°$$
$$m \angle A = 45°$$

Therefore, the measure of $\angle A$ is 45 degrees.   ◆

When you are trying to find out how much space there is inside a figure, you are looking for the area of that figure. Knowing the perimeter of your back yard enables you to build a fence around it, but the area is the number of square units to be covered by grass. Perimeter measurement is always given in linear units because it measures "the distance around" in one direction. Thus, perimeter answers are given in feet, inches, yards, meters, miles, and so on. Area measurements are always given in square units because they are found by measuring the number of squares necessary to cover the inside surface. It may help you to think of 1-foot-by-1-foot square floor tiles. Each of these tiles measures 1 square foot. (How many 1-square-foot floor tiles would it take to cover the floor of the room where you are sitting?) Area, then, is always given in square inches, square feet, square yards, square meters, and so on.

The area formula for a triangle is $A = 1/2\ bh$, where b represents the length of one side, and h represents the length of the altitude drawn from the opposite vertex. An **altitude** is a line segment drawn from a vertex, perpendicular to the opposite side. Look at Figure 14.

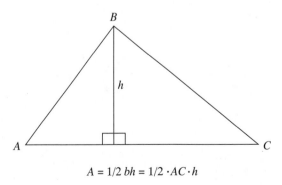

$$A = 1/2\ bh = 1/2 \cdot AC \cdot h$$

F I G U R E **14**

The base in this triangle would be the length of side $\overline{AC}$, and the altitude, or height h, is drawn from B perpendicular to $\overline{AC}$. Base does not always have to be the side that the triangle seems to be sitting on; rather, any side can be a base if the altitude is drawn to that side. Remember, perimeter is measured in linear units (inches, feet, meters) and area is measured in square units (square inches, square feet, square meters).

E X A M P L E   3   Find the area of $\triangle ABC$ in Figure 14 if $AC = 12$ feet and $h = 7$ feet.

Solution $A = 1/2\ bh$

$A = 1/2 \cdot AC \cdot h$

$A = 1/2 \cdot 12 \cdot 7$

$A = 6 \cdot 7$

$A = 42$

Therefore, the area of triangle ABC is 42 square feet.   ◆

Consider $\triangle PQR$ in Figure 15. $\overline{QR}$ is the base, and the altitude h is drawn perpendicular from vertex P to the extension of side $\overline{QR}$. In this case, the altitude is measured outside the triangle. (This happens in obtuse triangles.) To find the height, or altitude, in that situation, you extend the base side and measure the perpendicular distance from the vertex to that base.

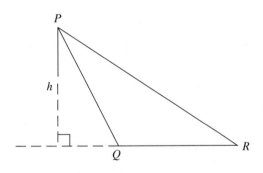

$A = 1/2\ bh = 1/2 \cdot QR \cdot h$

F I G U R E 15

E X A M P L E   4   Find the area of $\triangle PQR$ in Figure 15 if $QR = 36$ meters and $h = 9$ meters.

Solution $A = 1/2\ bh$

$A = 1/2 \cdot QR \cdot h$

$A = 1/2 \cdot 36 \cdot 9$

$A = 18 \cdot 9$

$A = 162$

Therefore, the area of triangle PQR is 162 square meters.   ◆

Because the base is the side to which an altitude is drawn, any side in a triangle can be a base. Figure 16 shows $\triangle PQR$ with an altitude drawn from Q to $\overline{PR}$. In this case, then, $\overline{PR}$ is the base.

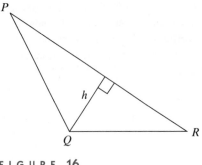

FIGURE 16

E X A M P L E 5 Find the perimeter and the area of △*DEF* in Figure 17. ∡*E* is a right angle.

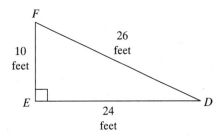

FIGURE 17

Solution We find the perimeter of a triangle by using the formula $P = a + b + c$.

$P = a + b + c$

$P = DE + EF + DF$

$P = 24 + 10 + 26$

$P = 60$

Therefore, the perimeter of triangle *DEF* is 60 feet.

The area formula for a triangle is $A = 1/2\ bh$. Because ∡*E* is a right angle, $\overline{DE}$ is perpendicular to $\overline{EF}$, ($\overline{DE} \perp \overline{EF}$). You can use $\overline{DE}$ as the altitude, and *EF* then is the base.

$A = 1/2\ bh$

$A = 1/2 \cdot DE \cdot EF$

$A = 1/2 \cdot 24 \cdot 10$

$A = 12 \cdot 10$

$A = 120$

Therefore, the area of triangle *DEF* is 120 square feet. ◆

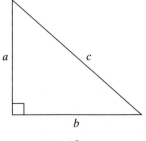

FIGURE 18

Right triangles have another special relationship. The lengths of their sides are related by the Pythagorean Theorem. This theorem was named for its author, Pythagorus. It is an extremely important and useful theorem. To understand this theorem, you need to know that the **hypotenuse** of a right triangle is the side opposite the right angle. The other sides in a right triangle are called **legs**. Because it is opposite the largest angle, the hypotenuse is always the longest side in a right triangle. Look at Figure 18, which shows a right triangle with legs *a* and *b* and hypotenuse *c*. The Pythagorean Theorem states the special relationship between the sides, or legs, of a right triangle and its hypotenuse.

Pythagorean Theorem

> The lengths of the legs in a right triangle, represented by a and b, and the length of its hypotenuse, represented by c, are related according to the formula
>
> $$a^2 + b^2 = c^2$$

To work problems using the Pythagorean Theorem, you will need to use the information in the Square Root Table on the inside back cover of this text. You may also want to review Section 2.2 before going on to try these problems.

E X A M P L E **6**   Find the length of the hypotenuse of $\triangle ABC$ if the legs have lengths of 3 yards and 4 yards.

Solution   Sketch a figure like the one in Figure 19.

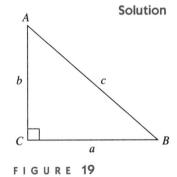

FIGURE **19**

Let $AC = b = 3$ yards.

Let $BC = a = 4$ yards.

Then $a^2 + b^2 = c^2$

$$4^2 + 3^2 = c^2$$

$$16 + 9 = c^2$$

$$25 = c^2$$

Because the value of c, not c^2, is needed, take the square root of both sides of the equation.

$$\sqrt{25} = \sqrt{c^2}$$

$$5 = c$$

The square root of 25 could be either $+5$ or -5, but because c represents a length, $+5$ is the only sensible answer. Therefore, the length of the hypotenuse, AB, is 5 yards. ◆

E X A M P L E **7**   Find the missing length of the leg in triangle GHI in Figure 20 if GH equals 8 feet and the hypotenuse is 12 feet. Round your answer to the nearest tenth of a foot.

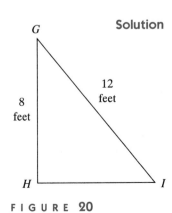

FIGURE **20**

Solution

$a^2 + b^2 = c^2$	Pythagorean Theorem.
$GH^2 + IH^2 = GI^2$	Substitution.
$8^2 + IH^2 = 12^2$	Substitution.
$64 + IH^2 = 144$	$8^2 = 64$ and $12^2 = 144$.
	$64 +$ what number $= 144$?
$IH^2 = 80$	$64 + 80 = 144$, so $IH^2 = 80$.
$\sqrt{IH^2} = \sqrt{80}$	Take the square root of both sides.
$IH \doteq 8.9$	See table on the inside back cover.

Therefore, the length of the hypotenuse is approximately 8.9 feet. ◆

8.3 ◆ Practice Problems

A. **Complete the following sentences by filling in the blanks.**

1. A(n) _____ triangle has three equal sides and three equal angles.

2. The two equal angles in an isosceles triangle are called _____ angles.

3. The corners in a triangle are called _____ .

4. A triangle with one right angle is called a _____ triangle.

5. A(n) _____ triangle has two equal sides.

6. The perimeter formula for a triangle whose sides have lengths a, b, and c is _____ .

7. A(n) _____ is the perpendicular line drawn from a vertex in a triangle to the opposite side.

8. The area formula for a triangle with base b and altitude h is _____ .

9. The sum of the angles in a triangle is always _____ .

10. The Pythagorean Theorem says that the sum of the squares of the two legs in a right triangle is equal to the square of the _____ .

11. If $\triangle ABC$ and $\triangle DEF$ are exact duplicates of each other, $\triangle ABC$ is said to be _____ to $\triangle DEF$.

12. _____ is the symbol for "is congruent to."

B. **Find the answer to each of the following problems.**

13. If in $\triangle PQR$, $m \angle P = 62°$ and $m \angle Q = 54°$, find the measure of $\angle R$.

14. If in $\triangle DEF$, $\angle E$ is a right angle and $m \angle F = 36°$, find the measure of $\angle D$.

15. In $\triangle KLM$, $m \angle K = 40°$ and $m \angle L = 65°$. Find the measure of $\angle M$.

16. In $\triangle DEF$, $m \angle D = 26°$ and $m \angle F = 22°$. Find the measure of $\angle E$.

17. Triangle XYZ is a right triangle with $\overline{XY} \perp \overline{YZ}$. Find the measure of $\angle Z$ if $m \angle X = 30°$.

18. In $\triangle GHI$, $\angle G$ is a right angle and $m \angle I = 18°$. Find the measure of $\angle H$.

C. **Find the answer to each of the following problems.**

19. Find the area of $\triangle PQR$ in Figure 21.

20. Find the perimeter of $\triangle PQR$ in Figure 21.

21. Find the perimeter of $\triangle KLM$ in Figure 22.

22. Find the area of $\triangle KLM$ in Figure 22.

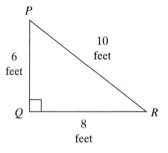

FIGURE 21

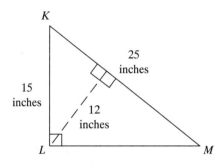

FIGURE 22

D. **Problems 23 through 26 refer to Figure 23. Round answers to the nearest tenth.**

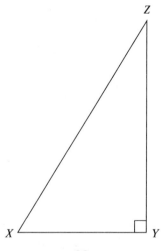

FIGURE 23

23. If $XY = 15$ meters, $XZ = 25$ meters, and $YZ = 20$ meters, find the perimeter of $\triangle XYZ$.

24. If $XY = 6$ feet and $ZY = 8$ feet, find the length of $\overline{XZ}$.

25. If $XY = 5$ inches and $YZ = 8$ inches, find the length of $\overline{XZ}$.

26. If $XY = 5$ inches and $YZ = 8$ inches, find the area of $\triangle XYZ$.

E. Problems 27 through 30 refer to Figure 24.

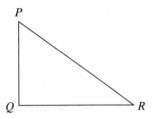

FIGURE **24**

27. If in $\triangle PQR$, $PQ = 24$ feet and $PR = 40$ feet, find the area of $\triangle PQR$.

28. If in $\triangle PQR$, $PR = 35$ inches and $QR = 20$ inches, find the perimeter of $\triangle PQR$.

29. Find the perimeter of $\triangle PQR$ using the information from Problem 27.

30. Find the area of $\triangle PQR$ using the information from Problem 28.

 F. Use your calculator to find the answers to the following problems. Round your answers to the nearest tenth. Problems 31 through 34 refer to Figure 25.

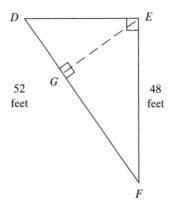

FIGURE **25**

31. Find the area of $\triangle DEF$.

32. Find the perimeter of $\triangle DEF$.

33. Find the length of $\overline{EG}$ using what you know from Problem 31.

34. Find the length of $\overline{DG}$.

◆ **8.4** ◆ Similarity and Congruence

Similar triangles are triangles that have the same shape but are of different size. The corresponding angles of two similar triangles have the same measure (are congruent), but the corresponding sides are not the same lengths—they are proportional. Two triangles are similar, ∼, if one is an enlargement or reduction of the other. Two triangles are congruent, ≅, if one is an exact duplicate of the other. In both similar and congruent triangles, the measures of the corresponding angles are the same; that is, the corresponding angles are congruent.

To better understand the ideas of similarity and congruence, think about your favorite photograph. Although the shape you are thinking about is probably not a triangle, similarity and congruence still apply. Suppose that the photo is a 4 × 5 snapshot. A friend also really likes the picture, so you have another copy of the photo made in the 4 × 5 size. Your photo and your friend's are congruent—exact duplicates of each other. Later you decide that you want to enlarge the photo and put it in an 8 × 10 frame. The enlargement is similar to the original photo but is not congruent to it. The sides in the enlargement are twice as long as the sides in the original photo. But all the angles in the original photo, in your friend's copy, and in the enlargement, are right angles and are therefore congruent. Figure 26 shows how the copy and the enlargement are related to the original. Perhaps this explanation will help you to understand the two concepts and to remember what similarity and congruence mean.

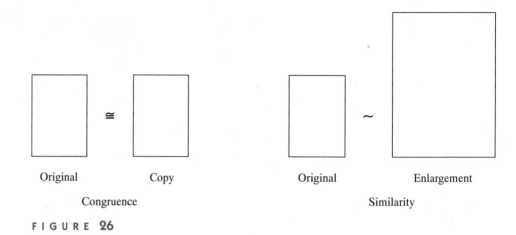

Original   Copy   Original   Enlargement

Congruence   Similarity

FIGURE 26

Figure 27 shows two similar triangles.

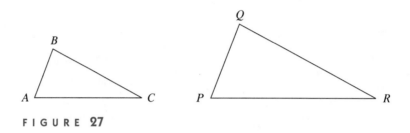

FIGURE 27

$\triangle ABC$ is similar to $\triangle PQR$; that is, $\triangle ABC \sim \triangle PQR$. We show the similarity of the corresponding angles by writing

$$\angle A \cong \angle P, \quad \angle B \cong \angle Q, \quad \text{and} \quad \angle C \cong \angle R$$

You may want to identify the congruent angles with tick marks as in Figure 28.

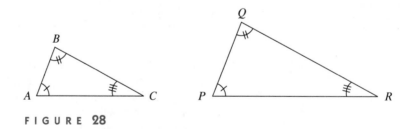

FIGURE 28

Because $\triangle ABC \sim \triangle PQR$, the corresponding sides are proportional. Therefore, the proportion showing the relationships between the lengths of the corresponding sides is written like this:

$$\frac{AB}{PQ} = \frac{BC}{QR} = \frac{AC}{PR}$$

EXAMPLE 1   Given that $\triangle DEF$ is similar to $\triangle KPT$, identify the congruent angles and write the proportion relating the lengths of the sides.

Solution   Making a sketch is always a good idea in solving a geometry problem. Although you do not know what the triangles actually look like, you do know how the corresponding sides need to be placed. Because $\triangle DEF \sim \triangle KPT$, angles D and K are in the same relative position, angles E and P are corresponding, and angles F and T need to match. The sketch might look like the one in Figure 29.

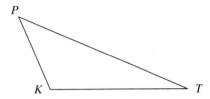
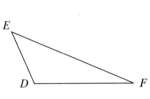

FIGURE **29**

Therefore, $\angle D \cong \angle K$, $\angle E \cong \angle P$, and $\angle F \cong \angle T$. The sides opposite these angles are corresponding sides and so are proportional. Therefore,

$$\frac{EF}{PT} = \frac{DF}{KT} = \frac{DE}{KP}$$ ◆

E X A M P L E 2 For the triangles shown in Figure 30, $\triangle ABC \sim \triangle PQR$, $BC = 12$ inches, $AC = 14$ inches, $PR = 42$ inches, and $PQ = 27$ inches. Find the length of $\overline{AB}$.

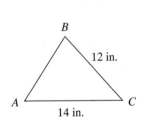
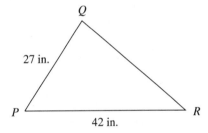

FIGURE **30**

Solution Sketch the triangles as in Figure 30, and fill in the given measurements.

Because the triangles are arranged in the same configuration, you will need to know the length of PQ (the corresponding side), as well as the values for one other pair of corresponding sides. The sketch reveals that the lengths of sides $\overline{AC}$ and $\overline{PR}$ are given. Use the two parts of the proportion that relate these sides.

$$\frac{AB}{PQ} = \frac{AC}{PR}$$

Fill in the known values: $AC = 14$, $PR = 42$, and $PQ = 27$.

$$\frac{AB}{27} = \frac{14}{42}$$

Solve the proportion. Compute the cross-products.

$42 \cdot AB = 27 \cdot 14$

$42 \cdot AB = 378$

Divide both sides by 42.

$$\frac{\cancel{42} \cdot AB}{\cancel{42}} = \frac{378}{42}$$

$AB = 9$ inches ◆

E X A M P L E 3   Using the values given in Example 2, find the length of $\overline{QR}$ (see Figure 30).

Solution   You will need the ratio of AC to PR and the ratio of BC to QR. Even though you now have a value for AB, it is safer to use the given information to do this second problem, because you might have made a mistake in getting AB. Use the parts of the proportion that contain the unknown and the two given values.

$$\frac{BC}{QR} = \frac{AC}{PR}$$

Substitute the given values: $BC = 12$, $AC = 14$, and $PR = 42$.

$$\frac{12}{QR} = \frac{14}{42}$$

$$14 \cdot QR = 504$$

Solve by dividing both sides by 14.

$$\frac{\cancel{14}QR}{\cancel{14}} = \frac{504}{14}$$

$$QR = 36 \text{ inches}$$ ◆

E X A M P L E 4 $\triangle XYZ$ and $\triangle DCE$, in Figure 31, are similar triangles in which $\angle X \cong \angle D$, $\angle Y \cong \angle C$, and $\angle Z \cong \angle E$. $YZ = 10$ feet, $XZ = 18$ feet, $CD = 6$ feet, and $CE = 4$ feet.

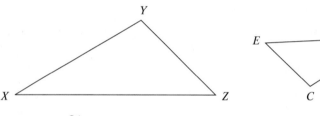

F I G U R E **31**

a. Find the length of $\overline{XY}$.
b. Find the length of $\overline{DE}$.

Solution   a. Before labeling the given parts, rotate $\triangle DCE$ so that the equal angles are in the same position in both triangles. $\angle C$ will need to be at the top, $\angle D$ to the left, and $\angle E$ to the right (Figure 32).

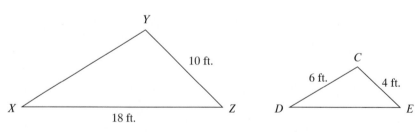

F I G U R E **32**

Now you can readily write the proportion for the corresponding sides.

$$\frac{XY}{CD} = \frac{YZ}{CE} = \frac{XZ}{DE}$$

YZ and *CE* are the pairs of corresponding sides that have both values given. Use that part of the proportion and the part with the side you are looking for. Then substitute *CD* = 6 feet, *YZ* = 10 feet, and *CE* = 4 feet.

$$\frac{XY}{CD} = \frac{YZ}{CE}$$

$$\frac{XY}{6} = \frac{10}{4}$$

$$4 \cdot XY = 60$$

$$\frac{\cancel{4} \cdot XY}{\cancel{4}} = \frac{60}{4}$$

$$XY = 15 \text{ feet}$$

b. Substitute *YZ* = 10, *CE* = 4, and *XZ* = 18.

$$\frac{YZ}{CE} = \frac{XZ}{DE}$$

$$\frac{10}{4} = \frac{18}{DE}$$

$$10 \cdot DE = 72$$

$$\frac{\cancel{10} \cdot DE}{\cancel{10}} = \frac{72}{10}$$

$$DE = 7.2 \text{ feet} \qquad\qquad\qquad \blacklozenge$$

E X A M P L E   **5**   In Figure 33, $\triangle ABC \sim \triangle DBE$. Given that *AB* = 12 feet, *AD* = 4 feet, and *DF* = 6 feet, find *AC*.

Solution   The first thing to do is separate the two triangles so you can see what you have to work with (Figure 34).

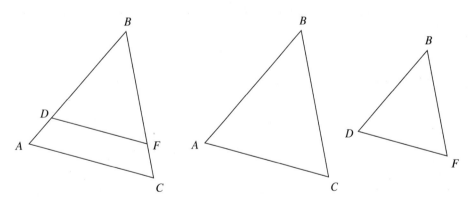

FIGURE **33**   FIGURE **34**

Because $\triangle ABC \sim \triangle DBF$, we know that $\angle A \cong \angle D$, $\angle B \cong \angle B$, and $\angle C \cong \angle F$. The proportion of the corresponding sides is

$$\frac{BC}{BF} = \frac{AC}{DF} = \frac{AB}{DB}$$

Write the proportion substituting the known values.

$$\frac{BC}{BF} = \frac{AC}{6} = \frac{12}{DB}$$

To find *AC*, you need *DF* and two other parts in the same ratio. Can you find *DB*? To find the length of *DB*, go back to the original sketch in Figure 31. You are given $AB = 12$ feet and $AD = 4$ feet. To find *DB* you need to subtract: $AB - AD = 12$ feet $- 4$ feet $= 8$ feet. Therefore, $DB = 8$ feet. Now use the last two ratios and solve the proportion.

$$\frac{AC}{6} = \frac{12}{8}$$

$$8 \cdot AC = 6 \cdot 12$$

$$\frac{\cancel{8} \cdot AC}{\cancel{8}} = \frac{72}{8}$$

$$AC = 9$$

Therefore, the length of side *AC* is 9 feet. ◆

8.4 ◆ Practice Problems

A. Complete the following sentences by filling in the blanks.

1. If $\triangle ABC$ and $\triangle DEF$ are exact duplicates of each other, $\triangle ABC$ is said to be _____ to $\triangle DEF$.

2. If the sides of $\triangle XYZ$ are three times as long as the sides of $\triangle KLM$, $\triangle XYZ$ is said to be _____ to $\triangle KLM$.

3. The ratio of corresponding sides in two congruent triangles is _____ .

4. Two equiangular triangles are always _____ and are sometimes _____ .

5. In two similar triangles, the corresponding sides are _____ .

6. In two similar triangles, the corresponding angles are _____ .

7. Two equilateral triangles are always _____ and are also always _____ .

8. In two congruent triangles, the corresponding sides are _____ .

B. Solve for the missing parts of the pairs of similar triangles shown in Figures 35 through 38. If necessary, round answers to tenths.

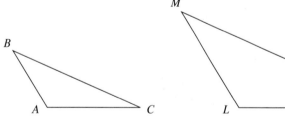

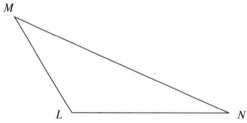

FIGURE **35**

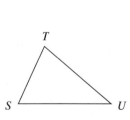

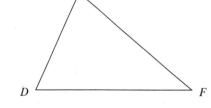

FIGURE **36**

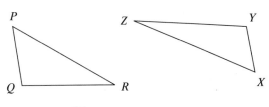

FIGURE 37

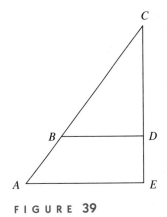

FIGURE 39

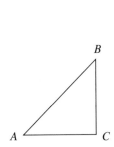

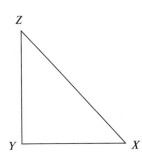

FIGURE 38

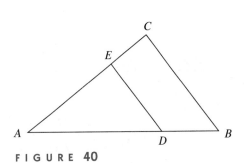

FIGURE 40

9. Use Figure 35, wherein △ABC and △LMN are similar, ∡A ≅ ∡L, ∡B ≅ ∡M, ∡C ≅ ∡N, AB = 6 yards, AC = 8 yards, MN = 18 yards, and LN = 12 yards. Find the length of $\overline{BC}$.

10. Use Figure 36, wherein △STU and △DEF are similar, ∡S ≅ ∡D, ∡T ≅ ∡E, ∡U ≅ ∡F, ST = 6 inches, TU = 4 inches, SU = 8 inches, and DF = 12 inches. Find DE.

11. Use Figure 35 and the information in Problem 9 to find the length of $\overline{ML}$.

12. Use Figure 36 and the information in Problem 10 to find the length of $\overline{EF}$.

13. In the similar triangles shown in Figure 37, ∡P ≅ ∡X, ∡Q ≅ ∡Y, ∡R ≅ ∡Z, PQ = 20 meters, QR = 40 meters, PR = 50 meters, and XZ = 45 meters. Find the length of $\overline{XY}$.

14. Use this information and Figure 38 to find the length of $\overline{AC}$: △ABC is similar to △XYZ, ∡A ≅ ∡X, ∡B ≅ ∡Z, ∡C ≅ ∡Y, AB = 5 feet, BC = 3 feet, XY = 12 feet, and XZ = 15 feet.

15. Use Figure 37 and the information in Problem 13 to find the length of $\overline{YZ}$.

16. Use Figure 38 and the information in Problem 14 to find the length of $\overline{YZ}$.

17. In Figure 39, △ACE is similar to △BCD, ∡C ≅ ∡C, ∡A ≅ ∡B, ∡E ≅ ∡D, AB = 5 yards, BC = 5 yards, BD = 4 yards, and CD = 4 yards. Find the length of $\overline{AE}$.

18. △ABC and △ADE in Figure 40 are similar. Also, ∡A ≅ ∡A, ∡E ≅ ∡C, ∡B ≅ ∡D, AE = 12 feet, EC = 8 feet, ED = 6 feet, and AD = 15 feet. Find the length of $\overline{BC}$.

19. Use Figure 39 and the information given in Problem 17 to find the length of $\overline{CE}$.

20. Use the information given in Problem 18 and Figure 40 to find the length of $\overline{AB}$.

C. Problems 21 through 30 refer to Figure 41, wherein △KJN is similar to △KLM, $\overline{KL} \perp \overline{LM}$, $\overline{KJ} \perp \overline{JN}$, KM = 40 yards, KJ = 20 yards, and JN = 15 yards.

21. Find the length of $\overline{KN}$.
22. Find the length of $\overline{NM}$.
23. Find the length of $\overline{LM}$.
24. Find the length of $\overline{KL}$.
25. Find the length of $\overline{JL}$.
26. Find the area of △KJN.
27. Find the area of △KLM.
28. What is the ratio of KM to KN?
29. What is the ratio of the perimeter of △KLM to the perimeter of △KJN?

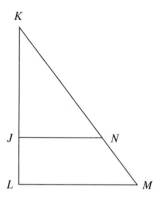

FIGURE 41

30. What is the ratio of the area of △KLM to the area of △KJN?

8.5 Quadrilaterals

Four-sided plane geometric figures are called **quadrilaterals**. There are plane geometric figures that have more than four sides too, of course. This whole group of plane figures with straight-line sides is called **polygons**. A stop sign, for instance, is a type of polygon called an **octagon**. The Defense Department in Washington, D.C., is housed in a building shaped like a polygon with five equal sides that is called, for this very reason, the **Pentagon**. Triangles are also a special kind of polygon. If all the sides in a polygon are the same length, all the angles have the same measure. Such a polygon is called a **regular polygon**. A square is a regular quadrilateral. An equilateral triangle is also a regular polygon. The stop sign we mentioned is a regular polygon. For the purposes of this section, we will study only four-sided polygons—quadrilaterals. Several special types of quadrilaterals are described in Chart 3. You need to be able to recognize all of them and to know what characteristics are particular to each type.

CHART 3

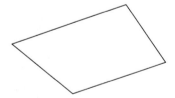

Quadrilateral: four sides.

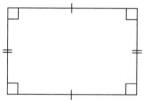

Rectangle: four sides, four right angles; opposite sides are parallel (∥) and congruent (≅).

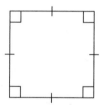

Square: four congruent sides, four right angles; opposite sides are parallel; all sides are congruent.

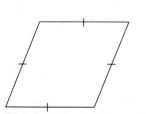

Rhombus: four congruent sides; opposite angles are congruent and parallel.

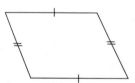

Parallelogram: opposite sides are parallel and congruent; opposite angles are congruent.

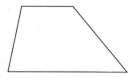

Trapezoid: only one pair of parallel sides called *bases*.

The perimeter of any geometric figure is the sum of its sides. The sum of the angles in any quadrilateral is 360°. This is true because any quadrilateral can be split into two triangles, each containing angles whose sum is 180°. Follow this "proof":

Look at Figure 42. It shows an irregular quadrilateral. The figure has four sides and four angles. A dashed line connects vertex A and vertex C. This line segment $\overline{AC}$ separates the quadrilateral into two triangles, $\triangle ABC$ and $\triangle ACD$. Because the sum of the angles in a triangle is equal to 180°, you know that

$$m \angle ABC + m \angle BCA + m \angle CAB = 180°$$
$$m \angle ADC + m \angle DCA + m \angle DAC = 180°$$

If you combine the two equations by adding them together, you get the statement

$$m \angle ABC + m \angle ADC + m \angle BCA + m \angle DCA + m \angle CAB +$$
$$m \angle DAC = 360°$$

Now look at $\angle BCA + \angle DCA$. Their sum is $\angle BCD$. Also, $\angle CAB + \angle DAC = \angle BAD$. Substitute these two equivalent angles into the foregoing combined statement.

$$m \angle ABC + m \angle ADC + \underline{m \angle BCA + m \angle DCA} + \underline{m \angle CAB + m \angle DAC}$$
$$= 360°$$

$$m \angle ABC + m \angle ADC + \qquad m \angle BCD \qquad + \qquad m \angle BAD \qquad = 360°$$

Looking at these four angles in Figure 42, you can see that the angles in the statement name the four angles in the quadrilateral.

$$m \angle B + m \angle D + m \angle C + m \angle A = 360°$$

Therefore, you have proved that the sum of the measures of the four angles in any quadrilateral is equal to 360°.

Now study these examples to see how you can use some of the information you have learned about quadrilaterals.

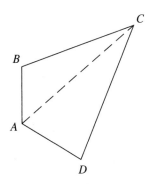

FIGURE **42**

E X A M P L E   1   Find the perimeter of the quadrilateral in Figure 43 if $a = 2$ feet, $b = 4$ feet, $c = 7$ feet, and $d = 6$ feet.

Solution

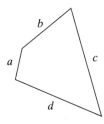

$$P = a + b + c + d$$
$$P = 2 + 4 + 7 + 6$$
$$P = 19$$

The perimeter is 19 feet.

FIGURE **43**

E X A M P L E   2   Find the measure of $\angle B$ as seen in Figure 44 if $m \angle A = 80°$, $m \angle C = 110°$, and $m \angle D = 105°$.

Solution

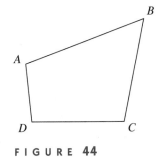

$$m \angle A + m \angle B + m \angle C + m \angle D = 360°$$
$$80° + m \angle B + 110° + 105° = 360°$$
$$m \angle B + 295° = 360°$$
$$m \angle B = 65°$$

Therefore, the measure of $\angle B$ is 65°.

FIGURE **44**

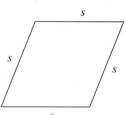

E X A M P L E   3   Find the perimeter of a rhombus if one of its sides is 12 meters.

Solution   First sketch a rhombus similar to the one shown in Figure 45.

$$P = s + s + s + s \qquad\qquad P = 4s$$
$$P = 12 + 12 + 12 + 12 \quad \text{or} \quad P = 4(12)$$
$$P = 48 \qquad\qquad\qquad\qquad P = 48$$

Therefore, the perimeter of this rhombus is 48 meters.   ◆

F I G U R E   45

There is only one area formula for triangles, but quadrilaterals have a different formula for each special shape. The area formulas for various quadrilaterals are given in Chart 4. Be sure to memorize these formulas and to remember that area is always given in square units.

C H A R T   4

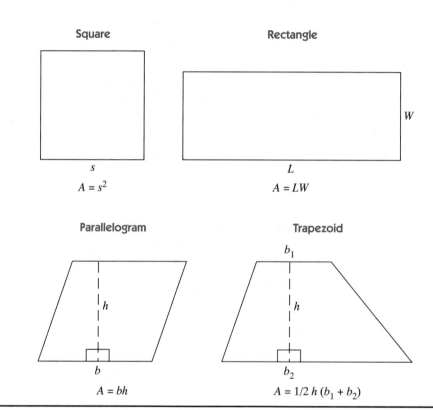

Area Formulas for Quadrilaterals

Square

$A = s^2$

Rectangle

$A = LW$

Parallelogram

$A = bh$

Trapezoid

$A = 1/2\, h\, (b_1 + b_2)$

Note: In the formula for the area of a trapezoid, the two bases are denoted by b, but b_1 is a different base from b_2. Subscript notation is often used in mathematics.

E X A M P L E   4   Find the area of a parallelogram whose base is 16 feet and whose height is 5 feet. (See Figure 46.)

Solution

$$A = bh$$
$$A = (16)(5)$$
$$A = 80$$

The area of this parallelogram is 80 square feet.   ◆

F I G U R E   46

E X A M P L E   **5**   Find the area of the trapezoid in Figure 47 if its height is 12 inches.

Solution

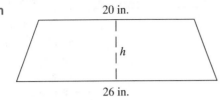

FIGURE 47

$A = 1/2\ h(b_1 + b_2)$

$A = 1/2 \cdot 12(20 + 26)$

$A = 1/2 \cdot 12(46)$

$A = 6(46)$

$A = 276$

The area of this trapezoid is 276 square inches.

8.5 ◆ Practice Problems

A. For each condition listed on the left, choose, from the list on the right, the names of all figures that always satisfy that condition.

Condition

1. four equal angles

2. four equal sides

3. two pairs of parallel sides

4. four sides

5. two pairs of equal sides

6. one pair of parallel sides

7. no right angles

8. four right angles

Name

A. quadrilateral

B. square

C. rectangle

D. parallelogram

E. rhombus

F. trapezoid

B. Solve each of the following problems.

9. Find the area of the rectangle in Figure 48.

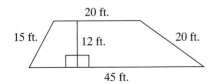

FIGURE 48

10. Find the perimeter of the rectangle in Figure 48.

11. Find the area of a square if one of its sides measures 18 inches.

12. Find the perimeter of a rhombus if one of its sides measures 23 meters.

13. Find the perimeter of the trapezoid in Figure 49.

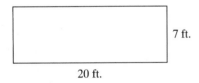

FIGURE 49

14. Find the area of the trapezoid in Figure 49. (Its altitude is 12 feet.)

15. Find the perimeter of the parallelogram in Figure 50.

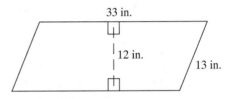

FIGURE 50

16. Find the area of the parallelogram in Figure 50.

17. Find the area of the trapezoid in Figure 51, whose altitude is 35 feet.

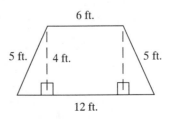

FIGURE 51

18. Find the perimeter of the trapezoid in Figure 51.

19. Find the area of the quadrilateral in Figure 52.

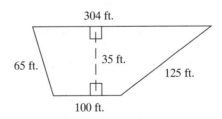

FIGURE 52

20. Find the area of the quadrilateral in Figure 53.

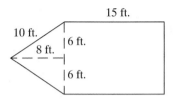

FIGURE **53**

21. Katie is landscaping her back yard and needs to determine how many square yards of sod to order. The yard's shape is shown in Figure 54. How many square yards of sod should she buy?

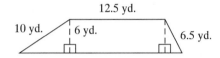

FIGURE **54**

22. Wright Rug is laying carpet in the Swenson's bedroom and adjoining hallway. How many square feet of carpeting are needed to cover rectangular areas that measure 14 feet by 11 feet and 4 feet by 29 feet?

23. How much edging will Katie need to enclose her back yard if its dimensions are as shown in Figure 54?

24. After the carpeting is laid in the areas described in Problem 22, new baseboard molding will be installed. Ignoring openings for doorways, find the number of linear feet of molding that the bedroom and hallway will require.

C. Use your calculator to help you to answer each of the following questions. If necessary, round your answers to the nearest hundredth.

25. Find the cost to carpet a banquet hall whose dimensions are 12.5 yards by 20 yards if the cost of the carpet is $17.85 per square yard.

26. Find the cost to lay parquet tile in an office entrance area if the dimensions of the entryway are 24 feet by 15 feet. The parquet tile sells for $6.35 a square foot.

27. How much more would it cost to cover the banquet hall in Problem 25 with the parquet tile described in Problem 26 instead of using carpet? (How many square feet are there in a square yard?)

28. How much less would it cost to carpet the office entryway if the company used the carpet from Problem 25 instead of the parquet flooring in Problem 26? (How many square feet are there in a square yard?)

29. Farmer Don is using a header 16 feet wide on his combine to cut his wheat field. If the field is 448 feet wide, how many passes must he make to cut the field?

30. A roll of wallpaper will cover approximately 35 square feet of wall. How many rolls of paper will be needed to cover a stage backdrop if the set measures 42 feet wide by 18 feet high?

8.6   Circles

Circles are different from other plane figures, because they don't have corners, or vertices, and their sides are curved lines rather than straight. For this reason, the vocabulary used in studying circles is different than that used with polygons. In Figure 55 the various parts of a circle are labeled.

When measuring the distance around the outside of a circle, you are measuring its **circumference**, not its perimeter. The surface inside a circle is still its area and is measured in square units.

By definition, a **circle** is a set of points that lie the same distance from a point called its **center**. The distance from the center to a point on the circle is called its **radius**, and the length of a line segment from one side of the circle through its center to the other side is called its **diameter**.

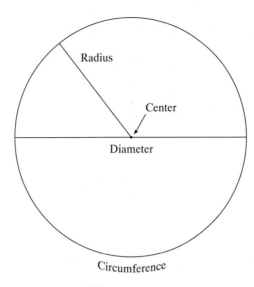

FIGURE 55

Both the formula for circumference and the one for the area of a circle use a number denoted by the Greek letter π. Any time you measure the circumference of a circle and divide that quantity by the length of the diameter of the circle, you get the same result. The value is approximately equal to 3.14. Because this result, 3.14, is the same for all circles, it was given a special name, pi (π). Whenever you need to solve a formula using π, you can substitute 3.14, 3 1/7, or 22/7 for π; all are acceptable approximations. Remember, though, that your answers are approximations and not exact answers.

The formula for the circumference of a circle is $C = \pi d$, or $C = 2\pi r$, where d represents the diameter and r is the radius. (The formulas are equivalent because the radius goes from the center to the circle, and two radii lying in a straight line would form a diameter.) The formula for the area of a circle is $A = \pi r^2$. Sometimes you will be asked to substitute the decimal or fraction value for π, and in other problems you will be able to choose which to use or to leave your answer in terms of π.

Formulas for a Circle

The formula for the circumference of a circle is given by $C = 2\pi r$, or $C = \pi d$, where r is the radius, and d is the diameter, and $\pi \doteq 3.14$ or $22/7$. ($\doteq$ means "is approximately equal to.")

The formula for the area of a circle is $A = \pi r^2$.

E X A M P L E 1 Find the circumference of a circle whose radius is 3 inches. Round your answer to the nearest tenth.

Solution $C = 2\pi r$

$C \doteq 2(3.14)(3)$

$C \doteq 6.28(3)$

$C \doteq 18.84$

The circumference is approximately 18.8 inches. ◆

E X A M P L E   **2**   Find the area of a circle with a radius of 3 inches. Round your answer to the nearest tenth.

Solution $A = \pi r^2$
$A \doteq (3.14)(3)^2$
$A \doteq (3.14)(9)$
$A \doteq 28.26$

The area is approximately 28.3 square inches.   ◆

E X A M P L E   **3**   Find the circumference of a circle whose diameter is 5 feet. Give the answer in fraction form.

Solution $C = \pi d$
$C \doteq (22/7)(5)$
$C \doteq (22/7)(5/1)$
$C \doteq 110/7$ or $15\ 5/7$

The circumference is approximately 15 5/7 feet.   ◆

E X A M P L E   **4**   Find the area of a circle whose diameter is 7 feet. Leave the answer in terms of π.

Solution   If the diameter is 7 feet, the radius is one-half of 7, or 7/2 feet.
$A = \pi r^2$
$A = \pi(7/2)^2$
$A = \pi(49/4)$
$A = 49/4\ \pi$ or $12\ 1/4\ \pi$

The area is approximately 49/4 π square feet, or 12 1/4 π square feet.   ◆

E X A M P L E   **5**   Find the area of the composite shape in Figure 56. The dimensions are shown on the figure. Use 3.14 as an approximation for π.

Solution   A **composite** shape is a shape that is formed when two or more geometric shapes are joined together. The shape in Figure 56 appears to be a rectangle with a **semicircle** (half of a circle) attached to its right side. You need to calculate the area of the rectangle and the area of the semicircle and then add those two areas together.

Area of rectangle = length times width
$A = LW$
$A = (28)(12)$
$A = 336$ square feet

Area of semicircle = $1/2\pi r^2$
$A \doteq (0.5)(3.14)(7^2)$
$A \doteq (0.5)(3.14)(49)$
$A \doteq 76.93$ square feet

Area of composite figure $\doteq 336 + 76.93$
$\doteq 412.93$ square feet   ◆

12 ft.

7 ft.

28 ft.

F I G U R E   **56**

When finding the area of composite figures, you will sometimes need to add and sometimes to subtract. You might also be asked to find the perimeter of a composite figure. In that case, you need to be careful not to count a common side twice. In the shape in Figure 56, you would not want to include the diameter of the semicircle, because it is also part of the length of one side of the rectangle. This type of problem is more challenging than some you will do. Composite figures are a little like puzzles.

8.6 ◆ Practice Problems

A. Answer each of the following questions. Use $\pi \doteq 3.14$.

1. What is the radius of a circle whose diameter is 9.3 feet?

2. What is the diameter of a circle whose radius is 16 inches?

3. Find the area of a circle whose radius is 10 meters.

4. Find the area of a circle whose radius is 4 feet.

5. Find the circumference of a circle whose diameter is 8 inches.

6. Find the circumference of a circle whose diameter is 12.2 meters.

7. Find the circumference of a circle whose radius is 3.5 feet.

8. Find the circumference of a circle whose radius is 10 feet.

9. Find the area of a circle whose diameter is 18 meters.

10. Find the area of a circle whose diameter is 12 inches.

B. In Problems 11 through 18, leave the answer in terms of π.

11. Find the circumference of a circle whose radius is 3.4 meters.

12. Find the circumference of a circle whose diameter is 5 1/2 inches.

13. Find the area of the circle in Problem 11.

14. Find the area of the circle in Problem 12.

15. Find the circumference of a circle whose diameter is 7 1/4 feet.

16. Find the circumference of a circle whose radius is 3 1/8 meters.

17. Find the area of the circle in Problem 15.

18. Find the area of the circle in Problem 16.

C. In Problems 19 through 22, give the answers correct to the nearest hundredth.

19. Nancy is finishing a circular pillow by sewing a ruffle around the edge. How much of this ruffled edging will she need if the pillow has a 22-inch diameter?

20. A gardener is ordering ivy plants to cover a circular planting area. He wants the plants to spread and fill in this circular area, which is 40 feet in diameter. How many plants will he need if each plant will cover 1 square foot?

21. If Nancy decides to make a second pillow from a contrasting fabric, how much material will she need to purchase to cover the front and back of a slightly smaller pillow whose diameter is 20 inches?

22. How much edging would be needed to enclose the planting area in Problem 20?

23. Find the perimeter of the shape in Figure 57.

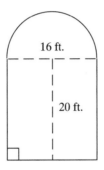

16 ft.

20 ft.

FIGURE 57

24. Find the area of the shape in Figure 57.

25. Find the area of the shape in Figure 58.

26. Find the perimeter of the shape in Figure 58.

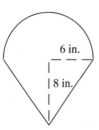

6 in.

8 in.

FIGURE 58

27. Find the area of the shape in Figure 59.

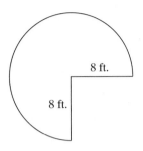

FIGURE 59

28. Find the area of the shaded portion in Figure 60.

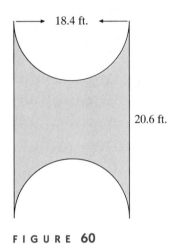

FIGURE 60

29. Find the area of the shaded portion in Figure 61.

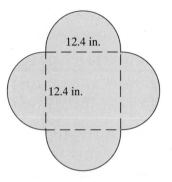

FIGURE 61

30. Find the area of the sidewalk around Malcolm's swimming pool if the dimensions are as shown in Figure 62.

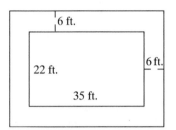

FIGURE 62

8.7 Three-Dimensional Figures

Up to this point, all the geometric figures you have studied could be drawn on a piece of paper—that is, on a plane. They were at most two-dimensional, having only length and width. Now we will consider three-dimensional figures—figures that have length, width, and height. Most of these figures get their names from the two-dimensional figures from which they come.

A **rectangular solid**, shaped like a shoe box, has six sides, has all right angles, and is described by its length, width, and height. A **cube** is a rectangular solid all of whose sides are squares of the same size. The **surface area** of a rectangular solid is the sum of the areas of all its surfaces. Volume is the measurement of the space inside a three-dimensional figure. Volume is measured in *cubic units*. The **volume of a rectangular solid** is given by the formula $V = LWH$. Figure 63 shows a rectangular solid and a special type of rectangular solid called a cube.

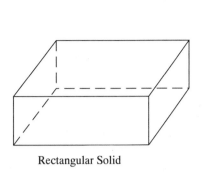
Rectangular Solid

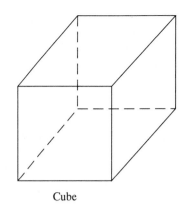
Cube

F I G U R E **63**

E X A M P L E **1** Find the volume of a rectangular solid whose length is 14 feet, whose height is 12 feet, and whose width is 3 feet.

Solution Because you are finding volume, use the formula $V = LWH$.

$V = (14)(3)(12)$

$V = (42)(12)$

$V = 504$

Therefore, the volume of this rectangular solid is 504 cubic feet. ◆

E X A M P L E **2** Find the surface area of the rectangular solid in Example 1.

Solution In order to find the surface area, you will need to compute the area of the six surfaces. There will be three pairs of matching surfaces: a top and a bottom, a front and a back, and two sides. Draw a sketch similar to the one in Figure 64 to help you determine each area. Because each surface is a rectangle, you will use the area formula $A = LW$.

The area of the top and of the bottom is found using $A = (14)(3)$, so each of those two areas measures 42 square feet.

The area of each end is found using $A = (12)(3)$, so each measures 36 square feet.

The area of the front and back is found using $A = (14)(12)$, so each of the two measures 168 square feet.

Therefore, the total surface area is $42 + 42 + 36 + 36 + 168 + 168$, or 492 square feet. ◆

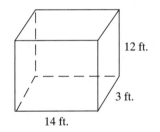

12 ft.

3 ft.

14 ft.

F I G U R E **64**

A **right circular cylinder** is shaped like a tin can. Its base and top are circles, and its sides are perpendicular to the bases (see Figure 65). You can think of the side of this figure as the label going around the can. What shape would the label be if you pulled it off the can? It would be a rectangle whose length is the distance around the can, the circumference of its top or bottom. The width of the label is equal to the height of the can. The total surface area (T) of a right circular cylinder is given by the formula $T = 2\pi r^2 + 2\pi rh$, where r is the length of a radius of a base, and h is the height of the cylinder. You are adding the area of the circular top and bottom $2(\pi r^2)$, to the area of the rectangular label, $2\pi r$ times h. The volume of a circular cylinder is given by the formula $V = \pi r^2 h$, where r is the measure of the radius of a base, and h is the measure of an altitude. Figure 65 shows the can and also the label and the top and bottom. The label of the can is a rectangle with height h and length the same as the circumference of the can, $2\pi r$. You can see by studying this figure how the formula for the surface area of a right circular cylinder was derived.

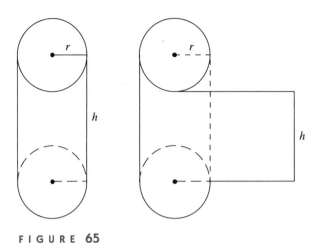

FIGURE 65

EXAMPLE 5   Find the surface area of a right circular cylinder whose radius is 3 inches and whose height is 4 inches. Leave the answer in terms of π.

Solution   The figure looks like a tin can. It has two circular ends, and its sides can be thought of as a rectangular label wrapped around it. The formula for the surface area will be used.

$T = 2\pi r^2 + 2\pi rh$

$T = 2\pi(3)^2 + 2\pi(3)(4)$

$T = 2\pi(9) + 2\pi(12)$

$T = 18\pi + 24\pi$

$T = 42\pi$

The total surface area is approximately equal to 42π square inches.   ◆

EXAMPLE 6   You must design the label for a new product to be sold in local stores. The product will be sold in cans that are 7 inches high and have a diameter of 5 inches. How much area do you have to work with when designing the label?

Solution   The question is asking for the surface area of the side of the can. In order to find the area of this rectangular piece of paper, you need to know its length and width. As was shown in Figure 65, the height of the can is the width of the rectangle, and the circumference of the top is the length of the rectangle. First, calculate the circumference of the can. Because you are talking about real-world measurement, you might want to use the decimal equivalent for π, 3.14.

$C = 2\pi r = \pi d,$

$C \doteq (3.14)(5)$

$C \doteq 15.7$ inches

The area formula for a rectangle is $A = LW$, and you have determined that $L = 15.7$ inches and $W = 7$ inches. Therefore,

$A = LW$

$A = (15.7)(7)$

$A = 109.9$ square inches

You will have approximately 110 square inches of space on which to design the label.   ◆

The final figure you will study is a sphere. A **sphere** is shaped like a globe or a soccer ball (Figure 66). The total surface area (T) of a sphere with a radius of r units is given by the formula $T = 4\pi r^2$. The volume (V) of a sphere is given by the formula $V = 4/3 \, \pi r^3$, where r is the measure of its radius.

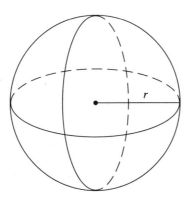

F I G U R E **66**

E X A M P L E **6**   What is the volume of a basketball whose diameter is 12 inches? Use $\pi \doteq 3.14$.

Solution   A basketball is a sphere, so you need to use the formula $V = 4/3 \, \pi r^3$. Be sure to find the radius because that is what the formula calls for. If the diameter is 12 inches, the radius is 6 inches.

$$V = 4/3 \, \pi r^3$$
$$V \doteq 4/3(3.14)(6)^3$$
$$V \doteq 4/3(3.14)(216)$$
$$V \doteq 288(3.14)$$
$$V \doteq 904.32$$

The volume of this basketball is approximately 904.32 cubic inches. ◆

Studying these three-dimensional figures should expand your ability to think in three dimensions. You need to be able to visualize what something looks like. Learn the correct names for these figures and keep a simple picture of each in your mind to help you work these problems. Chart 5 gives the formulas for the surface area and volume of each of these figures.

C H A R T **5**

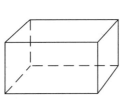 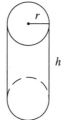 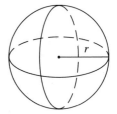

Figure:	Rectangular solid	Right circular cylinder	Sphere
Total Surface Area, T:	$T =$ sum of areas of sides	$T = 2\pi r^2 + 2\pi rh$	$T = 4\pi r^2$
Volume, V:	$V = LWH$	$V = \pi r^2 h$	$V = 4/3 \, \pi r^3$

8.7 ◆ Practice Problems

A. Answer each of the following questions. In Problems 1 through 12, leave the answers in terms of π where appropriate.

1. Find the volume of a cube that is 6 cm wide.

2. Find the surface area of the cube in Problem 1.

3. Find the surface area of a rectangular box that is 11 inches long, 8 inches wide, and 4 inches deep.

4. Find the volume of the box in Problem 3.

5. Find the surface area of a juice can if its radius is 2 inches and it is 10 inches high.

6. Find the volume of the juice can in Problem 5.

7. Find the volume of a basketball whose radius is 5 inches.

8. Find the surface area of a basketball whose radius is 6 inches.

9. Find the volume of a rectangular box if its length is 8 inches, its height is 5 inches, and its width is 3 inches.

10. Find the surface area of a juice can that is 10 inches high if its diameter is 8 inches.

11. Find the volume of the juice can in Problem 10.

12. Find the surface area of the box described in Problem 9.

B. Use your calculator to help you answer these questions. Round your answers to the nearest hundredth where appropriate.

13. Find the surface area of the carton in Figure 67.

14. Find the volume of the carton in Figure 67.

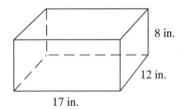

8 in.

12 in.

17 in.

F I G U R E **67**

15. How many gallons of water can be put into a can if the can is 1.6 feet high and has a diameter of 0.8 feet? (1 cubic foot of water is approximately 7.48 gallons.)

16. How many gallons of water could be stored in a drum that is 4 feet tall and has a diameter of 2.6 feet? (1 cubic foot of water is approximately 7.48 gallons.)

17. How many ounces of water could be stored in a can that is 14 inches high and has a diameter of 8 inches? (1 cubic inch of water weighs approximately 0.361 pound, and 1 ounce is 1/16 pound.)

18. How many ounces of water does a can hold if its height is 8 inches and its diameter is 14 inches? (Use the relationships given in Problem 17.)

19. The circumference of the earth is approximately 6,400 kilometers. What is its approximate area?

20. What is the approximate diameter of the earth if its circumference is 6,400 kilometers?

21. The circumference of a dime is approximately 5.652 centimeters. Find its diameter.

22. The circumference of a quarter is approximately 7.85 centimeters. Find its area.

23. Find the area of the dime in Problem 21.

24. Find the radius of the quarter in Problem 22.

25. What is the surface area of the solid shown in Figure 68? Round your answer to the nearest hundredth of a meter.

26. What is the volume of the solid shown in Figure 68?

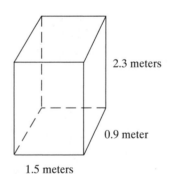

2.3 meters

0.9 meter

1.5 meters

F I G U R E **68**

8.8   Chapter Review

In studying this chapter, it is important for you to become familiar with the vocabulary of plane geometry. The types of angles, the kinds of triangles, the various other two- and three-dimensional figures should all be recognizable now. It is not necessary to have memorized all the formulas, but be sure you know where to find them. And you must know how to answer questions when you are given the formula and the specifications of the figure you are studying.

Now that you have completed this chapter, you should be able to:

1. Give working definitions of the terms angle, area, perimeter, acute, obtuse, right, scalene, perpendicular, parallel, intersecting, line segment, ray, complementary, supplementary, vertical, adjacent, congruent, similar, altitude, and hypotenuse.

2. Sketch each of the following figures: triangle, quadrilateral, trapezoid, rectangle, cube, rhombus, parallelogram, circle, and sphere.

3. Find missing measures of angles that appear in triangles or quadrilaterals or are formed by intersecting lines.

4. Use the Pythagorean Theorem to find the missing dimension of a right triangle.

5. Find missing measures in similar triangles.

6. Discuss congruence and be able to work problems where congruent figures are used.

7. Find the perimeter, area, volume, and surface area of geometric figures.

Review Problems

A. Fill in the blank in each of the following tatements.

1. If $m \angle ABC$ is 76°, $\angle ABC$ is a(n) _____ angle.

2. If $m \angle PQR$ is 102°, $\angle PQR$ is a(n) _____ angle.

3. The measure of $\angle XYZ$ is 180°, so $\angle XYZ$ is a(n) _____ angle.

4. $\angle ABC$ and $\angle CBD$ are complementary angles, so $m \angle ABC + m \angle CBD = $ _____ .

5. $\angle PRS$ and $\angle RST$ are supplementary angles, so $m \angle PRS + m \angle RST = $ _____ .

6. In $\triangle KLM$, $\angle L = 90°$, so $m \angle K + m \angle M = $ _____ .

7. In $\triangle XYZ$, $m \angle X = 42°$ and $m \angle Y = 106°$, so $m \angle Z = $ _____ .

8. Two angles are _____ if their measures are exactly the same.

9. A(n) _____ triangle has three equal sides.

10. A(n) _____ triangle has two equal sides.

11. A(n) _____ triangle has no equal sides.

12. Two triangles that are exact duplicates of each other are said to be _____ .

13. If the sides of one triangle are exactly half as long as the sides of another triangle and their shapes are the same, they are said to be _____ .

14. Area is always measured in _____ units.

15. Volume is always measured in _____ units.

B. Solve each of the following problems using Figure 69.

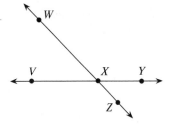

F I G U R E **69**

16. Find the measure of $\angle YXZ$ if $m \angle WXY = 122°$.

17. Find the measure of $\angle VXZ$ if $m \angle WXY = 122°$.

18. $\angle WXY$ and $\angle VXZ$ are _____ angles.

19. $\angle WXV$ and $\angle VXZ$ are _____ angles.

20. $m \angle WXV + m \angle VXZ = $ _____

C. **Answer the following questions using Figure 70.**

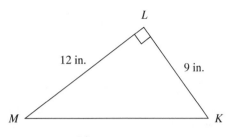

FIGURE 70

21. How long is $\overline{KM}$?
22. What is the perimeter of $\triangle KLM$?
23. What is the area of $\triangle KLM$?

D. **Using Figure 71, answer the following questions.**

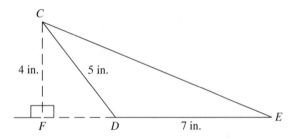

FIGURE 71

24. Find the area of $\triangle CDE$.
25. Find the length of $\overline{FD}$.
26. Find the length of $\overline{CE}$. (*Hint*: Use right triangle *CFE*.)
27. Find the perimeter of $\triangle CDF$.

E. **Using Figure 72, answer each of the following questions given that $\triangle ABC \sim \triangle DFE$.**

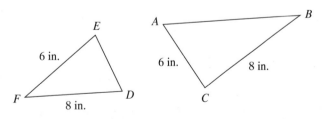

FIGURE 72

28. Find the length of $\overline{AB}$.
29. Find the length of $\overline{ED}$.

F. **Answer each of the following questions.**

30. Figure *ABCD* is a quadrilateral wherein $m \angle A = 25°$, $m \angle C = 110°$, and $m \angle D = 145°$. Find $m \angle B$.
31. Find the area of a trapezoid if its bases are 6 meters and 11 meters and its height is 8 meters.
32. Find the perimeter of a rectangle that is 14 yards long and 9 yards high.
33. Find the area of the rectangle in Problem 32.
34. Find the area of a circle whose diameter is 4 inches. (Use $\pi \doteq 3.14$.)
35. Find the circumference of the circle in Problem 34.
36. Find the volume of a cardboard packing crate whose dimensions are 3 feet by 2 1/2 feet by 4 feet.
37. Find the surface area of the crate in Problem 36.
38. Find the surface area of a cube whose edge measures 4 1/2 feet.

G. **Answer each of the following questions. Leave your answer in terms of π.**

39. Find the volume of an oil drum whose diameter is 2 1/2 feet and whose height is 3 1/2 feet.
40. Find the surface area of the oil drum in Problem 39.
41. Find the volume of a small soccer ball if its diameter is 9 inches.
42. Find the amount of leather needed to cover the soccer ball in Problem 41.

Chapter Test

Check with your instructor to see if you may use a list of the formulas in this chapter when you take the test, or if you must memorize the formulas.

1. Find the area of a rectangle whose length is 16 yards and whose width is 2 1/4 yards.

2. Find the volume of a cube that is 8 cm deep.

3. Find the area of the triangle in Figure 73.

FIGURE **73**

4. Find the circumference of a circle whose radius is 7 inches. (Use $\pi \doteq 3.14$.)

5. In $\triangle PQR$, $\angle R$ is a right angle. Find the length of $\overline{PQ}$ if $PR = 4$ feet and $QR = 6$ feet.

6. Line JKL intersects line MKN at point K. Name the angle that is congruent to $\angle LKN$.

7. $\angle R$ and $\angle S$ are supplementary angles. If $m \angle R = 38°$, what is the measure of $\angle S$?

8. If two angles are complementary and congruent, what is the measure of each angle?

9. $\triangle RST$ is a right triangle with $m \angle S = 90°$. Find the length of $\overline{RT}$ if $RS = 10$ feet and $ST = 24$ feet.

10. $\triangle ABC$ is similar to $\triangle DEF$. Find the length of $\overline{AB}$ if $BC = 12$ cm, $EF = 8$ cm, and $DE = 6$ cm.

11. In the parallelogram pictured in Figure 74, $\angle A \cong \angle C$, $\angle B \cong \angle D$, and $m \angle A = 76°$. Find the measure of $\angle B$.

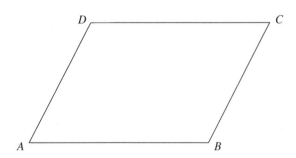

FIGURE **74**

12. Find the perimeter of the trapezoid in Figure 75.

13. Find the area of the trapezoid in Figure 75.

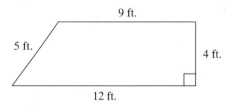

FIGURE **75**

14. Find the area of a circle whose diameter is 6.4 inches. Leave your answer in terms of π.

15. Find the surface area of the box in Figure 76.

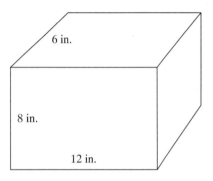

FIGURE **76**

16. Calculate the volume of the box in Figure 76.

17. Find the area of the label on a fruit drink can if the can is 9 inches high and has a radius of 2 inches. Round your answer to the nearest tenth.

18. How many pounds of grass seed are needed to re-seed a lawn if the dimensions of the lawn are 20 feet by 26 feet, and 1 pound of grass seed will cover 25 square feet?

19. Calculate the volume of a juice can whose radius is 1.5 inches and whose height is 10 inches. Leave the answer in terms of π.

20. A large beach ball has a diameter of 24 inches. What is the volume of this beach ball? (Use $\pi \doteq 3.14$.)

Thought-Provoking Problems

1. What reasons can you think of for studying some geometry in a fundamentals course?

2. List four things that you learned in this chapter that you did not know before. Which of those things are important to you?

3. Write a real-life problem wherein the sketch you would use would consist of two intersecting lines. Ask a question, and then calculate the answer to the question.

4. The formula for the area of a rectangle is $A = LW$. Sketch a rectangle, and draw a diagonal connecting two opposite corners. From the figure, write an explanation of the formula for the area of a triangle, $A = 1/2\ bh$.

5. $\triangle KPR$ is a right triangle with $\angle P = 90°$. If $KP = 2.6$ feet and $PR = 5.2$ feet, find the length of $\overline{KR}$. Round your answer to the nearest tenth of a foot. You may use your calculator if you wish.

6. Given $\triangle DEF$ as shown in Figure 77, find each of the following:
   a. Length of $\overline{DG}$
   b. Length of $\overline{EF}$
   c. Area of $\triangle DEF$
   d. Area of $\triangle DGE$
   e. Area of $\triangle GEF$
   f. How could you find the answer to part e by using parts c and d?

7. The Rockmans have a rectangular swimming pool in their back yard that is 25 feet wide and 40 feet long. The pool is surrounded by a concrete deck that is 4 feet wide. What is the area of this concrete sunning area? (*Hint*: Making a sketch might be very helpful.)

8. The Municipal Swimming Pool in Kokomo, Indiana, is circular and has a diving tower in the center. The distance from the center of the diving tower to the edge of the pool is 80 feet. The pool is surrounded by a concrete apron 12 feet wide. What is the area of this concrete apron? You may leave your answer in terms of π if you think that is easier, but giving the answer in square feet makes more sense in this situation.

9. A circular spa has a depth of 3.5 feet and a diameter of 5 feet. How many cubic feet of water does the spa hold? If the weight of a cubic foot of water is 62 pounds, what is the weight of the water in the spa?

10. Which is larger, a circular pizza with a diameter of 12 inches or a square pizza with a length of 12 inches? What is the difference between the areas of these two pizzas?

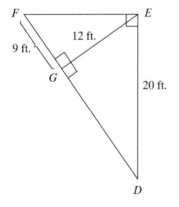

FIGURE **77**

The Why and What of Algebra

Arithmetic to Algebra

In this chapter, we will begin to tie arithmetic and algebra together. We will look at the differences between them and examine some applications that show why algebra is such a useful tool.

◆ **Learning Objectives**

When you have completed this chapter, you should be able to:

1. Recognize situations wherein using algebra might be helpful.
2. Recognize variables, expressions, and equations.
3. Simplify algebraic expressions by using the Order of Operations Agreement and by combining like terms.
4. Evaluate algebraic expressions.
5. Recognize the properties of real numbers.
6. Solve simple algebraic equations by inspection or intuition.

9.2   The "Why" of Algebra

You may think that you have never had any experience with algebra other than in a classroom. That, in fact, is probably not the case. Think about how you figure your grades in a class. If your grade for the term is based simply on the average of your scores on three tests, how do you calculate your grade? To find your average, you add the three scores together and divide the total by three. For instance, if your grades were 87, 95, and 94, you would figure the average this way:

$$87 + 95 + 94 = 276 \qquad 276 \div 3 = 92$$

Therefore, your average would be 92.

If your instructor wanted to find the averages for all the students in her class, she would have to figure each total separately and then divide by three. She might decide to use her computer to help her, instructing it to do the task over and over again with each student's grades. One way to do this would be to write a general statement that the computer could "understand" and would respond to by adding each set of three numbers and dividing the total by three to get the average. She would write a *formula*, using letters to keep it general, and the formula would give exactly those instructions. She might use this one:

$$A = \frac{a + b + c}{3}$$

where A stands for average and a, b, and c are the three scores. Any letter could be used to stand for the average, as long as what that letter represents is clearly defined, but A seems appropriate to stand for "average."

The use of letters and mathematical symbols to describe a process is **algebra**. Algebra and its formulas and symbols state in general terms what arithmetic numbers and symbols state in specific terms.

This isn't as complicated as it sounds. Think about the last vacation you planned. You knew approximately how fast you could travel in an hour (miles per hour) and about how many hours you could drive each day. Knowing that information, you could get an approximate idea of how many miles you could travel per day.

If you know you can drive about 50 miles in an hour, you know you can drive 100 miles in 2 hours and 150 miles in 3 hours, and so on. You can multiply the rate at which you travel by the number of hours you drive and find out how many miles you can travel. In general, the *distance* you can travel is equal to the *rate* at which you travel *times* your traveling *time*. In an algebraic formula,

$$\text{Distance} = \text{rate} \times \text{time}, \quad \text{or } D = rt$$

Again, you do not have to use D, r, and t, but they are the most logical choices to help you remember what the parts of the formula represent.

You could write this information in table form as well. In this example, the average rate at which you drive never changes, so the only things that vary are the time and the distance. The following table displays the relationship between the time and the distance traveled.

t (hours)	1	2	3	4	5	6	7
D (miles)	50	100	150	200	250	300	350

In Chapter 18 you will learn how to display this type of information on a two-dimensional graph called a Cartesian coordinate system. You could indicate the hours (t) on the horizontal scale and the miles or distance (D) on the vertical scale. Figure 1 shows how this graph might be drawn.

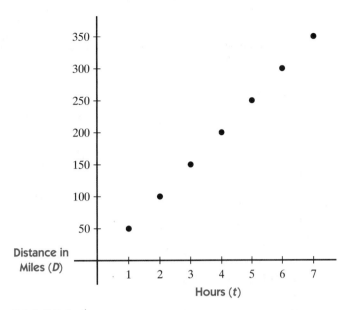

F I G U R E 1

In any situation in which formulas are used to describe in a general manner something that is always true, algebraic techniques are used. Think about a sale at a store. The sale price is equal to the regular price minus the discount. You could also say $S = P - D$, where S stands for the sale price, P stands for the regular price, and D stands for the amount of the discount.

Computers were mentioned earlier. Any time a computer is used to keep records, figure prices, or do any other repetitive tasks, the person programming the computer must communicate with it in mathematical, algebraic formulas.

Think of other situations wherein you do a task over and over. Could you use a general statement and write a formula that would describe what you are doing?

These examples do not, of course, tell the whole story about algebra, but they do form a basis for understanding the difference between arithmetic and algebra. Arithmetic deals with very specific situations, and algebra generalizes those situations. Remember that the use of a letter symbol (called a **variable**) instead of a number symbol (called a **numeral**) generalizes what you are doing and says that the relationship can be true in more than one instance.

Algebra

General statement of a relationship between quantities that is always true.

Sale price = Regular price − Discount
  $S = P - D$

Distance = rate × time
  $D = rt$

Arithmetic

Specific problem involving calculations to arrive at an answer.

Find the sale price of a $120 coat if a $30 discount is offered.

How far can you travel in 6 hours if you average 55 miles per hour?

In order to translate from words to symbols, you should become familiar with certain terms commonly used to indicate arithmetic operations. You need to recognize these terms and to know which operation each implies.

Sum   ——————→ addition

Difference ——————→ subtraction

Product ——————→ multiplication

Quotient ——————→ division

Remember these terms so that you can translate correctly from words to symbols when writing formulas.

Another common practice in algebra is to indicate multiplication by putting a number next to a letter, with no operation sign between them. This is done to avoid the confusion between the multiplication symbol $\times$ and the letter X. Multiplication can be indicated in these ways:

$$4 \cdot y \qquad 4(y) \qquad (4)(y), \quad \text{or, more commonly,} \quad 4y.$$

Try to write an algebraic sentence, or **formula**, for each of the following problems.

E X A M P L E 1   Write a formula for the distance, D, that you can travel at 55 miles per hour in a given number of hours, t.

Solution   In 1 hour, you can travel 55 miles.

In 2 hours, you can travel 110 miles.

In 3 hours, you can travel 165 miles.

From this pattern, it would seem you can multiply the number of hours by 55 to get the distance. Therefore, the formula is

$$D = 55t$$

(Remember that in algebra, writing a number next to a letter with no operation symbol in between means you are to multiply.)   ◆

E X A M P L E 2   Write a formula for the number of inches, I, in a given number of feet, represented by F.

Solution   You know that 1 foot has 12 inches.

You know that 2 feet have 24 inches.

You know that 3 feet have 36 inches.

From this pattern, it would seem you could multiply the number of feet you are given by 12 to change the measurement to inches. Therefore, the formula is

$$I = 12F$$ ◆

E X A M P L E 3   Using the formula in Example 2, find the number of inches in 9 feet.

Solution $I = 12F$ Write the formula.

$I = 12(9)$ Substitute the values and perform the indicated operations.

$I = 108$

There are 108 inches in 9 feet.   ◆

E X A M P L E 4   Write a formula for the volume of a box if the volume is equal to the product of its length, width, and height.

Solution   Volume should be represented by V, length by L, width by W, and height by H. Because *product* means to multiply, the formula is

$$V = LWH$$

No symbol is necessary between the L and W or between W and H.   ◆

E X A M P L E 5   Write a formula for the cost (C) of buying t theater tickets if the tickets cost $6.75 each. Then find the cost of 8 tickets.

Solution   You are looking for the cost of t tickets at $6.75 each. You could add $6.75 t times or multiply $6.75 times t. Because you don't have a value for t, the formula to use is

$$C = 6.75t$$

To find the cost of 8 tickets, substitute 8 in place of t and do the calculation.

$C = 6.75t$

$C = 6.75(8)$

$C = 54.00$

Therefore, the cost for 8 tickets is $54.00. ◆

E X A M P L E 6 Look at the information given in the following table. See if you can write a formula to show the relationship between the given values.

F	5	6	7	8	9	10
G	10	11	12	13	14	15

Solution If you consider the first vertical pair of values, you could multiply 5 by 2 to get 10, or you could add 5 to 5 to get 10. You have two choices: either multiply by 2 or add 5. Look at the second pair of values. Multiplying 6 by 2 does not give 11. Adding 5 to 6 does yield 11. Just to be sure, consider the third pair. If you add 5 to 7, you do get 12. Therefore, the relationship in the chart is such that if you add 5 to the top number, you get the bottom number.

5	6	7	8	9	10
+ 5	+ 5	+ 5	+ 5	+ 5	+ 5
10	11	12	13	14	15

The formula that makes this same statement is

Top number   plus   five   equals   bottom number.

$\quad\quad F \quad\quad + \quad 5 \quad\quad = \quad\quad\quad G$

If you "began with" the bottom values, you would be subtracting 5, so you would also be safe in writing

Bottom number   minus   five   equals   top number.

$\quad\quad\quad G \quad\quad\quad - \quad 5 \quad\quad = \quad\quad\quad F$ ◆

E X A M P L E 7 Look at the information given in the following table, and write a formula that shows the relationship between the values for R and the values for S.

R	12	16	20	24	28
S	3	4	5	6	7

Solution Top number   divided by   4   equals   bottom number.

$\quad\quad R \quad\quad\quad ÷ \quad\quad 4 \quad = \quad\quad\quad S$

You could also write

4   times   bottom number   equals   top number.

$\quad 4 \quad\quad · \quad\quad\quad S \quad\quad\quad = \quad\quad\quad R$

In this example, $R/4 = S$ or $4S = R$ would also be acceptable. Because an equality statement is true whether you read from left to right or from right to left, $S = R/4$ and $R = 4S$ are also acceptable. ◆

In this book, you will often be asked to write a formula that says in algebraic language what the specific numbers tell you to do. Don't be discouraged if this takes a lot of practice before you are comfortable doing it. It's not easy at first, but your skill will improve with effort and practice.

9.2 ◆ Practice Problems

A. Translate each of these statements into an algebraic formula. If you are not told what letters to use, choose letters that are appropriate to the quantities being represented.

1. The number of quarts (q) is equal to four times the number of gallons (g).

2. The number of yards (Y) is equal to the number of feet (f) divided by three.

3. The gasoline mileage for your car (M) is equal to the number of miles driven (m) divided by the number of gallons used (g). (*Hint*: Be careful when writing lowercase m and uppercase M. They stand for two different quantities.)

4. The average (A) of four grades is equal to the sum of the grades, a, b, c, and d, divided by four. (*Hint*: Be careful to distinguish between A and a. They stand for two different quantities.)

5. The cost (C) of postage stamps is equal to 25 cents times the number of stamps (s).

6. The cost (C) of the movie tickets is $4 times the number of adult tickets (a) added to $3 times the number of student tickets (s).

7. The total number of students in the class is equal to the number of males added to the number of females.

8. The cost of the trip is the sum of the food costs and the hotel costs. (What operation does *sum* indicate?)

9. The number of feet is equal to the quotient of the number of inches and twelve. (What operation is signaled by *quotient*?)

10. The product of the rate at which you travel and the time you travel gives you the distance you can go. (What operation does *product* call for?)

B. Look at the information given in each of the following tables, and write a formula for each.

11.
p	3	4	5	6	7
q	6	7	8	9	10

12.
j	5	4	3	2	1
k	15	12	9	6	3

13.
a	4	6	8	10	12
b	8	12	16	20	24

14.
g	7	6	5	4	2
h	5	4	3	2	0

15.
w	1	3	5	7	9
x	6	8	10	12	14

16.
e	12	10	8	6	4
f	6	5	4	3	2

C. Write an algebraic formula for each of the following situations, and then answer the question.

17. Write a formula for A if it is equal to the difference between R and S. Then find A when $R = 17$ and $S = 9$.

18. Write a formula for K if it is equal to the product of A and B. Then find K when $A = 13$ and $B = 12$.

19. Write a formula for L if it is equal to the quotient of W and Y. Then find L when $W = 171$ and $Y = 19$.

20. Write a formula for M if it is equal to the sum of K and L. Then find M when $K = 32$ and $L = 4$.

21. Write a formula for the distance you can drive in h hours at 52 miles per hour. Find how many miles you can travel in 7 hours.

22. Write a formula for the total cost of an article if its price is c cents and you need to pay t cents in tax. Now find the total cost of a pair of shoes that cost $21.95 if the tax is $1.32.

23. Write a formula for the number of chairs in the theater if there are r rows of s seats each. How many people can be seated in a theater that has 54 rows of 36 seats each?

24. Write a formula for the cost of p pairs of socks at $1.19 per pair. How much will you pay for 5 pairs of socks?

9.3 The "What" of Algebra

As we noted in the previous section, letters that stand for numerical values are called **variables**. That name makes sense because their values can change, or *vary*. The values of numbers do not change, so when they stand alone, they are called **constants**. When a number stands in front of a variable, indicating multiplication, it is called the **numerical coefficient**. When the variable is written alone—for example, Y—the numerical coefficient is understood to be 1. It can also be written as $1Y$. Any single variable or constant, or any grouping of variables or constants without an equals sign, is called an **expression**. Following are some examples of algebraic expressions:

$x + 3$ x is the variable; 1 is its coefficient.

            3 is the constant term.

            $x + 3$ is the algebraic expression.

$3y - 2$ y is the variable; 2 is the constant.

            3 is the numerical coefficient of y.

            $3y - 2$ is the expression.

$4 - 3c$ 4 is the constant; c is the variable.

            3 is the numerical coefficient of c.

            $4 - 3c$ is the expression.

$5xy$ 5 is the numerical coefficient.

            x and y are the variables.

            $5xy$ is the expression.

An expression in algebra is like a phrase in English—it contains only a part of an idea. The phrase "gone fishing" expresses a part of the idea, but it does not tell you who has gone fishing or where! In order to get the complete idea, you need to write a complete sentence. "Mike has gone fishing at the pond" conveys a whole idea. An algebraic sentence that completes a thought or makes a statement and has values on both sides of an equals sign is called an equation or a formula. Whenever there is a statement that has an equals sign ($=$) in it, this statement is called an **equation**. Special equations that involve more than one variable are often called **formulas**. And as we have said, an algebraic statement without an equals sign, such as $x - 4$, is called an **expression**.

Consider the following examples of equations or formulas:

$A = x + 3$

$14 = 3y - 2$

$4 - 3c = K$

$B = 5xy$

$P = a + b + c$

Note that each of these examples is a full "sentence" because it contains an equals sign. To distinguish between an equation and a formula, you need to consider the number of variables, or letters, in the sentence. Generally speaking, for our purposes here, an equation will involve only one variable, or letter, and a formula will involve more than one. Thus, $x + 3 = 7$ is an example of an equation, and $M = 3A - 2B + 4C$ is an example of a formula. (Later on, you will study equations that also have more than one variable.)

Let's look at the parts of an algebraic expression or equation that are called terms. A term can be made up of symbols, variables, or constants. In an expression or equation, **terms** are separated by addition or subtraction signs.

Consider this expression with one term: $5a$.

The 5 is called the *numerical coefficient*.

The a is called the *variable* or the *literal factor*.

Consider the one-term expression $7ab^2$.

The 7 is called the *numerical coefficient*.

The a and b are *variables* or *literal factors*.

The ^{2} is called the *power* or *exponent*.

Consider the expression $3y + 2$.

This expression has two terms, because $3y$ and 2 are separated by an addition sign.

The 3 is the *numerical coefficient* of the first term.

The y is the *variable* of the first term.

The 2 is the *constant* term in the expression.

Consider the formula $L = 4m + 3n$.

This formula has two terms because $4m$ and $3n$ are added together.

4 and 3 are the *numerical coefficients*.

m and n are the *variables*.

L is the *value* that results from the expression on the right.

Sometimes in an expression or an equation, there will be more than one term that has the same variable part. If you **combine** (add or subtract) those like-variable (same-variable) terms, you can write the equation or expression in simplified form. Consider the following expression, and decide for yourself how you might make it look simpler.

$$4a + 6a - 2a$$

Because all the terms have the same variable part—the letter a—you can combine those terms using the signs you are given. Work from left to right.

$$4a + 6a - 2a$$
$$10a - 2a$$
$$8a$$

Did you guess that answer correctly? If not, see where you made your error and try this one: $7y - 5y + 4y - 3y$. Remember to combine from left to right.

$$7y - 5y + 4y - 3y$$
$$2y + 4y - 3y$$
$$6y - 3y$$
$$3y$$

Like terms are terms that have the same variable part but not necessarily the same numerical coefficients. Like terms in any algebraic expression should be combined. Use the operation sign before each term, combine the numerical coefficients, and write the variable part unchanged. Remember that a variable term with no numerical coefficient showing is understood to have a coefficient of 1. You might wish to write that 1 in the expression when you are combining like terms.

What do you suppose you would do with a problem like this?

$$9W + 5Y - 3W + 2Y$$

You can combine like terms, but only like terms. (Which are they?) You can rearrange the terms in any order as long as the sign in front of each term moves with it.

$$9W + 5Y - 3W + 2Y$$
$$9W - 3W + 5Y + 2Y$$
$$6W + 7Y$$

You have to stop there because $6W$ and $7Y$ are not like terms. What is unusual about the following problem?

$$9P + 7R - 3P + R$$

R is a variable by itself. Its numerical coefficient is 1 because $1 \cdot R = R$.

$$9P - 3P + 7R + R$$
$$6P + 8R$$

Here is an example involving Y^2 instead of Y. Terms can be combined as long as their variable parts are exactly alike, but Y^2 is a different variable term from Y alone. The letter *and* any exponent parts must be *exactly* alike for the terms to be like terms.

$$8Y^2 + 5Y - 3Y^2 + 4Y$$
$$8Y^2 - 3Y^2 + 5Y + 4Y$$
$$5Y^2 + 9Y$$

Now try the following problem. Combine like terms.

$$3R + 2S - 5T$$

There aren't any like terms, so none can be combined. The expression already is in simplest form.

9.3 ◆ Practice Problems

A. Fill in the blanks.

1. In the formula $K = 3W + 7Y$,
   a. 3 and 7 are _____ .
   b. W and Y are _____ .
2. In the formula $A = 2c + 3d + 5$,
   a. 2 and 3 are _____ .
   b. c and d are _____ .
   c. 5 is a(n) _____ .
3. In the formula $4m + 3n + 6 = 8m$,
   a. the constant term is _____ .
   b. the variables are _____ .
   c. the numerical coefficients are _____ .
4. In the formula $4p + 5q + 7 = 12$,
   a. the variables are _____ .
   b. the numerical coefficients are _____ .
   c. the constants are _____ .
5. $a + b$ is called a(n) _____ .
6. $W = p + q$ is called a(n) _____ .
7. In $6a$, 6 is the _____ .
8. In $9y$, y is the _____ .
9. In $3a^2b$, ^{2} is called the _____ .
10. In WXY^3, ^{3} is called the _____ .

B. Combine like terms. Remember to combine from left to right.

11. $6w - 3w + 2w$
12. $9d - 5d + 4d$
13. $17x + 4x - 9x$
14. $6b^2 - 5b^2 + 2b^2$
15. $4a^3 - 3a^3 + a^3$
16. $9f + 3f - 5f$
17. $11P - P + 3P$
18. $14k^3 - 3k^3 - 5k^3$
19. $7c^2 + 4c^2 - 2c^2$
20. $16m - 9m - 5m$

C. Simplify each expression by combining like terms. Rearrange the order of the terms if it helps.

21. $6y + 4z + 5y + 3z$
22. $5W^2 + 2X^3 + 3W^2 + 4X^3$
23. $19f + 7g - 4f + 2g$
24. $9t - 6t + 8w - 4w$
25. $17a + 7b - 4a - 3b$
26. $23x + 7y - 8x + 3y$
27. $a^2 + 5b^2 + 6a^2 - b^2$
28. $24m - 7m + 2n + 8m$
29. $19b + 4c - 3c + 4c$
30. $11k - 4k + 6k - 3M$

9.4 ◆ **Evaluating Expressions and Formulas**

Review the Order of Operations Agreement, which was presented in Section 2.3. It is just as important in algebra as it was in arithmetic.

Order of Operations Agreement

> In any problem containing more than one mathematical operation, always follow these steps:
>
> 1. Simplify within parentheses or grouping symbols.
> 2. Simplify any expressions that involve exponents or roots.
> 3. Multiply or divide, in order, from left to right.
> 4. Add or subtract (combine), in order, from left to right.

In a moment, we will try some arithmetic examples as a review. In order to do these, however, you need to know how to simplify expressions with exponents, or powers. In the expression 3^2, which is read "three to the second power" or "three squared," the 3 is called the **base**. The ^{2} is the **exponent** or **power**.

3^2 means $3 \cdot 3$, or 9

4^3 means $4 \cdot 4 \cdot 4$, or 64

5^2 means $5 \cdot 5$, or 25

2^5 means $2 \cdot 2 \cdot 2 \cdot 2 \cdot 2$, or 32

The last two examples make it clear that 5^2 and 2^5 are not the same.

Exercises   Try these problems and see if you can simplify them correctly. The answers follow.

1. 2^3 **2.** 7^2 **3.** 5^3 **4.** 1^5 **5.** 3^3

Answers:

1. 8   **2.** 49   **3.** 125   **4.** 1   **5.** 27

In terms that contain several factors, the exponent applies only to its base, the number or letter right beside it. For example, $3(2)^4 = 3(2 \cdot 2 \cdot 2 \cdot 2) = 3(16) = 48$. Only 2 is raised to the fourth power, not the product of 3 and 2. In another example, $2^3(5) = (2 \cdot 2 \cdot 2)(5) = 8(5) = 40$; only 2 is raised to the third power.

Now you should be ready to try some simplifying problems using the Order of Operations Agreement and the definition of exponent, or power.

E X A M P L E 1   Simplify: $26 - 8 \times 2 + 9 \div 3$

Solution $26 - \underline{8 \times 2} + \underline{9 \div 3}$ Multiply or divide, left to right.

$\underline{26 - 16} + 3$ Add or subtract, left to right.

$\underline{10 + 3}$ Add or subtract, left to right.

13

E X A M P L E **2**   Simplify: $4^3 - 9 \times 2 + 6$

Solution $\underline{4^3} - 9 \times 2 + 6$ Exponents first.

$64 - \underline{9 \times 2} + 6$ Multiply or divide, left to right.

$\underline{64 - 18} + 6$ Add or subtract, left to right.

$\underline{46 + 6}$ Add or subtract, left to right.

52 ◆

E X A M P L E **3**   Simplify: $45 - 4(8 + 3)$

Solution $45 - 4(\underline{8 + 3})$ Parentheses first.

$45 - \underline{4(11)}$ Multiply or divide, left to right.

$\underline{45 - 44}$ Add or subtract, left to right.

1 ◆

Algebra and arithmetic come together when you are asked to find the value of a formula and are given specific values for the variables. A typical problem, for example, might be to find the value of K in $K = 5a + 3b$ when $a = 4$ and $b = 6$. First substitute the given values for a and b into the equation or formula. Then use the Order of Operations Agreement to simplify the resulting numerical expression. Replace a with 4 and b with 6 (remember that a number next to a letter means to multiply).

$K = 5a + 3b$

$K = 5(4) + 3(6)$ Multiply or divide.

$K = 20 + 18$ Add or subtract (combine).

$K = 38$

Using the same formula, find the value of K if $a = 7$ and $b = 4$. Substitute for a and b.

$K = 5a + 3b$

$K = 5(7) + 3(4)$ Multiply or divide.

$K = 35 + 12$ Add or subtract (combine).

$K = 47$

Any algebraic expression or equation can be evaluated by substituting the given numerical values and using the Order of Operations Agreement.

E X A M P L E **4**   Find the value of $P(Q - R)$ when $P = 8$, $Q = 6$, and $R = 2$.

Solution $P(Q - R)$ Substitute for P, Q, and R.

$= 8(6 - 2)$ Simplify within the parentheses first.

$= 8(4)$ Multiply.

$= 32$ ◆

E X A M P L E **5**   Find the value of $6x^3y$ when $x = 2$ and $y = 5$.

Solution $6x^3y$ Substitute for x and y.

$= 6(2)^3(5)$ Apply the rule for exponents.

$= 6(8)(5)$ Multiply.

$= 48(5)$ Multiply.

$= 240$ ◆

9.4 ◆ Practice Problems

A. Simplify each of the following expressions using the Order of Operations Agreement. Be sure to write out each step.

1. $16 + 3 \cdot 5$
2. $7 + 4 \cdot 3$
3. $4^2 - 5$
4. $3^4 + 2$
5. $12 + 3^2$
6. $2^3 - 3$
7. $14 - 3 \cdot 2$
8. $20 - 4 \cdot 3$
9. $4 + 5(3 - 1)$
10. $18 - 3(4 - 2)$
11. $7(8 + 3)$
12. $9(5 - 2)$
13. $19 - 5 + 4$
14. $26 - 9 - 4$
15. $(8 + 7)(3 + 2)$
16. $(7 - 3)(8 - 5)$
17. $12 + 3 \times 5 - 9$
18. $17 - 2 \times 4 + 5$
19. $30 - 6 \div 2 + 4$
20. $18 - 9 \div 3 + 4$

B. Find the value of each of the following expressions or formulas by substituting the given values for the variables and using the Order of Operations Agreement. Be sure to write out each step.

21. Find the value of $(a + b)(c + d)$ when $a = 4$, $b = 5$, $c = 2$, and $d = 6$.

22. Find the value of pqr when $p = 3$, $q = 4$, and $r = 6$.

23. Find the value of $ab - cd$ when $a = 7$, $b = 5$, $c = 3$, and $d = 4$.

24. Find the value of $2A^2B$ when $A = 3$ and $B = 4$.

25. Find the value of $3xy^2$ when $x = 4$ and $y = 3$.

26. Find the value of $K(P - R)$ when $K = 7$, $P = 8$, and $R = 3$.

27. Find the value of $(A - 3)(B - 9)$ when $A = 12$ and $B = 11$.

28. Find the value of $KL - M$ when $K = 5$, $L = 9$, and $M = 23$.

29. Find the value of $D + EF$ when $D = 7$, $E = 8$, and $F = 9$.

30. Find the value of AB^2C when $A = 3$, $B = 4$, and $C = 2$.

31. Find the value of K in $K = 3m - 4n$ when $m = 9$ and $n = 4$.

32. Find the value of P in $P = 5(R - S)$ when $R = 12$ and $S = 8$.

33. Find the value of A in $A = 4s^2t$ when $s = 2$ and $t = 3$.

34. Find the value of T in $T = 2(A + 4)(B - 5)$ when $A = 7$ and $B = 9$.

35. Find the value of R in $R = K - LM$ when $K = 23$, $L = 4$, and $M = 5$.

36. Find the value of f in $f = 8g - 4h$ when $g = 4$ and $h = 6$.

37. Find the value of g in $g = hk - 4j$ when $h = 9$, $k = 7$, and $j = 8$.

38. Find the value of m in $m = 3e^2 - 5f^2$ when $e = 5$ and $f = 3$.

39. Find the value of x in $x = 7w - 3y^2$ when $w = 8$ and $y = 2$.

40. Find the value of C in $C = 4f^3 - 2g$ when $f = 2$ and $g = 5$.

9.5 The Number System and Its Properties

The Order of Operations Agreement is, of course, not the only property of whole numbers that applies to algebra. As a matter of fact, most whole number properties can also be stated algebraically.

One notable difference that you might have seen when comparing an arithmetic text to an algebra text is the use of many letters (variables). You now know that algebra is a general way of saying the same things that arithmetic has said. In order not to have to restate properties every time we use them with different values, we state them generally—algebraically—using variables rather than numbers.

Let's first look at the set of real numbers. This group includes almost every kind of number that you will encounter in beginning algebra. Once you have studied this system, you can be confident that, because of the way the properties are stated, they apply to all real numbers. Here, then, are all the kinds of numbers that make up the real number system.

Natural Numbers   These are also called counting numbers. From their names, you can guess that they include 1, 2, 3, 4, and 5 and go on forever. (Is there ever a largest number?) The set, or collection, of natural numbers is written this way:

$$\{1, 2, 3, 4, 5, \ldots\}$$

Whole Numbers   This set of numbers includes the natural numbers but it has one more member, zero. When the thinkers who developed our number system tried to show in nature what happened when they had five twigs and the wind came along and blew them all away, they realized there was a problem. There was no symbol for 5 take away 5. Because what remained was a void, "a hole," they wrote the symbol 0 to show that. The set, or collection, of whole numbers is written this way:

$$\{0, 1, 2, 3, 4, 5, \ldots\}$$

Integers   The set of integers includes all the whole numbers and their opposites. You can have two pencils, $+2$, or be missing two pencils, -2. You can have \$5, $+5$, or owe \$5, -5. The set of integers is written this way:

$$\{\ldots -3, -2, -1, 0, 1, 2, 3, \ldots\}$$

This set extends indefinitely in both directions.

Rational Numbers   What happens when you divide 8 by 2? You get 4. What happens when you divide 9 by 2? So far in the sets of numbers we have described, there is no way to show that result. This difficulty led to the development of the set of rational numbers, sometimes called fractions. Hence $9 \div 2 = 9/2$. The rationals include all numbers that can be written in the form a/b, where a and b are integers and b does not equal 0. (Why can't we have $a/0$?) This set of numbers includes the whole numbers because 6 can be written as $6/1$, -26 as $-26/1$, and 0 as $0/7$. The set of rationals also includes all repeating or terminating decimals; they too can be written in a/b form. This collection of numbers includes far too many numbers to show by example, so it is described as the set of all numbers that can be written in the form a/b, where a and b are integers, where $b \neq 0$. Using set notation, we express the set of all rational numbers like this:

$$\{a/b \mid a \text{ and } b \text{ are integers}, b \neq 0\}$$

This notation is read "the set of all numbers represented by a divided by b, such that a and b are integers and b is not equal to zero."

Irrational Numbers   This set of numbers is made up of numbers that cannot be written in a/b form. These are decimal numbers that cannot be written as fractions—for example, 2.010010001 .... There is a pattern, but the decimal is neither terminating nor repeating. The irrational numbers also include pi, π, the Greek letter that represents the quotient of the circumference of any circle divided by its diameter. π is a non-terminating, non-repeating decimal. Numbers that are the square roots of quantities that are not perfect squares ($\sqrt{2}$, $\sqrt{5}$, and $\sqrt{17}$, for example) are also irrational numbers.

Real Numbers   When you put all of these numbers together, you have the set of real numbers. If you were to draw a line resembling a ruler but with zero in the middle of the scale, the real numbers would be represented by *all the points* that make up that line, not just the labeled points.

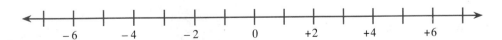

The line would also include values such as 3 1/2, -1.5, π, and $\sqrt{6}$, but for the sake of clarity, only the integer values are usually labeled on the scale. The numbers on the number line that fall to the left of the zero are called *negative numbers*. If you think in terms of temperature, those numbers would represent readings below zero degrees. If you lived in the Swiss Alps in the middle of winter, you might well see a temperature of $-12°$. The numbers that fall to the right of the zero on the number line are *positive numbers*. On a winter day in Pittsburgh, Pennsylvania, a temperature of $+12°$ is to be expected. Negative 12 (-12) and positive 12 ($+12$) are **opposites**. They are the same distance from zero but in opposite directions. When we compare any two values on a number line, we find that the one to the left is the smaller (just as on a thermometer, the lower number on the scale is the colder temperature).

If you think about the logical need for different types of numbers, and consider what you know about their history, you can easily remember the real number system. When students begin to learn about signed numbers (so-called because such numbers have a numerical value and a positive or negative sign), they learn what these numbers are and how to do calculations with them. You will learn about these signed numbers in the next chapter. Then you will become familiar with some of the processes for solving equations and finding solutions to word problems by using algebra. Once you are comfortable with these processes, you will be ready to expand your knowledge to include other topics in algebra.

Now, let's review the properties you studied in Chapter 1 and show that they apply to all real numbers.

Commutative Property of Addition

For all real numbers represented by a and b,

$$a + b = b + a$$

For example,	$7 + 6 = 6 + 7$
because	$13 = 13$
and	$13\ 1/2 + 7\ 1/4 = 7\ 1/4 + 13\ 1/2$
because	$20\ 3/4 = 20\ 3/4$

Commutative Property of Multiplication

For all real numbers represented by a and b,

$$a \cdot b = b \cdot a$$

For example,	$9 \cdot 5 = 5 \cdot 9$
because	$45 = 45$
and	$(6.2)(1.4) = (1.4)(6.2)$
because	$8.68 = 8.68$

Associative Property of Addition

For all real numbers represented by a, b, and c,

$$a + (b + c) = (a + b) + c$$

Consider	$2.6 + (3.5 + 4.9) = (2.6 + 3.5) + 4.9$
	$2.6 + 8.4 = 6.1 + 4.9$
	$11.0 = 11.0$

Associative Property of Multiplication

For all real numbers represented by a, b, and c,

$$a(bc) = (ab)c$$

For example,　$1/2(3/4 \cdot 5/8) = (1/2 \cdot 3/4)5/8$

$$1/2(15/32) = (3/8)(5/8)$$

$$15/64 = 15/64$$

Addition Property of Zero

For all real numbers represented by a,

$$a + 0 = a$$

You know that　　　$8 + 0 = 8$

and　　　$2.7 + 0 = 2.7$

Multiplication Property of Zero

For all real numbers represented by a, $a \neq 0$,

$$a \cdot 0 = 0$$

You know that　　　$12 \cdot 0 = 0$

and　　　$5/8 \cdot 0 = 0$

Multiplication Property of One

For all real numbers represented by a,

$$a \cdot 1 = a$$

You know that　　　$14 \cdot 1 = 14$

and　　　$2\ 3/4 \cdot 1 = 2\ 3/4$

Division by One

For all real numbers represented by a,

$$a \div 1 = a$$

You know that　　　$9 \div 1 = 9$

and　　　$1/2 \div 1 = 1/2$

You may want to go back and review Section 1.5 if these next two properties are not clear to you.

Division by Zero

For all real numbers represented by a,

$$a \div 0 \text{ is undefined}$$

It has no real number answer.

Division into Zero

For all real numbers represented by a, $a \neq 0$,

$$0/a = 0$$

Distributive Principle

For all real numbers represented by a, b, and c,

$$a(b + c) = ab + ac$$

and

$$a(b - c) = ab - ac$$

Consider

$$2(3 + 5) = 2 \cdot 3 + 2 \cdot 5$$
$$2(8) = \quad 6 \; + \; 10$$
$$16 = \qquad 16$$

and

$$1/2(7/10 - 3/5) = 1/2(7/10) - 1/2(3/5)$$
$$1/2(7/10 - 6/10) = 7/20 - 3/10$$
$$1/2(1/10) = 7/20 - 6/20$$
$$1/20 = 1/20$$

To these properties that we studied before, we can add four more that we have used but have not yet expressed in so many words.

Closure Property for Addition

For any two real numbers represented by a and b, the sum of a and b, $a + b$, is also a real number.

Closure Property for Subtraction

For any two real numbers represented by a and b, the difference between a and b, $a - b$, is also a real number.

Closure Property for Multiplication

For any two real numbers represented by a and b, the product of a and b, ab, is also a real number.

Closure Property for Division

For any two real numbers represented by a and b, $b \neq 0$, the quotient of a divided by b, $a \div b$, is also a real number.

This seems like a lot of theoretical information, but if you think about it, you will realize that it is these rules that have enabled you to do the mathematics that you have been doing all along.

9.5 ◆ Practice Problems

Match each of the statements in Column I with the property that it illustrates from Column II. You may use a property more than once.

Column I

1. $8 \times 1 = 1 \times 8$
2. $12 \div 0$ is undefined.
3. $3 + (4 + 9) = (3 + 4) + 9$
4. $4(9 - 3) = 4(9) - 4(3)$
5. $15 \cdot 1 = 15$
6. $9 \div 1 = 9$
7. $14 + 0 = 0 + 14$
8. $(4 + 8) + 6 = 4 + (8 + 6)$
9. $0 \div 4 = 0$
10. $5(2 \cdot 3) = (5 \cdot 2)3$
11. $7(6 - 4) = 7(6) - 7(4)$
12. $0 = 4 \cdot 0$
13. When a and b are real numbers and $ab = c$, c is a real number.
14. $5 \cdot 9 = 9 \cdot 5$
15. $6 + 0 = 6$
16. $6(9 + 5) = 6(9) + 6(5)$
17. $3(7 \cdot 6) = (3 \cdot 7)6$
18. When a and b are real numbers and $a + b = c$, c is a real number.
19. $7(3 + 5) = 7(3) + 7(5)$
20. $0 + 3 = 3 + 0$

Column II

a. Commutative Property of Addition
b. Commutative Property of Multiplication
c. Associative Property of Addition
d. Associative Property of Multiplication
e. Addition Property of Zero
f. Multiplication Property of Zero
g. Multiplication Property of One
h. Division By Zero
i. Division Into Zero
j. Division By One
k. Distributive Principle
l. Closure Property for Addition
m. Closure Property for Multiplication

9.6 Solving Simple Equations and Inequalities

Suppose you are given a simple equation—for example, the statement $y + 5 = 12$. You know that y is a variable and can take on any value. When you talk about "solving the equation," you are trying to find the one value that, when it is substituted for y, will give a true statement. You could try a lot of different possibilities and, by *trial and error*, by *intuition*, or by *inspection*, find the solution. To solve $y + 5 = 12$, you could ask yourself intuitively, "What number must be added to 5 to get 12?" You might consider several possibilities. Because there are no restrictions placed on y, the value you try for y could be any number.

$$y + 5 = 12$$

Let $y = 4$.	Is $4 + 5 = 12$ a true statement? No.
Let $y = 9$.	Is $9 + 5 = 12$ a true statement? No.
Let $y = 0.4$.	Is $0.4 + 5 = 12$ a true statement? No.
Let $y = 2\ 1/2$.	Is $2\ 1/2 + 5 = 12$ a true statement? No.
Let $y = 7$.	Is $7 + 5 = 12$ a true statement? Yes.

Thus 7 is the *solution* of the equation. It is the one value for *y* that makes the statement true. The value that makes the statement true is called the **solution**. It can be written in set notation, {7}, and 7 can be called the **solution set**.

Try to solve these simple equations by inspection. Look at the equation and take a guess. Use your intuition. Then check your answer by substituting the number for the variable and simplifying. If you get a true statement, or an **identity** (that is, the number on the right equals the number on the left), then the solution you have found is the correct one.

Exercises   1. $A - 5 = 3$ 2. $B + 4 = 9$ 3. $7C = 63$ 4. $D \div 5 = 3$

5. $2E + 6 = 14$

Answers:

1. $A = 8$; Check: $3 = 3$ 2. $B = 5$; Check: $9 = 9$

3. $C = 9$; Check: $63 = 63$ 4. $D = 15$; Check: $3 = 3$

5. $E = 4$; Check: $14 = 14$

It is not always possible or convenient to solve all equations by inspection or intuition. You will learn methods for solving more complex equations in Chapter 11. These methods will make solving equations more like solving a puzzle.

Suppose that instead of the equation $y + 5 = 12$, you were given the inequality $y + 5 > 12$. This is read "*y* plus five is greater than twelve." Begin by solving the equation $y + 5 = 12$. By intuition and inspection, you know that $y = 7$ will make the equation true. Now try to solve the inequality by substituting 7 and two values greater than 7 and two values less than 7. This should give you a pretty good idea of which values will work.

$y + 5 > 12$

Try 5.	$5 + 5 > 12 \longrightarrow 10 > 12$	False.
Try 6.	$6 + 5 > 12 \longrightarrow 11 > 12$	False.
Try 7.	$7 + 5 > 12 \longrightarrow 12 > 12$	False, but it's close.
Try 8.	$8 + 5 > 12 \longrightarrow 13 > 12$	True.
Try 9.	$9 + 5 > 12 \longrightarrow 14 > 12$	True.

You can see that any whole number greater than 7 will satisfy the inequality. In fact, because $7\,1/2 + 5 > 12$, or $12\,1/2 > 12$, you can correctly assume that every real number greater than 7 will make the inequality true.

$y + 5 > 12$ is true whenever $y > 7$.

Consider $K - 2 < 6$. This is read "*K* decreased by two is less than six." You can see by inspection that $K - 2 = 6$ is true when $K = 8$, so try 8 and values on either side of 8.

$K - 2 < 6$

Try 6.	$6 - 2 < 6 \longrightarrow 4 < 6$	True.
Try 7.	$7 - 2 < 6 \longrightarrow 5 < 6$	True.
Try 8.	$8 - 2 < 6 \longrightarrow 6 < 6$	False, but it's close.
Try 9.	$9 - 2 < 6 \longrightarrow 7 < 6$	False.
Try 10.	$10 - 2 < 6 \longrightarrow 8 < 6$	False.

You can see by inspection that $K - 2 < 6$ is true for the whole numbers less than 8. In fact,

$K - 2 < 6$ is true whenever $K < 8$.

Now try $d + 2 \geq 9$. This is read "the sum of *d* and two is greater than or equal to nine." You know by inspection that $d + 2 = 9$ is true when $d = 7$. Therefore, try 7 and two numbers on either side of 7.

$$d + 2 \geq 9$$

Try 5. $5 + 2 \geq 9 \longrightarrow 7 \geq 9$ False.

Try 6. $6 + 2 \geq 9 \longrightarrow 8 \geq 9$ False.

Try 7. $7 + 2 \geq 9 \longrightarrow 9 \geq 9$ True, because $9 \geq 9$ says that 9 is *greater than or equal to 9*.

Try 8. $8 + 2 \geq 9 \longrightarrow 10 \geq 9$ True.

Try 9. $9 + 2 \geq 9 \longrightarrow 11 \geq 9$ True.

You can see by inspection that $d + 2 \geq 9$ is true for the whole numbers greater than or equal to 7. In fact, it is true for all real numbers greater than or equal to 7.

$d + 2 \geq 9$ is true whenever $d \geq 7$.

Study these examples and see how some inequalities can be solved by inspection.

E X A M P L E   1   Solve: $g + 3 < 8$

Solution   You can see by inspection that $g + 3 = 8$ when $g = 5$, so try 5 and values greater than and less than 5.

Try 3. $3 + 3 < 8 \longrightarrow 6 < 8$ True.

Try 4. $4 + 3 < 8 \longrightarrow 7 < 8$ True.

Try 5. $5 + 3 < 8 \longrightarrow 8 < 8$ False.

Try 6. $6 + 3 < 8 \longrightarrow 9 < 8$ False.

Try 7. $7 + 3 < 8 \longrightarrow 10 < 8$ False.

Therefore, $g + 3 < 8$ is true whenever $g < 5$.   ◆

E X A M P L E   2   Solve: $6a \geq 42$

Solution   You know by inspection that $6a = 42$ when $a = 7$, so try 7 and values greater than and less than 7.

Try 5. $6(5) \geq 42 \longrightarrow 30 \geq 42$ False.

Try 6. $6(6) \geq 42 \longrightarrow 36 \geq 42$ False.

Try 7. $6(7) \geq 42 \longrightarrow 42 \geq 42$ True.

Try 8. $6(8) \geq 42 \longrightarrow 48 \geq 42$ True.

Try 9. $6(9) \geq 42 \longrightarrow 54 \geq 42$ True.

Therefore, $6a \geq 42$ is true whenever $a \geq 7$.   ◆

In a later chapter, you will learn how to solve more complicated inequalities, but an intuitive idea of what they are and how they work is all you need for now.

9.6 ◆ Practice Problems

A. Fill in the blanks.

1. The statement $y - 9 = 4$ is called a(n) _____ .

2. You can solve simple equations by inspection or by _____ .

3. The value that makes the equation true is called the _____ .

4. In $A + 6 = 7$, _____ is the solution.

5. In the equation $b - 5 = 4$, _____ is the value that makes the statement true when it is substituted in *place* of b.

6. 17 _____ (is or is not?) the solution of $25 = c + 8$.

7. 9 makes $Y - 8 = 17$ a(n) _____ statement.

8. If $D + 4 = 7$, 5 _____ (is or is not?) the solution.

B. Solve each of the following equations by "look-ing," by inspection, or by intuition. Show a check for each equation.

9. $y + 8 = 17$ 10. $2 + n = 11$
11. $M \div 3 = 5$ 12. $4P = 28$
13. $12 = T - 5$ 14. $12 = L - 4$
15. $17 = N + 5$ 16. $8 = r \div 3$
17. $11 = 21 - K$ 18. $6E = 42$
19. $12 = 8 + C$ 20. $B \div 4 = 3$

C. Solve each of the following inequalities by in-spection. Give your answer in statement form. (See Examples 1 and 2, page 255.)

21. $x + 3 < 9$ 22. $c + 5 > 8$
23. $y - 2 > 10$ 24. $3d \geq 12$
25. $4a \leq 20$ 26. $k \div 4 < 24$
27. $30 \geq a + 10$ 28. $14 \leq y + 5$
29. $m \div 2 > 5$ 30. $n - 4 > 7$

9.7 Chapter Review

Now that you have completed this chapter, you should be able to:

1. Write an algebraic formula from a given description.
2. Write an algebraic formula from information given in a table.
3. Recognize the following algebraic quantities: variable, constant, numerical coefficient, base, power or exponent, algebraic expression, equation, inequality, and solution set.
4. Combine like terms.
5. Use the Order of Operations Agreement to simplify numerical expressions.
6. Find the value of a given algebraic expression or formula.
7. Solve a simple algebraic equation by intuition or inspection.
8. Solve a simple inequality by inspection.

Review Problems

A. Write a formula for each of the following situations.

1. The number of inches (i) in Y yards is equal to the product of 36 and Y.
2. The value (V) of s stamps at 45 cents each is equal to the product of 45 and s.
3. The cost (c) of 9 movie tickets at D dollars each is equal to 9 times D.
4. The tax (t) on an item if it is equal to 5% (0.05) times the price (p).
5. The number of students (s) in a class is equal to the sum of p, those who are present, and a, those who are absent.
6. The number of hours (h) is equal to the quotient of the number of minutes (m) and 60.
7. The rate (r) you can drive is equal to the quotient of the distance (d) and the travel time (t).
8. The sale price, S, of an item is equal to the difference between the regular price, R, and the discount, D.

9. The total cost of your vacation (V) is equal to the sum of the transportation costs (t) and the food costs (f).
10. The number of chocolates (c) in a box is equal to the product of r rows and p pieces in each row.

B. Write a formula from the information given in each of the following tables. There are two options for your answer, and either one is correct.

11.
K	9	8	7	6
L	3	2	1	0

12.
E	12	14	16	18
F	4	6	8	10

13.
A	9	12	15	18
B	3	4	5	6

14.
Y	8	12	18	22
A	4	6	9	11

15.

P	1	2	3	4
R	4	8	12	16

16.

M	3	7	11	15
N	7	11	15	19

17.

C	9	11	13	15
D	11	13	15	17

18.

E	2	5	8	11
F	6	15	24	33

19.

T	5	10	15	20
V	10	15	20	25

20.

S	12	20	28	36
W	3	5	7	9

C. Identify the parts asked for in each expression or equation.

21. the variables in $6x - 3y$

22. the constants in $4a - 5 + 3b + 7$

23. the numerical coefficient(s) in $7x^2 - 3x + 2$

24. the exponent(s) in $3a^4 - 6b^2$

25. the literal factor(s) in $7p^3r$

26. the base(s) in $-8a^4b^5$

27. the solution set of $3 + y = 9$

28. the power(s) in $14 + 5x^3$

29. the numerical coefficient(s) in $5a + 3b + c$

30. the numerical coefficient of p

D. Combine like terms in each of the following problems.

31. $5W + 7Y - 3W - 3Y$ **32.** $6a + 5a - 3a$

33. $9c - 6c + 8c$ **34.** $4a + 3b - 2c$

35. $3a^2 + 4a^2 - 2a^2$ **36.** $9d^2 - 4d^2 + 3d^2$

37. $4B - B + 2B$ **38.** $12Y - 3Y + Y$

39. $6w - 3x + 2y$ **40.** $9y^2 - 4y^2 + 6y$

41. $3y^2 + 6y^2 - y^2$ **42.** $8c^2 - 4c^2 + 3c^2$

43. $7a^2 - 5a^2 + 2a^2$

44. $6x^2 + 4x - 2x^2 + 5x$

45. $8t^3 + 3t^2 - 4t^3 + 8t^2$

E. Simplify the following numerical expressions using the Order of Operations Agreement.

46. $9 - 4 + 3$ **47.** $12 + 3(8 - 2)$

48. $14 - 2(6 - 2)$ **49.** $15 - 6 + 3$

50. $24 - 3 \times 2 + 8 \div 2$ **51.** $(7 - 4)(8 \times 2)$

52. $(11 + 3)(9 \div 3)$ **53.** $3^5 - 12 \div 3 + 9$

54. $9 \div 3 \times 2$ **55.** $6 \times 4 \div 3$

56. $2^5 + 3$ **57.** $4(2)^3$

58. $5^2 + 3 \times 2 - 4$ **59.** $8 + 3(9 - 5)$

60. $6 + 3^2 - 4$

F. Find the value of each expression or formula by substituting the given values and using the Order of Operations Agreement.

61. Evaluate $(P - T)(R + S)$ when $P = 7$, $T = 3$, $R = 5$, and $S = 2$.

62. Evaluate $5A^2B$ when $A = 4$ and $B = 2$.

63. Evaluate $3CD^2$ when $C = 5$ and $D = 2$.

64. Evaluate $K + 6(M - L)$ when $K = 5$, $M = 8$, and $L = 3$.

65. Evaluate M in $M = 4R - 3S$ when $R = 5$ and $S = 2$.

66. Evaluate L in $L = 5A + B - 2C$ when $A = 4$, $B = 6$, and $C = 5$.

67. Evaluate R in $R = 3S^2T$ when $S = 4$ and $T = 2$.

68. Evaluate K in $K = 3W - 4Y$ when $Y = 5$ and $W = 7$.

69. Evaluate r in $r = 5p^3 - 3q^2$ when $p = 2$ and $q = 3$.

70. Evaluate d in $d = 4a^2 - 3b^3$ when $a = 3$ and $b = 2$.

G. Find the solution for each equation by inspection or intuition.

71. $X - 9 = 4$ **72.** $16 - a = 9$

73. $f + 5 = 11$ **74.** $28 = 4Y$

75. $13 - B = 5$ **76.** $19 - K = 4$

77. $m/3 = 5$ **78.** $18 = 3Y$

79. $2y + 3 = 11$ **80.** $4a - 5 = 7$

H. Find the solution for each inequality by inspection. Write your answer in statement form. (See Examples 1 and 2 in Section 9.6, page 255.)

81. $g - 7 > 2$ **82.** $y + 2 < 8$

83. $y \div 6 < 3$ **84.** $3b \geq 15$

85. $2d \leq 12$ **86.** $m - 4 > 5$

87. $40 > m + 10$ **88.** $k \div 3 \leq 9$

89. $x + 7 \geq 11$ **90.** $18 < a + 4$

Chapter Test

In Problems 1 through 6, write a formula for each situation.

1. The number of students present (p) is equal to the difference between the number of students enrolled (s) and the number absent (a).

2. The total price of an item (P) is equal to the sum of its cost (c) and the sales tax (t).

3. The number of seconds (s) in m minutes is equal to the product of 60 and the number of minutes (m).

4. The class average (A) on the last test is equal to the quotient of the total points earned (P) and the number of students (S) who took the test.

5.

X	8	10	12	16
Y	4	6	8	12

6.

G	3	5	7	9
H	18	30	42	54

7. Combine like terms: $7a - 3a + a$

8. Combine like terms: $16X + 4Y - 5X + 3Y$

9. Simplify: $(6 + 7)(18 \div 3)$

10. Simplify: $45 - 4(12 - 5)$

11. Simplify: $3^2 + 12 \div 2 - 4$

12. Simplify: $26 - 5[16 - 3(7 - 3)]$

13. Evaluate $(K - P)(K + P)$ when $K = 7$ and $P = 4$.

14. Evaluate $5A^2B$ when $A = 3$ and $B = 2$.

15. Evaluate $7R - 3S - 4T$ when $R = 5$, $S = 6$, and $T = 3$.

16. Find the value of r in $r = 3s^3t$ when $s = 2$ and $t = 4$.

17. Solve: $M - 5 = 6$

18. Solve: $4 = K \div 12$

19. Solve: $Y + 3 < 11$

20. Solve: $4p \geq 20$

Thought-Provoking Problems

1. Explain in three or four sentences some of the differences between arithmetic and algebra.

2. Explain the difference between an expression and an equation.

3. Why is the commutative property stated only for addition and multiplication? Are subtraction and division commutative? Why or why not?

4. Find the value of K in

$$K = 4a^2 - 3b[2c - d(7 - e)]$$

when $a = 6$, $b = 5$, $c = 7$, $d = 2$, and $e = 3$.

5. Solve the following equations by inspection and explain your answers.
   a. $X + 7 = 2 + X + 5$
   b. $m + 3 = 4 + m$

6. Inequalities have more than one answer or solution (unlike the equations you looked at in this chapter). Such solutions are often represented by shading the values that make the inequality true. To get started, replace the inequality symbol by an equals sign and solve the equation. If you want to include the starting point, use a solid dot for that point. If you don't want to include the starting point, use an open circle around the point. Then test values on either side of that number to see which part of the number line should be shaded.
   a. $x + 4 < 7$
   b. $y - 2 > 3$
   c. $3 \leq y - 1$
   d. $5 \geq x + 2$

7. Solve each of the following inequalities by inspection, and explain your answer. What would the number-line representation look like?
   a. $9 + K > K + 9$
   b. $Y + 12 < 20 + Y - 8$
   c. $N + 4 \geq N + 4$
   d. $B - 5 \leq B - 5$

8. Explain why we cannot divide by zero.

9. Draw a number line and show, by using a large dot and a letter, approximately where each of the following values would be located on the line.
   a. 2^3
   b. $17/15$
   c. $\sqrt{11}$

10

Equations

10.1 Introduction

In Chapter 9, you found solutions to simple equations by inspection or intuition. These methods can be used only when you are trying to find the solution to very simple equations. In this chapter, we will develop procedures to solve more complex equations. You will use these same procedures when you go on to other algebra courses and in science and technical courses where solving equations is required.

Later in the chapter you will be solving equations that contain decimals and fractions. In order to do this successfully, you must be very competent with these kinds of numbers. If you are unsure of these skills, do the Skills Checks in Appendix C. Check your answers, and for any errors you made, go back to the indicated sections in the text and review that material at this time.

Learning Objectives

When you have completed this chapter, you should be able to:

1. Solve one-step equations.
2. Solve two-step equations.
3. Solve equations that require combining like terms.
4. Solve equations with unknown terms on both sides of the equals sign.
5. Solve equations that contain fraction or decimal numbers.
6. Understand the meaning of formulas, formula substitution, and rearrangement.

10.2   Solving One-Step Equations

In this section, you will learn how to solve equations such as

$$x + 5 = 7 \qquad y - 3 = 8 \qquad 7t = 28 \qquad W/4 = 12$$

To solve these equations, you rearrange the terms to get the variable alone on one side of the equals sign and the constant term on the other side. In each of these "one-step" equations, the variable has one operation performed on it. You will "undo" that operation by performing the opposite operation.

Addition $\leftrightarrow$ subtraction

Multiplication $\leftrightarrow$ division

To maintain equality, you must do the same operation on *both sides* of the equals sign. An equation is like a balance scale (Figure 1). If you add something to one pan of a scale that is in balance, but not to the other, the scale is no longer in balance. If you add the same quantity to both sides, however, the balance is maintained. The same theory applies to solving equations.

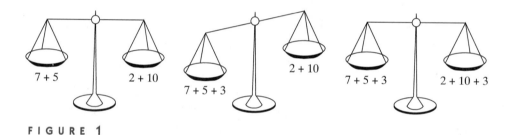

| 7 + 5 | 2 + 10 | | 7 + 5 + 3 | 2 + 10 | | 7 + 5 + 3 | 2 + 10 + 3 |

FIGURE 1

This process of doing addition or subtraction on both sides of the equation is an application of a property of real numbers called the Addition or Subtraction Property of Equality.

Addition or Subtraction Property of Equality

For all real numbers a, b, and c

$a = b$ if and only if $a + c = b + c$ and

$a = b$ if and only if $a - c = b - c$

In the first sample equation, $x + 5 = 7$, the variable x has 5 added to it. To "undo" that operation, we must do the opposite operation: subtraction. Five must be subtracted from both sides of the equation, as illustrated in Figure 2.

This means that the value for x that will give a true statement in $x + 5 = 7$ is 2. Try it.

$$x + 5 = 7$$
$$2 + 5 = 7$$
$$7 = 7 \quad \text{True.}$$

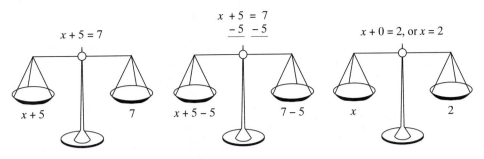

FIGURE 2

This last process is the one used to check the answer to any equation. You should develop the habit of checking *every* equation that you solve. Substitute the answer back into the *original* equation, and simplify the arithmetic problem that has been created. If the answers on both sides come out to be the same value that equation is called an **identity** statement, and then you know that your answer is correct. If you do not get the same value on both sides, then you need to go back and redo the problem. It is not often that there is a way you can check answers on homework or on a test and know for certain that they are correct! Use the opportunity whenever you can—and that should be every time you solve an equation.

The solution to the second sample problem is similar. In $y - 3 = 8$, the variable y has 3 subtracted from it. To solve and get y all alone, we must do the opposite operation: three must be added to both sides.

$$\begin{aligned} y - 3 &= 8 \\ \underline{+\,3} \quad &\underline{+\,3} \\ y + 0 &= 11 \\ y &= 11 \end{aligned}$$

Check: $y - 3 = 8$
  $11 - 3 = 8$
    $8 = 8$ True.

Here are more solutions to equations like these. Study every step and try to understand why each process (opposite operation, addition or subtraction property of equality, simplifying, and so forth) was performed.

E X A M P L E   1   Solve: $15 + T = 26$.

Solution
$$\begin{aligned} 15 + T &= 26 \\ \underline{-15} \qquad &\underline{-15} \\ T &= 11 \end{aligned}$$

Check
$$\begin{aligned} 15 + T &= 26 \\ 15 + 11 &= 26 \\ 26 &= 26 \quad \text{True.} \end{aligned}$$

◆

E X A M P L E   2   Solve: $13 = M - 9$.

Solution
$$\begin{aligned} 13 &= M - 9 \\ \underline{+\,9} \quad &\underline{+\,9} \\ 22 &= M \end{aligned}$$

Check
$$\begin{aligned} 13 &= M - 9 \\ 13 &= 22 - 9 \\ 13 &= 13 \quad \text{True.} \end{aligned}$$

◆

Exercises Try these problems. Remember to check each answer.

1. $X + 7 = 13$ 2. $B - 5 = 6$ 3. $14 = J - 9$
4. $8 + G = 12$ 5. $13 + P = 21$ 6. $W + 4 = 9$

Answers:

1. $X = 6$; check: $13 = 13$ 2. $B = 11$; check: $6 = 6$
3. $23 = J$; check: $14 = 14$ 4. $G = 4$; check: $12 = 12$
5. $P = 8$; check: $21 = 21$ 6. $W = 5$; check: $9 = 9$

In the third sample problem on page 260, $7t = 28$, the operation performed on t is multiplication (a number and variable written side by side indicate multiplication). Because division is the opposite operation, divide both sides by the numerical coefficient 7. This will get t all alone with a coefficient of 1. Use the fractional form to indicate the division.

$$7t = 28$$
$$\frac{7t}{7} = \frac{28}{7} \quad \text{(Simplify: } 7/7 = 1, 28/7 = 4)$$
$$1t = 4 \text{ or } t = 4$$

Check: $7t = 28$
$7(4) = 28$
$28 = 28$ True.

Remember, it is not necessary to write the coefficient 1 before a variable. Any literal term standing alone is understood to have a coefficient of 1.

In the last sample problem, $W/4 = 12$, the variable W is being divided by 4. Do the opposite operation to both sides of the equation: multiply both sides by 4.

$$\frac{W}{4} = 12$$
$$\frac{(\overset{1}{\cancel{4}})W}{\underset{1}{\cancel{4}}} = (4)12$$
$$1W = 48 \text{ or } W = 48$$

Check: $\dfrac{W}{4} = 12$
$\dfrac{48}{4} = 12$
$12 = 12$ True.

The two previous sample problems utilize an important property of equations:

Multiplication or Division Property of Equality

For all real numbers a, b, and c,

$a = b$ if and only if $ac = bc$ and
$a = b$ if and only if $a/c = b/c$, where $c \neq 0$.

Here are two more solutions to one-step equations. Study every step and try to understand how each process (opposite operation, do to both sides, simplify, and so forth) works.

EXAMPLE 3 Solve: $27 = 3N$

Solution
$$27 = 3N \qquad \text{Divide by the coefficient of } N.$$
$$\frac{27}{3} = \frac{3N}{3}$$
$$9 = N$$

Check $27 = 3N$

$27 = 3(9)$

$27 = 27$ True.

$\blacklozenge$

E X A M P L E 4 Solve: $\dfrac{T}{4} = 24$

Solution $\dfrac{T}{4} = 24$

$\dfrac{\overset{1}{(\cancel{4})}T}{\underset{1}{\cancel{4}}} = (4)24$

$T = 96$

Check $\dfrac{T}{4} = 24$

$\dfrac{96}{4} = 24$

$24 = 24$ True.

$\blacklozenge$

Exercises   Try these problems. Remember to check each answer.

1. $\dfrac{H}{5} = 35$ **2.** $32 = 8G$ **3.** $45 = 9R$

4. $\dfrac{M}{4} = 20$ **5.** $12P = 84$ **6.** $90 = \dfrac{F}{6}$

Answers:

1. $H = 175$, check: $35 = 35$ **2.** $4 = G$, check: $32 = 32$
3. $R = 5$, check: $45 = 45$ **4.** $M = 80$, check: $20 = 20$
5. $P = 7$, check: $84 = 84$ **6.** $F = 540$, check: $90 = 90$

Included in this discussion of one-step equations should be equations such as $2/3\ Y = 12$ (this might also be written $2Y/3 = 12$). The variable Y is being multiplied by $2/3$, so to "undo" the multiplication, you must divide both sides of the equation by $2/3$. Review complex fractions (Section 3.5).

$$\frac{2}{3}Y = 12$$

$$\frac{\dfrac{2}{3}Y}{\dfrac{2}{3}} = \frac{12}{\dfrac{2}{3}}$$

Rewrite each side as division.

$$\frac{2}{3}Y \div \frac{2}{3} = 12 \div \frac{2}{3}$$

Invert and multiply; reducing first by common factors could be done.

$$\frac{\overset{1}{\cancel{2}}}{\underset{1}{\cancel{3}}}Y\frac{\overset{1}{(\cancel{3})}}{\underset{1}{(\cancel{2})}} = \frac{\overset{6}{\cancel{12}}}{1}\frac{(3)}{\underset{1}{(\cancel{2})}}$$

$$Y = 18$$

You can shorten the process and eliminate the need to write the equation as a division problem. Simply multiply both sides by the reciprocal of the fraction. The reciprocal of 2/3 is 3/2.

$$\frac{2}{3}Y = 12$$

$$\frac{(3)}{(2)}\frac{2}{3}Y = \frac{(3)}{(2)}\frac{12}{1}$$

Reducing by common factors could be done.

$$\frac{(\overset{1}{\cancel{3}})}{(\underset{1}{\cancel{2}})}\frac{\overset{1}{\cancel{2}}}{\underset{1}{\cancel{3}}}Y = \frac{(3)}{(\underset{1}{\cancel{2}})}\frac{\overset{6}{\cancel{12}}}{1}$$

$$Y = 18$$

Check
$$\frac{2}{3}Y = 12$$

$$\frac{2}{3}(18) = 12$$

$$\frac{36}{3} = 12$$

$$12 = 12 \quad \text{True.}$$

You can solve any equation of the form $2/3Y = 12$ by using the short cut. Multiply both sides of the equation by the reciprocal of the coefficient of the unknown term. Look at the steps in each of these examples before doing problems yourself.

E X A M P L E 5 Solve: $\dfrac{4}{5}t = \dfrac{3}{20}$

Solution
$$\frac{4}{5}t = \frac{3}{20}$$

$$\frac{(5)}{(4)}\frac{4}{5}t = \frac{(5)}{(4)}\frac{3}{20}$$ Multiply both sides by the reciprocal 4/5 and simplify.

$$\frac{(\overset{1}{\cancel{5}})}{(\underset{1}{\cancel{4}})}\frac{\overset{1}{\cancel{4}}}{\underset{1}{\cancel{5}}}t = \frac{(\overset{1}{\cancel{5}})}{(4)}\frac{3}{\underset{4}{\cancel{20}}}$$

$$t = \frac{3}{16}$$

Check
$$\frac{4}{5}t = \frac{3}{20}$$

$$\frac{4}{5}\frac{(3)}{(16)} = \frac{3}{20}$$ Substitute 3/16 for t.

$$\frac{\overset{1}{\cancel{4}}}{5}\frac{(3)}{(\underset{4}{\cancel{16}})} = \frac{3}{20} \quad \text{or} \quad \frac{\overset{1}{\cancel{2}}\cdot\overset{1}{\cancel{2}}\cdot 3}{5\cdot\underset{1}{\cancel{2}}\cdot\underset{1}{\cancel{2}}\cdot 2\cdot 2} = \frac{3}{20}$$

$$\frac{3}{20} = \frac{3}{20} \quad \text{True.}$$ ◆

E X A M P L E **6** Solve: $\dfrac{3X}{10} = 32$

Solution $\dfrac{3X}{10} = 32$

$\dfrac{(10)}{(3)} \dfrac{3X}{10} = \dfrac{(10)}{(3)} \dfrac{32}{1}$ Multiply both sides by the reciprocal of 3/10.

$\dfrac{30X}{30} = \dfrac{320}{3}$

$X = \dfrac{320}{3}$ or 106 2/3

Check $\dfrac{3X}{10} = 32$ $\dfrac{3X}{10}$ is the same is $\dfrac{3}{10} X$.

$\dfrac{3}{10} \dfrac{(320)}{(3)} = 32$ Replace X with 320/3.

$\dfrac{960}{30} = 32$

$32 = 32$ True.

E X A M P L E **7** Solve: $1\dfrac{2}{3} K = 30$

Solution $1\dfrac{2}{3} K = 30$

$\dfrac{5}{3} K = 30$ Change the mixed number to an improper fraction.

$\dfrac{(3)}{(5)} \dfrac{5}{3} K = \dfrac{(3)}{(5)} \dfrac{30}{1}$ Multiply by the reciprocal.

$K = 18$

Check $1\dfrac{2}{3} K = 30$

$\dfrac{(5)}{(3)} \dfrac{18}{1} = 30$ Substitute 18 for K.

$\dfrac{90}{3} = 30$

$30 = 30$ True.

Exercises Try these problems. Be sure to check your answers.

1. $\dfrac{4}{9} P = 60$ **2.** $4\dfrac{2}{3} S = 56$ **3.** $1\dfrac{4}{5} Y = \dfrac{6}{5}$ **4.** $\dfrac{2}{3} = \dfrac{4}{9} M$

Answers:

1. $P = 135$; check: $60 = 60$ **2.** $S = 12$; check: $56 = 56$
3. $Y = 2/3$; check: $6/5 = 6/5$ **4.** $M = 3/2$; check: $2/3 = 2/3$

10.2 ◆ Practice Problems

Solve and check.

1. $B + 4 = 9$
2. $t + 3 = 15$
3. $8 + W = 12$
4. $11 + R = 13$
5. $m - 3 = 7$
6. $B - 9 = 7$
7. $6 = t - 3$
8. $15 = C - 4$
9. $12 = 3N$
10. $5q = 20$
11. $5s = 30$
12. $9T = 36$
13. $\dfrac{C}{3} = 6$
14. $\dfrac{d}{4} = 5$
15. $9 = \dfrac{P}{3}$
16. $7 = \dfrac{B}{4}$
17. $\dfrac{2y}{3} = \dfrac{5}{6}$
18. $\dfrac{3T}{4} = 12$
19. $7 = 1\dfrac{3}{4}X$
20. $10 = 3\dfrac{1}{3}R$
21. $3W = 21$
22. $9 = t - 2$

23. $\dfrac{1}{2}Y = 6$
24. $\dfrac{X}{8} = 3$
25. $S - 8 = 12$
26. $9A = 27$
27. $B - 14 = 3$
28. $\dfrac{3}{4}T = \dfrac{15}{24}$
29. $\dfrac{m}{7} = \dfrac{5}{9}$
30. $9 + A = 12$
31. $1\dfrac{5}{6}X = 22$
32. $4r = 15$
33. $10 = 3t$
34. $2\dfrac{3}{4}Y = \dfrac{33}{64}$
35. $\dfrac{W}{4} = 20$
36. $\dfrac{y}{5} = \dfrac{3}{8}$
37. $11 + A = 14$
38. $\dfrac{3Y}{4} = \dfrac{5}{8}$
39. $8 = 8 + R$
40. $56 = 8 + M$

10.3 ◆ Two-Step Equations

Solving equations such as $3X + 5 = 20$ and $11 = 2Y - 7$ requires two different operations. In the first equation, $3X + 5 = 20$, the variable X is multiplied by 3 *and* has 5 added to it. In $11 = 2Y - 7$, the variable Y is multiplied by 2 and has 7 subtracted from it. In such equations, we "undo" the addition or subtraction first and then do the multiplication or division. The thing to remember is that every step makes the problem simpler.

E X A M P L E   1   Solve: $3X + 5 = 20$

Solution
$$3X + 5 = 20 \qquad \text{Subtract 5 from both sides.}$$
$$\underline{- 5 \quad - 5}$$
$$3X = 15$$

Now a one-step equation remains.

$$\frac{3X}{3} = \frac{15}{3} \qquad \text{Divide by 3.}$$
$$X = 5$$

Check
$$3X + 5 = 20$$
$$3(5) + 5 = 20$$
$$15 + 5 = 20$$
$$20 = 20 \qquad \text{True.}$$

◆

E X A M P L E 2   Solve: $11 = 2Y - 7$

Solution
$$11 = 2Y - 7 \qquad \text{Add 7 to both sides.}$$
$$\underline{+\ 7 \qquad\quad +\ 7}$$
$$18 = 2Y \qquad \text{Divide by 2.}$$
$$\frac{18}{2} = \frac{2Y}{2}$$
$$9 = Y$$

Check
$$11 = 2Y - 7$$
$$11 = 2(9) - 7$$
$$11 = 18 - 7$$
$$11 = 11 \qquad \text{True.}$$

◆

E X A M P L E 3   Solve: $6A - 8 = 5$

Solution
$$6A - 8 = 5 \qquad \text{Add 8 to both sides.}$$
$$\underline{+8 \quad +8}$$
$$6A = 13 \qquad \text{Because } A \text{ must have a coefficient of 1, divide}$$
$$\text{both sides by 6.}$$
$$\frac{6A}{6} = \frac{13}{6}$$
$$A = \frac{13}{6} \text{ or } 2\frac{1}{6}$$

Check
$$6A - 8 = 5$$
$$6\left(\frac{13}{6}\right) - 8 = 5$$
$$13 - 8 = 5$$
$$5 = 5 \quad \text{True.}$$

◆

E X A M P L E 4   Solve: $7m + 4 = 18$

Solution
$$7m + 4 = 18$$
$$\underline{-4 \quad -4}$$
$$7m = 14$$
$$\frac{7m}{7} = \frac{14}{7}$$
$$m = 2$$

Check
$$7m + 4 = 18$$
$$7(2) + 4 = 18$$
$$14 + 4 = 18$$
$$18 = 18 \quad \text{True.}$$

◆

The presence of fraction or decimal numbers in the equation does not change the strategy for solving two-step equations. "Undo" addition or subtraction first; then "undo" multiplication or division. Use the rules of fraction and decimal arithmetic operations. (You may need to review parts of Chapters 3, 4, and 5.)

EXAMPLE **5**   Solve: $\dfrac{H}{2} - 15 = 3$

Solution $\dfrac{H}{2} - 15 = \quad 3$ Add 15 to both sides.

$\underline{\quad +15 \quad\quad +15\quad}$

$\dfrac{H}{2} \quad\quad = \quad 18$

$(2)\dfrac{H}{2} = (2)18$ Multiply by 2.

$H = 36$

Check $\dfrac{H}{2} - 15 = 3$

$\dfrac{36}{2} - 15 = 3$

$18 - 15 = 3$

$3 = 3$ True.

EXAMPLE **6**   Solve: $\dfrac{2}{3}R + 4 = 12$

Solution $\dfrac{2}{3}R + 4 = \quad 12$ Subtract 4 from both sides.

$\underline{\quad -4 \quad\quad -4\quad}$

$\dfrac{2}{3}R \quad\quad = \quad 8$

$\dfrac{(3)2}{(2)3}R = \dfrac{(3)(8)}{(2)1}$ Multiply both sides by the reciprocal of 2/3. You could divide out common factors.

$\dfrac{(\cancel{3})\cancel{2}}{(\cancel{2})\cancel{3}}R = \dfrac{(3)(\cancel{8})^{4}}{(\cancel{2})1}$

$R = 12$

Check $\dfrac{2}{3}R + 4 = 12$

$\dfrac{2(\cancel{12})^{4}}{\cancel{3} \cdot 1} + 4 = 12$

$8 + 4 = 12$

$12 = 12$ True.

EXAMPLE **7**   Solve: $\dfrac{2}{3}Y + \dfrac{1}{2} = \dfrac{5}{6}$

Solution $\dfrac{2}{3}Y + \dfrac{1}{2} = \dfrac{5}{6}$

$$\frac{2}{3}Y + \frac{1}{2} = \frac{5}{6}$$

Subtract 1/2 from both sides, using the common-denominator form, $1/2 = 3/6$.

$$-\frac{1}{2} \qquad -\frac{1}{2}$$

$$\frac{2}{3}Y + \frac{1}{2} = \frac{5}{6}$$

$$-\frac{1}{2} \qquad -\frac{3}{6}$$

$$\frac{2}{3}Y = \frac{2}{6}$$

$$\frac{3(2)}{2(3)}Y = \frac{3(2)}{2(6)}$$

Multiply both sides by the reciprocal of 2/3. Then divide common factors or multiply and reduce.

$$\frac{\overset{1}{\cancel{3}}(\overset{1}{\cancel{2}})}{\underset{1}{\cancel{2}}(\underset{1}{\cancel{3}})}Y = \frac{\overset{1}{\cancel{3}}(\overset{1}{\cancel{2}})}{\underset{1}{\cancel{2}}(\underset{2}{\cancel{6}})}$$

$$Y = \frac{1}{2}$$

Check $\dfrac{2}{3}Y + \dfrac{1}{2} = \dfrac{5}{6}$

$$\frac{\cancel{2}(1)}{3(\overset{}{\underset{1}{\cancel{2}}})} + \frac{1}{2} = \frac{5}{6}$$

$$\frac{2}{6} + \frac{1}{2} = \frac{5}{6}$$

$$\frac{2}{6} + \frac{3}{6} = \frac{5}{6}$$

$$\frac{5}{6} = \frac{5}{6} \quad \text{True.}$$

E X A M P L E 8   Solve: $14 = 0.8R + 1.2$

Solution $14.0 = 0.8R + 1.2$ Subtract 1.2 from both sides.

$$\underline{-\ 1.2 \qquad\qquad -\ 1.2}$$

$$12.8 = 0.8R$$ Divide by 0.8.

$$\frac{12.8}{0.8} = \frac{0.8R}{0.8}$$

$$16 = R$$

Check $14 = 0.8R + 1.2$

$$14 = 0.8(16) + 1.2$$

$$14 = 12.8 + 1.2$$

$$14 = 14 \quad \text{True.}$$

E X A M P L E **9**   Solve: $0.6X - 0.003 = 0.021$

Solution

$$0.6X - 0.003 = 0.021$$

$$\underline{+\ 0.003\quad +\ 0.003}$$

$$0.6X \qquad\qquad = 0.024$$

$$\frac{0.6X}{0.6} = \frac{0.024}{0.6}$$

$$X = 0.04$$

Add 0.003 to both sides.

Divide by 0.6.

Check

$$0.6X - 0.003 = 0.021$$

$$0.6(0.04) - 0.003 = 0.021$$

$$0.024 - 0.003 = 0.021$$

$$0.021 = 0.021 \quad \text{True.}$$

◆

Exercises Try these problems. If necessary, use the examples as guides. Remember to check your answers.

1. $3H - 5 = 25$
2. $8M + 6 = 62$
3. $\frac{5}{6}X + 4 = 19$

4. $2.2 = 2X - 0.4$
5. $16 = 3Y - 2$
6. $4T + 3 = 23$

Answers:

1. $H = 10$; check: $25 = 25$
2. $M = 7$; check: $62 = 62$
3. $X = 18$; check: $19 = 19$
4. $X = 1.3$; check: $2.2 = 2.2$
5. $Y = 6$; check: $16 = 16$
6. $T = 5$; check: $23 = 23$

10.3 ◆ Practice Problems

Solve and check.

1. $2T - 7 = 11$
2. $2W - 8 = 10$
3. $4F + 1 = 25$
4. $3M + 5 = 17$
5. $8 + 5Y = 38$
6. $9 + 3T = 27$
7. $1.2P - 8 = 1.6$
8. $11N - 4 = 7$
9. $5T + 6 = 41$
10. $3R - 1.66 = 0.17$
11. $9C + 12 = 48$
12. $0.06K + 1.1 = 7.7$
13. $7X + 3 = 3$
14. $6T + 7 = 7$
15. $0.2W - 1.6 = 2.4$
16. $3M - 21 = 15$
17. $56 + \frac{5F}{6} = 206$
18. $0.43 + 0.3F = 0.73$

19. $4 + \frac{5N}{2} = 44$
20. $9 + \frac{7Y}{3} = 30$
21. $2T - 6 = 0$
22. $7P - 16 = 0$
23. $8K + 4 = 13$
24. $9R + 5 = 15$
25. $\frac{1}{2}F - 7 = 17$
26. $\frac{1}{4}Y - 1 = 15$
27. $5A - \frac{2}{3} = 3$
28. $64 = 5C - 6$
29. $0.8 = 0.16 + 4W$
30. $\frac{5}{6} + 3R = \frac{7}{2}$

10.4 Equations That Require Combining Like Terms

An equation such as $6X - 3 + 2X = 29 - 8$ looks much more complex than the two-step equations that you just solved. But actually, after you simplify, it will be a two-step equation. To simplify, combine like terms on each side of the equals sign. This concept was introduced in Chapter 9.

E X A M P L E **1**   Solve: $6X - 3 + 2X = 29 - 8$

Solution $6X - 3 + 2X = 29 - 8$ $6X$ and $2X$ are like terms. 29 and 8 are like terms. Combine the like terms, using the operation sign that appears in front of each term:

$$6X + 2X \qquad 29 - 8$$

$$8X - 3 = \quad 21$$

This should now look familiar. It's a two-step equation.

$$\underline{+3 = +\ 3}$$

$$8X \quad = \quad 24$$

$$\frac{8X}{8} \quad = \quad \frac{24}{8}$$

$$X = 3$$

Check   When checking an equation, *always start with the original* equation, not one of the simplified forms. You may have made a mistake in getting to that step, and you will not find such an error if you don't take your answer into the original equation.

$$6X - 3 + 2X = 29 - 8$$
$$6(3) - 3 + 2(3) = 29 - 8$$
$$18 - 3 + 6 = 29 - 8$$
$$15 + 6 = 21$$
$$21 = 21 \quad \text{True.}$$

◆

Note that for the first time the check answer, $21 = 21$, is not a number that was carried down from the original equation. This happens in equations that have more than one term or variable on each side of the equals sign.

E X A M P L E **2**   Solve: $K + 2K + 3K = 180$

Solution $K + 2K + 3K = 180$ Combine like terms:

$$6K = 180 \qquad\qquad 1K + 2K + 3K = 6K$$

$$\frac{6K}{6} = \frac{180}{6}$$

$$K = 30$$

Check $K + 2K + 3K = 180$
$$30 + 2(30) + 3(30) = 180$$
$$(30) + (60) + (90) = 180$$
$$180 = 180 \quad \text{True.}$$

◆

E X A M P L E **3**   Solve: $7m + 9 - 3m + 4 = 27 + 6$

Solution $7m + 9 - 3m + 4 = 27 + 6$ $7m$ and $3m$, 9 and 4, and 27 and 6 are like terms. Combine like terms: $7m - 3m = 4m$, $9 + 4 = 13$, and $27 + 6 = 33$.

$$4m + 13 = \quad 33$$

$$\underline{-13 = -13}$$

$$4m \qquad = \quad 20$$

$$\frac{4m}{4} \qquad = \quad \frac{20}{4}$$

$$m = 5$$

Check $\qquad 7m + 9 - 3m + 4 = 27 + 6$

$$7(5) + 9 - 3(5) + 4 = 27 + 6$$
$$35 + 9 - 15 + 4 = 27 + 6$$
$$44 - 15 + 4 = 33$$
$$29 + 4 = 33$$
$$33 = 33 \quad \text{True.} \qquad \blacklozenge$$

When solving equations, always look first to see if terms can be combined on either side of the equal sign. If so, do this simplifying first before "undoing" any operations.

Exercises Try these equations before doing the practice problems. Follow the steps shown in the examples if you need to. Be sure to check your answers.

1. $11X + 4X + X = 64$ **2.** $14 + 5R + 16 = 50$
3. $T + 7 + 2T - 2 = 22 + 4$ **4.** $M + 8 + 5M + 2 = 40$

Answers:

1. $X = 4$; check: $64 = 64$ **2.** $R = 4$; check: $50 = 50$
3. $T = 7$; check: $26 = 26$ **4.** $M = 5$; check: $40 = 40$

10.4 ◆ Practice Problems

A. Combine any like terms first; then solve and check.

1. $5T + 3T = 16$ **2.** $11M + 2M = 26$
3. $6F + 3 - 4F = 13$ **4.** $5X + 5 - 2X = 17$
5. $14 + 6R + 11 = 43$ **6.** $9 + 3Y + 4 = 28$
7. $240 = 4n + 2n$ **8.** $190 = 14R + 5R$
9. $12T + 6 - 4T = 22$ **10.** $17W + 4 - 4W = 43$
11. $3M - 9 = 24 - 6$ **12.** $5Y - 7 = 38 - 5$
13. $6r - 3r = 12 + 9$ **14.** $8t - 3t = 19 - 4$
15. $13M - 8M = 34 - 9$ **16.** $22 - 7 = 38W - 8W$

B. Solve and check.

17. $4 + 28 = 10Y - 2Y$
18. $75 + 5 = 25K + 15K$
19. $14T - 4T = 30$
20. $80 = 55k - 15k$
21. $12M + M = 52$
22. $Y + 7Y = 8$
23. $7R + 2 - 3R + 6 = 21 + 3$
24. $6w + 3 - 2w + 5 = 36$
25. $240 = C + 19C$

26. $180 = X + 17X$
27. $12x - 8x + 12 = 28$
28. $5r + 2r + 2 = 23$
29. $6t + 6 + 4t = 16$
30. $9n + 3 - 6n = 15$
31. $8C + 12 - 9 = 15 - 4$
32. $7A + 4 - 3A = 9 + 3$
33. $9R - 6R = 36$
34. $54 - 18 = 6T - 3 + 7T$
35. $26T + 4 - 22T + 5 = 15$
36. $19M - 11M = 24$
37. $84 - 78 = 13M + 6 - 7M$
38. $8A - 7 + 11A = 12$
39. $8N + 15 + 9N - 6 = 62 - 53$
40. $47 - 8 = 24 + 8W + 15 - 3W$

10.5 Equations with Unknown Terms on Both Sides

The next equations you will solve in this chapter are those that have variable terms on both sides of the equals sign. In solving equations such as:

$$3Y + 4 = Y + 12 \quad \text{and} \quad 16 - 7T = 3T - 4$$

You are somewhat restricted in how you can proceed with these equations until you have studied Chapter 12 on signed numbers. For now you must work with positive numbers so look at the numerical coefficients of the variable terms, and picture their positions on a number line to determine which is *smaller*. The number that is farther to the left on the line is the smaller number.

In the first sample problem, 1 would be farther to the left than 3, so $1Y$ is the smaller variable term. You should start solving the equation by eliminating that smaller term. You will have to subtract $1Y$ from Y to eliminate it on the right side. Do the same to the other side of the equation, subtracting $1Y$ from $3Y$. (Adding $1Y$ to Y would give $2Y$, not the $0Y$ that you want.)

$$
\begin{array}{rcl}
3Y + 4 = & & Y + 12 \\
- 1Y & & - 1Y \\
\hline
2Y + 4 = & & 0Y + 12
\end{array}
$$

It's now a familiar two-step equation.

$$
\begin{array}{rcl}
2Y + 4 = & 12 \\
- 4 & - 4 \\
\hline
2Y \quad = & 8 \\
\dfrac{2Y}{2} = \dfrac{8}{2} \\
Y = 4
\end{array}
$$

Check
$$
\begin{array}{rcl}
3Y + 4 &=& Y + 12 \\
3(4) + 4 &=& 4 + 12 \\
12 + 4 &=& 4 + 12 \\
16 &=& 16 \quad \text{True.}
\end{array}
$$

In the second sample problem, look at the variable terms with the sign that is in front of them, $-7T$ and $3T$. 3 is assumed to be $+3$ and is to the right of the zero on the number line. The other variable coefficient, -7, is to the left of the zero and is therefore smaller than $+3$. You should add its opposite to both sides of the equation. (**Opposites** have the same number value but different signs. Opposites always add to zero.)

$$
\begin{array}{rcl}
16 - 7T = & & 3T - 4 \\
+ 7T & & + 7T \\
\hline
16 \quad = & & 10T - 4
\end{array}
$$

A two-step equation remains.

$$16 = 10T - 4$$
$$\underline{+\ 4 \qquad\qquad +\ 4}$$
$$20 = 10T$$
$$\frac{20}{10} = \frac{10T}{10}$$
$$2 = T$$

Check $16 - 7T = 3T - 4$
$$16 - 7(2) = 3(2) - 4$$
$$16 - 14 = 6 - 4$$
$$2 = 2 \quad \text{True.}$$

E X A M P L E 1   Solve: $11N - 5 = 12 - 6N$

Solution $11N - 5 = 12 - 6N$ $-6N$ is smaller than $+11N$. Add $+6N$ to both sides.
$$\underline{+6N \qquad\qquad +\ 6N}$$
$$17N - 5 = 12$$
$$\underline{+\ 5 = +5}$$
$$17N \qquad = 17$$
$$\frac{17N}{17} = \frac{17}{17}$$
$$N = 1$$

Check $11N - 5 = 12 - 6N$
$$11(1) - 5 = 12 - 6(1)$$
$$11 - 5 = 12 - 6$$
$$6 = 6 \quad \text{True.}$$

◆

In some problems, you might have to combine like terms first. Follow through this example carefully. Remember that no matter how complicated an equation looks at first, each step you do makes it easier.

E X A M P L E 2   Solve: $3W + 5 + W - 3 = 18 + 2W - 6$

Solution $3W + 5 + W - 3 = 18 + 2W - 6$ Combine $3W + W$, $5 - 3$, and $18 - 6$. Which is smaller, $4W$ or $2W$?
$$4W + 2 = 12 + 2W$$
$$\underline{-\ 2W \qquad\qquad -\ 2W}$$
$$2W + 2 = 12$$
$$\underline{-\ 2 \quad -2}$$
$$2W \qquad = 10$$
$$\frac{2W}{2} = \frac{10}{2}$$
$$W = 5$$

Check $3W + 5 + W - 3 = 18 + 2W - 6$ Use the original equation.

$3(5) + 5 + 5 - 3 = 18 + 2(5) - 6$

$15 + 5 + 5 - 3 = 18 + 10 - 6$

$20 + 5 - 3 = 28 - 6$

$25 - 3 = 22$

$22 = 22$ True.   ◆

Exercises   Here are a few equations for you to try. Follow the steps of the examples as you solve these equations. Check your answers in the original equation first. Then look at the answers that follow if you need more proof that you are correct.

1. $3W - 7 = 8 - 2W$ 2. $9M - 16 = 6 - 2M$

3. $6X + 4 = X + 14$ 4. $5N + 3 = 2N + 30$

Answers:

1. $W = 3$; check: $2 = 2$ 2. $M = 2$; check: $2 = 2$

3. $X = 2$; check: $16 = 16$ 4. $N = 9$; check: $48 = 48$

In this chapter, you began by solving one-step equations and gradually learned to solve those that require three and four steps. The following list summarizes what you have learned.

Steps for Solving Equations

1. Combine like terms on each side of the equals sign.
2. If unknowns occur on both sides, start by eliminating the smaller variable term.
3. Add or subtract variable terms and numerical terms.
4. Do multiplication or division.
5. Check your answer in the original equation.

10.5 ◆ Practice Problems

A. Solve and check.

1. $9R = 33 - 2R$ 2. $7M = 15 + 2M$

3. $3 - t = 8t$ 4. $5P = 2 + P$

5. $11W = 36 - 7W$ 6. $3Y + 6 = 4Y$

7. $4k + 8 = 12k$ 8. $5c = 21 + 2c$

9. $6m = 36 - 3m$ 10. $8X + 5 = 4X + 13$

11. $6T + 2 = T + 17$ 12. $7N + 4 = 6N + 7$

13. $5w - 4 = 2w + 5$ 14. $9C - 10 = 3C + 2$

15. $12Y - 2 = 9Y + 7$ 16. $13A - 1 = 4A + 17$

B. Solve and check.

17. $15x - 22 = 4x + 11$ 18. $7a - 5 = 2a + 20$

19. $3W + 1 = 11 - 2W$ 20. $n - 2 = 6 - 3n$

21. $2x - 3 = 9 - 2x$ 22. $4y - 5 = 16 - 3y$

23. $5a + 7 = 2a + 7$ 24. $4y + 8 = y + 8$

25. $10x + 1 = 6x + 10 - 1$

26. $5r - 3 = 7 + 4r - 6$

27. $11C - 4 = 7 + 4C + 10$

28. $5w + 6w - 1 = 3w + 15$

29. $4k + 8 = 20 + k$

30. $3Y + 6 = 55 - 4Y$

C. Solve and check. These equations include all types of problems found in Sections 10.2 through 10.5.

31. $b - 6 = 27$ 32. $6W + 9 = 45$

33. $\dfrac{y}{3} - 8 = 2$ 34. $3\frac{1}{2}P - 9 = 45$

35. $3a + 7 = 28$ 36. $2 = y - 8$

37. $3T - 25 = 8$ 38. $2X - 1 = 7$

39. $35 = 10y - 45$ 40. $4M + 7 = 7M - 20$

41. $3c + 7 = 8c - 18$

42. $9w + 4 - 4w = 3w + 24$

43. $2a - 7 = 7$ **44.** $24 = 18c + 21$

45. $\dfrac{x}{5} = 6$ **46.** $\dfrac{m}{6} = 0$

47. $6Z - 17 = 8Z + 10 - 5Z$

48. $4 + 3X = 13$

49. $24R + 25 = 73$ **50.** $\dfrac{3}{5}B + 7 = 12$

◆ 10.6 ◆   Solving Equations with Fractions and Decimals

In this section, you will learn how to solve equations that contain fraction or decimal numbers. The process for solving these equations includes the same steps as those you followed in Sections 10.2 through 10.5.

If you are to succeed with this section, it is important that you have good skills in fraction and decimal operations. If you have not reviewed or tested yourself on these skills, as was suggested at the beginning of the chapter, then do so now before going any farther (see the Skills Check, Appendix C).

In Section 10.3, you solved some equations that contain fractions and decimals by working with those numbers and doing the same procedures you would have done with whole numbers. In this section you will learn an alternative method that eliminates the fraction or decimal numbers and thus enables you to convert these equations to equations containing only whole numbers.

The new method has you multiply *every* term in the equation on *both sides* of the equals sign by the same value. Let's explore that idea first. Solve each of the following equations:

$$X + 2 = 6 \qquad 2X + 4 = 12 \qquad 5X + 10 = 30$$

Did you get $X = 4$ as the solution to each one? In fact, all three represent the same equation. See what happens when each term of $X + 2 = 6$ is multiplied by 2 or by 5.

$$X + 2 = 6 \text{ multiplied by 2:}\quad 2(X) + 2(2) = 2(6)$$
$$2X + 4 = 12$$
$$X + 2 = 6 \text{ multiplied by 5:}\quad 5(X) + 5(2) = 5(6)$$
$$5X + 10 = 30$$

The fact that all three equations have a solution of $X = 4$ is not what makes them **equivalent equations**. They are equivalent because each of the equations is just a variation of the first equation. That is, each term of the first equation was multiplied by the same value to produce the second and third equations. Division can be used on each term as well, though you will not see such an example in this chapter.

Equivalent Equations

> If every term in an equation is multiplied or divided by the same value, the resulting equation is equivalent to the original and has the same solution.

This process is similar to the balance illustration (Figure 1 in Section 10.2), which showed that adding or subtracting the same number on both sides of the equation maintains the balance. Multiplying or dividing every term on both sides of the equation by the same number also maintains the balance or equality. It produces an equivalent equation—one with a different appearance but the same solution. This is another application of the Multiplication or Division Property of Equality.

E X A M P L E 1 Are $Y - 2 = 5$ and $3Y - 6 = 15$ equivalent equations?

Solution Yes, because when each term in $Y - 2 = 5$ is multiplied by 3, $3(Y) - 3(2) = 3(5)$, it becomes $3Y - 6 = 15$. ◆

E X A M P L E 2 Are $6M - 15 = 21$ and $2M - 5 = 8$ equivalent equations?

Solution Inspection reveals that the second equation has a smaller coefficient with the variable. You could multiply $2M$ by 3 to get $6M$. Then multiply every term in the equation by 3 to see if you get an equation identical to the first.

$$(3)2M - (3)5 = (3)8$$
$$6M - 15 = 24 \qquad \text{This is not exactly the same as } 6M - 15 = 21, \text{ so the}$$
$$\text{equations are not equivalent.} \quad ◆$$

In solving a fraction or decimal equation by this new method, you must multiply through the equation by a value that eliminates the denominators or the decimal places. When all terms in a fraction or decimal equation are multiplied by the right value, the resulting equation contains only whole number terms and coefficients. In fraction equations, you multiply by the least common denominator (LCD) of all the denominators in the equation.

E X A M P L E 3 Solve for Y: $\dfrac{2}{3}Y + \dfrac{1}{2} = \dfrac{5}{6}$

Solution $\dfrac{2}{3}Y + \dfrac{1}{2} = \dfrac{5}{6}$

The least common denominator of the denominators 3, 2, and 6 is 6. Multiply every term by 6 to eliminate the fractions and thus produce an equation that you can easily solve using steps you already know.

$$\frac{6(2)}{1\ 3}Y + \frac{6(1)}{1\ 2} = \frac{6(5)}{1\ 6}$$

$$\frac{\overset{2}{\cancel{6}}(2)}{1\ \underset{1}{\cancel{3}}}Y + \frac{\overset{3}{\cancel{6}}(1)}{1\ \underset{1}{\cancel{2}}} = \frac{\overset{1}{\cancel{6}}(5)}{1\ \underset{1}{\cancel{6}}} \qquad \text{Dividing out common factors could be used to simplify the terms.}$$

$$4Y + 3 = 5 \qquad \text{Now solve the whole number equation.}$$
$$\underline{ - 3 \quad - 3}$$
$$4Y = 2$$
$$\frac{4Y}{4} = \frac{2}{4}$$
$$Y = \frac{1}{2}$$

Check $\dfrac{2}{3}Y + \dfrac{1}{2} = \dfrac{5}{6}$ \qquad Always start with the original equation.

$$\frac{\overset{1}{\cancel{2}}(1)}{3(\underset{1}{\cancel{2}})} + \frac{1}{2} = \frac{5}{6}$$

$$\frac{1}{3} + \frac{1}{2} = \frac{5}{6}$$

$$\frac{2}{6} + \frac{3}{6} = \frac{5}{6}$$

$$\frac{5}{6} = \frac{5}{6} \quad \text{True.} \qquad \blacklozenge$$

E X A M P L E **4** Solve: $A + \frac{2}{3}A = \frac{4}{9}$

Solution Multiply every term by the LCD, 9. Even the whole number term $(1A)$ must be multiplied by 9.

$$A + \frac{2}{3}A = \frac{4}{9}$$

$$9(A) + \frac{\overset{3}{\cancel{9}}(2)}{1(\cancel{3})}A = \frac{\overset{1}{\cancel{9}}(4)}{1(\cancel{9})}$$

$$9A + 6A = 4$$

$$15A = 4$$

$$\frac{15}{15}A = \frac{4}{15}$$

$$A = \frac{4}{15}$$

Check $A + \frac{2}{3}A = \frac{4}{9}$

$$\frac{4}{15} + \frac{2(4)}{3(15)} = \frac{4}{9}$$

$$\frac{4}{15} + \frac{8}{45} = \frac{4}{9}$$

$$\frac{12}{45} + \frac{8}{45} = \frac{4}{9}$$

$$\frac{20}{45} = \frac{4}{9}$$

$$\frac{4}{9} = \frac{4}{9} \quad \text{True.} \qquad \blacklozenge$$

In decimal equations, the method of eliminating decimal places involves multiplying by 10, 100, 1,000, and so forth, depending on the greatest number of decimal places in any term of the equation. Review the short cut for doing this multiplication in Section 5.4.

E X A M P L E **5** Solve for X: $0.6X - 0.003 = 0.021$

Solution $0.6X - 0.003 = 0.021$

The greatest number of decimal places in this problem is 3, so multiply each term by 1,000 to eliminate the decimals and produce a whole number equation. If two were the greatest number of decimal places, you would use 100; if 1, you would use 10.

$$1000(0.6X) - 1000(0.003) = 1000(0.021)$$

$$
\begin{aligned}
600X - 3 &= 21 \qquad \text{Now solve this whole number equation.}\\
+3 \quad &\ +3\\
\hline
600X &= 24\\
\frac{600X}{600} &= \frac{24}{600} \qquad \text{Always divide } by \text{ the coefficient of the}\\
&\qquad\qquad\quad \text{variable term.}\\
X &= 0.04
\end{aligned}
$$

Check $0.6X - 0.003 = 0.021$ Use the original equation.

$0.6(0.04) - 0.003 = 0.021$

$0.024 - 0.003 = 0.021$

$0.021 = 0.021$ True.   ◆

E X A M P L E 6   Solve: $4.2B + 8.6 - 3.1B = 14.1$

Solution   Multiply by 10 to eliminate the decimals.

$$10(4.2B) + 10(8.6) - 10(3.1B) = 10(14.1)$$

$$
\begin{aligned}
42B + 86 - 31B &= 141\\
11B + 86 &= 141\\
-86 \quad &\ -86\\
\hline
11B &= 55\\
\frac{11B}{11} &= \frac{55}{11}\\
B &= 5
\end{aligned}
$$

Check $4.2B + 8.6 - 3.1B = 14.1$

$4.2(5) + 8.6 - 3.1(5) = 14.1$

$21.0 + 8.6 - 15.5 = 14.1$

$29.6 - 15.5 = 14.1$

$14.1 = 14.1$ True.   ◆

You have now experienced two methods for solving decimal and fraction equations: one where you work with those numbers and one where you eliminate them. As you solve the following problems, try using both methods and then decide which works better for you or which looks easier to use for that particular problem. Be sure to check your answers in the original equation.

Exercises   **1.** $2W + \dfrac{1}{3} = \dfrac{7}{3}$ **2.** $7 = 0.4Y + 0.6$

3. $3.6 = 3.42 + 0.006X$ **4.** $\dfrac{7}{2} = \dfrac{2}{5}X + 3$

Answers:

1. $W = 1$; check: $7/3 = 7/3$ **2.** $Y = 16$; check: $7 = 7$

3. $X = 30$; check: $3.6 = 3.6$ **4.** $X = 5/4$; check: $7/2 = 7/2$

10.6 ◆ Practice Problems

A. Determine whether the pairs of equations are equivalent. Use the procedures shown in Examples 1 and 2 in this section.

1. $3Y - 6 = 15$ and $Y - 2 = 5$

2. $7C + 14 = 35$ and $C + 2 = 5$

3. $2W - 9 = 12$ and $12W - 54 = 84$

4. $3P - 4 = 15$ and $21P - 28 = 90$

5. $T + 6 = 15$ and $75 = 5T + 30$

6. $18V + 12 = 24$ and $4 = 2 + 3V$

B. Solve these equations using the method introduced in this section. Multiply every term by the value that will clear the denominators or decimal points. Check each answer.

7. $\dfrac{3}{4}K - \dfrac{1}{3} = \dfrac{1}{2}$

8. $12 + 0.4M = 14.32$

9. $2.4N - 0.08 = 7.12$

10. $\dfrac{3}{8}P + \dfrac{3}{2} = 6$

11. $0.2W - 1.6 = 2.4$

12. $\dfrac{3M}{4} - \dfrac{21}{5} = 15$

13. $56 + \dfrac{5F}{6} = 206$

14. $0.3F + 0.43 = 0.73$

C. Solve the following equations. You don't have to use the same method for all problems. Check your answers.

15. $\dfrac{1}{3}X + 1 = \dfrac{7}{2}$

16. $28 = 1.4Y + 2.1$

17. $1.4K - 3.5 = 1.12$

18. $\dfrac{1}{2}A - 3 = \dfrac{3}{4}$

19. $\dfrac{3}{4}N + 7 = \dfrac{15}{2}$

20. $0.735 + 2.4W = 12.9W$

21. $3.92 + 5.6X = 7.2X$

22. $\dfrac{3}{7}C + \dfrac{1}{4} = 1$

23. $2M - \dfrac{3}{4} = \dfrac{13}{4}$

24. $8.7P = 3.9P + 16.8$

25. $2.7X + 9 = 4.5X$

26. $3Y - \dfrac{5}{6} = \dfrac{13}{6}$

27. $0.2N + 1.3 = 1.5 - 0.6N$

28. $3\dfrac{3}{4}W + 1\dfrac{1}{2}W = 7$

29. $1\dfrac{5}{6}A + 2\dfrac{2}{3}A = 6$

30. $2.4X + 0.94 = 12.42 - 5.8X$

D. Solve and check using a calculator. If necessary, round answers to the nearest hundredth.

31. $83.6y - 19.8 = 63.5$

32. $\dfrac{0.7R}{0.72} + 6.91 = 11.3$

33. $\dfrac{4.3}{18.46} = 0.94t$

34. $1.82m = 6.56 - 0.86m$

10.7 Formulas

FIGURE 3

A **formula** is a general statement of how several quantities are related, so formulas usually contain more than one variable. The formula $P = 2L + 2W$, which you have seen before, is used to find the perimeter P of any rectangle of length L and width W (see Figure 3). The formula $A = LW$ is used to find the area A of any rectangle.

These formulas show the general relationships that apply to all rectangles of any size. If you are given dimensions (measurements) of a specific rectangle, then you can find the perimeter (distance around the figure) or area (surface of the figure) by substituting those known values into the formula and solving the resulting equation.

E X A M P L E 1 Find the perimeter of the rectangle in Figure 4 using the formula $P = 2L + 2W$.

Solution

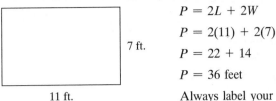

7 ft.

11 ft.

F I G U R E 4

$P = 2L + 2W$

$P = 2(11) + 2(7)$

$P = 22 + 14$

$P = 36$ feet

Always label your answers with the correct units.

Check $P = 2L + 2W$

$36 = 2(11) + 2(7)$

$36 = 22 + 14$

$36 = 36$ True. ◆

E X A M P L E 2 Find the area of the rectangle in Figure 4 using the formula $A = LW$.

Solution $A = LW$

$A = (11)(7)$

$A = 77$ square feet

Note: When you multiply feet × feet, the resulting units are square feet. Area is *always* expressed in square units.

Check $A = LW$

$77 = (11)(7)$

$77 = 77$ True. ◆

E X A M P L E 3 The perimeter of a rectangle is 98 centimeters and the width is 12 centimeters. Find the length, using $P = 2L + 2W$.

Solution $P = 2L + 2W$

$98 = 2L + 2(12)$ Substitute the known values and simplify.

This is now an equation with one variable (L). Solve it as a two-step equation.

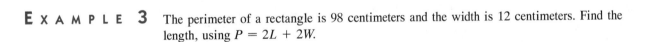

$98 = 2L + 24$

$\underline{- 24 \qquad\quad - 24}$

$74 = 2L$

$\dfrac{74}{2} = \dfrac{2L}{2}$

$37 = L$ The length is 37 centimeters.

Check $P = 2L + 2W$

$98 = 2(37) + 2(12)$

$98 = 74 + 24$

$98 = 98$ True. ◆

Another approach to Example 3 would be to rearrange or solve the formula for L before substituting the given values. The formula $P = 2L + 2W$ is solved for P, because P is alone on one side of the equals sign. To solve that equation for L, add or subtract terms *on both sides* in such a way that $2L$ ends up alone on one side. Then divide both sides by 2 to leave L. After L is isolated, substitute the given values and simplify.

Solution

$$P \quad\quad = 2L + 2W \quad\quad \text{Solve for } L.$$

$$\underline{\quad - 2W \quad\quad - 2W \quad}$$

$$P - 2W = 2L$$

$$\frac{P - 2W}{2} = \frac{2L}{2}$$

$$\frac{P - 2W}{2} = L \quad\quad \text{Substitute given values.}$$

$$\frac{98 - 2(12)}{2} = L$$

$$\frac{98 - 24}{2} = L$$

$$\frac{74}{2} = L$$

$$37 = L \quad\quad \text{The length is 37 centimeters.} \quad\quad\blacklozenge$$

E X A M P L E   4   Solve $V = \dfrac{1}{3}bh$ for b.

Solution

$$V = \frac{1}{3}bh \quad\quad \text{Multiply both sides by 3 to eliminate the fraction.}$$

$$(3)V = \frac{\overset{1}{(\cancel{3})}}{1}\frac{1}{\cancel{3}}bh$$

$$3V = bh \quad\quad \text{Divide both sides by } h.$$

$$\frac{3V}{h} = \frac{bh}{h}$$

$$\frac{3V}{h} = b \quad\quad\quad\quad\quad\quad\quad\quad\quad\quad\quad\quad\quad\quad\quad\blacklozenge$$

E X A M P L E   5   Find the circumference of (distance around) the circle in Figure 5 using $C = 2\pi r$ and $r = 8$ inches. π is a numerical constant that equals approximately 3.14.

Solution

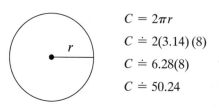

$$C = 2\pi r$$

$$C \doteq 2(3.14)(8)$$

$$C \doteq 6.28(8)$$

$$C \doteq 50.24 \quad\quad \text{The circumference is approximately 50.24 inches.}$$

F I G U R E   5

Check $C = 2\pi r$
$50.24 \doteq 2(3.14)(8)$
$50.24 \doteq 6.28(8)$
$50.24 \doteq 50.24$ True.   ◆

E X A M P L E 6 Find the area of the circle in Figure 5 using $A = \pi r^2$, $r = 8$ inches, and $\pi \doteq 3.14$.

Solution $A = \pi r^2$
$A \doteq (3.14)(8)^2$
$A \doteq (3.14)(64)$
$A \doteq 200.96$ The area is approximately 200.96 square inches.

Check $A = \pi r^2$
$200.96 \doteq (3.14)(8)^2$
$200.96 \doteq (3.14)(64)$
$200.96 \doteq 200.96$ True.   ◆

E X A M P L E 7 The circumference of a circle is 28.26 meters. Find the radius. Use $\pi \doteq 3.14$.

Solution You can substitute first and then solve for r:
$C = 2\pi r$ Substitute.
$28.26 = 2(3.14)r$ Simplify.
$28.26 \doteq 6.28r$ Solve for r.
$$\frac{28.26}{6.28} \doteq \frac{6.28}{6.28}r$$
$4.5 \doteq r$ The radius is approximately 4.5 meters.

Solution Or you can solve for r first and then substitute:
$C = 2\pi r$ Divide by 2π.
$$\frac{C}{2\pi} = \frac{2\pi r}{2\pi}$$
$$\frac{C}{2\pi} = r$$ Substitute the given values.
$$\frac{28.26}{2(3.14)} \doteq r$$
$4.5 \doteq r$ The radius is approximately 4.5 meters.

Check $C = 2\pi r$
$28.26 \doteq 2(3.14)(4.5)$
$28.26 \doteq 6.28(4.5)$
$28.26 \doteq 28.26$ True.   ◆

There are many formulas used in business, science, and mathematics with which you may be familiar. The formula for finding simple interest is $I = PRT$ (I is interest, P is principal, R is rate, and T is time). $S = C + M$ shows the relationship among the selling price (S) of an item, the cost (C) of the item, and the markup (M) needed to cover expenses and profit.

EXAMPLE 8 Find the amount of markup on an item that sells for $114.50 and that costs $69.79. Use $S = C + M$.

Solution

$$S = C + M$$
$$114.50 = 69.79 + M$$
$$\underline{-69.79 -69.79}$$
$$44.71 = M$$

The markup is $44.71.

Check

$$S = C + M$$
$$114.50 = 69.79 + 44.71$$
$$114.50 = 114.50 \quad \text{True.}$$ ◆

EXAMPLE 9 Find the amount that was invested (principal) at 8% interest for 1 year if $125.00 interest was earned. Use $I = PRT$.

Solution

$$I = PRT$$
$$125 = P(8\%)(1) \qquad \text{Change 8\% to a decimal.}$$
$$125 = P(0.08)(1)$$
$$125 = 0.08P$$
$$\frac{125}{0.08} = \frac{0.08P}{0.08}$$
$$\$1562.50 = P$$

Check

$$I = PRT$$
$$125 = (1562.50)(0.08)(1)$$
$$125 = 125.00(1)$$
$$125 = 125 \quad \text{True.}$$ ◆

EXAMPLE 10 What selling price should be placed on an item that cost $19.36 if the manager requires a markup of $2.17? Use $S = C + M$.

Solution

$$S = C + M$$
$$S = 19.36 + 2.17$$
$$S = \$21.53$$

Check

$$S = C + M$$
$$21.53 = 19.36 + 2.17$$
$$21.53 = 21.53 \quad \text{True.}$$ ◆

10.7 ◆ Practice Problems

A. Substitute the given values into each formula and solve for the unknown. Label answers with correct units, and check your answers. If necessary, round answers to the nearest hundredth.

1. The formula for perimeter of a square is $P = 4s$. Find the perimeter if $s = 6$ feet.

2. The formula for the perimeter of a regular hexagon (six-sided figure) is $P = 6s$. Find the perimeter if the side (s) is 11 feet.

3. Find the perimeter of a rectangle ($P = 2L + 2W$) if the length is 5 centimeters and the width is 3 1/2 centimeters.

4. Find the perimeter of a rectangle ($P = 2L + 2W$) if the length is 11 1/2 yards and the width is 5 1/4 yards.

5. The formula for the perimeter of a rectangle is $P = 2L + 2W$. Find the width if the perimeter is 156 inches and the length is 28 inches.

6. The formula for the perimeter of a rectangle is $P = 2L + 2W$. Find the length if the perimeter is 96 meters and the width is 12 meters.

7. Find the circumference of a circle whose radius is 19 feet. Use $C = 2\pi r$ and $\pi \doteq 3.14$.

8. Use the circumference formula, $C = 2\pi r$ ($\pi \doteq 3.14$), to find the circumference of a circle that has a radius of 2.4 inches.

9. Find the area of the rectangle in Problem 3. Use $A = LW$.

10. Find the area of the rectangle in Problem 4. Use $A = LW$.

11. Find the area of the square in Problem 1. Use $A = s^2$.

12. Find the area of a square ($A = s^2$) if $s = 1.2$ centimeters.

13. Find the area of the circle in Problem 7. Use $A = \pi r^2$ and $\pi \doteq 3.14$.

14. Find the area of the circle in Problem 8. Use $A = \pi r^2$ and $\pi \doteq 3.14$.

B.   Use the formula $I = PRT$ to solve for the unknown quantity. Percent answers should be rounded to the nearest tenth of a percent.

15. Find the amount of interest earned if $10,000 is invested at 10% annual interest for 3 years.

16. How much interest must be paid if $5,000 is borrowed at 8% annual interest for 2 years?

17. How long did it take to repay a loan if $500 was borrowed at 13% interest and the interest charged was $260?

18. How much money was borrowed at 15% interest if $90 in interest was charged for the 3-year period?

19. What is the rate of annual interest if $1,600 earns $416 in 2 years?

20. How long does it take to earn $240 when $400 is invested at 12% annual interest?

21. How much money was invested at 11.5% annual interest if it earned $172.50 in 3 years?

22. What is the annual rate of interest on a charge account that collects $19.80 in interest on a balance of $110?

23. The formula $A = P + I$ is used to compute the total amount (A) of an investment or loan. Using this formula and the simple interest formula, $I = PRT$, find the amount of an investment at the end of 3 years if $P = \$12,500$ and $R = 12.5\%$.

24. Find the total amount to be repaid at the end of 2 years if $10,000 is borrowed at 11.6%. The formula $A = P + I$ gives the total amount (A) of an investment or loan. Use this formula and the simple interest formula, $I = PRT$.

C.   Using the given formula and information, solve for the unknown quantity.

25. Formula: $S = C + M$. Find S if $C = \$117.30$ and $M = \$40.00$.

26. Formula: $S = C + M$. Find M if $S = \$19.99$ and $C = \$10.00$.

27. Formula: $A = 1/2 bh$. Find the area (A) if $b = 13$ inches and $h = 10$ inches.

28. Formula: $P = a + b + c$. Find P if $a = 6$ feet, $b = 7.5$ feet, and $c = 11$ feet.

29. Formula: $P = a + b + c$. Find b if $P = 65$ centimeters, $c = 25$ centimeters, and $a = 28$ centimeters.

30. Formula: $P = 2L + 2W$. Find L if $P = 124$ yards and $W = 27$ yards.

31. Formula: $C = 2\pi r$. Find r if $C = 43.96$ inches and $\pi \doteq 3.14$.

32. Formula: $C = 2\pi r$. Find r if $C = 56.52$ centimeters and $\pi \doteq 3.14$.

D.   Solve each formula or equation for the stated variable.

33. $V = \dfrac{1}{3} bh$ for h

34. $I = PRT$ for T

35. $I = PRT$ for P

36. $y = mx + b$ for m

37. $y = mx + b$ for x

38. $A = P + PRT$ for R

39. $A = P + PRT$ for T

40. $E = IR$ for I

41. $E = IR$ for R

42. $V = \dfrac{1}{3} \pi r^2 h$ for r^2

43. $V = \dfrac{1}{3} \pi r^2 h$ for h

44. $F = \dfrac{9}{5} C + 32$ for C

45. $V = \dfrac{4}{3} \pi r^3$ for r^3

10.8 Chapter Review

Solving an equation means finding the value for the variable that produces a true statement. The steps for solving an equation are as follows:

1. If desired, eliminate fractions or decimal numbers by applying the Multiplication or Division Property of Equality to create an equivalent equation.

2. Combine like terms on each side of the equals sign.

3. If unknowns occur on both sides of the equals sign, start by eliminating the smaller variable term.

4. Add or subtract variable terms and numerical terms by using the Addition or Subtraction Property of Equality.

5. Do multiplication or division.

6. Check the answer in the *original* equation.

Formulas are equations that are general statements about how several quantities are related. You can substitute known values into the formula and then solve the resulting equation, or you can rearrange first and then substitute.

This chapter looks very long, but don't be too concerned about that as you prepare for the test. Each section in the chapter just added a step to the beginning of the process of solving equations. Once you learned the first sections of material, you used them in every other section. With that much practice, those earlier steps have become almost second nature to you!

Also, this material contains a confidence-building factor. Because you can check each answer, you know that you are doing things correctly. This is very reassuring, especially at test time! Many people think of solving equations as "doing algebra." Thus, your success with this material should make you feel even more confident about your ability to learn algebra.

Review Problems

A. Solve and check.

1. $3X = 12$

2. $N + 5 = 7$

3. $F - 19 = 23$

4. $\frac{3}{4}C = 12$

5. $9 = H + 4$

6. $20 = 4Y$

7. $2A + 4 = 16$

8. $\frac{T}{9} = 4$

9. $11 = 5W - 9$

10. $J + 3.9 = 10.4$

11. $c - 1.37 = 9.44$

12. $R - 7 = 20$

13. $\frac{2}{3}Y = 24$

14. $\frac{F}{12} = 12$

15. $14 = 8M - 2$

16. $3N - 8 = 19$

17. $32.1 = 7S + 4.1$

18. $\frac{5}{6}X + 4 = 24$

19. $\frac{3}{4}X + 6 = 24$

20. $8.9 = 12W + 1.7$

21. $9P = 24$

22. $2X - 5 = 16$

23. $1.1Y + 0.3 = 3.6$

24. $\frac{4}{5}Y + 3 = 20$

25. $19C - 12 = 64$

26. $11W = 20$

27. $A + \frac{3}{4}A = 6$

28. $2X + 4 - X = 29$

29. $3S - 4 + S = 18$

30. $T + \frac{2}{3}T = 13$

31. $3F - 5 = F + 3$

32. $8R + 4 = 5R + 12$

33. $5Y - 4 = 22 - 6$

34. $180 = 4N + 2N$

35. $24 = 14R + 10R$

36. $4M - 8 = 20 - 6$

37. $\frac{5y}{6} = \frac{2y}{9} + 3$

38. $1.8K = 3 - 0.7K$

39. $1.1 + 0.45M = M$

40. $2 - \frac{2}{3}T = \frac{5}{6} + \frac{1}{2}T$

41. $\frac{5}{8}Y = \frac{2}{3}$ **42.** $3C + 16 = 6C - 2$

43. $8 - w + 6 = 7w - 2$ **44.** $11 = n - 1$

45. $8 = 5 + \frac{3}{4}R$ **46.** $6 - 2M = 8M - 14$

47. $17R - 5R + R = 45 - 6$

48. $3A - 21 - A - 1 = 0$

49. $3y - 16 = 4 + y$ **50.** $17 = \frac{5m}{2} - 1$

B. Substitute the given values into each formula and solve for the unknown. If necessary, round answers to the nearest hundredth.

51. If $r = 50$ miles per hour and $D = 375$ miles, find t in $D = rt$.

52. Find I in $I = PRT$ if $1,500 is invested for 4 years at 7.5%.

53. $P = 2L + 2W$. Find P if $L = 22$ feet and $W = 13$ feet.

54. In $A = LW$, find W if $L = 10$ yards and $A = 95$ square yards.

55. If $360 in interest is paid on a loan of $2,000 after 3 years, what annual interest rate was charged? Use $I = PRT$.

56. Find L in $P = 2L + 2W$ if P is 80 meters and W is 14 meters.

57. Find M in $S = C + M$ if S is $6.49 and C is $4.15.

58. If $20.90 in interest is earned on a 2-year investment at 5.5% a year, what amount (principal) was originally deposited? Use $I = PRT$.

59. Use $A = 1/2bh$ to find b when A is 36 square inches and h is 9 inches.

60. Use $A = \pi r^2$ to find the area of a circle whose radius is 8 feet. (Use $\pi \doteq 3.14$.)

61. Use $C = 2\pi r$ to find the radius of a circle whose circumference is 150.72 centimeters. (Use $\pi \doteq 3.14$.)

62. Use $A = 1/2bh$ to find h when A is 24 square feet and b is 12 feet.

63. Find the area of a circle if $r = 9$ meters and $\pi \doteq 3.14$. Use $A = \pi r^2$.

64. Use $C = 2\pi r$ to find the radius of a circle whose circumference is 15.7 inches. (Use $\pi \doteq 3.14$.)

C. Solve each equation for the variable indicated.

65. $P = 2L + 2W$ for W **66.** $P = a + b + c$ for c

67. $V = LWH$ for H **68.** $V = LWH$ for L

69. $\frac{2x}{3} + 3y = 12$ for x **70.** $\frac{2x}{3} + 3y = 12$ for y

Chapter Test

Solve and check.

1. $3R = 81$ **2.** $5R + 3 = 38$

3. $2C - 7 + C = 35$ **4.** $\frac{w}{4} = 12$

5. $21 + 2A = 5A$ **6.** $0.4m + 12 = 14.32$

7. $\frac{1}{2}B - 3 = \frac{3}{4}$ **8.** $9R - 6R = 24 - 18$

9. $11w - 8 = 3$ **10.** $\frac{3}{5}T - \frac{1}{3} = 1$

11. $A + \frac{2}{3}A = 25$ **12.** $4Y - 1 = 11$

13. $c + 17 = 2 + 6c$

14. $6N - 1 + 5N = 3N + 15$

15. $\frac{3T}{8} = 4$

16. $0.6M + 1.4 + 0.04M = 14.2$

17. $12P - 2 = 9P + 7$ **18.** $Y + 7Y = 8$

19. Find the length of a rectangle if the perimeter is 92 yards and the width is 14 yards. Use $P = 2L + 2W$.

20. Using the formula $C = 2\pi r$ ($\pi \doteq 3.14$), find the circumference of a circle that has a radius of 4 inches.

Thought-Provoking Problems

1. Explain the relationship (similarity or difference) between the steps in the Order of Operations Agreement and the steps for solving equations.

2. Why does the short cut of multiplying both sides of the equation by the reciprocal of the fraction coefficient work in an equation such as $2/3\ X = 6$? After all, you were told to "undo" multiplication by dividing.

3. Have your feelings about your ability to do algebra changed at all now that you have finished this chapter? If so, in what ways? If not, how *do* you feel?

4. What property describes the steps illustrated in each of the following equations?

$$\frac{x}{4} = 7 \qquad Y - 5 = \quad 8$$

$$(4)\frac{x}{4} = (4)7 \qquad \begin{array}{c} Y - 5 = \quad 8 \\ \underline{+5 \qquad +5} \end{array}$$

5. Use your calculator to solve and check each of the following equations.

   a. $2.026 = 1.48Y + 0.25$

   b. $1.42K - 3.449 = 1.308$

   c. $2.9X + 8.03 = 12.42 - 5.88X$

   d. $\dfrac{0.445m}{0.89} - 6.72 = 1.775$

6. Use your calculator to find the value of A in each of the following formulas.

   a. Find A in $A = \pi r^2$ when $r = 1.97$ inches. Use $\pi = 3.14$ if your calculator does not have a π key.

   b. Find A in $A = \sqrt{\dfrac{BC}{4}}$ when $B = 2.6$ and $C = 1.15$. If your calculator does not have a square root key, estimate your answer using the Table of Square Roots on the inside back cover.

   c. Find A in $Z = 3X + 4A$ when $Z = 55.6$ and $X = 0.28$.

7. Find X in $3MX - 2X = NX + 4P$ for each set of values given below.

   a. $M = 3 \qquad N = 2 \qquad P = 1$

   b. $M = 5/6 \qquad N = 1/6 \qquad P = 2/3$

   c. $M = 1.5 \qquad N = 0.6 \qquad P = 3.8$

8. Using the equation-solving skills you have learned in this chapter, solve each of the following formulas for the variable indicated.

   a. $S = \dfrac{a}{1 - r}$ for a

   b. $P = 3a + 2b - 5c$ for c

   c. $PV = nRT$ for R

   d. $A = \dfrac{B - C}{T}$ for B

11

Word Problems

11.1 Introduction

Often when students in a math class hear that the next topic is word problems, there is a loud chorus of groans. Cries of "Ugh!" and "I hate word problems!" are heard all around the room. We will try, in this chapter, to change some of those feelings. Word problems are not very exciting and often seem contrived. After all, who cares how many nickels or dimes are in Michael's piggy bank? But perhaps if you understand what skills are being developed in studying application problems, they will become more meaningful to you. Every time you need to make a choice, you look at all the factors that affect this decision. You consider the pros and cons. If it is a major decision—to buy a house, to marry, to move, or where to invest money—you think long and hard about it. You look at all the angles. You weigh both sides of the decision and then make your best choice. That gathering of information, outlining of options, and logical considering of both sides are the same process you follow in solving word problems. The situations are not so complex, but the logical reasoning is the same. If you try to keep this in mind when you begin to study word problems, they will be more manageable. They are not unlike a jigsaw puzzle. You put the pieces together in some logical order and produce a finished product.

 **Learning Objectives**

When you have completed this chapter, you should be able to:

1. Translate words and expressions into algebraic symbols.
2. Translate word sentences into algebraic equations.
3. Write the correct equation from word problems of the following types:
   a. Number problems
   b. Consecutive-number problems
   c. Perimeter problems
   d. Coin or mixture problems
4. Solve the equation you have written and check your answer.

11.2

Words to Symbols

You are now comfortable with solving equations in one variable. In this chapter, you will learn to write equations from word problems. This skill is important because there are many situations in algebra, business math, and science courses, and also at work, where writing and solving an equation will enable you to find the solution to a given problem.

The first skill you need is that of translating expressions given in words to expressions that consist of variables and symbols.

E X A M P L E   **1**   If the price of a movie ticket is $5, then the price of two tickets could be expressed as $5(2) and the cost of 10 tickets as $5(10). The cost of an unknown number of tickets (n) would be expressed as $5(n)$, or $5n$.   ◆

E X A M P L E   **2**   If Uri is 23 years old now, you could express his age in 5 years as $23 + 5$ or 28 years. Eight years ago he would have been $23 - 8$ or 15 years old. If Maria won't tell you her age now, you can still express her age in 5 years as $Y + 5$ or her age 8 years ago as $Y - 8$, where $Y = $ Maria's age now.   ◆

E X A M P L E   **3**   If the temperature in Houston is D degrees, a 6-degree rise in temperature could be expressed as $D + 6$. A 10-degree fall in temperature would be $D - 10$.   ◆

When changing from words to symbols, look for key words that indicate which mathematical operation you should use. Many of these words were listed in Section 1.6.

Add	Subtract	Multiply	Divide
sum	difference	product	quotient
plus	less	times	divided
total	less than	total	split
increase	decrease	in all	each
more than	reduce	twice	shared
raise	remain	part of	average
combined	larger	altogether	ratio
altogether	fewer	area	per
in all	taller	volume	equal parts
	other *-er* words		

Use these key words to help you write the following word phrases as algebraic expressions. N will be used to represent the number in each example, but any letter, capital or lowercase, could have been chosen.

Word Expression	Algebraic Expression
1. six more than the number	$N + 6$
2. difference between a number and 8	$N - 8$
3. product of 12 and a number	$12N$ is the preferred form; but $12(N)$, $(12)(N)$, and $(12)N$ are also acceptable.
4. 15 divided by a number	$15/N$ or $15 \div N$
5. a number decreased by 11	$N - 11$
6. 11 *less than* a number	$N - 11$
7. 11 less a number	$11 - N$
8. a number divided into 3 equal parts	$N/3$ or $N \div 3$
9. 2/3 of a number	$2/3N$ or $2N/3$
10. 16% of a number	$0.16N$

A few of the algebraic expressions listed here need some explanation. In expression 1 in the list, $N + 6$ is the direct translation, though $6 + N$ would also be correct because of the Commutative Property of Addition. A variable represents any real number, so the properties of real numbers apply to variable expressions. In the multiplication examples (expressions 3, 9, and 10), although the Commutative Property does apply, the numerical coefficient should always be written before the literal factor ($12N$, not $N12$). This will avoid confusion later when exponents become involved: $N12$ could be misread as N^{12}.

The subtraction and division examples (expressions 2, 4, 5, 7, and 8) are not commutative and therefore *must* be written in the order given. Exceptions to writing in the order given occur only in expressions 1 and 6. In "six more than the number," the exact translation is $N + 6$. In "11 *less than* a number," the number (N) has to have 11 *subtracted from it*. Be aware that any time "less than" or "more than" is used, the order of the terms in the algebraic expression is reversed from the order that appears in the word expression. You may want to underline or circle "less than" and "more than" whenever they appear in a problem so that you will remember to write the expression correctly.

11.2 ◆ Practice Problems

A. Write the algebraic expression for each of the following.

1. One can of tuna costs 68 cents. Express the cost of x cans.

2. Pineapples cost $1.98 each. Express the cost of k pineapples.

3. Jerome is M years old now. Express his age 5 years ago.

4. Quick Key calculators cost $4 less than they did three years ago. Express the cost now if they cost L dollars three years ago.

5. R represents the total rainfall for the past four months. Represent the average rainfall for any month.

6. An estate of T dollars is equally divided among six heirs. Represent each person's inheritance.

7. Greg got P points on his term paper. Express Marie's grade if she got 11 points more than Greg.

8. In a company's benefits package, one health insurance plan costs $14 more per month than another. The less expensive policy costs P. Represent the monthly cost of the other policy.

B. Match each expression in Column I with the correct phrase in Column II.

Column I

9. $3X + 6$
10. $M - 12$
11. $1/2 F$
12. $12 - P$
13. $5N + 13$
14. $\dfrac{2}{3}T$
15. $J + 6$
16. $4/B$
17. $Y/4$
18. $12R$

Column II

a. half of a number
b. two-thirds of a number
c. the product of 12 and a number
d. the sum of three times a number and six
e. twelve less a number
f. twelve less than a number
g. the sum of a number and six
h. the quotient of a number and four
i. the quotient of four and a number
j. five times a number plus thirteen

C. Write the algebraic expression for each of the following. Use _T_ to represent the unknown quantity.

19. eleven more than a number

20. the difference between a number and 69

21. a number decreased by ten

22. the sum of thirteen and a number

23. twice a number

24. one-fourth of a number

25. seventeen divided by a number

26. the quotient of a number and four

27. two-thirds of a number increased by 12

28. twice a number decreased by 14

29. eight less than a number

30. eight less a number

31. seventeen more than a number

32. three-eighths of a number

33. twenty less than a number

34. eighty more than a number

35. the difference between eight and a number

36. three times a number increased by twelve

37. four less than twice a number

38. fifteen less than four times a number

39. eighteen more than a number

40. the difference between six and twice a number

11.3 Word Sentences to Algebraic Sentences

You can now take algebraic expressions and make them part of a full sentence. These sentences in algebra are **equations**. When forming algebraic sentences (equations) by converting from words to symbols, first state what variable you will use and what it represents. Key words such as "equals," "is the same as," "the result is," "was," and "are" will translate into the equals sign in the equation. In the following table, N is used to represent the number in each example.

Word Sentence	Algebraic Equation
The sum of a number and sixteen is twenty-seven.	$N + 16 = 27$
Thirty-six is equal to the product of nine and some number.	$36 = 9N$
The difference of a number and seventeen is equal to sixty-three.	$N - 17 = 63$
Eight is equal to the quotient of ninety-six and a number	$8 = 96/N$
The product of six and some number is equal to the sum of that number and forty-five.	$6N = N + 45$
The product of five and what number equals thirty?	$5N = 30$

Of course, once you have written the equation from the words, you will need to solve it and check your answer. The steps for solving equations follow. Review them before going on. The only difference between this list and the list that was given in Chapter 10 is in the statement about checking. This change will be discussed after the next example.

Steps for Solving Equations

1. If desired, eliminate fraction or decimal numbers by applying the Multiplication or Division Property of Equality to create an equivalent equation.

2. Combine like terms on each side of the equals sign.

3. If unknowns occur on both sides of the equals sign, start by eliminating the smaller variable term.

4. Add or subtract variable terms and numerical terms to isolate the variable.

5. Do multiplication or division to obtain the variable coefficient of positive 1.

6. Check your answer in the words of the problem.

E X A M P L E   1   If a number is increased by 12, the result is 29. Find the number.

Equation   Let N = the number

Number	increased by 12	result is	29.
N	$+ 12$	$=$	29

Solution

$$N + 12 = 29$$
$$\underline{-12 \quad -12}$$
$$N = 17$$

Check

$$N + 12 = 29$$
$$17 + 12 = 29 \qquad 29 = 29 \quad \text{True.} \qquad \blacklozenge$$

Note: This kind of check is not sufficient for word problems, because it only checks the solution to the equation. The most common error in doing word problems is in not writing the correct equation. This check does not expose that kind of error. Instead, check a word problem by substituting the answer into the wording of the problem.

If a number is increased by 12 the result is 29.

If 17 is increased by 12, the result is 29.

$17 + 12 = 29$ True. $\blacklozenge$

E X A M P L E   2   Six less than twice a number is 10. Find the number.

Equation   Let N = number

Twice a number	6 less than	is	10.
$2N$	-6	$=$	10

Solution

$$2N - 6 = 10$$
$$\underline{+6 \quad +6}$$
$$2N = 16$$
$$\frac{2N}{2} = \frac{16}{2}$$
$$N = 8$$

Check   Six less than twice a number is 10.

Twice (8) less 6 is $2(8) - 6 = (16) - 6 = 10$ True. $\blacklozenge$

E X A M P L E   3   One number is three times the other number. Their sum is 48. Find the numbers.

Equation   This time there are two different numbers, but they must be represented by the same variable. In "One number is three times the other number," the "other number" is the one you know the least about, so call it N.

Let N = "other" number

One number is "3 times the other number." Three times the "other number" (N) is $3N$.

Let $3N$ = one number

Now that you have defined the two numbers, use them to write the equation.

Their sum	is	48.
$3N + N$	$=$	48

Solution $3N + N = 48$

$$4N = 48$$

$$\frac{4N}{4} = \frac{48}{4}$$

$$N = 12$$

Look at the list above the equation. Because N = "other" number, it equals 12. The first number is $3N = 3(12) = 36$. Be sure to give all the answers asked for! The problem said to "Find the *numbers*," so you must give both answers to be correct. The answers are 12 and 36.

Check   One number is three times the other: $36 = 3(12)$. Their sum is 48: $12 + 36 = 48$. True. Both parts of the problem are true when the solution is substituted.   ◆

E X A M P L E 4   The net receipts of a rock concert were divided equally among the five band members. If each member received $525, what were the net receipts?

Equation   Let R = net receipts

Net receipts	divided equally among five	is	525.
R	$\div 5$	$=$	525

Solution $\dfrac{R}{5} = 525$

$$\frac{(5)R}{5} = (5)525$$

$$R = \$2{,}625$$

Check   If total receipts are divided by 5, the result is 525. $2,625/5 does equal $525.   ◆

E X A M P L E 5   Together, Marc and Mike purchased a hamburger franchise for $72,000. Mike's part was $18,000 more than Marc's. How much did each invest?

Equation   Let D = Marc's amount

$$D + 18{,}000 = \text{Mike's amount}$$

Marc	and	Mike	is	72,000.
D	$+$	$D + 18{,}000$	$=$	72,000

Solution $D + D + 18{,}000 = 72{,}000$

$$2D + 18{,}000 = 72{,}000$$

$$\underline{-\ 18{,}000 \qquad -18{,}000}$$

$$2D = 54{,}000$$

$$\frac{2D}{2} = \frac{54{,}000}{2}$$

$$D = 27{,}000$$

Marc's amount = $27,000

Mike's amount = $D + 18{,}000 = \$45{,}000$

Check $45,000 is $18,000 more than $27,000. $45,000 + $27,000 does equal $72,000.   ◆

EXAMPLE 6 One number is 6 more than another number. The sum of the larger number and twice the smaller number is 27. What are the numbers?

Equation Let N = "another" number = smaller number

$N + 6$ = first number = larger number

Larger number	sum	twice smaller number	is	27.
$N + 6$	$+$	$2N$	$=$	27

Solution
$$N + 6 + 2N = 27$$
$$3N + 6 = 27$$
$$\frac{-6 \quad -6}{}$$
$$3N = 21$$
$$\frac{3N}{3} = \frac{21}{3}$$
$$N = 7$$

The smaller number = 7, and the larger number is $N + 6 = 13$.

Check One number is 6 more than the other. $7 + 6 = 13$. The sum of the larger and twice the smaller is 27.

$13 + 2(7) = 27$. $13 + 14 = 27$. True. ◆

EXAMPLE 7 Four-fifths of the students in a class received a grade of "C" or better. If 28 students received a "C" or better, how many students are in the class?

Equation Let S = number of students in the class

"Part of" means to multiply.

$$\frac{4}{5}S = 28$$

Multiply by the reciprocal and divide common factors.

$$\frac{(5)4}{(4)5}S = \frac{(5)28}{(4)(1)}$$

$$\frac{(5)4}{(4)5}S = \frac{(5)28}{(4)(1)}$$

$$S = 35 \text{ students}$$

Check Four-fifths of the students are 28 students. 4/5 of 35 = 4/5 × 35 = 28. True. ◆

11.3 ◆ Practice Problems

A. Write an algebraic equation for each of the following problems. Then solve it and check your answer.

1. Seventeen more than a number is 39. Find the number.

2. A number increased by 21 is 54. What is the number?

3. Eleven times a number is 99. What is the number?

4. The product of 16 and a number is 48. Find the number.

5. The quotient of a number and 4 is 9. Find the number.

6. When a number is divided by 12, the result is 4. What is the number?

7. Four less than a number is 13. Find the number.

8. Eight less than a number is 37. Find the number.

B. Write an algebraic equation, and then solve and check.

9. Six more than twice a number is 24. Find the number.

10. If the product of 3 and a number is increased by 12, the result is 48. What is the number?

11. Half a number decreased by 6 is equal to 48. Find the number.

12. Sixteen less than two-thirds of a number is equal to 56. Find the number.

13. One number is 4 less than another number, and their sum is 20. Find the numbers.

14. One number is 11 less than another number, and their sum is 55. Find the numbers.

15. Four more than twice a number is equal to the number increased by 16. Find the number.

16. Six times a number is equal to twice the number increased by 12. Find the number.

C. Write the algebraic equation for each word problem that follows. Then solve the equation and check your answer.

17. Two-thirds of the students in a class completed their assignment on schedule. If 16 students finished on time, how many students are in the class?

18. Three-fourths of the students in this school drive their own cars to campus each day. If 7,200 students drive, how many students come to campus each day?

19. One number is 4 times another number. The difference between the larger number and the smaller number is 54. Find the numbers.

20. One number is 5 times another. The difference between the larger and the smaller is 84. Find the numbers.

21. The Cardinals' star running back scored one-fourth of the team's total points in this week's game. He scored 12 points. What was the team's score in the game?

22. Ron deposits two-thirds of his paycheck in the bank each week. If he deposited $144 this week, what were his total earnings?

23. A woman earned $40 more than twice what her sister earned. Together they earned $580. Find each woman's salary.

24. During a six-month diet contest, Sue's weight loss was 2 pounds more than twice what Jean's weight loss was. Together they lost 26 pounds. What was Sue's total?

25. One number is four times another number. When twice the smaller number is subtracted from the larger number, the result is 28. What are the numbers?

26. The total cost of buying a used car is $9,816. If Alexandria pays $600 down and makes 48 monthly payments to purchase the car, how much will she pay each month?

27. Max received a car repair bill for $245. If the bill included $80 for parts, and if labor costs $45 per hour, how long did the mechanic work on the car?

28. This week Mrs. Chang worked six hours less than Mrs. Alvarez. Together they worked 70 hours. How many hours did Mrs. Chang work during this period?

29. The total cost of the VCR that Manuella purchased was $179. If she paid $47 down and made twelve monthly payments to finish paying for it, how much was each monthly payment?

30. The sum of two numbers is 52, and one of them is three times the other. What are the numbers?

11.4 Consecutive-Number Problems

The next three sections of this chapter will show you how to set up and solve some special kinds of word problems. You will find these problems easier to deal with if you follow the procedures suggested. They will help you to organize the information so that nothing is left out of the translation from words into equations.

Consecutive means that things follow one another successively without interruption. **Consecutive numbers or integers** are numbers that follow one another, such as 13, 14, 15 or 36, 37, 38, 39. In these sets of consecutive numbers, each number is 1 larger than the previous number.

In 13, 14, 15, $14 = 13 + 1$

$15 = 14 + 1$ or $(13 + 1) + 1 = 13 + 2$.

In general terms, if you don't know the numbers but you do know that they are consecutive, then let

n = 1st number

$n + 1$ = 2nd consecutive number

$n + 2$ = 3rd consecutive number

$n + 3$ = 4th consecutive number

and so on

Use this notation to write equations for consecutive-number problems.

E X A M P L E 1   The sum of three consecutive numbers is 69. What are the numbers?

Equation   Let n = 1st number

$n + 1$ = 2nd number

$n + 2$ = 3rd number

1st number	+	2nd number	+	3rd number	is	69.
n	+	$n + 1$	+	$n + 2$	=	69

Solution
$$n + n + 1 + n + 2 = 69$$
$$3n + 3 = 69$$
$$\underline{-3 \quad -3}$$
$$3n = 66$$
$$\frac{3n}{3} = \frac{66}{3}$$
$$n = 22, \ n + 1 = 23, \ n + 2 = 24$$

The numbers are 22, 23, and 24.

Check   The sum of three consecutive numbers is 69. 22, 23, and 24 are consecutive numbers. $22 + 23 + 24 = 69$. The sum is 69.   True.   ◆

E X A M P L E 2   The sum of the first and third of three consecutive numbers is 96. Find all three numbers.

Equation   Let n = 1st number

$n + 1$ = 2nd number

$n + 2$ = 3rd number

Though all three numbers are asked for in the answer, only two are used in the equation.

first	sum	third	is	96.
n	+	$n + 2$	=	96

Solution
$$n + n + 2 = 96$$
$$2n + 2 = 96$$
$$\underline{-2 \quad -2}$$
$$2n = 94$$
$$\frac{2n}{2} = \frac{94}{2}$$
$$n = 47, \ n + 1 = 48, \ n + 2 = 49$$

The numbers are 47, 48, and 49.

Check   The numbers are consecutive: 47, 48, and 49.

The sum of the 1st and 3rd is 96. $47 + 49 = 96$.   True.   ◆

E X A M P L E   **3**   The sum of four consecutive numbers is 102. What are the four numbers?

Equation   Let n = 1st number

$n + 1$ = 2nd number

$n + 2$ = 3rd number

$n + 3$ = 4th number

$$\underline{\text{1st}} + \underline{\text{2nd}} + \underline{\text{3rd}} + \underline{\text{4th}} \quad \text{is} \quad 102.$$
$$n \; + \; n+1 \; + \; n+2 \; + \; n+3 \; = \; 102$$

Solution
$$n + n + 1 + n + 2 + n + 3 = 102$$
$$4n + 6 = 102$$
$$\underline{ -6 \qquad -6}$$
$$4n \quad = \quad 96$$
$$\frac{4n}{4} = \frac{96}{4}$$
$$n = 24$$

$n = 24$, $n + 1 = 24 + 1 = 25$, $n + 2 = 24 + 2 = 26$, $n + 3 = 24 + 3 = 27$

The numbers are 24, 25, 26, and 27.

Check   24, 25, 26, and 27 are consecutive numbers.

$$(24) + (25) + (26) + (27) = 102$$
$$102 = 102 \quad \text{True.}$$ ◆

Problems may also be concerned with consecutive *even* numbers (for example, 18, 20, 22) or consecutive *odd* numbers (for example, 33, 35). Note that in either situation, the next number is *2 more than* the previous number. Therefore, the notation that might be used in any problem involving *consecutive even or consecutive odd* numbers is:

Let n = 1st consecutive even/odd number

$n + 2$ = 2nd consecutive even/odd number

$n + 4$ = 3rd consecutive even/odd number

Don't be confused by the 2 and 4 notation. They are used so you will skip the number that comes between your consecutive even or odd numbers, *not* because they are even numbers.

E X A M P L E   **4**   The sum of four consecutive even numbers is 28. Find the numbers.

Equation   Let n = 1st consecutive even number

$n + 2$ = 2nd consecutive even number

$n + 4$ = 3rd consecutive even number

$n + 6$ = 4th consecutive even number

$$\underline{\text{1st}} \quad + \quad \underline{\text{2nd}} \quad + \quad \underline{\text{3rd}} \quad + \quad \underline{\text{4th}} \quad \text{is} \quad \underline{28}.$$
$$n \quad + \quad n + 2 \quad + \quad n + 4 \quad + \quad n + 6 \quad = \quad 28$$

Solution
$$n + n + 2 + n + 4 + n + 6 = 28$$
$$4n + 12 = 28$$
$$\frac{-12 \qquad -12}{4n \qquad = \quad 16}$$
$$\frac{4n}{4} = \frac{16}{4}$$
$$n = 4, n + 2 = 6, n + 4 = 8, n + 6 = 10$$

The numbers are 4, 6, 8, and 10.

You need to list all four answers, because the problem tells you to find the *numbers*.

Check The numbers 4, 6, 8, 10 are consecutive even numbers. $4 + 6 + 8 + 10 = 28$. The sum is 28. True. ◆

E X A M P L E **5** The sum of the first and third of three consecutive odd numbers is 130. Find all of the numbers.

Equation Let n = 1st consecutive odd number

$n + 2$ = 2nd consecutive odd number

$n + 4$ = 3rd consecutive odd number

$$\underline{\text{1st}} \quad + \quad \underline{\text{3rd}} \quad = \quad \underline{130}$$
$$n \quad + \quad n + 4 \quad = \quad 130$$

Solution
$$n + n + 4 = 130$$
$$2n + 4 = 130$$
$$\frac{-4 \qquad -4}{2n \qquad = 126}$$
$$\frac{2n}{2} = \frac{126}{2}$$
$$n = 63$$
$$n = 63, n + 2 = 63 + 2 = 65, n + 4 = 63 + 4 = 67$$

The numbers are 63, 65, and 67.

Even though the equation used only two numbers, you must give all three answers.

Check The numbers 63, 65, 67 are consecutive odd numbers. The sum of the first and third is 130. $(63) + (67) = 130$. $130 = 130$ True. ◆

11.4 ◆ Practice Problems

A. Write an expression for each of the following preliminary steps for consecutive-number problems. They do not need equations.

1. If 15 is the first of three consecutive numbers, what is the next number?

2. If 47 is the first of three consecutive numbers, what is the next number?

3. If 69 is the first of three consecutive odd numbers, what are the next two numbers?

4. If 128 is the first of three consecutive even numbers, what are the next two numbers?

5. If 16 is the first of three consecutive numbers, what are the next two numbers?

6. If 91 is the first of four consecutive numbers, what are the next three numbers?

7. If 44 is the first of two consecutive even numbers, what is the next number?

8. If 25 is the first of two consecutive odd numbers, what is the next number?

9. If N represents the first of three consecutive numbers, and N equals 29, what is the value of $N + 1$ and of $N + 2$?

10. If N represents the first of two consecutive numbers, and N equals 11, what is the value of $N + 1$ and of $N + 2$?

11. What is always the difference between the largest and smallest of any three consecutive even numbers?

12. What is the difference between the first and third of any three consecutive odd numbers?

B. For each of the following problems, write the equation, solve it, and then check your answer(s).

13. The sum of three consecutive numbers is 153. Find the numbers.

14. The sum of two consecutive numbers is 93. What are the numbers?

15. The sum of two consecutive even numbers is 90. What are the numbers?

16. The sum of two consecutive odd numbers is 152. Find the numbers.

17. Find four consecutive numbers whose sum is 178.

18. The sum of four consecutive numbers is 450. What are the numbers?

19. The sum of the first and third of three consecutive numbers is 146. Find all three numbers.

20. The sum of the second and third of three consecutive numbers is 89. Find all three numbers.

21. The sum of the largest and smallest of three consecutive even numbers is 68. What are the three numbers?

22. Find three consecutive odd numbers such that the sum of the smallest and largest is 114.

23. The sum of the first two of three consecutive numbers is equal to the third number. What are the numbers?

24. Find three consecutive even numbers such that the sum of the first two numbers is equal to the third.

25. The sum of the second and twice the first of two consecutive numbers is equal to 28. What are the two numbers?

26. The sum of the largest and three times the smallest of three consecutive numbers is 38. Find the three numbers.

27. The sum of the third and twice the first of three consecutive even numbers is 16. Find the three numbers.

28. If four times the first of three consecutive odd numbers is added to the third, the sum is 179. Find the three numbers.

11.5   Perimeter Problems

You are now ready to solve some word problems involving simple geometric figures. You have worked with these in previous chapters. The *perimeter* of a figure is the total distance around it—that is, the sum of all its sides. Figure 1 shows where the different perimeter formulas came from. Having a picture or diagram to work with is often very helpful.

You will not have to memorize any formulas to do this work. You can write the perimeter formula for any figure by adding all the sides together.

Rectangle

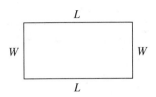

$$P = L + W + L + W$$
or
$$P = 2L + 2W$$

Square

$$P = s + s + s + s$$
or
$$P = 4s$$

Triangle

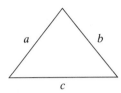

$$P = a + b + c$$

Irregular Figure

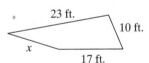

$$P = a + b + c + d + e$$

F I G U R E 1

E X A M P L E 1 The perimeter of Figure 2 is 57 feet. Find the length of x. Write a formula and an equation, and then solve the equation for x.

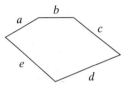

F I G U R E 2

Equation The formula for the perimeter of a four-sided figure is $P = a + b + c + d$.

Solution Substitute the values in the drawing for a, b, c, and d.

$$P = a + b + c + d$$
$$57 = 23 + 17 + 10 + x$$ Combine the numerical values and then isolate the variable.

$$57 = 50 + x$$
$$\underline{-50 \quad -50}$$
$$7 = x$$
$$x = 7 \text{ feet}$$

Check $23 + 17 + 10 + 7$ does equal 57.

In the problems you will do in this section, the actual lengths of the sides are not known. However, you will be given some information about the dimensions. You will represent one of the lengths (the dimension you know the least about) by a variable. Then you will use the other information that is given to express the other sides. Making a drawing of the figure and labeling its parts will help. Study the examples that follow.

E X A M P L E 2   The length of a rectangle is 6 feet longer than the width. The perimeter is 60 feet. Find the dimensions.

Equation   Let x = width (the dimension you know the least about)

$x + 6$ = length (the length is 6 feet longer than the width)

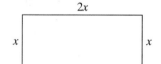

x + 6

x   x

x + 6

F I G U R E 3

Draw a sketch such as Figure 3, and label the sides.

| $P =$ | L | $+ W +$ | L | $+ W$ | Write the formula. |

$60 = \underline{x + 6} + \underline{x} + \underline{x + 6} + \underline{x}$ Substitute for P, L, and W.

Solution $60 = x + 6 + x + x + 6 + x$

$60 = 4x + 12$ Combine like terms.

$\underline{-12 \qquad -12}$

$48 = 4x$

$\dfrac{48}{4} = \dfrac{4x}{4}$

$12 = x$

Because x represents the width, the width is 12 feet. Because $x + 6$ represents the length, the length is $12 + 6 = 18$ feet. Be sure to include the units of measurement in the answer and to give *both* dimensions (one length and one width).

Check $P = L + W + L + W$

$60 = 18 + 12 + 18 + 12$

$60 = 60$ True.

And the length, 18 feet, is 6 feet more than the width of 12 feet.   ◆

E X A M P L E 3   The length of a rectangle is twice the width. The perimeter is 66 yards. Find the length.

Equation   Let x = width (you know the least about it)

$2x$ = length (twice the width)

$2x$

x   x

$2x$

F I G U R E 4

Refer to the sketch in Figure 4.

$P = L \ + W + L + W$

$66 = \underline{2x} + \ \underline{x} + \underline{2x} + \underline{x}$

Solution $66 = 2x + x + 2x + x$

$66 = 6x$

$\dfrac{66}{6} = \dfrac{6x}{6}$

$11 = x$

Because length = $2x$, the length is 22 yards.

Check $P = L + W + L + W$

$66 = 22 + 11 + 22 + 11$

$66 = 66$ True. And 22 is twice 11.   ◆

E X A M P L E **4** The perimeter of a triangle is 55 inches. Two of the sides are equal, and the third side is 5 inches less than twice the length of one of the equal sides. Find the dimensions of the triangle.

Equation Let x = length of one equal side

x = length of other equal side

$2x - 5$ = length of third side (5 inches less than twice the length of one of the equal sides)

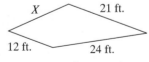

x x

$2x - 5$

FIGURE **5**

Refer to Figure 5.

$$P = a + b + \quad c$$
$$55 = \underline{x} + \underline{x} + \underline{2x - 5}$$

Solution
$$55 = x + x + 2x - 5$$
$$55 = 4x - 5$$
$$\underline{+\ 5 \qquad\quad +\ 5}$$
$$60 = 4x$$
$$\frac{60}{4} = \frac{4x}{4}$$
$$15 = x$$

Because x represents the length of the equal sides, each of them is 15 inches. The third side is represented by $2x - 5$, so it is $2(15) - 5 = 30 - 5 = 25$ inches. The sides are 15 inches, 15 inches, and 25 inches.

Check $P = a + b + c$
$$55 = 15 + 15 + 25$$
$$55 = 55 \quad \text{True.}$$

Two of the sides are equal, and the third is 5 inches less than twice the length of one of the equal sides. ◆

11.5 ◆ Practice Problems

A. For each of these problems, write a formula and an equation. Then solve the equation and check your answer.

1. The perimeter of Figure 6 is 75 feet. Find X.

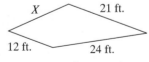

X 21 ft.

12 ft. 24 ft.

FIGURE **6**

2. The perimeter of Figure 7 is 33 inches. Find X.

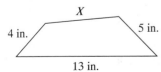

X

4 in. 5 in.

13 in.

FIGURE **7**

3. Find X if the perimeter of the triangle in Figure 8 is 48 meters.

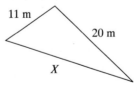

11 m 20 m

X

FIGURE **8**

4. If the perimeter of the triangle in Figure 9 is 25 feet, what is the length of *X*?

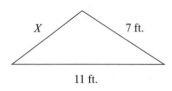

FIGURE 9

5. What is the length of each of the two equal sides of the triangle in Figure 10 if the perimeter of the triangle is 58 inches?

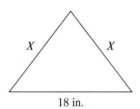

FIGURE 10

6. The perimeter of the triangle in Figure 11 is 200 inches. What is the length of each equal side?

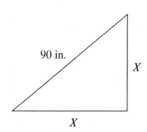

FIGURE 11

7. If the perimeter of Figure 12, an octagon, is 176 feet, what is the length of *X*?

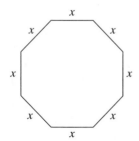

FIGURE 12

8. Find *X* if the perimeter of the hexagon in Figure 13 is 144 feet.

FIGURE 13

B.   For each problem, write a formula and an equation. Then solve the equation and check your answer(s). A sketch might be helpful.

9. The perimeter of a square is 96 meters. What is the length of each side?

10. A triangle with three equal sides (an *equilateral* triangle) has a perimeter of 135 feet. What is the length of each side?

11. The length of a rectangle is 6 inches more than the width. The perimeter is 132 inches. Find the dimensions.

12. The length of a rectangle is 20 centimeters more than the width. The perimeter is 96 centimeters. Find the dimensions.

13. The length of a rectangle is twice the width. The perimeter is 24 yards. Find the length.

14. The length of a rectangle is four times the width. The perimeter is 40 inches. Find the length.

15. The length of a rectangle is 6 feet more than twice the width. The perimeter is 84 feet. Find the dimensions.

16. The length of a rectangle is 4 inches more than three times the width, and the perimeter is 88 inches. Find the dimensions.

17. The perimeter of a triangle is 37 meters. Two of the sides are equal, and the third side is 7 meters less than twice the length of one of the equal sides. Find the dimensions.

18. Find the lengths of the sides of a triangle that has two equal sides and a third side that is 6 yards less than twice the length of one of the equal sides. Its perimeter is 66 yards.

19. Find the lengths of the sides of a triangle if the first side is 5 inches longer than the second side, and the third side is twice the length of the second side. Its perimeter is 41 inches.

20. One side of a triangle is 7 feet longer than the second side. The third side is 6 feet longer than twice the length of the second side. Find the sides of the triangle if its perimeter is 37 feet.

11.6 Coin Problems and Mixture Problems

Another type of problem you will solve in this chapter includes coin problems and mixture problems. In both, several like items (for example, coins, stamps, or theater tickets), each of different value, are combined to give a total cost or a total value. It is important to remember that the total value cannot be found by just combining the numbers of each item. The value or cost of each item must also be used. It is not just the fact that you have four coins that gives you $1.00. The fact that each of those coins is worth $0.25, or 4(0.25), is also essential.

As you work coin problems and mixture problems, you may find it easier to write the correct equation if you organize your information in a box like Chart 1. This should help prevent you from leaving important information out of the equation.

CHART 1

Items	Value of Each	Number of Each	Value of Each Kind
			Total Value

EXAMPLE 1 Your son's collection bag for his paper route contains quarters and dimes. There are nine times as many quarters as dimes. The total amount in the bag is $23.50. Find the number of quarters and dimes.

Equation Fill in the information that is given: Items are dimes and quarters, and the total amount is $23.50. From outside the problem, you know that each dime has a value of $0.10 and each quarter a value of $0.25. Because the number of dimes is the quantity you know the least about, call it x. That means the number of quarters must be $9x$. Place all the information in the boxes shown in Chart 2.

CHART 2

Items	Value of Each	Number of Each	Value of Each Kind
Dimes	0.10	x	
Quarters	0.25	$9x$	
			Total Value
			23.50

The last box, *Value of Each Kind*, will be the *product* of the quantities in the *Value of Each* box and the *Number of Each* box.

CHART **3**

Items	Value of Each	Number of Each	Value of Each Kind
Dimes	0.10	x	$0.10x$
Quarters	0.25	$9x$	$0.25(9x)$
			Total Value
			23.50

Now, using the *Value of Each Kind* boxes (Chart 3), you can write the equation. The sum of the *value of dimes* and the *value of quarters* is equal to the *Total Value*.

$$0.10x + 0.25(9x) = 23.50$$

Solution

$$0.10x + 0.25(9x) = 23.50$$
$$0.10x + 2.25x = 23.50$$
$$100(0.10x) + 100(2.25x) = 100(23.50)$$
$$10x + 225x = 2350$$
$$235x = 2350$$
$$\frac{235x}{235} = \frac{2350}{235}$$
$$x = 10$$

Take this answer back into the *Number of Each* box. Because x is the number of dimes and $9x$ is the number of quarters, there are 10 dimes and 90 quarters.

Check

$$10 \text{ dimes} = 10(0.10) = \$\ 1.00$$
$$90 \text{ quarters} = 90(0.25) = \underline{\$22.50}$$
$$\$23.50$$

And there are 9 times more quarters than dimes. ◆

E X A M P L E 2 The ticket booth for tonight's basketball game collected $675. The Cougars had three times as many students as nonstudents attend the game. If student tickets cost $2.00 and nonstudent tickets cost $3.00, how many of each kind of ticket were sold?

Equation Because the number of nonstudent tickets is what you know the least about, call it *N*. Draw a box like Chart 4, and fill in the information that is given.

CHART **4**

Items	Value of Each	Number of Each	Value of Each Kind
			Total Value

Your information box should look like Chart 5.

CHART 5

Items	Value of Each	Number of Each	Value of Each Kind
Student tickets	2.00	$3N$	$2.00(3N)$
Nonstudent tickets	3.00	N	$3.00N$
			Total Value 675.00

The equation is the sum of the *Value of Each Kind* boxes, set equal to the *Total Value*, or the amount collected.

$$2.00(3N) + 3.00N = 675.00$$

Solution $$2.00(3N) + 3.00N = 675.00$$
$$6.00N + 3.00N = 675.00$$
$$9.00N = 675.00 \text{ or}$$
$$9N = 675$$
$$\frac{9N}{9} = \frac{675}{9}$$
$$N = 75$$

Therefore, 75 nonstudent tickets and 3(75), or 225, student tickets were sold.

Check $$75(3.00) = \$225$$
$$3(75)(2.00) = \underline{\$450}$$
$$\text{Total} = \$675$$

And there were three times as many student tickets as nonstudent tickets sold.   ◆

E X A M P L E 3   While emptying his vending machine, Jerome noticed that he had five times as many dimes as quarters and twice as many nickels as dimes. How many of each coin did he have if there was a total of $37.50 in the machine?

Equation   Because the number of quarters is what you know the least about, call it X. Draw a box and fill in the rest of the information from the problem. Your box should look like Chart 6, but you can abbreviate the headings of the columns if you wish.

CHART 6

Items	Value of Each	Number of Each	Value of Each Kind
Dimes	0.10	$5X$	$0.10(5X)$
Quarters	0.25	X	$0.25(X)$
Nickels	0.05	$2(5X)$	$0.05(2)(5X)$
			Total Value 37.50

$$0.10(5X) + 0.25(X) + 0.05(2)(5X) = 37.50$$

Solution

$$0.10(5X) + 0.25(X) + 0.05(2)(5X) = 37.50$$

$$0.50X + 0.25X + 0.50X = 37.50$$

$$1.25X = 37.50$$

$$\frac{1.25X}{1.25} = \frac{37.50}{1.25}$$

$$X = 30$$

Therefore, there are 30 quarters; $5(x) = 5(30) = 150$ dimes; and $2(5X) = 2(5)(30) = 300$ nickels.

Check

$$30 \text{ quarters} = 30(0.25) = \$\ 7.50$$

$$150 \text{ dimes} = 150(0.10) = \$15.00$$

$$\underline{300 \text{ nickels} = 300(0.05) = \$15.00}$$

$$\text{Total} = \$37.50$$

And there are five times as many dimes as quarters and twice as many nickels as dimes.

◆

11.6 ◆ Practice Problems

A. Write the algebraic expression for each of the following. These are not entire problems that require an equation and a solution. They are the beginning steps used to solve mixture or coin problems.

1. There are twice as many nickels as dimes in a bank, and N represents the number of dimes. Express the number of nickels.

2. There are three times as many nickels as pennies in a child's piggy bank. Represent the value of nickels if there are X pennies.

3. Represent the value of the nickels and the value of the dimes in Problem 1.

4. Represent the value of the nickels and the value of the pennies in Problem 2.

5. What arithmetic operation would you perform with the values of the coins in Problem 3 if you wanted to find the total amount of money in the bank?

6. What arithmetic operation would you perform with the values of the coins in Problem 4 if you wanted to find the total amount of money in the piggy bank?

7. Using the expressions you wrote in Problems 1, 3, and 5, fill in Chart 7, labeling the columns as they are labeled in the examples in this section.

CHART 7

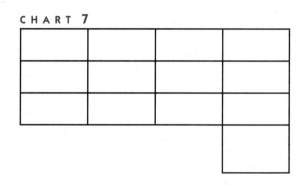

8. Using the expressions you wrote in Problems 2, 4, and 6, fill in Chart 8, labeling the columns as they are labeled in the examples in this section.

CHART 8

B. Using an information box, write an equation and solve each of the following problems.

9. If the bank in Problem 1 contains $2.40, how many dimes and nickels are there?

10. If the piggy bank in Problem 2 contains $2.88, how many nickels and pennies are there?

11. Niko sets off the alarm system when passing through an airport security gate. Upon emptying his pockets, he finds equal numbers of quarters and pennies that total $1.56 in value. How many of each coin does he have?

12. The toll collector at the Riverview exit notices that she has the same number of dimes as quarters in her box. If there is a total of $9.80 in the box, and she has only dimes and quarters, how many of each coin does she have?

13. When cleaning out her purse, Jan found $2.10 in loose change. There were eight times as many dimes as quarters. How many of each coin did she have?

14. Emptying his pants pockets one evening, Steve finds $1.00 in change. In the pile are five times as many nickels as quarters. How many nickels are there?

15. The afternoon ticket sales for the Plaza Movie Theater totaled $269.50. Twice as many children's tickets as adults' tickets were sold. How many of each kind of ticket were sold if children's tickets cost $1.50 and adults' tickets cost $2.50?

16. Elena collects the money from her co-workers and goes to the Sub House to get lunches for all of them. She collects $17.90 and has orders for twice as many sandwiches as drinks. If sub sandwiches cost $3.98 and large drinks cost $0.99, how many of each is she buying?

17. Joanna had to buy twice as many textbooks as workbooks for her nursing classes next quarter. Workbooks cost $12.50 each, and textbooks cost $38.00 each. If her total bill is $177.00, how many textbooks did she buy?

18. The college bookstore charges $1.95 for notebooks and $0.85 for pens. When Cheryl bought twice as many notebooks as pens, her bill was $9.50. How many of each item did she purchase?

19. Franz has half as many quarters as half-dollars. All together he has $13.75. How many quarters does he have?

20. Juan noticed that his Salvation Army kettle had only dollar bills and quarters in it. There was $14.00 in the kettle. If the number of dollar bills was one-fourth the number of quarters, how many dollar bills were there?

11.7 ◆ Chapter Review

In this chapter, you have learned and practiced how to change words to symbols and word sentences to equations. When solving special kinds of word problems, remember these facts and suggestions:

1. Read the problem carefully to see what is given and what is to be found.

2. Use a variable to represent one of the items, often the one you know the least about. Then represent the other items in terms of this variable.

3. Consecutive numbers can be represented as n, $n + 1$, $n + 2$, and so on.

4. Consecutive even or odd numbers can be represented as n, $n + 2$, $n + 4$, and so on.

5. Perimeter is the sum of all the sides of a figure.

6. In coin problems and mixture problems, you must consider both the *value of each item* and the *number* of each item. (Making a chart helps to organize the information.)

7. Check your answers in the words of the problem, not just in the equation.

8. Be sure to give all answers that are requested and to label answers with the correct units when appropriate.

Review Problems

A. Match each phrase in Column I with the correct algebraic expression in Column II.

I		II
1. six more than X		**a.** $2X$
2. twice X		**b.** $6/X$
3. the difference between 6 and X		**c.** $6X$
4. six less than X		**d.** $1/2\, X$
5. the quotient of 6 and X		**e.** $X + 6$
6. the product of 6 and X		**f.** $X - 6$
7. half of X		**g.** $6 - X$
8. the quotient of X and 6		**h.** $X/6$

B. Match each sentence in Column I with the correct algebraic equation in Column II.

I		II
9. Eleven more than a number equals 19.		**i.** $n + 4 = 19$
		j. $n - 4 = 19$
10. When a number is decreased by 11, the result is 19.		**k.** $2n = 19$
		l. $n + 11 = 19$
11. Two-thirds of a number is 12.		**m.** $1/2\, n = 3$
12. The quotient of a number and 12 is 3.		**n.** $2/3\, n = 12$
		o. $n - 11 = 19$
13. When a number is increased by 4, the result is 19.		**p.** $2/3\, n - 4 = 12$
		q. $n/12 = 3$
14. Twice a number is 19.		**r.** $2n + 6 = 12$
15. The difference between a number and 4 is 19.		**s.** $2/3\, n + 4 = 12$

16. Half of a number is 3.

17. Six more than twice a number is 12.

18. Two-thirds of a number decreased by 4 is 12.

C. For each problem, write an equation and then solve and check your answer(s).

19. Twice a number increased by 7 is 25. Find the number.

20. The difference between a number and 4 is 13. What is the number?

21. The sum of three consecutive numbers is 87. Find the numbers.

22. The perimeter of a rectangle is 64 feet. The length is 2 feet more than the width. Find the dimensions.

23. Four less than half a number is 16. Find the number.

24. Find three consecutive numbers whose sum is 174.

25. The length of a rectangle is 12 inches more than the width. Find the dimensions if the perimeter is 52 inches.

26. Two-thirds a number increased by 17 is 35. Find the number.

27. A pile of dimes and nickels contains equal numbers of the two coins. The value of the coins is $1.95. Find how many of each coin there are.

28. Equal numbers of adults' and children's tickets were sold for the movie *Cinderella*. Find the number of children's tickets sold if the total receipts were $225.00. Adults' tickets cost $3.00, and children's tickets cost $2.00.

29. One number is four times the other, and their sum is 35. Find the numbers.

30. Twice a number increased by 11 is 25. Find the number.

31. The sum of the first and third of three consecutive even numbers is 72. Find all the numbers.

32. A church collection basket contains as many dollar bills as quarters. The basket contains $21.25. How many quarters are there?

33. One number is 5 more than the other. The sum of the larger and three times the smaller is 37. Find the larger number.

34. The sum of the largest and smallest of three consecutive odd numbers is 58. Find all the numbers.

35. The length of a rectangle is 6 inches more than the width. The perimeter is 48 inches. Find the dimensions.

36. The perimeter of a square is 96 inches. Find the length of a side.

37. The sum of three consecutive numbers is 60. Find the three numbers.

38. One number is three times the other number. The sum of the larger and twice the smaller is 30. Find the numbers.

39. The perimeter of a triangle is 81 feet. All three sides have the same length (it is an equilateral triangle). Find the measure of each side.

40. The sum of two consecutive numbers is 19. Find the numbers.

41. Eight less than three-fourths a number is 10. Find the number.

42. In the college cafeteria, pizza sells for $0.99 a slice and chicken sandwiches for $1.49 each. During a one-hour period, $31.23 was collected when twice as many pizza slices as sandwiches were sold. How many slices of pizza were purchased?

43. In a triangle, the first side is twice the length of the second side. The third side is 4 inches longer than the second side. Find the length of the longest side if the perimeter is 40 inches.

44. Six more than three-eighths of a number is 12. Find the number.

45. The coffee vending machine in the student lounge contains the same number of quarters as dimes. If they total $15.75, how many of each coin are there?

46. The length of a rectangle is three times the width. The perimeter is 72 meters. Find the dimensions.

47. Lee Ann purchased twice as many letter stamps as postcard stamps when she went to the Post Office. Letter stamps cost $0.25, and postcard stamps cost $0.15. If she spent $6.50, how many of each did she purchase?

48. Find the lengths of the sides of a triangle if two sides are equal, and the third side is two feet shorter than one of the equal sides. The perimeter is 37 feet.

Chapter Test

Write an algebraic expression for each of the following.

1. A number increased by seven

2. Twice a number

3. Four less than three times a number

4. Doug has Q quarters. Mark has half as many quarters as Doug. Express Mark's number of quarters.

5. The average annual rainfall in Austin is R inches. The average annual rainfall in Tacoma is five inches more than four times the Austin average. Express Tacoma's annual rainfall.

6. Three-fourths a number increased by seven

7. The sum of two consecutive odd numbers

8. The value of N dimes

In each of Problems 9 through 20, write an equation and solve it.

Six less than twice a number is 20. What is the number?

9. Equation:

10. Solution:

Find three consecutive even numbers whose sum is 102.

11. Equation:

12. Solution:

Five-sixths of the students were in history class today. If 35 students were present, how many students are enrolled in the class?

13. Equation:

14. Solution:

The length of a rectangle is 8 meters more than the width. The perimeter is 180 meters. What are the dimensions?

15. Equation:

16. Solution:

The Whitestone Bridge toll collector noticed that she had twice as many quarters as dollar bills and had no other coins. If she had a total of $43.50, how many quarters did she have?

17. Equation:

18. Solution:

The sum of the first and third of three consecutive numbers is 24. What are the three numbers?

19. Equation:

20. Solution:

Thought-Provoking Problems

1. Has your attitude toward solving word problems undergone any change as a result of your work in this chapter? Why or why not?

2. Write a consecutive-number problem. Be sure that the problem is clearly stated and that it can be solved using the algebraic techniques of Chapters 10 and 11.

3. The perimeter of a triangle is 91 feet. If the length of the second side is twice the length of the first side, and the third side is 1 foot longer than the second side, what are the lengths of the three sides?

4. The sum of the three angles of any triangle is always 180 degrees. If the first angle is half the measure of the second angle, and the third angle is 12 degrees more than the first, what are the measurements of the three angles?

5. Leonard has 200 feet of chain-link fencing to enclose his rectangular back yard. If he uses his house for one of the lengths, what will be the dimensions of the fenced yard if the length is 8 feet more than twice the width?

6. The width of a rectangle is one-third the measure of the length. If each dimension is increased by 12 meters, the perimeter is doubled. What are the dimensions of the original rectangle?

7. Use the techniques of a coin problem to solve the following problems.
   a. A woman invests $1000 in a certificate of deposit that is paying 9% annual interest. She also invests some money in a high-risk stock that yields 12% annual interest. If after 1 year she earns $480 in interest, what amount did she invest in the stock? (*Hint*: principal × rate = interest for 1 year.)
   b. A company retirement plan has $24,000 invested in a high-return venture paying 20% annual interest. How much should it invest in a project that is yielding a 15% return in order to earn $20,400 annually from the combined investments?

Signed Numbers

12.1 ◆ A Suggested Approach

The ability to do calculations with signed numbers is a fundamental math skill. Often such calculations are one of the first topics studied in an algebra course. What signed numbers are and what they mean are important concepts to grasp. One way to accomplish this is to begin by memorizing the rules, but be sure to keep reading through the explanations of those rules. The more problems you do, the more familiar the processes will become. Soon the reasoning behind the rules may become clearer too.

◆ **Learning Objectives**

When you have completed this chapter, you should be able to:

1. Translate verbal expressions into signed-number expressions.
2. Graph signed numbers on a number line.
3. Add, subtract, multiply, and divide signed numbers.
4. Make use of the Order of Operations Agreement in expressions with signed numbers.
5. Simplify signed-number expressions that include exponents.
6. Evaluate algebraic expressions using signed numbers.

12.2 What are Signed Numbers?

In the beginning, you might think of signed numbers as "temperature" numbers. Everyone knows the difference between 12 degrees above zero ($+12$) and 12 degrees below zero (-12). Money is also a practical application using signed numbers. Many people have had the experience of having money in a checking account (a positive balance) and overdrawing that account by writing a check that is larger than the amount of money in the account (leaving a negative balance).

Rather than a vertical scale like a thermometer, a horizontal number line is often used to illustrate signed numbers.

The zero is placed approximately in the middle, and it is thought of as the beginning point for all calculations. It separates the positive numbers on its right from the negative numbers on its left. Like a ruler, a number line is divided into units, and each unit is the same length. A number line extends without end in both directions, so arrows are used at each end of the line. Each signed number has distance and direction. For example,

$+3$ is 3 units to the right of zero.

-5 is 5 units to the left of zero.

0 is in the middle and has no sign.

From the number line that was shown, it would seem that the signed numbers can be only integers. **Integers** are the positive and negative whole numbers and zero. Examples of integers that are also signed numbers are $+5$, -6, 0, and -8. Fractions and decimals are not included in the group of numbers called integers. We could graph $+2\ 3/4$, -1.8, $-2/3$, and $+11.76$, because they are signed numbers, but for the sake of clarity on the number line, we usually indicate only the integer values on the scale. On the following number line, -4, -1.5, $+1$, and $+2\ 1/4$ are graphed.

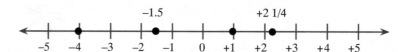

Because in many contexts (such as football and finances), it is customary to indicate a gain as a good thing, *a positive*, and to indicate a loss as a bad thing, *a negative*, it is easy to represent gains and losses as signed numbers. On a vertical graph, such as a thermometer, the positive values are above zero and the negative values are below zero. Applying these ideas to a map, where we could measure both north–south and east–west, north and east would be positive, and south and west would be negative.

Exercises Represent each of the following expressions by a signed number.

1. four miles east
2. six feet below ground
3. eight degrees above zero
4. twenty-three miles north
5. a loss of twelve dollars
6. seventeen and one-half miles west
7. eighty-two miles south
8. ten points ahead in a basketball game

Answers:

1. +4   **2.** −6   **3.** +8   **4.** +23   **5.** −12   **6.** −17 1/2

7. −82   **8.** +10

Consider the relative positions of +5 and +3 on the number line.

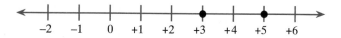

You know that +5 is larger than +3 and is to the right of +3 on the number line. Using inequality notation, you would write +5 > +3 or +3 < +5.

Consider the relative positions of +2 and −2 on the number line.

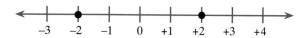

Because +2 is to the right of −2 on the number line, +2 must be the larger number. So +2 > −2 or −2 < +2.

Consider the relative positions of −4 and −10 on the number line.

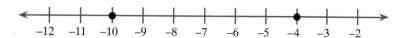

Because −4 is to the right of −10 on the number line, −4 is larger than −10. You would write −4 > −10 or −10 < −4.

Whenever you need to know which of two numbers is larger, picture their positions on the number line: the larger one is always to the right. You can also think of them as temperature numbers; the larger number is the "warmer" number. For example, +2 is to the right of −2, and 2 degrees above zero is warmer than 2 degrees below zero. Four degrees below zero, −4, is cold but it is still warmer than 10 degrees below zero, −10. On the number line, −4 is to the right of −10.

Another concept that students usually study when they are learning about signed numbers is absolute value. Consider the relative positions of +3 and −3 on the number line.

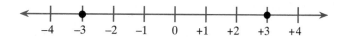

Each of these integers is three units from zero. We say that the absolute value of +3, which is written $|+3|$, is equal to 3. **Absolute value** is the distance that a given number is from zero. Distance is always positive (you can't walk a negative two miles), the absolute value of a number is the positive value associated with it. The absolute value of −5, $|-5|$, is equal to +5. Similarly, $|-2.5| = 2.5$ and $|3\ 1/4| = 3\ 1/4$. Study these examples.

E X A M P L E   **1**   Find the value of $|-7|$.

Solution   Because −7 is 7 units from zero,
$|-7| = 7$ ◆

E X A M P L E   **2**   Find the value of $|+6|$.

Solution   Because +6 is 6 units from zero,
$|+6| = 6$ ◆

E X A M P L E   **3**   Find the value of $|-2.6|$.

Solution   Because -2.6 is 2.6 units from zero,
$$|-2.6| = 2.6$$
◆

E X A M P L E   **4**   Find the value of $|0|$.

Solution   Because there is no distance between 0 and 0,
$$|0| = 0$$
◆

E X A M P L E   **5**   Find the value of $|-3| + |+5|$.

Solution   Because $|-3| = 3$ and $|+5| = 5$,
$$|-3| + |+5| = 3 + 5 = 8$$
◆

Look again at the number line.

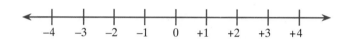

Note that $+3$ and -3 are each 3 units from zero. These two numbers are "opposites" of each other. They both have an absolute value of 3, but they lie on opposite sides of zero. Numbers that have the same absolute value but opposite signs are called **opposites**.

Now, you have three different ways of reading -3:

-3 can tell you to subtract three,

-3 can mean the signed integer, "negative three," and

-3 can mean "the opposite of three."

You will be able to tell from the problems themselves which meaning applies.

12.2 ◆ Practice Problems

A.   Represent each of the following descriptions as a signed number.

1. 4 steps down

2. 350 feet below sea level

3. winning $26

4. a drop of 18 degrees

5. 276 miles east

6. 143 miles north

7. 5,280 feet above sea level

8. a deposit of $125

9. 75 degrees above zero

10. losing by 14 points

B.   Make a number line and graph each of the following signed numbers.

11. -5	12. $+3\ 1/2$
13. $+2$	14. -5.25
15. 0	16. $-5/8$
17. $-3\ 1/3$	18. $+6$
19. $+4.2$	20. $+1.5$

C.   Write each of the following statements using inequality notation. Use $>$ or $<$ to fill in the blank.

21. -4 _____ -1	22. $+8$ _____ -8
23. -5 _____ -6	24. $+9$ _____ $+12$

25. +5 _____ +6   **26.** −8 _____ +8

27. 0 _____ −6   **28.** +7 _____ 0

37. $|5| + |-3|$ **38.** $|-2| + |-7|$

39. $|-12| - |+10|$ **40.** $|-8| - |+3|$

D. Simplify each of the following absolute-value statements.

29. $|-6|$ **30.** $|0.4|$

31. $|+9|$ **32.** $|-4|$

33. $|0|$ **34.** $|+13|$

35. $|-1.4|$ **36.** $|-1 \ 1/4|$

E. Use your calculator to simplify each of the following expressions.

41. $|-2.63| + |+5.91|$ **42.** $|+28.35| - |-19.261|$

43. $|+27.3| - |-17.45|$ **44.** $|-37.6| - |+29.834|$

45. $|-13.95| - |0|$ **46.** $|0| + |-21.45|$

12.3 Adding Signed Numbers

The addition of signed numbers can be shown easily by making use of the number line. The following graph shows the addition of (+3) and (+4). Begin at zero and travel 3 units in the positive direction. From that point, move 4 more units in the positive direction.

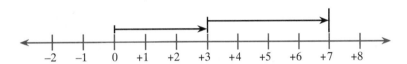

The number line shows what you already know, that

$$(+3) + (+4) = +7$$

Because the positive signs get in the way, you can simply write $3 + 4 = 7$. But if you mean negative 3, you must write -3.

Consider $(-5) + (-1)$.

Therefore, $(-5) + (-1) = -6$.

It makes sense that if you are combining positive numbers, you are moving farther in the positive direction, and your answer will be positive. Likewise, if you are combining negative values, you are moving farther in the negative direction, and your answer will be negative.

When combining numbers with like signs, we add the distances (absolute values), measured from starting point to ending point, and use the common sign. For example,

$$(-6) + (-8) = -14$$

$$(+7) + (+5) = +12$$

$$(-3) + (-8) + (-4) = -15$$

$$11 + 12 + 6 + 5 = 34$$

The next question is what happens when you are adding (combining) and the signs are different? Look at these examples, which were worked out using the number line, and see if you can figure out the rule.

Consider $(-5) + (+3)$. Starting at zero, you will move 5 spaces to the left (in the negative direction), and then, from that point, you will move 3 spaces to the right (in the positive direction).

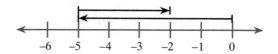

Therefore, $(-5) + (+3) = -2$.

Consider $(+4) + (-7)$.

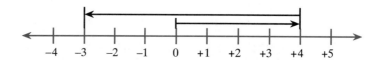

Therefore, $(+4) + (-7) = -3$.

Consider $(-3) + (+8)$.

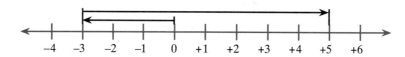

Therefore, $(-3) + (+8) = +5$.

Consider $(-6) + (+6)$.

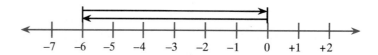

Therefore, $(-6) + (+6) = 0$.

These examples indicate that *when combining numbers of different signs, we find the answer by taking the difference between the two distances (absolute values) and using the sign of the longer distance.* The "longer distance" is the one with the larger absolute value. $|-5| = 5$ and $|+3| = 3$; we use the negative sign because 5 is greater than 3. Although the number line makes it easier to visualize what is going on when you combine signed numbers, you will not want to use it all the time. You need to learn the rules and be able to apply them.

Consider the following sample problems.

$(-5) + (+3) = -2$

The difference between 5 and 3 is 2, and 5 is the longer distance so you use its sign, $-$.

$(+4) + (-7) = -3$

The difference between 4 and 7 is 3, and $|-7| > |+4|$ so you use the sign from -7.

$(-3) + (+8) = +5$

The difference between 3 and 8 is 5, and $|+8| > |-3|$ so you use the sign from $+8$.

$(-6) + (+6) = 0$

The difference is 0 and it has no sign. *Note*: -6 and $+6$ are called *opposites*, because they are the same number of units from zero but in opposite directions. The sum of two opposites is always zero.

Study the following examples and see how the rule applies.

1. $(-11) + (+4) = -7$ The difference is 7, and $|-11| > |+4|$.
2. $(+12) + (-8) = +4$ The difference is 4, and $|+12| > |-8|$.
3. $(-7) + (+9) = +2$ The difference is 2, and $|+9| > |-7|$.
4. $(+9) + (-9) = 0$ The difference is zero, and $|+9| = |-9|$ because $9 = 9$.

You can now formally write the rules for the addition of signed numbers, which is also called *combining signed numbers*.

Combining Signed Numbers

If the numbers being added have the same sign, combine the absolute values and use that sign.

If the numbers being added have different signs, find the difference between the absolute values, and use the sign of the number with the larger absolute value.

Exercises   Try these problems to see if you understand how to combine (add) signed numbers.

1. $(-6) + (-9)$ 2. $(+4) + (-7)$ 3. $(-3) + (+8)$
4. $(-11) + (+11)$ 5. $(+7) + (+4)$

Answers:

1. -15 2. -3 3. $+5$ 4. 0 5. $+11$

The properties of real numbers that you studied in Chapter 9 also apply, of course, to signed numbers. The Commutative Property for Addition applies to addition with signed numbers.

For all real numbers represented by a and b,

$$a + b = b + a$$
$$(-5) + (+3) = (+3) + (-5) \qquad (+4) + (-7) = (-7) + (+4)$$
$$-2 = -2 \qquad\qquad -3 = -3$$

The Associative Property for Addition also applies to calculations with signed numbers.

For all real numbers represented by a, b, and c,

$$(a + b) + c = a + (b + c)$$
$$[(+9) + (-12)] + (-4) = (+9) + [(-12) + (-4)]$$
$$[-3] + (-4) = (+9) + [-16]$$
$$-7 = -7$$
$$[(-5) + (-8)] + (+2) = (-5) + [(-8) + (+2)]$$
$$[-13] + (+2) = (-5) + [(-6)]$$
$$-11 = -11$$

You will find that by making use of these two properties, you can combine signed numbers in any order that seems most convenient to you, as long as the sign and the number are rearranged together. For instance, because it is easier to combine numbers with like signs, rearrange the following problem so that the like-signed terms are together.

$$(-8) + (+7) + (+3) + (-5)$$
$$= \underline{(-8) + (-5)} + \underline{(+7) + (+3)}$$
$$= \qquad \underline{(-13)} \quad + \quad \underline{(+10)}$$
$$= \qquad\qquad -3$$

If you were to do the same problem combining from left to right as the order of operations suggests, it would look like this.

$$\underline{(-8) + (+7)} + (+3) + (-5)$$
$$= \underline{\quad (-1) \quad + (+3)} + (-5)$$
$$= \underline{\quad\quad (+2) \quad\quad + (-5)}$$
$$= \quad\quad\quad\quad -3$$

The answers are the same no matter which approach you use.

12.3 ◆ Practice Problems

A. Simplify each of the following expressions by combining the signed numbers.

1. $(+4) + (-7)$
2. $(-6) + (+11)$
3. $(-8) + (-3)$
4. $(+7) + (+3)$
5. $(-11) + (+4)$
6. $(-2) + (-8)$
7. $(+6\ 1/2) + (+3\ 1/2)$
8. $(-7) + (+4\ 1/2)$
9. $0 + (-5)$
10. $(-14) + (+14)$
11. $(+3) + (-3)$
12. $0 + (-7)$
13. $(-8) + (-5) + (-3)$
14. $(-7) + (-6) + (-2)$
15. $(+6) + [(-9) + (-3)]$
16. $(+4) + [(-9) + (+3)]$
17. $(-4) + (0) + (-6)$
18. $(+8) + (-9) + 0$
19. $[(+4) + (-6)] + (+9)$
20. $[(-8) + (-4)] + (+9)$

B. Write each of the following descriptions as a signed-number expression, and compute its solution. Label your answers where possible.

21. A loss of 14 points followed by a gain of 10 points.
22. A gain of 7 pounds followed by a loss of 12 pounds.
23. Driving up 250 feet from 150 feet below sea level.
24. Putting $142 in your checking account and then writing a check for $87.35.

25. You drive 23 miles east and then turn around and drive 40 miles west. How far and in what direction are you from your original position?
26. You fly 125 miles north and then turn around and fly 50 miles south. How far and in what direction are you from your original position?

C. Simplify each of the following expressions.

27. $12 + (-3\ 2/3)$
28. $(-8) + (+11\ 2/5)$
29. $(-9\ 3/4) + (-5\ 1/8) + (2\ 7/8)$
30. $(-12\ 5/8) + (+4\ 3/4) + (-2\ 1/2)$

D. Use your calculator to simplify each of the following addition problems.

31. $(-2.673) + (+9.85)$
32. $(+14.892) + (-29.37)$
33. $(-31.4) + (+2.695) + (+17.43)$
34. $(-42.906) + (-18.479)$
35. $(+113.457) + (-206.384)$
36. $(+27.863) + (-19.095) + (-36.52)$

12.4 Subtracting Signed Numbers

Looking at the number line again may help you discover the rules used in subtracting signed numbers.

Consider $(+3) - (+5)$.

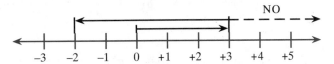

First go 3 spaces to the right of zero. From that point you cannot go 5 more spaces to the right, because that would be adding $+5$ to $+3$. You want to move 5 spaces, but you need to change direction because you are subtracting and doing the operation that is opposite to addition. You will go 5 spaces to the left.

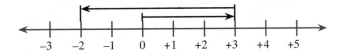

Therefore, $(+3) - (+5) = -2$.

Consider $(-4) - (-7)$.

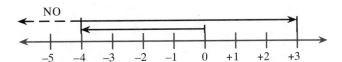

First go 4 spaces to the left of zero. From there you can't go 7 more spaces to the left, because that would be adding -7 to -4, so you need to change direction and go 7 spaces to the right.

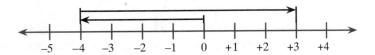

Therefore, $(-4) - (-7) = +3$.

If you look *only* at the number line for $(+3) - (+5)$, the arrows make it appear that you are adding (-5) to $(+3)$.

$$\overset{\text{S}}{(+3)} - \overset{\text{A}}{(+5)} \text{ looks like } (+3) + (-5).$$

If you look *only* at the graph of $(-4) - (-7)$, it would appear that you are adding $+7$ to -4.

$$\overset{\text{S}}{(-4)} - \overset{\text{A}}{(-7)} \text{ looks like } (-4) + (+7).$$

From these observations, you can formalize the rule for subtracting signed numbers.

Subtracting Signed Numbers

> To subtract signed numbers, change the sign of the number being subtracted, and change the operation to addition. Symbolically, this may be written
>
> $$a - b = a + (-b)$$

Consider the following true statements:

      S   A

1. $(-6) - (+3) = (-6) + (-3)$ Now the signs are the same, so you add the
                $= -9$ distances and use the sign they share.

      S   A

2. $(+9) - (+4) = (+9) + (-4)$ Now the signs are different, so you find the
                $= +5$ difference between the absolute values and
                                         use the sign of the number with the larger
                                         absolute value.

        S   A

3. $(+14) - (+18) = (+14) + (-18) = -4$

       S   A

4. $(-8) - (-12) = (-8) + (+12) = +4$

       S   A

5. $(-9) - (-9) = (-9) + (+9) = 0$

Exercises   Try these problems to see if you understand the subtraction rule.

1. $(+8) - (+3)$ 2. $(+7) - (-12)$ 3. $(-4) - (-9)$

4. $(-6) - (+6)$ 5. $(-6) - (-6)$ 6. $(+7) - (+11)$

Answers:

1. $+5$ 2. $+19$ 3. $+5$ 4. -12 5. 0 6. -4

    You might also find it helpful to know that when you try to find the answer to the question "What is $7 - 4$?" you are also asking "How far and in what direction is it from 4 to 7?" Beginning at 4 and moving to 7 is moving 3 spaces in a positive direction, and $7 - 4$ is equal to $+3$. Anytime you are finding the value of $a - b$, you can ask, "How far and in what direction is it from b to a?" Does this make a problem like $(-9) - (-3)$ easier to understand? "How far and in what direction is it from (-3) to (-9)?" It is 6 spaces in a negative direction. Therefore, you know that $(-9) - (-3) = -6$. Of course, you get the same result when you work the problem by applying the rule for subtraction.

          S

    $(-9) - (-3)$

          A

  $= (-9) + (+3)$

  $= -6$

    There are two popular short cuts for the rule for subtraction. One is "change the sign and add." The other is "add the opposite." It is fine to use these short cuts as long as you understand what changes to make. Most students find it helpful to write S over the subtraction symbol in the original statement of the problem and to write A once they have made the changes. This technique will be useful in the next chapters as well.

12.4 ◆ Practice Problems

A. **Subtract, following these examples:**

$$
\begin{array}{cc}
S & S \\
(-8) - (+6) & (+11) - (-3) \\
A & A \\
= (-8) + (-6) & = (+11) + (+3) \\
= -14 & = +14 \text{ or } 14
\end{array}
$$

1. $(-4) - (-3)$ 2. $(+8) - (+5)$

3. $(+6) - (+9)$ 4. $(-11) - (-3)$

5. $(-12) - (+6)$ 6. $(-4) - (+6)$

7. $(+7) - (-3)$ 8. $(+12) - (-2)$

9. $0 - (+6)$ 10. $0 - (-8)$

B. **Subtract, following these examples:**

$$
\begin{array}{cccc}
-8 & -8 & +11 & +11 \\
S\ -6 & A\ +6 & S\ -\ 4 & A\ +\ 4 \\
\hline
& -2 & & +15
\end{array}
$$

11. $\begin{array}{r} +6 \\ S\ +3 \\ \hline \end{array}$ 12. $\begin{array}{r} -8 \\ S\ -2 \\ \hline \end{array}$

13. $\begin{array}{r} -\ 9 \\ S\ -11 \\ \hline \end{array}$ 14. $\begin{array}{r} -6 \\ S\ +9 \\ \hline \end{array}$

15. $\begin{array}{r} +12 \\ S\ -\ 5 \\ \hline \end{array}$ 16. $\begin{array}{r} +11 \\ S\ +15 \\ \hline \end{array}$

17. $\begin{array}{r} -14 \\ S\ +\ 9 \\ \hline \end{array}$ 18. $\begin{array}{r} +7 \\ S\ -9 \\ \hline \end{array}$

C. **Simplify each expression, following the examples given. First change all subtraction operations to addition operations, and then combine the resulting signed numbers.**

$$
\begin{array}{cc}
S & S \\
(+6) - (+9) + (-8) & (-11) + (+6) - (-5) \\
A & A \\
= (+6) + (-9) + (-8) & = (-11) + (+6) + (+5) \\
= (-3) + (-8) & = (-5) + (+5) \\
= -11 & = 0
\end{array}
$$

19. $(-11) - (-5) - (+6)$

20. $(+12) - (+4) - (-8)$

21. $(+20) + (-3) - (+8)$

22. $(+14) + (-11) - (+4)$

23. $(-15) - (-2) + (-6)$

24. $(-7) - (-9) + (-8)$

25. $(-9) - (+4) - (-8)$

26. $(-11) - (+6) - (-7)$

27. $(-5\ 3/4) - (-2\ 1/2)$

28. $(+8\ 6/7) - (+9\ 5/8)$

29. $(-3\ 1/5) + (-2\ 7/10) - (+9\ 3/5)$

30. $(-2\ 3/4) - (-3\ 5/12) - (+2\ 1/2)$

D. **Use your calculator to simplify each of the following expressions.**

31. $(-24.95) - (-13.268)$

32. $(+175.863) - (-205.93)$

33. $(-39.84) - (+42.765)$

34. $(+492.76) - (+508.375)$

35. $(-21.83) - (-19.276) + (-14.95)$

36. $(+123.9) - (+247.3) - (+142.8)$

12.5 Multiplying Signed Numbers

The rules for multiplying signed numbers are fairly simple to develop. Consider the following pattern:

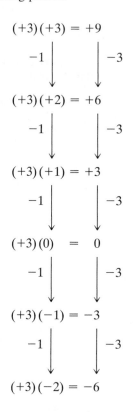

$(+3)(+3) = +9$

$(+3)(+2) = +6$ Note that a decrease of 1 in a factor produces a decrease of 3 in the answer (product).

$(+3)(+1) = +3$ A decrease of 1 in a factor produces a decrease of 3 in the answer.

$(+3)(0) = 0$ A decrease of 1 in a factor produces a decrease of 3 in the answer.

$(+3)(-1) = -3$ This must be correct because a decrease of 1 in a factor produces a decrease of 3 in the answer.

$(+3)(-2) = -6$ A decrease of 1 in a factor again produces a decrease of 3 in the answer.

You already know that $3 \times 3 = 9$ and that $3 \times 0 = 0$, so you must have established a good working pattern to go by. From this pattern, you can see that the following rules hold true.

$$+ \quad \times \quad +$$

A positive $\times$ a positive $=$ a positive.

$$+ \quad \times \quad -$$

A positive $\times$ a negative $=$ a negative.

You know that multiplication is a commutative operation. For example, $6 \times 7 = 7 \times 6$. Therefore, if $(+3)(-2) = -6$, $(-2)(+3)$ must also equal -6. From this statement it would seem that

$$- \quad \times \quad +$$

A negative times a positive $=$ a negative.

You have only one case left to consider. What is a negative times a negative equal to? Let us again look at a pattern. You accept that $(-2)(+3) = -6$. Begin there.

$$(-2)(+3) = -6$$

A negative $\times$ a positive = a negative.

$$-1 \downarrow \quad \downarrow +2$$

$$(-2)(+2) = -4$$

Because you are multiplying by -2 this time, it would seem that a decrease of 1 in a factor produces an increase of 2 in the answer. You know that -4 is larger than -6.

$$-1 \downarrow \quad \downarrow +2$$

$$(-2)(+1) = -2$$

A decrease of 1 in a factor produces an increase of 2 in the answer. (You know that -2 is larger than -4.)

$$-1 \downarrow \quad \downarrow +2$$

$$(-2)(0) = 0$$

A decrease of 1 in a factor produces an increase of 2 in the answer. Also, you know that any number times 0 equals 0.

$$-1 \downarrow \quad \downarrow +2$$

$$(-2)(-1) = +2$$

A decrease of 1 in the factor produces an increase of 2 in the answer.

$$-1 \downarrow \quad \downarrow +2$$

$$(-2)(-2) = +4$$

Given the pattern you have observed, this must be true.

It would seem from the pattern that

$$- \quad \times \quad -$$

A negative times a negative = a positive.

We can now summarize the rules for multiplication of signed numbers.

Multiplying Signed Numbers

When two factors with like signs are multiplied, the answer is positive.
When two factors with unlike signs are multiplied, the answer is negative.

You can summarize these rules by using symbols as well.

$$+ \times + = +$$
$$+ \times - = -$$
$$- \times + = -$$
$$- \times - = +$$

Look at the following correct sample problems.

$$(-8)(-2) = +16 \qquad (-7)(+4) = -28$$
$$(+6)(+5) = +30 \qquad (+4)(-3) = -12$$
$$(+7)(0) = 0 \qquad (0)(-8) = 0$$
$$(-1/4)(+24) = -6 \qquad (+2.3)(-1.4) = -3.22$$

Exercises Try these problems to see if you understand the basic multiplication rules.

1. $(-8)(+3)$ **2.** $(-2)(-5)$ **3.** $(+4)(+9)$

4. $(-7)(0)$ **5.** $(+9)(-5)$

Answers:

1. -24 **2.** $+10$ **3.** $+36$ **4.** 0 **5.** -45

If a multiplication problem has more than two factors, you have two choices. (1) You can multiply from left to right, following the order of operations and dealing with the signs as you figure each individual product. (2) Or you can determine the sign of the answer first and then multiply the quantities together.

Look at these problems, paying attention to the number of negative factors in each and the answer that you get.

$(-3)(+4) = -12$ 1 negative factor = negative answer

$(-3)(-5) = +15$ 2 negative factors = positive answer

$(-4)(-3)(-2) = -24$ 3 negative factors = negative answer

$(-2)(-2)(-2)(-2) = +16$ 4 negative factors = positive answer

It would seem that *an even number of negative factors produces a positive answer. An odd number of negative factors produces a negative answer.* This is true, and it enables you to determine the sign of the product (answer) first and then to multiply without worrying about the signs. Consider the following problems that are done both ways, and decide which you prefer. Either method is fine.

E X A M P L E **1** Simplify: $(-6)(+2)(-5)$

Solution **Method 1**   **Method 2**

$(-6)(+2)(-5)$ $(-6)(+2)(-5)$ Two negative factors, so the answer will be positive.

$= (-12)(-5)$ $= +(12)(5)$
$= +60$ $= +60$ ◆

E X A M P L E **2** Simplify: $(+6)(-5)(+3)$

Solution **Method 1**   **Method 2**

$(+6)(-5)(+3)$ $(+6)(-5)(+3)$ One negative factor, so the answer will be negative.

$= (-30)(+3)$ $= -(30)(3)$
$= -90$ $= -90$ ◆

E X A M P L E **3** Simplify: $(-2)(-3)(-4)$

Solution **Method 1**   **Method 2**

$(-2)(-3)(-4)$ $(-2)(-3)(-4)$ Odd number of negative factors, so the answer will be negative.

$= (+6)(-4)$ $= -(6)(4)$
$= -24$ $= -24$ ◆

E X A M P L E **4**   Simplify: $(+3)(-4)(-2)(+1)$

Solution   **Method 1**   **Method 2**

$(+3)(-4)(-2)(+1)$ $(+3)(-4)(-2)(+1)$ Even number of nega-
                                                         tive factors, so the
                                                         answer will be positive.

$= (-12)(-2)(+1)$ $= +(3)(4)(2)(1)$
$= (+24)(+1)$ $= +(12)(2)(1)$
$= +24$ $= +(24)(1)$
                           $= +24$ ◆

Normally it is easier to determine the sign of the product first and then multiply.

When exponents are involved, the rules for counting negative signs are a real help. Remember that in multiplication, an even number of negative factors produces a positive answer, whereas an odd number of negative factors produces a negative answer. And exponents are just a short-cut notation for repeated multiplication. Study the following examples.

E X A M P L E **5**   Simplify: $(-2)^5$

Solution $(-2)^5 = (-2)(-2)(-2)(-2)(-2)$
$= (+4)(-2)(-2)(-2)$
$= (-8)(-2)(-2)$
$= (+16)(-2)$
$= -32$

Or $(-2)^5$ has an odd number of negative signs, and will therefore give a negative answer.
$(-2)^5 = $ negative $(2)^5$
$= -32$ ◆

E X A M P L E **6**   Simplify: $(-1)^4$

Solution $(-1)^4 = (-1)(-1)(-1)(-1)$
$= (+1)(-1)(-1)$
$= (-1)(-1)$
$= +1$

Or $(-1)^4$ has an even number of negative factors and will therefore produce a positive answer.
$(-1)^4 = $ positive $(1)^4$
$= +1$ ◆

Of course, any number of positive factors will always produce a positive answer.

Study the next two examples to see the difference between $(-3)^2$, *negative three raised to the second power*, and -3^2, the negative of *3 raised to the second power*.

E X A M P L E **7**   Simplify: $(-3)^2$

Solution $(-3)^2 = (-3)(-3)$
$= +9$ ◆

E X A M P L E   **8**   Simplify: -3^2

Solution $-3^2 = -(3)(3)$
$= -(9)$
$= -9$ ◆

An exponent, or power, applies only to its base. In Example 8, only the 3 is the base. In Example 7, the base is negative 3, (-3). The parentheses make the difference in these two examples. Be very careful when working problems like these.

E X A M P L E   **9**   Simplify: -3^4

Solution -3^4 means to take the opposite of 3 raised to the fourth power. $3^4 = 81$, and the opposite of 81 is negative 81. Therefore, $-3^4 = -81$.   ◆

E X A M P L E   **10**   Simplify: $(-3)^4$

Solution   In this problem the base is -3, so that whole quantity is raised to the fourth power.
$(-3)^4 = (-3)(-3)(-3)(-3) = 81$ ◆

E X A M P L E   **11**   Simplify: $-22 - 3(-8)$

Solution $-22 - 3(-8)$ Multiply $3(-8)$ before you subtract.
$= -22 - (-24)$
$= -22 + (+24)$
$= +2$ ◆

E X A M P L E   **12**   Simplify using the Order of Operations Agreement: $(-5)^2 - 2[7 - (-2)]$

Solution $(-5)^2 - 2[7 - (-2)]$
$= (-5)^2 - 2[7 + (+2)]$
$= (-5)^2 - 2[9]$
$= 25 - 18$
$= 7$ ◆

12.5   ◆   Practice Problems

A. Find the product indicated in each of the following exercises.

1. $(+11)(+5)$
2. $(+7)(+9)$
3. $(-8)(+6)$
4. $(-7)(+4)$
5. $(-18)(-3)$
6. $(-22)(-6)$
7. $(+4)(-9)$
8. $(+8)(-3)$
9. $(-8)(0)(-4)$
10. $(-8)(-5)(0)$
11. $(-3)(-2)(-5)$
12. $(-6)(-2)(-3)$
13. $(+7)(-2)(+3)$
14. $(+9)(-3)(+2)$
15. $(-4)(-6)(0)$
16. $(-7)(0)(+6)$
17. $(-1)(-2)(-3)(-4)$
18. $(+2)(-3)(-5)(+4)$
19. $(+2)(-2)(-2)(+3)$
20. $(-1)(-1)(5)(+1)$

B. Simplify each of the following expressions using the Order of Operations Agreement (Section 12.4) and the rules for signed numbers. Remember to use S and A where they will help.

21. $23 - (+4)(-2)$
22. $37 - (-3)(+8)$
23. $17 + (-3)(+6)$
24. $12 + (-7)(+3)$

25. $(-6)(+2) + (-3)(+4)$

26. $(+7)(-2) + (-3)(+6)$

27. $(-6)(-2) - (-3)(-5)$

28. $(-8)(-4) - (-6)(-5)$

29. $12 - 6(3 - 8)$

30. $22 - 11(2 - 6)$

C. Simplify each of the following expressions involving exponents.

31. -3^2 **32.** -2^4

33. -4^3 **34.** -1^5

35. $(-2)^3$ **36.** $(+5)^2$

37. $(+2)^4$ **38.** $(-1)^6$

39. $(-1)^5(-2)^2$ **40.** $(-2)^3(-3)^2$

41. $(+3)^3(-2)^2$ **42.** $(-4)^3(+2)^3$

43. $(-1)^7 - 3(-2)^4$ **44.** $4(+2)^3 - (-3)^3$

D. Simplify each of the following expressions involving fractions by using the Order of Operations Agreement.

45. $(-1/2)(5/9) - (2/3)(-7/9)$

46. $(-1\ 3/4) - (2/3)(9/8)$

47. $(-7\ 1/4) - (-5/8)(7/2)$

48. $(2/5)^2 - 7\ 9/10$

49. $(-3/4)^2 + (-4\ 3/8)$

50. $(7/12)(-2/3) - (-5\ 5/6)$

E. Use your calculator to simplify each of the following expressions. Round answers to the nearest hundredth.

51. $(-2.45)^3$ **52.** $(-7.59)^4$

53. $(0.76)^2(-24.5)^3$ **54.** $(-16.4)^3(-2.45)^2$

55. $(-19.3)(-4.76)^3$ **56.** $(-9.1)^2(-17.65)$

57. $(17.3)(19.02) - (-16.3)(4.5)$

58. $(-21.4)(0.52) - (17.4)(-11.1)$

59. $(-12.4)(-2.35)(-17.8)$

60. $(-13.5)(+21.6)(-12.7)$

61. $27.5 - 16.2(8.75 - 11.3)$

62. $31.9 + 12.4(16.5 - 19.2)$

63. $14.27 + 3.9(-15.4 + 7.59)$

64. $42.7 - 24.3(18.4 - 27.1)$

12.6 Dividing Signed Numbers

The division process for signed numbers is a simple one. The rules for division are exactly the same as those for multiplication. Looking at the process we use to check a division problem will help you understand the rules. You know that to check a division problem you multiply the quotient by the divisor and add on the remainder, if there is one. For example,

To check $28 \div 7 = 4$, we see if 4 times $7 = 28$.

The signs for 28, 7, and 4 are all positive, so

Positive $\div$ positive = positive

Consider now what happens when you divide $+28$ by (-7).

$(+28) \div (-7) = ?$

? times (-7) must equal $(+28)$, so ? must be (-4) because you know that (-4) times $(-7) = +28$. Therefore,

$(+28) \div (-7) = -4$ because (-4) times $(-7) = +28$.

Positive $\div$ negative = negative

Consider now what happens when you divide (-28) by $(+7)$.

$$(-28) \div (+7) = ?$$

? times $(+7)$ must equal (-28), so ? must be (-4) because you know that (-4) times $(+7) = -28$. Therefore,

$$(-28) \div (+7) = -4 \text{ because } (-4) \text{ times } (+7) = -28.$$

Negative $\div$ positive $=$ negative

The only case left to look at is $(-28) \div (-7)$.

$$(-28) \div (-7) = ?$$

The check tells you that ? times (-7) must equal (-28), so ? must be $+4$ because $(+4)$ times $(-7) = -28$.

$$(-28) \div (-7) = +4 \text{ because } (+4) \text{ times } (-7) = -28$$

Negative $\div$ negative $=$ positive

These statements can be summarized in the rules that follow.

Dividing Signed Numbers

In division, two like signs produce a positive quotient (answer). Unlike signs produce a negative quotient.

For example,

$$(-16) \div (+4) = -4 \qquad (+28) \div (+4) = +7$$
$$(-20) \div (-5) = +4 \qquad (-42) \div (-7) = +6$$
$$(+12) \div (-2) = -6 \qquad (+84) \div (-7) = -12$$
$$(-8/3) \div (-5/4) = (-8/3) \times (-4/5) = +32/15$$
$$\frac{-6}{+3} = -2 \qquad \frac{+12}{+4} = +3 \qquad \frac{-9}{-2} = \frac{9}{2} \text{ or } 4\ 1/2$$

Exercises Try these division problems to check your understanding of the division rules.

1. $(-12) \div (+6)$ **2.** $(+18) \div (-2)$ **3.** $(+40) \div (+4)$

4. $(-36) \div (-9)$ **5.** $(0) \div (-3)$

Answers:

1. -2 **2.** -9 **3.** $+10$ **4.** $+4$ **5.** 0

When multiplication and division are combined within the same problem, the **number of negative signs rule** applies. *An even number of negative signs gives a positive answer, and an odd number of negative signs gives a negative answer.*

Consider these examples.

E X A M P L E 1 Simplify: $\dfrac{(-8)(+5)}{-10}$

Solution $\dfrac{(-8)(+5)}{-10} = \dfrac{-40}{-10} = +4$

Because this problem is *all* multiplication or division, you could first count the number of negative signs to determine the sign of the answer, and then calculate the numbers.

$$\frac{(-8)(+5)}{-10} = 4 \qquad \qquad 2 \text{ negatives} \longrightarrow \text{positive answer}$$

◆

E X A M P L E **2** Show the two ways of simplifying $\dfrac{(-6)(-4)(-8)}{(-3)(-2)}$.

Solution $\dfrac{(-6)(-4)(-8)}{(-3)(-2)} = \dfrac{(+24)(-8)}{+6} = \dfrac{-192}{+6} = -32$

Or, because there are 5 negative numbers in this problem and the operations are all either multiplication or division, the result will be negative (5 is an odd number).

$-\dfrac{\overset{2}{\cancel{(6)}}\,\overset{2}{\cancel{(4)}}\,(8)}{\underset{1}{\cancel{(3)}}\,\underset{1}{\cancel{(2)}}} = -32$

Therefore, the answer is -32.   ◆

When you have more than one operation in a problem, follow the Order of Operations Agreement (Section 9.4). Note that the fraction bar acts as a grouping symbol. You need to simplify the numerator and denominator separately and then perform the indicated division.

E X A M P L E **3** Simplify: $\dfrac{(-4)(-3)}{-2} - \dfrac{(-2)(+6)}{+3}$

Solution $\dfrac{(-4)(-3)}{-2} - \dfrac{(-2)(+6)}{+3}$

$= \dfrac{(+12)}{-2} - \dfrac{(-12)}{+3}$

$\overset{S}{= (-6) - (-4)}$

$\overset{A}{= (-6) + (+4)}$

$= -2$ ◆

E X A M P L E **4** Simplify: $\dfrac{(-20) + (-5)}{(-20) - (-5)}$

Solution $\dfrac{(-20) + (-5)}{(-20) - (-5)}$

$= \dfrac{(-20) + (-5)}{(-20) + (+5)}$

$= \dfrac{-25}{-15}$

$= \dfrac{+5}{3}$ or $+1\ 2/3$ ◆

12.6 ◆ Practice Problems

A. Simplify each of the following expressions.

1. $(-10) \div (+5)$ **2.** $(+20) \div (-4)$

3. $(+8) \div (-2)$ **4.** $(-12) \div (+3)$

5. $-16/-4$ **6.** $-30/-6$

7. $\dfrac{+8}{-20}$ **8.** $\dfrac{+15}{-20}$

9. $\dfrac{-12}{+5}$ **10.** $\dfrac{-14}{+3}$

11. $\dfrac{(-3)(-12)}{+4}$ **12.** $\dfrac{(-5)(-8)}{+10}$

13. $(-6\ 2/3) \div (+1\ 1/3)$ **14.** $(-5\ 1/3) \div (+2\ 2/3)$

15. $(+2.4) \div (-0.06)$ **16.** $(+3.5) \div (-0.07)$

17. $\dfrac{-2 + 8}{-3}$ **18.** $\dfrac{-9 + 3}{+2}$

19. $\dfrac{(-7) - (-3)}{(-7) + (+3)}$ **20.** $\dfrac{(-8) - (-2)}{(-8) + (-2)}$

B.   Simplify using the rules for signed numbers and the Order of Operations Agreement (Section 12.4).

21. $\dfrac{-8}{-2} \div \dfrac{+6}{+2}$ 22. $\dfrac{-12}{-3} \div \dfrac{-20}{+2}$

23. $\dfrac{-12}{-4} \div \dfrac{-10}{+8}$ 24. $\dfrac{(-8)(-3)(+4)}{(-6)(+10)}$

25. $-15 - \dfrac{+12}{-4}$ 26. $18 - \dfrac{+10}{-5}$

27. $-14 - \dfrac{-8}{+4}$ 28. $-10 - \dfrac{+12}{-3}$

29. $(-3)(-8) - (+6)(-2)$

30. $(-7)(+6) - (-5)(-4)$

31. $(-7)(+8) - (-3)(-2)$

32. $(-9)(+5) - (-4)(-5)$

33. $(-2)^4 \div (+2)^3$ 34. $(-3)^5 \div (+3)^3$

35. $(-3)^6 \div (-3)^3$ 36. $(-4)^3 \div (-2)^5$

37. $\dfrac{(-5)(-4)(-3)}{(-6)(-10)}$ 38. $\dfrac{-18}{+6} \div \dfrac{+9}{-3}$

39. $\dfrac{17 + (-5)(+2)}{-6 + 4}$ 40. $\dfrac{21 + (-3)(+7)}{-4 + 5}$

12.7   Evaluating Expressions Using Signed Numbers

You have substituted numbers into formulas in Chapter 9 and in earlier chapters. You will now do the same thing to evaluate expressions using signed numbers. Review the Order of Operations Agreement first.

Order of Operations Agreement

When simplifying a mathematical expression involving more than one operation, follow these steps:

1.   Simplify within parentheses or grouping symbols.
2.   Simplify any expressions that involve exponents or roots.
3.   Multiply or divide, in order, from left to right.
4.   Add or subtract (combine), in order, from left to right.

Study the following examples. Note that when you substitute a value for a variable, it is a good idea to use parentheses to keep the sign with the number. This practice often eliminates a lot of confusion. You can remove the parentheses as you work through the simplification steps.

E X A M P L E   1   Evaluate abc when $a = -2$, $b = +3$, and $c = -4$.

Solution abc

$= (-2)(+3)(-4)$

$= (-6)(-4)$

$= +24$

E X A M P L E   2   Evaluate $4b^2$ when $b = +3$.

Solution $4b^2$

$= 4(+3)^2$

$= 4(+9)$

$= +36$

E X A M P L E 3 Evaluate $bd - ac$ when $a = -2$, $b = +3$, $c = -4$, and $d = -1$.

Solution $bd - ac$

$$\overset{S}{= (+3)(-1) - (-2)(-4)}$$ Do not rewrite the subtraction until the multiplication has been done.

$$\overset{S}{= (-3) - (+8)}$$

$$\overset{A}{= (-3) + (-8)}$$

$$= -11$$ ◆

E X A M P L E 4 Evaluate $a + 2b - 3c$ when $a = -2$, $b = +3$, and $c = -4$.

Solution $a + 2b - 3c$

$$\overset{S}{= (-2) + 2(+3) - 3(-4)}$$

$$\overset{S}{= (-2) + (+6) - (-12)}$$

$$\overset{A}{= (-2) + (+6) + (+12)}$$

$$= (+4) + (+12)$$

$$= +16$$ ◆

E X A M P L E 5 Evaluate $\dfrac{a^2 - b^2}{cd}$ when $a = -2$, $b = +3$, $c = -4$, and $d = -1$.

Solution $\dfrac{a^2 - b^2}{cd}$

$$\overset{S}{= \dfrac{(-2)^2 - (+3)^2}{(-4)(-1)}}$$

$$\overset{S}{= \dfrac{4 - (+9)}{+4}}$$

$$\overset{A}{= \dfrac{4 + (-9)}{+4}}$$

$$= \dfrac{-5}{+4} \text{ or } -\dfrac{5}{4}$$

A negative divided by a positive gives a negative answer, so the answer can be written as either $\dfrac{-5}{+4}$ or $-\dfrac{5}{4}$. Usual algebra practice is not to leave a negative value in the denominator, so if the answer had been $\dfrac{+5}{-4}$, you would have written $-\dfrac{5}{4}$.   ◆

E X A M P L E **6** Evaluate $a - 3c$ when $a = -2$ and $c = -4$.

Solution

$a - 3c$

$$\overset{S}{= (-2) - 3(-4)}$$

$$\overset{S}{= (-2) - (-12)}$$

$$\overset{A}{= (-2) + (+12)}$$

$$= +10 \qquad\qquad\qquad\qquad\qquad\qquad\qquad ◆$$

Look at Example 6 again. You can consider the -3 as negative 3 and multiply -4 by -3, writing the product's sign as the operation sign. Then combine the results.

$a - 3c$	Substitute -2 for a and -4 for c.
$(-2) - 3(-4)$	Multiply $(-3)(-4)$.
$= -2 + 12$	Now combine.
$= +10$	

Treating this minus sign as a direction sign instead of as an operation sign is a short cut. It usually saves steps. Consider Example 4, this time using the short cut.

$a + 2b - 3c$	Substitute.
$(-2) + 2(+3) - 3(-4)$	Multiply $(+2)(+3)$ and $(-3)(-4)$.
$= -2 + 6 + 12$	Combine from left to right.
$= +4 + 12$	
$= +16$	

Follow through these last two examples to be sure that you understand this alternative. It is usually shorter, and it always works.

E X A M P L E **7** Evaluate $cb - 5d$ when $b = +3$, $c = -4$, and $d = -1$.

Solution

$cb - 5d$

$$= (-4)(+3) - 5(-1)$$

$$= -12 + 5$$

$$= -7 \qquad\qquad\qquad\qquad\qquad\qquad\qquad ◆$$

E X A M P L E **8** Evaluate $-5c - 4b + 2a$ when $a = -2$, $b = +3$, and $c = -4$.

Solution

$-5c - 4b + 2a$

$$= (-5)(-4) - 4(+3) + 2(-2)$$

$$= +20 - 12 - 4$$

$$= 8 - 4$$

$$= +4 \qquad\qquad\qquad\qquad\qquad\qquad\qquad ◆$$

12.7 ◆ Practice Problems

A. Find the value of each of the following expressions when $a = -2$, $b = +4$, $c = -3$, $d = -1$, and $e = 0$.

1. $a + b + c$
2. $d + bc$
3. $b + ad$
4. $a - b - c$
5. abc
6. bde
7. $2a - 3b$
8. $3c - 2d$
9. $a + 3b - 4c$
10. $a + 2b - 3c$
11. $4a - 5c$
12. $2b - 3d$
13. $3e - 4c$
14. $6e + 2d$
15. $a^2 c$
16. $b^2 d$
17. $5c^2$
18. $-3d^2$
19. $4a^3$
20. $2c^3$
21. $\dfrac{ab}{cd}$
22. $\dfrac{bc}{ad}$
23. $\dfrac{ab + cd}{c}$
24. $\dfrac{bd - ac}{d}$
25. $a^2 + 2b$
26. $c^2 - 2b$
27. $2b^2 - 3b + c$
28. $3a^2 + 2a - b$
29. $\dfrac{a^2 - b^2}{a^2 + b^2}$
30. $\dfrac{c^2 + d^2}{c^2 - d^2}$

B. Find the value of each of the following expressions when $w = 0.7$, $x = 1/4$, $y = -2/5$, and $z = -2.5$.

31. $wx - yz$
32. $x - 7z$
33. $3x + 10y$
34. $yz - 2wx$
35. xyz
36. $w + x + y$
37. $-2x + 3z$
38. $3w - 2z + y$
39. $3y - 2x^2 - 4z$
40. $4w - 20y^2$

12.8 ◈ Chapter Review

Now that you have completed this chapter, you should be able to:

1. Describe or talk about these terms and expressions: integer, absolute value, negative number, positive number, the odd and even rules for signs in multiplication, exponent, and division problems.

2. Plot numbers on a number line.

3. Translate verbal expressions into signed-number expressions and simplify them when possible.

4. Use the following rules for operations with signed numbers:

Combining Signed Numbers

If the numbers being added have the same sign, combine their absolute values and use the common sign.

If the numbers being added have different signs, find the difference between the absolute values and use the sign of the number with the larger absolute value.

Subtracting Signed Numbers

To subtract signed numbers, change the sign of the number being subtracted, and change the operation to addition.

Multiplying Signed Numbers

In the multiplication of two factors with like signs, the answer is positive.

In the multiplication of two factors with unlike signs, the answer is negative.

Dividing Signed Numbers

In division, two like signs produce a positive quotient (answer). Unlike signs produce a negative quotient.

5. Figure out the sign of the answer in a multiplication or division problem by using the odd and even rules for signs.

6. Substitute signed numbers into algebraic expressions.

7. Use the Order of Operations Agreement to simplify expressions involving signed numbers.

Review Problems

A. Complete the following sentences by filling in the blanks.

1. A signed number has both _____ and _____ .

2. 0 has _____ sign.

3. −6 is _____ than −4. (less than or greater than?)

4. −3 is _____ than −8. (less than or greater than?)

5. A gain of 6 pounds is represented by _____ .

6. A loss of $14 is represented by _____ .

7. When adding or combining numbers with like signs, _____ .

8. When adding or combining numbers with unlike signs, _____ .

9. When multiplying signed numbers, you can determine that the answer is positive if there is an _____ number of negative signs in the problem.

10. When dividing signed numbers, you can determine that the answer is negative if there is an _____ number of negative signs in the problem.

11. $(-2)^4$ will have a _____ answer.

12. $(+3)^5$ will have a _____ answer.

13. $(-5)^3$ will have a _____ answer.

14. $(-1)^7 = $ _____ .

15. -3^4 will have a _____ answer.

16. -2^5 will have a _____ answer.

17. What are the four steps in the Order of Operations Agreement?

18. In $-4p^3$, if p has the value −2, take −2 to the _____ power first, and then _____ by −4.

19. In evaluating $+3a - 2b$ when $a = +4$ and $b = +5$, it is easier to think of _____ times +5, rather than thinking of +2 times +5, before subtracting.

20. Zero times any quantity equals _____ .

B. Graph the following numbers on a number line.

21. −3   22. +2 1/2   23. +1 1/4

24. 0   25. 3   26. −5

27. +1/3   28. −5/8   29. −1.5

30. +4.2

C. Perform the indicated operations.

31. $-6 + 9$ 32. $4 - 8$

33. $(-3)(5)$ 34. $(-2)(-4)$

35. $(-45) \div (+9)$ 36. $(-18) \div (-3)$

37. $(-6)(0) + (-2)(+3)$ 38. $(4)(-3) - (2)(-1)$

39. $-8 - 4 + 5 - 2$ 40. $-11 + 6 - 5 - 4$

41. $+8 - (+11)$ 42. $-9 - (-3)$

43. $\dfrac{-9}{+3}$ 44. $\dfrac{+12}{-3}$

45. -2^4 46. -5^2

47. $(-2)^2(-1)^5$ 48. $(-3)^3(+2)^3$

49. $-3(-2) - 2(4)$ 50. $5(-1) - 3(-2)$

51. $\dfrac{-12}{-6} \div \dfrac{+3}{-2}$ 52. $\dfrac{-30}{+6} \div \dfrac{+12}{-2}$

53. -3^2 54. -2^6

55. $(-4)^2 \div (-3)^3$ 56. $(+5)^3 \div (-2)^4$

57. $-3/4(+12) - 1/2(-6)$

58. $2/3(-12) - 3/4(-16)$

59. $(1/2)^3(-2/5)^2$ 60. $(3/4)^2(-2/3)^3$

61. $(-5/8)(-3/10)$ 62. $(-4/5)(-1/2)(-5/8)$

63. $-5\ 1/2 + 3\ 3/4$ 64. $-6\ 2/3 + 7\ 1/2$

65. $-6 + 4\ 2/5$ 66. $-11 - 2\ 3/4$

67. $(-2.4)(+3.01)$ 68. $(-1.7)(-2.05)$

69. $\dfrac{(-6) + (-3)(+2)}{(-4)(-3)}$ 70. $\dfrac{8 + (-4)(+5)}{(2)(-6)}$

D. Evaluate each of the following expressions when w = -2, x = +3, y = 0, and z = -4.

71. xyz

72. wxz

73. $x + y - z$

74. $w + x - z$

75. $wx - wy$

76. $wy - wz$

77. $2w + 3x$

78. $-4w + 2z$

79. $-4x^2$

80. $-2z^3$

81. $-4xz$

82. $5wx$

83. $-3z - 2y$

84. $-6w + 5z$

85. $\dfrac{wx}{z}$

86. $\dfrac{wz}{x}$

87. $\dfrac{x + w}{x - w}$

88. $\dfrac{x + z}{x - z}$

89. $\dfrac{w^2 - x^2}{w^2 + x^2}$

90. $\dfrac{x^2 - z^2}{x^2 + z^2}$

Chapter Test

I. Find the value of each of the following expressions.

1. $-8 + 5$

2. $(-3) - (-7)$

3. $(-3)(-4)(-2)$

4. $\dfrac{+21}{-3}$

5. $(-4)^3$

6. $2(-8) + 3(-4)$

7. $(+6) - (+11)$

8. $(-5) + (-2) - (+3)$

9. $(-24) \div (-3)$

10. $(-2)^3(+3)^2$

11. $21 - 4[(-2) + (-3)]$

12. $3(-6) - 2(+5)$

13. $-4(+3) - 2(-3)$

14. $14 - 3[(-2) - (-4)]$

II. Evaluate each of the following expressions when d = +3, e = -2, f = -1, and g = 0.

15. def

16. $d - ef$

17. $fg + df$

18. $-2f^2 - 3f$

19. $\dfrac{d + e}{d - e}$

20. $-6f^5 + e$

Thought-Provoking Problems

1. Write an algebraic expression that describes each of the following situations, and then find the answers to the questions.

   a. Suppose you started out in the Rocky Mountains at an elevation of 11,200 feet and drove to Denver (the Mile High City) at an elevation of 5,280 feet. Describe the change in elevation in terms of a signed number.

   b. Your bank account has a balance of $270. You write two checks, one for $125 and another for $82. Then you make a deposit of $165 and write one more check for $94. What is your ending balance after all of these transactions?

   c. A stock opens the week at 31 1/4 points. It experiences four straight days of 5/8-point gains and then loses 1 1/2 points on Friday. What is the value of the stock at the close of the week?

2. Describe $(-8) + (+6) + (-3)$ as movement on the number line beginning at zero.

3. Explain the meaning of absolute value. Why can it never be negative? Find the value of

   a. $-|2|$ b. $|6| - |-4|$ c. $|4 - 6|$

4. Find the value of each of the following expressions when $a = -2$, $b = 3$, $c = -1$, and $d = 0$.

   a. $|a|$ b. $-|c|$ c. $|a| - |b|$

   d. $a - |c|$ e. $|d| - c$ f. $|c| - |b| + |a|$

   g. $(|a|)(|b|)$ h. $(|c|)(|d|)$ i. $a - |a|$

   j. $a + |a|$

5. Insert the correct symbol ($>$, $<$, or $=$) in each of the following expressions.

   a. $|-2|$ _____ -2 b. -16 _____ 16

   c. $|5|$ _____ $|-6|$ d. $|18|$ _____ $|-18|$

   e. $|0|$ _____ -12 f. $|-7|$ _____ $|-9|$

   g. 8 _____ $-|-8|$

6. Rank the members of each of the following groups in order from the least to the greatest.

   a. $|-7|, |+2|, |0|, |-3|, |+4|$

   b. $|-2\,1/2|, |+2\,1/4|, |-5|, |+6|, |-3|$

   c. $|+9|, |-10|, |+3|, |-7|, |-12|$

7. If $5(4)$ means five fours, or $4 + 4 + 4 + 4 + 4$, explain why $3(-2) = -6$.

8. Using your calculator, simplify each of the following expressions.

   a. $7(-8) - (-3)(+9)$

   b. $(-3)^4 + (-5)^5$

   c. $(0.26)^2(-1.4)^3$

   d. $(-3)(+7)(-9)(-15)(+12)$

   e. $|-25| \cdot |+17| \cdot |0|$

   f. $|-4.2| + |-8|$

   g. $|17.5| - |-12|$

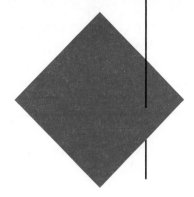

Cumulative Review, Chapters 1–12

Find the answer to each of the following problems. Show all your work. Reduce all fraction answers to lowest terms.

In Problems 1 through 5, add.

1. $11.56 + 3.8 + 18$
2. $11 \; 4/9 + 7 \; 2/3$
3. $5x + 8y + x + 3y$
4. $(-15) + (8) + (-3)$
5. $29 + (-17)$

In Problems 6 through 10, subtract.

6. $(-15) - (-12)$
7. $0 - 8 - (-5)$
8. $4 \; 1/4 - 2 \; 5/6$
9. $18 - 3.56$
10. $12a - 3a - 18a$

In Problems 11 through 15, multiply.

11. 0.176×100
12. $(-14)(-3)$
13. $5 \; 2/3 \times 2 \; 1/8$
14. Find $4/5$ of 36.
15. $(-2)(3)(-1)(2)$

In Problems 16 through 20, divide.

16. $(-32) \div (-8)$
17. $\dfrac{1.68}{0.04}$
18. $2 \; 5/8 \div (-3)$
19. $\dfrac{-54}{9}$
20. $0.8\overline{)24}$

Simplify Problems 21 and 22 using the Order of Operations Agreement.

21. $3 + 2[6 - 2(4 \div 2)]$
22. $\dfrac{25}{36} \div \left(\dfrac{2}{3}\right)^2 \div \dfrac{5}{9}$

Find the value for each of the expressions in Problems 23 through 25 by substituting the given values.

23. $ab^2 - ac$ when $a = 2$, $b = -1$, and $c = -3$
24. $\dfrac{2bc}{b^2 - c^2}$ when $a = 2$, $b = -1$, and $c = -3$
25. $2ab - 4c + 3ac$ when $a = 2$, $b = -1$, and $c = -3$

In Problems 26 through 28, write an algebraic expression for each of the English phrases.

26. the difference between a number and six
27. seven more than twice a number
28. the product of a number and eight

In Problems 29 and 30, write an algebraic equation for each of the English sentences.

29. The quotient of 15 and a number is -5.
30. The total cost (C) is equal to the product of the number of tickets (N) and $5.85.

Solve each of the following equations and inequalities.

31. $7Y = 56$

32. $\dfrac{15}{N} = \dfrac{75}{13}$

33. 16% of 90 is what number?

34. $3x + 5 = 20$

35. $\dfrac{3}{5}M + 6 = \dfrac{2}{3}$

36. $6A - 2 + 3A = 59 - 7$

37. $5.6W + 0.12 = 8W$

38. 18 is what percent of 90?

39. $2X > 14$

40. $4N - 5 = 2N + 9$

Solve Problems 41 and 42 as requested.

41. Solve for W in $P = 2L + 2W$ when $P = 100$ inches and $L = 29$ inches.

42. Solve for M in $S = C + M$ when $S = \$129.95$ and $C = \$74.25$.

In Problems 43 through 50, find the answer(s) to each of the questions asked. You may need to write an algebraic equation or a proportion to help you find the solution(s).

43. Sales tax on a \$28 item is \$1.54. What is the sales tax rate (percent)?

44. The sum of the three angles of any triangle is 180°. If one angle measures 39° and the second angle measures 78°, what is the measure of the third angle?

45. The sum of three consecutive odd integers is -75. What are the three integers?

46. Elena earns \$2.25 per hour when she babysits for the O'Toole children. If she babysat for 6 1/2 hours, how much money did she earn?

47. If Tom can drive 341 miles in 5 1/2 hours, how far can he drive in 1 hour?

48. A 10 3/4-ounce can of soup will be mixed with an equal amount of water, heated, and then divided into five equal servings. How many ounces will there be in each serving?

49. Using the formula $A = \pi r^2$, find the area of a circle whose radius is 10 inches. Use $\pi \doteq 3.14$.

50. Find the dimensions of a rectangle whose perimeter is 110 yards if the length is 4 times the width.

13

Expressions, Equations, and Inequalities

13.1   Introduction

To solve more complex first-degree equations, you will need to bring together many of the concepts you mastered in Chapters 11 and 12 as well as several new ones. You will also use many of these concepts again to solve inequalities.

Learning Objectives

When you have completed this chapter, you should be able to:

1. Simplify expressions using the Distributive Principle.
2. Apply the Order of Operations Agreement to simplify expressions that include more than one set of grouping symbols.
3. Follow the Steps to Solve Equations procedures to solve and check first-degree equations that contain grouping symbols.
4. Solve first-degree inequalities, and graph the solution sets on number lines.

13.2   Distributive Principle

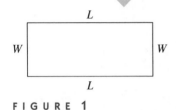

FIGURE 1

Do you remember the formula for finding the perimeter of a rectangle (Figure 1)? It was stated in several different ways:

$$P = L + W + L + W$$

or

$$P = 2L + 2W$$

It could also have been written

$$P = 2(L + W)$$

Let us compare the last two forms of the formula and check to see that they are equivalent. If $L = 8$ feet and $W = 5$ feet, then

$P = 2(L + W)$	or	$P = 2L + 2W$
$P = 2(8 + 5)$		$P = 2(8) + 2(5)$
$P = 2(13)$		$P = 16 + 10$
$P = 26$ feet		$P = 26$ feet

You should conclude that $2(L + W)$ and $2L + 2W$ are the same; they are equivalent statements.

$$2(L + W) = 2L + 2W$$

This example illustrates a property of numbers called the **Distributive Principle**. This rule concerns multiplication of a sum or difference. The Distributive Principle indicates that you get the same answer when you multiply first and then add (or subtract) as you get when you add (or subtract) first and then multiply. Read the examples that follow, remembering that when you multiply, you must multiply every term inside the parentheses by the factor that is outside.

E X A M P L E 1   Are $3(7 + 4)$ and $3(7) + 3(4)$ equal?

Solution   Add. $3(7 + 4)$ $3(7) + 3(4)$ Multiply.
Multiply. $3(11)$ $21 + 12$ Add.
                      33   33

Therefore, $3(7 + 4) = 3(7) + 3(4)$.   ◆

E X A M P L E 2   Are $-5(4 + 9 - 3)$ and $(-5)(4) + (-5)(9) - (-5)(3)$ equal?

Solution   Add. $-5(4 + 9 - 3)$ $(-5)(4) + (-5)(9) - (-5)(3)$ Multiply.
Subtract. $-5(13 - 3)$ $-20 + (-45) - (-15)$ Add.
Multiply. $-5(10)$ $(-65) - (-15)$ Subtract.
                  -50 $(-65) + (+15)$ Combine.
                                        -50

Therefore, the expressions are equivalent.   ◆

Distributive Principle

> For any numbers represented by a, b, and c,
>
> $$a(b + c) = ab + ac$$
>
> and
>
> $$a(b - c) = ab - ac$$

Here are two samples of problems that make use of the Distributive Principle.

$$3(7 - 5) \qquad -5(6 + 9)$$
$$= 3(7) - 3(5) \qquad = -5(6) + (-5)(9)$$
$$= 21 - 15 \qquad = -30 + (-45)$$
$$= 6 \qquad = -75$$

In many algebraic expressions involving the Distributive Principle, you will not be able to do the addition or subtraction first because the terms will not be like terms. In $-3(x + 2)$, the x and the 2 cannot be added together; therefore, the only operation that can be done is multiplication. $-3(x + 2) = (-3)(x) + (-3)(2) = -3x + (-6)$. Using the definition of subtraction, you can write this expression in a simpler form as $-3x - 6$. You can use a short cut when performing the distributive multiplication. When multiplying $-3(x + 2)$, treat the operation sign with the 2 as if it were a signed-number sign. You will be multiplying $(-3)(+x)$ and $(-3)(+2)$. You should get $-3x - 6$, using the signed-number answer with 6 as if it were the operation sign. It sounds more confusing than it is! The following examples illustrate the short cut.

E X A M P L E   3   Simplify: $6(5 - 2x)$

Solution $6(5 - 2x)$ $\qquad (+6)(+5)$ and $(+6)(-2x)$
$= 30 - 12x$ ◆

E X A M P L E   4   Simplify: $-11(-4y + 5)$

Solution $-11(-4y + 5)$ $\qquad -11(-4y)$ and $-11(+5)$
$= 44y - 55$ ◆

E X A M P L E   5   Simplify: $8(5P - 3R + 7T - 4)$

Solution $8(5P - 3R + 7T - 4)$ $\qquad 8(5P)$, $8(-3R)$, $8(+7T)$, and $8(-4)$
$= 40P - 24R + 56T - 32$ ◆

E X A M P L E   6   Simplify: $-(x - 5)$

Solution $-(x - 5)$ $\qquad$ Write a 1 before the parentheses and multiply through by -1:
$= -1(x - 5)$ $\qquad (-1)(+x)$ and $(-1)(-5)$
$= -1x + 5$
or $-x + 5$ ◆

In the examples and practice problems in this section, you are *simplifying* algebraic expressions. Keep in mind that they are expressions, not equations. Equations can be solved; expressions (no equals sign) can only be simplified. We simplify by removing grouping symbols and combining like terms.

$3(y + 2)$ is not the same as $3(y + 2) = 15$.

$3(y + 2) = 15$ can be solved to find a value for y, whereas $3(y + 2)$ can only be simplified.

13.2 ◆ Practice Problems

A. Are these expressions equivalent? Prove your answer by simplifying each problem and showing whether the answers are the same. Use the following example.

$$3(2 + 5) \quad \text{and} \quad 3(2) + 3(5)$$
$$3(7) \qquad\qquad 6 + 15$$
$$21 \qquad\qquad\quad 21$$

Therefore, $3(2 + 5) = 3(2) + 3(5)$ because $21 = 21$.

1. $-11(1 + 3)$ and $-11(1) + (-11)(3)$
2. $5(8 - 5)$ and $5(8) - 5(-5)$
3. $4(10 - 6)$ and $4(10) - 4(6)$
4. $-9(7 + 5)$ and $(-9)(7) + (-9)(5)$
5. $\dfrac{2}{3}(9 - 6)$ and $\dfrac{2}{3}(9) - \dfrac{2}{3}(6)$
6. $\dfrac{3}{4}(16 - 12)$ and $\dfrac{3}{4}(16) - \dfrac{3}{4}(12)$
7. $-3(8 - 11)$ and $-3(8) - (-3)(8)$
8. $-2(15 - 24)$ and $(-2)(15) - (-2)(24)$
9. $0.4(1.7 - 1.2)$ and $0.4(1.7) - (0.4)(1.2)$
10. $1.5(2.1 - 3.6)$ and $1.5(2.1) - 1.5(3.6)$

B. Simplify the following expressions using the Distributive Principle.

11. $-4(7 - 9)$
12. $3(5 + 6)$
13. $\dfrac{1}{2}(4 + 6)$
14. $-5(9 - 13)$
15. $7(4 + 5)$
16. $\dfrac{1}{4}(8 + 12)$
17. $3(X + 2)$
18. $-2(y + 4)$
19. $-5(m + 4)$
20. $7(R + 5)$
21. $6(K - 2)$
22. $-5(X - 3)$
23. $-8(Y - 4)$
24. $15(B - 3)$
25. $17(2x - 3)$
26. $8(9M - 5)$
27. $3.2(5N + 2.1P)$
28. $-11(4S - 3R + 6)$

29. $-12(3A - 5B + 2C)$
30. $1.1(4.2G - 3.2H)$
31. $\dfrac{3}{8}(24A - 16C)$
32. $\dfrac{5}{6}(18x - 30y)$
33. $-\dfrac{1}{3}(12M + 9N)$
34. $-\dfrac{1}{5}(25A + 90B)$

C. Use your calculator to simplify the following expressions.

35. $0.65(3.8F - 0.06 + 29.5F)$
36. $384(1,952X + 641 - 923X)$
37. $-5.6(11.2 - 18.2W + 39.8X)$
38. $0.07(7.5C + 2.8 - 11.7E)$
39. $915(3,497M + 9,567N)$
40. $-52.7(16.7X - 10.9)$

13.3 Simplifying Expressions That Contain Grouping Symbols

Grouping symbols such as parentheses, ( ), and brackets, [ ], are used in algebraic expressions or equations to show that terms are related to each other or that they have the same operation performed on them. For example, in $3(x + 2)$, the parentheses indicate that both x and 2 are to be multiplied by 3. In $6 - (Y + 4)$, the parentheses show that the entire quantity $Y + 4$ must be subtracted from 6. Without the parentheses in this example, $6 - Y + 4$, only the Y would be subtracted and the 4 would be added. Any time you wish to multiply, subtract, or divide a quantity that contains more than one term, you *must* enclose that quantity in parentheses.

EXAMPLE 1 Is $3(x + 2)$ equal to $3x + 6$?

Solution $3(x + 2)$ is equal to $3x + 6$, because the parentheses say to multiply everything inside by 3. Without the parentheses, $3x + 2$, only the x would be multiplied by 3. The answers would not be the same.

Therefore, $3(x + 2) \neq 3x + 2$.

EXAMPLE 2 Is $6 - (y + 4)$ equal to $2 - y$?

Solution $6 - (y + 4)$ is the same as $2 - y$ because:

$= 6 - [(+y) + (+4)]$ Rewrite subtraction.

$= 6 + (-y) + (-4)$ Rearrange terms.

$= 6 + (-4) + (-y)$ Combine like terms.

$= 2 + (-y)$ Definition of subtraction.

$= 2 - y$

Therefore, $6 - (y + 4) = 2 - y$. If the parentheses had not been there, the problem would have been $6 - y + 4$, which is equal to $10 - y$.

In Example 2, the process of doing the subtraction was very awkward. You can get the same result by applying the Distributive Principle. Because multiplying by -1 and subtracting both result in changed signs, you can write a 1 in front of $(y - 4)$ and treat it as a distributive problem.

$6 - (y + 4)$

$= 6 - 1(y + 4)$ Multiply by -1: $(-1)(y)$ and $(-1)(+4)$.

$= 6 - y - 4$ Combine like terms.

$= 2 - y$

This process enables you to remove the parentheses and make the sign changes required in subtraction, but it is less complicated to do. Compare these processes:

Rewrite Subtraction Process   **Distributive Process**

$$\overset{\text{S}}{7 - (3x - 4)}$$

$$7 - (3x - 4)$$

$$= 7 - \overset{\text{A}}{[3x + (-4)]}$$

$$= 7 - 1(3x - 4)$$

$$= 7 - \overset{\text{S}}{[3x + (-4)]}$$

$$= 7 - 3x + 4$$

$$= 7 + \overset{\text{A}}{(-3x) + (+4)}$$

$$= 7 + 4 - 3x$$

$$= 7 + (+4) + (-3x)$$

$$= 11 - 3x$$

$$= 11 + (-3x)$$

$$= 11 - 3x$$

In the examples that follow, the Distributive Principle is used to remove parentheses, and then the like terms are combined. In Section 12.7, you learned that you can combine terms by using the operation signs as direction signs. In the rest of the examples, this short cut is used when like terms are combined.

EXAMPLE 3   Simplify: $5(A - B) + 4(A + B)$

Solution $5(A - B) + 4(A + B)$ Multiply $5(A)$, $5(-B)$, $4(A)$, and $4(B)$.

$$= 5A - 5B + 4A + 4B$$ Combine like terms: $5A$ and $+4A$, $-5B$ and $+4B$.

$$= 9A - 1B \quad \text{or} \quad 9A - B$$ ◆

EXAMPLE 4   Simplify: $4(8x - 3) - 2(5 + 2x)$

Solution $4(8x - 3) - 2(5 + 2x)$ Multiply $4(8x)$, $4(-3)$, $-2(5)$, and $-2(2x)$.

$$= 32x - 12 - 10 - 4x$$ Combine like terms.

$$= 28x - 22$$ ◆

EXAMPLE 5   Simplify: $4(y - 2) - (6y + 5)$

Solution $4(y - 2) - (6y + 5)$ Write 1 before the second parentheses.

$$= 4(y - 2) - 1(6y + 5)$$ Multiply $4(y)$, $4(-2)$, $-1(6y)$, and $-1(+5)$.

$$= 4y - 8 - 6y - 5$$ Combine terms.

$$= -2y - 13$$ ◆

EXAMPLE 6   Simplify: $3 - 2(5A - 4)$

Solution $3 - 2(5A - 4)$ Multiply $(-2)(5A)$ and $(-2)(-4)$.

$$= 3 - 10A + 8$$ Combine like terms.

$$= 11 - 10A$$ ◆

E X A M P L E **7**   Simplify: $3M + (5M - 4)$

Solution $3M + (5M - 4)$ Write in 1 and multiply through by $+1$.

$= 3M + 1(5M - 4)$
$= 3M + 5M - 4$ Combine like terms.
$= 8M - 4$ ◆

E X A M P L E **8**   Simplify: $9(2z - 6) - 4(3z - 8)$

Solution $9(2z - 6) - 4(3z - 8)$
$= 18z - 54 - 12z + 32$
$= 6z - 22$ ◆

E X A M P L E **9**   Simplify: $5(3r + 4 - 8r) - (11 + 2r)$

Solution $5(3r + 4 - 8r) - (11 + 2r)$
$= 5(3r + 4 - 8r) - 1(11 + 2r)$
$= 15r + 20 - 40r - 11 - 2r$
$= -27r + 9$ ◆

Exercises   Simplify these problems and check your answers below.

1. $15 - (m + 3)$ **2.** $2(x - 3) + 5(2x + 4)$
3. $-4(B - 8) - 2(5B + 4 - 7B)$

Answers:

1. $-m + 12$ **2.** $12x + 14$ **3.** 24

Expressions that contain grouping symbols within another set of grouping symbols require the use of different kinds of symbols so that you can tell where one group begins and the other ends. In addition to parentheses ( ), brackets [ ] and braces { } are often used.

When simplifying expressions that contain grouping symbols within other grouping symbols, always begin with the *innermost* set of symbols and work outward.

$7 - [4n - 5(n + 6)]$ First eliminate the parentheses by multiplying through by -5.

$= 7 - 1[4n - 5n - 30]$ Eliminate the brackets by multiplying through by -1.
$= 7 - 4n + 5n + 30$ Combine like terms.
$= 37 + n$

As you simplify expressions like this, be sure to rewrite all parts of the expression that you have not worked with yet. Also, in the previous problem, you could combine like terms after the first step and have a simpler expression to multiply by -1.

$7 - [4n - 5(n + 6)]$ Multiply by -5 to eliminate parentheses first.
$= 7 - [4n - 5n - 30]$ Combine inside the brackets.
$= 7 - [-n - 30]$ Now multiply by -1.

$= 7 - 1[-n - 30]$
$= 7 + n + 30$ Combine like terms.
$= 37 + n$

E X A M P L E   **10**   Simplify: $5[7x + 4(3x - 6)]$

Solution

$5[7x + 4(3x - 6)]$ Multiply by $+4$.
$= 5[7x + 12x - 24]$

Combine inside next.   or   Multiply by 5 next.
$= 5[19x - 24]$ $= 35x + 60x - 120$
$= 95x - 120$ $= 95x - 120$

The result is the same.   ◆

E X A M P L E   **11**   Simplify: $6 - \{12[y + 3(2y - 4)]\}$

Solution

$6 - \{12[y + 3(2y - 4)]\}$ Multiply by $+3$ to remove parentheses first.
$= 6 - \{12[y + 6y - 12]\}$
$= 6 - \{12[7y - 12]\}$ or $= 6 - \{12y + 72y - 144\}$
$= 6 - \{84y - 144\}$ $= 6 - 12y - 72y + 144$
$= 6 - 84y + 144$ $= 150 - 84y$
$= 150 - 84y$ ◆

13.3 ◆ Practice Problems

Simplify the following expressions by removing the grouping symbols and combining like terms.

1. $11 - 6(x + 2)$
2. $2(y + 4) + 3(2y - 5)$
3. $3(2m - 5) + 4(m + 1)$
4. $7 - 5(k - 3)$
5. $5C - (C + 2)$
6. $2(3A - B) - (A - B)$
7. $-3(y + 2) - 2(y - 5)$
8. $6R - (2R - 4)$
9. $5(s + 2) - (s - 3)$
10. $-2(T - 5) - 4(2T + 3)$
11. $3X - 2(X - 4)$
12. $5 - 2(w + 2)$
13. $-2(m - 4) + 3(2m - 1)$
14. $2(x + 4) - (2x - 1)$
15. $8B - (B + 2)$
16. $6V - (2 - 5V)$
17. $15 - 2(y - 3)$
18. $-3(R + 1) - 5(2R - 3)$
19. $9 - (6c + 4)$
20. $18 - (7 + w)$
21. $3[4m - 2(m + 3)]$
22. $7 - 2[4N - (3 + 2N)]$
23. $9 - [6 - (S + 3)]$
24. $2[7Y - 3(2Y - 1)]$
25. $12 + [5F - 3(F + 2)]$
26. $6[3(2A + 1)]$
27. $2[3(4x - 1)]$
28. $15 + [3N - 2(4N - 3)]$
29. $11 - [5(w + 4) - 6]$
30. $5[7(R - 4) + 9]$
31. $-2[3y - (5y - 2)]$
32. $2A - 3[A - 2(4 - A)]$
33. $6(2K - 7) - 3(3 - 2K)$
34. $12(y - 2) + 2(3y - 7)$
35. $4(a - 3b) - 3(a + b)$
36. $-3[2x + (x - 7)]$
37. $-5[3g + 2(5 - g)]$
38. $3(a - 2b) - 5(4a + 7b)$
39. $2(y - 3x) + 2(7x - y)$
40. $-2[3H - (5H - 4)] + 3H$
41. $2\{4[c - 2(c - 3)]\}$
42. $3\{2[p + 2(p + 7)]\}$
43. $-3\{6p - 2[2p - (p + 1)]\}$
44. $-2\{7x - 2[3x + 2(x + 4)]\}$
45. $-\{2w - 3[w - 2(4 - w)]\}$
46. $-\{-5a + 2[3a - 7(3 - 2a)]\}$

13.4   Equations with Grouping Symbols

To solve more complex first-degree equations, you will use the concepts you learned in the two previous sections. You will also have to recall the procedures for solving equations. All of the following processes will be applied in solving the equations in this chapter.

1. When simplifying expressions that have more than one set of grouping symbols, begin by simplifying the innermost set and then work outward.

2. In expressions such as $-5(x - 2)$, apply the Distributive Principle and multiply each term inside the parentheses by -5.

3. After eliminating all grouping symbols, follow these Steps to Solve Equations:

   a. If desired, multiply every term in the equation by a number that will eliminate the fractions or decimals.

   b. Combine like terms on each side of the equals sign.

   c. Do steps that require addition or subtraction of variable and constant terms.

   d. Do steps that require multiplication or division.

   e. Always check answers by substituting the answer into the *original* equation.

4. Apply appropriate signed-number operation rules to all the foregoing steps.

Study the examples that follow to see where each step is used.

E X A M P L E 1   Solve and check: $3(x + 2) + 5 = 38$

Solution

$$3(x + 2) + 5 = 38 \qquad \text{Distributive Principle.}$$

$$3x + 6 + 5 = 38 \qquad \text{Combine like terms.}$$

$$3x + 11 = 38 \qquad \text{Do addition or subtraction.}$$

$$\underline{\; -11 \quad -11}$$

$$3x = 27$$

$$\div 3 \left| \frac{3x}{3} = \frac{27}{3} \right| \div 3 \qquad \text{Do multiplication or division.}$$

$$x = 9$$

Check

$$3(x + 2) + 5 = 38$$

$$3(9 + 2) + 5 = 38$$

$$3(11) + 5 = 38$$

$$33 + 5 = 38$$

$$38 = 38 \quad \text{True.}$$

◆

E X A M P L E **2**   Solve and check: $-24 = 3[2t - 4(t - 6)]$

Solution $-24 = 3[2t - 4(t - 6)]$ Remove parentheses by multiplying by -4.

$-24 = 3[2t - 4t + 24]$ Remove brackets by multiplying by 3.

$-24 = 6t - 12t + 72$ Combine like terms.

$-24 = -6t + 72$ Subtract 72.

$\dfrac{-72 \qquad\ -72}{-96 = -6t}$ Divide by -6.

$\dfrac{-96}{-6} = \dfrac{-6t}{-6}$

$16 = t$

Check $-24 = 3[2t - 4(t - 6)]$

$-24 = 3[2(16) - 4(16 - 6)]$

$-24 = 3[32 - 4(10)]$

$-24 = 3[32 - 40]$

$-24 = 3[-8]$

$-24 = -24$ True.

◆

When solving an equation, be sure that the final step shows the variable with a positive one ($+1$) coefficient. In the previous problem, both sides were divided by the coefficient and its sign: $-6t/-6$ and $-96/-6$. This resulted in $(+1)t = 16$, but with the $+1$ understood. Never leave the coefficient of the variable as anything but $+1$. You have not finished solving until it is positive one. Follow through Example 3 to see this process.

E X A M P L E **3**   Solve and check: $2(2k + 1) - 5(k - 2) = 36$

Solution $2(2k + 1) - 5(k - 2) = \quad 36$

$4k + 2 - 5k + 10 = \quad 36$

$-k + 12 = \quad 36$

$\dfrac{\qquad\quad -12 \quad -12}{-k \qquad = \quad 24}$

You are not finished because the coefficient of k is -1.

$\dfrac{-1k}{-1} = \dfrac{24}{-1}$

$k = -24$

Check $2(2k + 1) - 5(k - 2) = 36$

$2[2(-24) + 1] - 5(-24 - 2) = 36$

$2[-48 + 1] - 5(-26) = 36$

$2[-47] - 5(-26) = 36$

$-94 + 130 = 36$

$36 = 36$ True.

◆

E X A M P L E　4　Solve and check: $7x - (x + 2) = -2(x - 7)$

Solution
$$7x - (x + 2) = -2(x - 7)$$
$$7x - 1(x + 2) = -2(x - 7)$$
$$7x - x - 2 = -2x + 14$$
$$6x - 2 = -2x + 14$$

> You could eliminate any one of the terms now. This is just one choice.

$$\underline{+2x \qquad\quad +2x}$$
$$8x - 2 = \qquad\quad 14$$
$$\underline{ + 2 \qquad\quad + 2}$$
$$8x \quad = \qquad\quad 16$$
$$\frac{8x}{8} = \frac{16}{8}$$
$$x = 2$$

Check
$$7x - (x + 2) = -2(x - 7)$$
$$7(2) - (2 + 2) = -2(2 - 7)$$
$$14 - 4 = -2(-5)$$
$$10 = 10 \quad \text{True.}$$

◆

E X A M P L E　5　Solve and check: $2(y - 4) = 7 + 2y$

Solution
$$2(y - 4) = 7 + 2y$$
$$2y - 8 = 7 + 2y$$
$$\underline{-2y \qquad\qquad -2y}$$
$$-8 = 7 \quad \text{No solution}$$

Because this can never be true, $-8 \neq 7$, there is no solution to this problem. When the variable disappears from the problem and a false statement (such as $-8 = 7$) is left, the equation has no solution. In other words, it has an empty solution set, which is represented as $\{\ \}$ or ϕ (null set).　◆

E X A M P L E　6　Solve and check: $2x + 4 = 2(2 + x)$

Solution
$$2x + 4 = 2(2 + x)$$
$$2x + 4 = 4 + 2x$$
$$\underline{-2x \qquad\qquad -2x}$$
$$4 = 4$$

When the variable disappears and a true statement (an identity such as $4 = 4$) results, then any and all real numbers could be the solution to the equation. The solution is stated as "all real numbers."

Test a few values such as $x = 1$ and $x = -8$.

$$2x + 4 = 2(2 + x) \qquad\qquad 2x + 4 = 2(2 + x)$$
$$2(1) + 4 = 2(2 + 1) \qquad\qquad 2(-8) + 4 = 2[2 + (-8)]$$
$$2 + 4 = 2(3) \qquad\qquad -16 + 4 = 2[-6]$$
$$6 = 6 \qquad\qquad -12 = -12$$

Any real number would produce a true statement.　◆

E X A M P L E   **7**   Solve and check: $2/3(9R - 3) + 5 = R - 4$

Solution $2/3(9R - 3) + 5 = R - 4$

You can either eliminate the denominator or work with the fractions. This example is solved with the fractions, because when you distribute by $2/3$, $2/3(9R)$ and $2/3(-3)$ produce coefficient and constant terms that are whole numbers.

$$2/3(9R - 3) + 5 = \quad R - 4$$
$$6R - 2 + 5 = \quad R - 4$$
$$6R + 3 = \quad R - 4$$
$$\underline{- R \qquad\quad - R}$$
$$5R + 3 = - 4$$
$$\underline{- 3 \quad - 3}$$
$$5R \quad = - 7$$
$$\frac{5R}{5} = \frac{-7}{5}$$
$$R = -7/5 \text{ or } -1.4$$

Check   These checks are more difficult when fractions are involved, but don't give up! You have the skills to do them correctly.

$$2/3(9R - 3) + 5 = R - 4$$
$$2/3[9(-7/5) - 3] + 5 = -7/5 - 4$$
$$2/3[-63/5 - 3] + 5 = -7/5 - 4$$

Before going further, change the whole numbers to fractions with a denominator of 5: $3/1 = 15/5$, $4/1 = 20/5$, and $5/1 = 25/5$.

$$2/3[-63/5 - 15/5] + 5 = -7/5 - 20/5$$
$$2/3[-78/5] + 5 = -27/5$$
$$-52/5 + 25/5 = -27/5$$
$$-27/5 = -27/5 \quad \text{True.}$$

◆

13.4   ◆   Practice Problems

A. Solve each equation and check your answer by substituting into the *original* equation.

1. $3(x + 2) = 18$
2. $9 + 2(T + 1) = 27$
3. $5 - (C - 2) = -4$
4. $5(y - 1) = 40$
5. $24 = -5(P - 5)$
6. $14 - (2A - 1) = 7$
7. $8 + 3(N + 1) = 38$
8. $6 + 2/3(p + 8) = 6$
9. $12 = 2(A - 3)$
10. $2(x - 1) = 3(2x + 3)$
11. $-3(M + 2) = -12$
12. $1 - (2c + 5) = 10$
13. $8(w - 2) = w + 5$
14. $-6 = 3(B - 4)$
15. $1 + 3(c - 1) = 10$
16. $4 - 2(K + 1) = 16$
17. $6(N - 3) = 3(N + 1)$
18. $-5(r + 2) = 25$
19. $1/2 + 3/4(a + 6) = 29$
20. $7(5m + 3) = -49$

B. Solve and check. A calculator will be helpful, especially when you are doing the checks.

21. $18 = 3[3R - 2(R + 5)]$
22. $4(x + 2) + 5 = -35$
23. $3(y + 5) + 8 = 44$
24. $8(R - 2) = 4(3 + 2R)$
25. $-12 = 2[K - 3(K + 2)]$
26. $8(K - 2) - (2K + 2) = -24$
27. $6T - 1 = 3(2T + 5)$
28. $-2[5M + 2(3M - 1)] = 70$
29. $21 = -3[2A + 3(2A + 3)]$
30. $4(Y - 2) - 3(2Y + 5) = -1$
31. $12x - 1/2(4x - 6) = 28$
32. $7a - (3a - 4) = 12 - 4a$

33. $7m - 4(2m - 3) = 3(11 + 2m)$

34. $3/4(8y - 4) = 3(2y - 1)$

35. $6R - 2(R - 2) = 4(R - 1)$

36. $5 - 1/3(9 - 6Y) = 16 - 5Y$

37. $2T - 5 = 4(3T + 1) - 19$

38. $-36 = -2(8T - 1) + 3(T - 4)$

39. $1.6 = 0.5(3.2R + 4.8)$

40. $6Y - 2(5 + 3Y) = 15$

41. $11m = 5(m + 2) - (1 - 6m)$

42. $0.3(4.5x + 0.6) = 5.85(x - 2)$

43. $1/3(6x - 15) = 2x - 5$

44. $2m + 3 = 8m - 3(2m - 1)$

13.5 Solving Inequalities

Let's review the definition of inequality that was introduced in Section 9.6. On the number line, larger numbers are farther to the right than smaller numbers. Therefore, $8 > 3$ (8 is greater than 3) because 8 is farther to the right than 3, and $-6 < 3$ because 3 is farther to the right than -6.

The points on the number line correspond to all the *real* numbers, not just to the integers shown. The set of real numbers is made up of the *rational numbers* (integers, and fractions with integer numerators and denominators) and the *irrational numbers* (numbers such as $\sqrt{2}$, π, and $\sqrt{5}$ that are non-repeating and non-terminating in decimal form). Examples of rational numbers include -6, $4/5$, and $2.333\ldots$, and examples of irrational numbers include $\sqrt{2} = 1.4142135\ldots$ and $\pi = 3.14159265\ldots$.

An inequality such as $x > 3$ includes *all* real numbers (all points on the number line) greater than 3. This is illustrated by:

The open circle shows that 3 is not included in the answer (solution set), but all numbers to the right of that point are included.

If the statement had been $x \geq 3$, the graph would have had a filled-in circle around 3.

The line with an arrowhead extending to the right means that it continues in that direction forever. There are an infinite number of answers to the problem. Unlike the first-degree equations that have only one answer, first-degree inequalities have many answers. Graphing on the number line is one way to show the solution set. Another way to show it is to use set notation. Let's go back to the statement $x > 3$. In set notation, the solution is $(x \mid x > 3)$. This is read "the set of all real numbers represented by x such that x is greater than 3."

Examine the following examples of algebraic inequalities, solution set notation, and graphs.

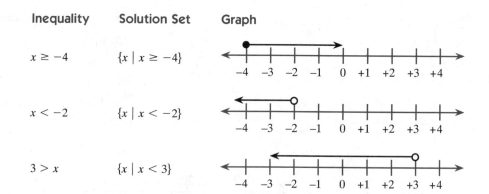

Inequality	Solution Set	Graph
$x \geq -4$	$\{x \mid x \geq -4\}$	
$x < -2$	$\{x \mid x < -2\}$	
$3 > x$	$\{x \mid x < 3\}$	

In the last example, did you notice that $3 > x$ and $x < 3$ express the same relationship? That is, "3 is greater than x" and "x is less than 3" are graphed the same way. You may find it easier to turn around inequalities such as $3 > x$ when you work with them. The inequality $5 < y$ might be easier to work with if it is stated as $y > 5$.

Both set notation and graphs will be required when you solve the algebraic inequalities in this section. To understand how to solve inequalities, look at the solutions to the equation and the inequality that follow. Do you see any difference in procedure?

Equation	Inequality
$2x + 5 = 17$	$2x + 5 > 17$
$\underline{-5 \quad -5}$	$\underline{-5 \quad -5}$
$2x = 12$	$2x > 12$
$\dfrac{2x}{2} = \dfrac{12}{2}$	$\dfrac{2x}{2} > \dfrac{12}{2}$
$x = 6$	$x > 6$

There was no difference in the steps for the equation and the inequality. The same number, 5, was subtracted from both sides of the symbol, and then both sides were divided by 2, the coefficient of the variable term.

The addition and subtraction properties are the same for equations and inequalities.

Addition–Subtraction Property of Equality

For all real numbers a, b, and c, if $a = b$, then

$$a + c = b + c$$

and

$$a - c = b - c$$

Addition–Subtraction Property of Inequality

For all real numbers a, b, and c, if $a > b$, then

$$a + c > b + c$$

and

$$a - c > b - c$$

The following arithmetic (no variables) examples show that these addition and subtraction properties for inequalities are true.

$$6 > 3 \qquad\qquad\qquad 14 < 20$$
$$6 + 2 > 3 + 2 \qquad\qquad 14 - 5 < 20 - 5$$
$$8 > 5 \quad \text{True.} \qquad\qquad 9 < 15 \quad \text{True.}$$

E X A M P L E   **1**   Solve $7 > x + 5$ and graph the solution.

Solution

$$7 > x + 5$$
$$\underline{-5 \qquad -5}$$
$$2 > x \qquad\qquad \{x \mid x < 2\}$$

Remember that if 2 is greater than x, then x is also less than 2. $2 > x$ and $x < 2$ are the same.

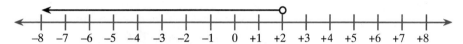

Check   You would not check the solution by substituting 2 into the original problem because x does not equal 2. Instead, pick any value that is *less than* 2, substitute it into the original inequality, and see if you get a true statement. Because you can choose any value in the solution set, look for one that is easy to use.

$$7 > x + 5 \qquad \text{The solution was } x < 2, \text{ so try } 0.$$
$$7 > 0 + 5$$
$$7 > 5 \quad \text{True.}$$ ◆

E X A M P L E   **2**   Solve $x + 2 > -3$ and graph the solution.

Solution $x + 2 > -3$
$$\underline{-2 \qquad -2}$$
$$x \qquad > -5 \qquad\qquad \{x \mid x > -5\}$$

Check $x + 2 > -3$ Use 0 to check, because $0 > -5$.
$$0 + 2 > -3$$
$$2 > -3 \quad \text{True.}$$ ◆

Now let's look at some more arithmetic examples that involve multiplication and division to see if there is a pattern.

$$4 > -2 \qquad\qquad\qquad 15 < 35$$

Multiply both sides by 3.   Divide both sides by 5.
$$3(4) > 3(-2) \qquad\qquad\qquad 15/5 < 35/5$$
$$12 > -6 \quad \text{True.} \qquad\qquad\qquad 3 < 7 \quad \text{True.}$$

$$4 > -2 \qquad\qquad\qquad 15 < 35$$

Multiply both sides by -3.   Divide both sides by -5.
$$-3(4) > -3(-2) \qquad\qquad\qquad 15/(-5) < 35/(-5)$$
$$-12 > +6 \quad \text{False.} \qquad\qquad -3 < -7 \quad \text{False.}$$

It appears that multiplying or dividing both sides of an inequality by a *positive* number does not affect the sense, or direction, of the inequality. However, multiplying or dividing by a *negative* number produces a false statement. In fact, the symbols in those answers are just the opposite of what they should be to give a true statement.

$$-12 > +6 \quad \text{False.} \qquad\qquad -3 < -7 \quad \text{False.}$$

$$\text{But } -12 < +6 \quad \text{True.} \qquad \text{But } -3 > -7 \quad \text{True.}$$

It is important to remember that multiplying or dividing both sides of an inequality by a negative number reverses the direction of the inequality.

Multiplication and Division Properties of Inequalities

> **1.** For all real numbers a, b, and c, with $c > 0$:
>   If $a > b$, then $ac > bc$.
>   If $a > b$, then $a/c > b/c$.
> **2.** For all real numbers a, b, and c, with $c < 0$:
>   If $a > b$, then $ac < bc$.
>   If $a > b$, then $a/c < b/c$.

Do you see that the difference between the initial statements is whether c is a positive (> 0) or a negative (< 0) number? These properties could have been expressed with $\geq$, $<$, or $\leq$ in the $a > b$ statement. For example,

For all real numbers a, b, and c, with $c < 0$:

If $a \leq b$, then $ac \geq bc$.

E X A M P L E **3**   Solve and graph: $4x - 3 \leq 15$

Solution

$$
\begin{array}{rcl}
4x - 3 & \leq & 15 \\
\underline{+\ 3} & & \underline{+\ 3} \\
4x & \leq & 18 \\
\dfrac{4x}{4} & \leq & \dfrac{18}{4} \\
x & \leq & 4.5 \qquad \{x \mid x \leq 4.5\}
\end{array}
$$

Check

$$
\begin{array}{rl}
4x - 3 \leq 15 & \qquad \text{Let } x = 0. \\
4(0) - 3 \leq 15 & \\
0 - 3 \leq 15 & \\
-3 \leq 15 & \quad \text{True.}
\end{array}
$$

E X A M P L E **4**   Solve and graph: $5 - 3x > 11$

Solution $5 - 3x >\ \ 11$

$\underline{-5 \qquad\quad -\ 5}$

$-3x >\ \ \ 6$

$\dfrac{-3x}{-3} < \dfrac{6}{-3}$ Divide by -3, which reverses the symbol.

$x < -2$ $\{x \mid x < -2\}$

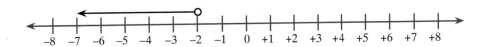

Check $5 - 3x > 11$ Let $x = -4$.

$5 - 3(-4) > 11$

$5 + 12 > 11$

$17 > 11$ True.

To solve more complex-looking inequalities, we use the same procedures we apply to equations that require more steps. Be especially careful with steps that require multiplication or division of *both* sides of the inequality by a negative number.

E X A M P L E **5**   Solve and graph: $3x + 6 > x - 4$

Solution $3x + 6 >\ \ x -\ \ 4$

$\underline{-\ x \qquad\quad -x}$

$2x + 6 >\ \ \ \ -\ 4$

$\underline{\ -\ 6 \qquad\quad -\ 6}$

$2x \quad\ > \qquad -10$

$\dfrac{2x}{2} > \dfrac{-10}{2}$

$x > -5$ $\{x \mid x > -5\}$

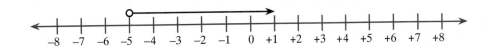

Check $3x + 6 > x - 4$ Try $x = 0$.

$3(0) + 6 > 0 - 4$

$0 + 6 > 0 - 4$

$6 > -4$ True.

E X A M P L E 6   Solve and graph: $-5(x - 3) + 3(x + 1) < 14$

Solution

$$-5(x - 3) + 3(x + 1) < 14$$
$$-5x + 15 + 3x + 3 < 14$$
$$-2x + 18 < 14$$
$$\underline{-18 \quad -18}$$
$$-2x < -4$$
$$\frac{-2x}{-2} > \frac{-4}{-2}$$
$$x > 2 \qquad\qquad \{x \mid x > 2\}$$

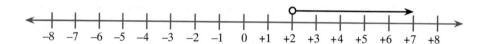

Check $-5(x - 3) + 3(x + 1) < 14$ Try $x = 4$.
$$-5(4 - 3) + 3(4 + 1) < 14$$
$$-5(1) + 3(5) < 14$$
$$-5 + 15 < 14$$
$$10 < 14 \quad \text{True.}$$

◆

E X A M P L E 7   Solve and graph: $\dfrac{1}{2}x + \dfrac{2}{3} < \dfrac{5}{6}$

Solution

$$\frac{1}{2}x + \frac{2}{3} < \frac{5}{6} \qquad\qquad \text{Multiply every term by the LCD of 6.}$$
$$6\left(\frac{1}{2}x\right) + 6\left(\frac{2}{3}\right) < 6\left(\frac{5}{6}\right)$$
$$3x + 4 < 5$$
$$\underline{-4 \quad -4}$$
$$3x < 1$$
$$\frac{3x}{3} < \frac{1}{3}$$
$$x < \frac{1}{3} \qquad\qquad \{x \mid x < 1/3\}$$

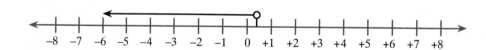

Check $\dfrac{1}{2}x + \dfrac{2}{3} < \dfrac{5}{6}$ Use $x = 0$.

$$\frac{1}{2}(0) + \frac{2}{3} < \frac{5}{6}$$

$$0 + \frac{2}{3} < \frac{5}{6} \quad \text{True, because } 2/3 = 4/6, \text{ and } 4/6 < 5/6.$$

◆

E X A M P L E **8** Graph the inequality $-1 < x \le 5$.

Solution This can be separated into two inequalities that you then graph on the same number line, including only the points that satisfy *both* parts.

$$-1 < x \quad \text{and} \quad x \le 5$$

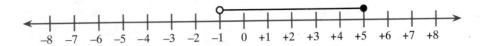

Check Try a point included in the graph, such as 0.

$$-1 < x \le 5, \quad -1 < 0 \le 5 \quad \text{True.}$$

◆

An inequality such as $0 \le x + 3 < 6$ is actually two inequalities, and the solution must satisfy both of them: $0 \le x + 3$ and $x + 3 < 6$. You could solve the two inequalities separately and then combine the solutions on the number line, including only the points that satisfy *both* parts of the inequality. Another method is to solve them at the same time, working on all parts of the inequality at the same time. Both methods are demonstrated in the example that follows.

E X A M P L E **9** Solve and graph: $0 < x + 3 < 6$

Solution Separate the inequalities.

$$0 < x + 3 \quad \text{and} \quad x + 3 < 6$$

Solve as if they were equations.

$$
\begin{array}{ll}
0 < x + 3 & x + 3 < 6 \\
\underline{-3 \quad -3} & \underline{-3 \quad -3} \\
-3 < x & x \quad < 3
\end{array}
$$

Combine the solutions on the number line, using only the points that satisfy both.

$$\{x \mid -3 < x < 3\}$$

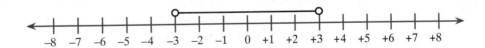

Solution Work with all parts at the same time.

$$
\begin{array}{ll}
0 < x + 3 < 6 & \text{Subtract 3 from all parts to isolate } x. \\
\underline{-3 \quad -3 \quad -3} & \\
-3 < x \quad < 3 & \{x \mid -3 < x < 3\}
\end{array}
$$

The graph must show all points that are greater than -3 *and are also* less than 3.

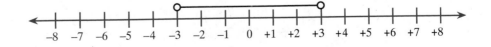

Check $0 < x + 3 < 6$ Try $x = 0$.

$0 < 0 + 3 < 6$

$0 < 3 < 6$ True, because 3 is greater than 0 *and* is also less than 6.   ◆

E X A M P L E   10   Solve and graph: $-4 \le 2(x + 3) < 1$

Solution $-4 \le 2(x + 3) < \quad 1$ Distribute 2.

$-4 \le 2x \;\; + 6 < \quad 1$ Subtract 6 from all parts.

$$\underline{\begin{array}{ccc} -\;6 & -\;6 & -6 \end{array}}$$

$-10 \le 2x \qquad < -5$

$\dfrac{-10}{2} \le \dfrac{2x}{2} \qquad < \dfrac{-5}{2}$ Divide all parts by 2.

$-5 \le x \qquad\quad < \dfrac{-5}{2}$ $\{x \,|\, -5 \le x < -5/2\}$

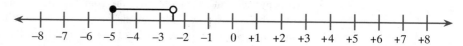

The graph shows all the points that are greater than or equal to -5 *and are also* less than $-5/2$.

Check $-4 \le 2(x + 3) < 1$ Let $x = -4$.

$-4 \le 2[(-4) + 3] < 1$

$-4 \le 2[-1] < 1$

$-4 \le -2 < 1$ True, because -2 is greater than or equal to -4 *and* is also less than 1.   ◆

E X A M P L E   11   Solve and graph: $-9 \le -2x + 7 < 13$

Solution   Separate the inequalities.

$-9 \le -2x + 7 \qquad\qquad -2x + 7 < \quad 13$ Subtract 7.

$$\underline{\begin{array}{cc} -7 & -\;7 \end{array}} \qquad\qquad \underline{\begin{array}{cc} -\;7 & -\;7 \end{array}}$$

$-16 \le -2x \qquad\qquad\quad -2x \quad < \quad 6$

$\dfrac{-16}{-2} \ge \dfrac{-2x}{-2} \qquad\qquad \dfrac{-2x}{-2} \quad > \dfrac{6}{-2}$ Divide by -2.

Remember to *reverse* both inequality symbols, because you are dividing by a *negative* number.

$8 \ge x \qquad\quad x > -3 \qquad \{x \,|\, 8 \ge x > -3\}$

The graph must show all values that are less than or equal to 8 *and are also* greater than -3.

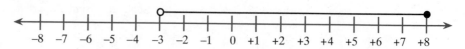

Check   Try a value between 8 and -3, such as 0.

$-9 \le -2x + 7 < 13$

$-9 \le -2(0) + 7 < 13$

$-9 \le 0 + 7 < 13$

$-9 \le 7 < 13$ Both parts are true.   ◆

13.5 ◆ Practice Problems

A. Graph each inequality on a number line.

1. $x < 0$ 2. $x \le -1$

3. $x \le -2$ 4. $x > 0$

5. $3 < x$ 6. $4 < x$

7. $-4 < x \le 4$ 8. $-3 < x < 3$

9. $0 < x \le 5$ 10. $-1 \le x < 4$

11. $-3 \le x \le 1$ 12. $-4 \le x \le 0$

B. Solve each inequality, state the answer in set notation (such as $\{x \mid x < 2\}$), and graph the solution on a number line.

13. $x - 3 \le 4$ 14. $5 - x > 4$

15. $2x + 5 > 17$ 16. $x + 7 \le 2$

17. $7 - x > 3$ 18. $4x - 3 \ge 17$

19. $3 - x < 3x - 13$ 20. $x + 2 < 3x - 10$

21. $5(2x - 1) \le 6(2x - 1)$

22. $x - (4x - 9) \ge 1 - 2(x + 4)$

23. $7 - (-6 - x) \ge x - (2x + 5)$

24. $-2(3x + 1) < 4(x - 6)$

25. $-7 \le 2x + 3 < 9$

26. $-5 < 3x + 1 < 7$

27. $2(x + 2) \ge -(x + 5)$

28. $x/4 + 1 \le 5(1 - x/4)$

29. $5(x/3 + 1) > x/3 - 1$

30. $2(2x - 1) < 6(x + 1)$

31. $-2 < 3x + 5 < 8$ 32. $-5 \le 2x - 1 < 3$

33. $3 \le 4 - x < 5$ 34. $-4 \le 8 - x \le 0$

35. $-4 \le \dfrac{2}{3}x - 2 \le 4$ 36. $-5 < \dfrac{3}{4}x + 1 < 3$

13.6 ◆ Chapter Review

There have been a number of new concepts to master in this chapter, but many of them were based on material you have already learned. You should recognize the differences between simplifying an expression and solving an equation.

Now that you have completed this chapter, you should be able to:

1. Simplify expressions or solve equations using the Distributive Principle.

2. Simplify expressions or solve equations that contain grouping symbols.

3. Solve inequalities, state the answers in set notation, and graph the solutions on number lines.

Review Problems

A. Simplify these expressions.

1. $7(8 - 3y)$ 2. $-2(x + 4)$

3. $-3(b + 6)$ 4. $4(9 - 2a)$

5. $1/4(8x + 12)$ 6. $2(b + 3) - (b - 4)$

7. $5(a + 9) - (2a + 3)$ 8. $2/3(9x - 24)$

9. $6 - [3r - 4(r + 5)]$

10. $7(6p - 4) + 3(5p + 1)$

11. $2(n - 5) + 5(3n + 1)$

12. $10c - 4[c + 2(5c + 1)]$

13. $3[2(5p + 3)]$

14. $5(3x - 2y + 4z - 1)$

15. $3(2x - 4y + 5)$ 16. $4[3(5a - 2)]$

17. $4[t - (t - 3)]$

18. $2\{5[2d + 3(5d - 2)]\}$

19. $10 - \{2[3c + 5(2c - 1)]\}$

20. $-2[3x - |(x - 5)]$

21. $2y - \{3[y + 2(2y + 3)]\}$

22. $-3\{2[m - (m - 5)]\}$

23. $-2\{5[3d - 2(d + 4)]\}$

24. $2\{-3[2b - (b + 4)]\}$

B. Solve and check each of these equations.

25. $6x - 2 = 7x + 5$ 26. $10 - 4y = 16 - y$

27. $2m - 4 = 6m$ 28. $2(5n - 4) = 22$

29. $3(2p - 5) = 45$ **30.** $2c - 10 = 7c$

31. $5 - (-2a + 9) = 2(a - 1)$

32. $3x - 7 = 5(2x + 7)$

33. $2y - 5 = 3(4y + 5)$

34. $4p + 3 = 7 - (-8p + 5)$

35. $7 - 2(r - 2) = 5r - 3$

36. $7 - 2(2m + 5) = -3m + 4$

37. $10[4 - (4x - 8)] = 4(10 - 6x)$

38. $2(3y + 5) - 1 = 6y + 1$

39. $-8[2y - 4(4y - 6) + 1] = 4(y + 4)$

40. $3[2 - 4(a - 1)] = 2(a - 5)$

41. $-2[4 - (3b + 2)] = 3b + 5$

42. $5 + 3[1 + 2(2x - 3)] = 2(x + 5)$

C. **Solve each inequality, express the solution in set notation, and graph the solution on a number line.**

43. $-x \geq -2$ **44.** $4 < x - 1$

45. $7 < x + 5$ **46.** $-x \geq 0$

47. $6(x + 1) > 7 - x$ **48.** $-4 < x - 3 \leq 1$

49. $0 < x + 2 < 3$ **50.** $4 - 3x \leq 2(x + 1)$

Chapter Test

A. **Simplify each expression completely.**

1. $6(x + 2)$

2. $5 - (2n - 1)$

3. $2[8y - 2(3y + 2)]$

4. $-2(R - 3) - 4(2R + 5)$

B. **Solve and check each equation.**

5. $3(x + 1) = 33$

6. $6 - (t + 1) = 13$

7. $24 = -2(p - 5)$

8. $-48 = 3[2m - 4(m - 60)]$

9. $28 - (2A - 1) = 7 - 2(A + 1)$

10. $2[K - 3(K + 2)] = -24$

11. $-40 = 5(T + 1)$

12. $8 + 3(F + 1) = 38$

13. $-4 = 5 - (c - 2)$

14. $3(2y + 5) - 4(y - 2) = -1$

15. $5[2 - (2w - 4)] = 2(5 - 3w)$

C. **Solve each inequality, express the solution in set notation, and graph the solution on a number line.**

16. $2x + 3 > 5$ **17.** $-x \leq 3$

18. $3 - 4x \leq -13$ **19.** $-14 \leq 4x + 6 < 18$

20. $2x - 10 > 7x$

Thought-Provoking Problems

1. In your own words, list the sequential steps for solving equations such as those that appear in Section 13.4.

2. Explain the difference between these solution sets: $\{0\}$ and $\{\ \}$.

3. Write an inequality for each problem. Then solve it and graph the solution.

   a. Seven more than twice a number is greater than 14. Find the numbers that satisfy these conditions.

   b. Eight less than twice a number is less than or equal to three times the number. Find the numbers that satisfy this relationship.

4. Solve these formulas for the letter indicated.

   a. $P = 2(L + W)$ for W

   b. $C = \dfrac{5}{9}(F - 32)$ for F

   c. $A = \dfrac{1}{2}h(b_1 + b_2)$ for b_1

   The subscripts (the 1 in b_1 and the 2 in b_2) are only for identification. They show that the bases (b) are not the same measure. The subscripts do not affect the operations.

5. Solve and graph. A calculator might be useful.

  a. $0.2(x + 30) + 0.08x > 65$

  b. $7.5y - (2.3y + 0.8) \le 4 + 3.2y$

  c. $p \ge 3.8 + 0.16p$

6. Simplify completely.

  a. $6 - \{-2[3x - (4x - 1)]\}$

  b. $-5y - 2[2x - 4(x + 7)] - 6$

  c. $4m - 2[2n - (n - 2m)] + 3n$

7. *Compound* inequality statements use the word *and* or *or* to connect the two parts. Such statements using *and* are *conjunctions*; they mean that the solution set must satisfy *both* inequalities. When *or* is used, the statement is a *disjunction*, which means that the solution set must satisfy *at least one* of the inequalities. Use this information to graph each of the following compound inequalities and to identify whether it is a conjunction or a disjunction.

  a. $x > 2$ and $x < 4$

  b. $x < -2$ or $x > 3$

  c. $x \ge -3$ or $x < 1$

  d. $x \ge 3$ and $x < -2$

8. Discuss the differences among the following:

  $5x + 3$

  $5x + 3 = 0$

  $5x + 3 > 0$

  $5x + 3 \le 0$

More Word Problems

  ## Introduction

In this chapter, you will solve application problems by writing equations that involve parentheses, the Distributive Principle, and signed numbers. Except for the added complexity of the equations, most of the types of word problems will be familiar because you worked with them in Chapter 11. They include consecutive-number, perimeter, and coin or mixture problems. Two new kinds of problems, involving motion and age, will be introduced. You will also work with word problems that require an inequality statement rather than an equation.

When you have completed this chapter, you should be able to:

1. Write equations for these types of word problems: general, consecutive-integer, perimeter, coin or mixture, motion, and age.

2. Solve the equation.

3. Give the answer(s) requested.

4. Check the solution in the words of the problem.

5. Find the solutions to word problems that require working with an inequality (instead of an equation).

Learning Objectives

14.2 General Word Problems

In this section, you will solve general word problems. Sometimes these seem more difficult than later problems because the approach does not seem structured (no charts, no sketches). But the structure is still there—it's just not quite as visible. For every word problem:

1. Read the problem carefully to see what is given and what is needed.

2. State what variable you will use and what it represents.

3. Use the variable to represent the item you know the least about; then define the other items in the problem in terms of that same variable.

4. If desired, organize the information in a chart, in a list, or on a sketch.

5. Translate the words of the problem into an equation.

6. Solve the equation.

7. State all the answers requested (include labels if they are needed).

8. Check the answer(s) in the words of the problem.

Read the following examples carefully and look for the procedures that were just listed.

E X A M P L E 1 One number is 12 more than another number. The sum of twice the larger number and three times the smaller number is 64. Find the numbers.

Equation Let n = smaller number (because the problem describes the larger number)

$n + 12$ = larger number

Twice the larger no.	sum	three times smaller no.	is	64.
$2(n + 12)$	$+$	$3n$	$=$	64

Because the larger number, $n + 12$, is an expression with more than one term, you must enclose it in parentheses so that both parts are multiplied by 2. If you left out the parentheses and wrote $2n + 12$, only n would be multiplied by 2.

Solution

$$2(n + 12) + 3n = 64 \qquad \text{Distributive Principle.}$$
$$2n + 24 + 3n = 64 \qquad \text{Combine like terms.}$$
$$24 + 5n = 64 \qquad \text{Do addition or subtraction.}$$
$$\underline{-24 \qquad\qquad -24}$$
$$5n = 40 \qquad \text{Do multiplication or division.}$$
$$\frac{5n}{5} = \frac{40}{5}$$
$$n = 8$$
$$n + 12 = 20 \qquad \text{The numbers are 8 and 20.}$$

Check

$$\text{Twice the larger} = 2(20) = 40$$
$$\text{Three times the smaller} = 3(8) = \underline{24}$$
$$\text{Sum} = 64$$

$$20 \text{ is } 12 \text{ more than } 8 \quad \text{True.}$$

◆

E X A M P L E **2** One number is 3 less than another number. The difference between the larger number and twice the smaller number is -16. Find the numbers.

Equation Let n = one of the numbers = larger (because the problem describes the smaller number)

$n - 3$ = the other number = smaller

Difference (subtraction) must be written in the order given.

$\underline{\text{Larger}}\ \underline{\text{difference}}\ \underline{\text{twice the smaller}}\ \underline{\text{is}}\ \underline{-16.}$

$n-2(n-3)=-16$

Solution $n - 2(n - 3) = -16$

$n - 2n + 6 = -16$

$-n + 6 = -16$

$\underline{-6-6}$

$-n = -22$

$\dfrac{-n}{-1} = \dfrac{-22}{-1}$

$n = 22$

$n - 3 = 19$ The numbers are 22 and 19.

Check 19 is 3 less than 22.

Larger $- 2$(smaller) $= -16$

$22 - 2(19) = 22 - 38 = -16$ True. ◆

E X A M P L E **3** Sarah, Laura, and Kristen work in the learning center. Last week Sarah worked 4 hours less than Kristen, and Laura worked twice as many hours as Sarah. If the three women worked a total of 52 hours, how many hours did Sarah work?

Equation Look at two people at a time. Because Sarah worked 4 hours less than Kristen, you know the least about Kristen's time. Let Kristen's time be x. Sarah's time would then be $x - 4$. Now that you have represented these two items, you can multiply your statement of Sarah's time by 2 to represent Laura's time because Laura worked twice as many hours as Sarah.

Let x = Kristen's hours

$x - 4$ = Sarah's hours

$2(x - 4)$ = Laura's hours

52 = Total hours worked

$x + (x - 4) + 2(x - 4) = 52$

Solution $x + x - 4 + 2x - 8 = 52$

$4x - 12 = 52$

$\underline{+12+12}$

$4x = 64$

$4x/4 = 64/4$

$x = 16$

The only answer requested is Sarah's time:

$x - 4 = 16 - 4 = 12$ hours

Check   See if the times add up to 52:

Kristen: $x = 16$

Sarah: $x - 4 = 16 - 4 = 12$

Laura: $2(x - 4) = 2(16 - 4) = 2(12) = 24$

Total time: $16 + 12 + 24 = 52$ hours.   True.

Sarah's time, 12 hours, is 4 less than Kristen's 16 hours. Laura's 24 hours is twice Sarah's time: $2(12) = 24$.   True.   ◆

E X A M P L E 4   One number is four more than twice the other number. The difference between the smaller number and the larger number is -23. What are the numbers?

Equation   Let $n =$ one of the numbers

$2n + 4 =$ the other number

Do not split the information. "Four more than twice the other number" is all describing one quantity.

Now determine which is the smaller number.

Let $n =$ smaller number

$2n + 4 =$ larger number

Difference (subtraction) is written in the given order.

$$\underline{\text{Smaller}} - \underline{\text{larger}} \quad \text{is } -23.$$
$$n \quad - (2n + 4) = -23$$

Solution

$$n - (2n + 4) = -23$$
$$n - 1(2n + 4) = -23$$
$$n - 2n - 4 = -23$$
$$-n - 4 = -23$$
$$\underline{\quad\quad + 4 \quad + 4}$$
$$-n \quad\quad = -19$$
$$\frac{-n}{-1} \quad\quad \frac{-19}{-1}$$
$$n = 19$$

$n =$ smaller number $= 19$

$2n + 4 =$ larger number $= 2(19) + 4 = 42$

Check   Four more than twice the smaller number is 42.

$2(19) + 4 = 38 + 4 = 42$

Smaller $-$ larger $= 19 - 42 = -23$ True.   ◆

14.2 ◆ Practice Problems

Write an equation for each word problem. Then solve the equation and check the answer.

1. Mary, Diane, and Jean are waitresses at Pop's Pizza Parlor. Last week Mary earned $19 less than Jean, and Diane earned twice as much as Mary. Altogether they earned $163. How much was each woman's salary?

2. The sum of two numbers is 84. The larger number is 12 less than twice the smaller. What is each number?

3. One number is 5 less than another number. If three times the smaller number is subtracted from twice the larger number, the result is -7. Find the numbers.

4. Three students took a total of 39 hours to complete a group project for their hydraulics class. Tom worked 4 hours less than Ted. Terry worked three times as much as Tom. How much time did each student contribute to the project?

5. The height of the World Trade Center is 123 feet less than twice the height of the Prudential Building. The difference in height between the World Trade Center in New York and the Prudential Building in Boston is 627 feet. How tall is the World Trade Center?

6. A student bought a notebook, three pens, and a candy bar at the bookstore. Each pen costs $0.54 more than the candy bar, and the notebook cost twice as much as a pen. The total cost was $5.10 (before tax). How much did the notebook cost?

7. A nurse's aide worked two more hours on Monday than on Tuesday, and twice as many hours on Wednesday as on Monday. If he worked a total of 42 hours, how many did he work each day?

8. One number is 8 less than another number. When twice the smaller number is subtracted from the larger number, the result is −10. What are the numbers?

9. A student bought textbooks for her history, chemistry, and mathematics courses. The mathematics book cost $5 more than the history text. The price of the chemistry book was twice the price of the mathematics book. The total bill was $90. Find the price of the chemistry book.

10. Bob, Connie, and Lorraine all work as receptionists in the learning center. During the last pay period they earned a total of $464. Find out how much Lorraine earned if Bob earned $16 more than Connie, and Lorraine earned twice as much as Bob.

11. The sum of three numbers is 64. The largest number is twice the smallest, and the smallest is 8 less than the middle number. Find the three numbers.

12. Three stone masons worked a total of 31 hours to construct a retaining wall. James worked 5 more hours than Joe, and Jerry worked twice as long as James. How many hours did Jerry work?

13. Raoul works part-time in the Financial Aid Office. He worked 3 more hours on Tuesday than on Wednesday, and he worked twice as many hours on Friday as on Tuesday. If he worked a total of 27 hours on those three days, how many hours did he work on Tuesday?

14. One number is 11 more than another number. The difference between twice the smaller and three times the larger number is −47. What are the numbers?

15. The difference between twice a number and 14 is equal to three times the sum of the number and 3. What is the number?

16. Five times the difference between twice a number and 1 is equal to the sum of twice the number and 3. What is the number?

17. Three times the difference between five times a number and 1 is equal to four times the sum of the number and 2. What is the number?

18. Five times the sum of a number and 1 is equal to twice the difference between the number and 2. What is the number?

19. Six times the difference between a number and 3 is equal to the difference between the number and 3. What is the number?

20. When the sum of a number and 3 is subtracted from twice the number, the result is 15. What is the number?

14.3 Consecutive-Integer Problems

To solve these consecutive-integer problems, you will use the same notation that you used in Section 11.4. In order to include both positive and negative whole numbers and zero, we will use the term *integer* instead of *number*.

Consecutive Integers	Consecutive Even/Odd Integers
Let n = 1st consecutive integer	Let n = 1st consecutive even/odd integer
$n + 1$ = 2nd consecutive integer	$n + 2$ = 2nd consecutive even/odd integer
$n + 2$ = 3rd consecutive integer	$n + 4$ = 3rd consecutive even/odd integer
and so on.	and so on.

The only new thing in these problems is the use of the Distributive Principle. Be sure to read the problem carefully, because although you might not use all the numbers in the equation, you will have to give all of them in the answer. These steps are demonstrated in the examples that follow.

EXAMPLE 1   The sum of three consecutive integers is -75. Find the integers.

Equation   Let $n = $ 1st consecutive integer

$n + 1 = $ 2nd consecutive integer

$n + 2 = $ 3rd consecutive integer

$\underline{\text{Sum of three consecutive integers}}$ $\underline{\text{is}}$ $\underline{-75.}$

$\qquad n + n + 1 + n + 2 \qquad = -75$

Solution $3n + 3 = -75$

$\qquad \underline{-3 \quad -3}$

$\qquad 3n = -78$

$\qquad \dfrac{3n}{3} = \dfrac{-78}{3}$

$\qquad n = -26$

$n + 1 = -25$

$n + 2 = -24$

Check $-26, -25,$ and -24 are consecutive integers. Note that -26 is smaller than -24 because it is farther to the left on the number line.

$(-26) + (-25) + (-24) = -75$ True.   ◆

EXAMPLE 2   The sum of the smallest and twice the largest of three consecutive even integers is 56. Find the three integers.

Equation   Let $n = $ 1st consecutive even integer

$n + 2 = $ 2nd consecutive even integer

$n + 4 = $ 3rd consecutive even integer

$\underline{\text{Smallest}}$ $\underline{\text{sum}}$ $\underline{\text{twice the largest}}$ $\underline{\text{is}}$ $\underline{56.}$

$\qquad n \qquad + \qquad 2(n + 4) \qquad = 56$

Solution $n + 2(n + 4) = 56$

$\qquad n + 2n + 8 = 56$

$\qquad 3n + 8 = 56$

$\qquad \underline{-8 \quad -8}$

$\qquad 3n \quad = 48$

$\qquad \dfrac{3n}{3} = \dfrac{48}{3}$

$\qquad n = 16$

$n + 2 = 18$

$n + 4 = 20$

Check 16, 18, and 20 are consecutive even integers.

The smallest plus twice the largest equals 56.

$16 + 2(20) = 16 + 40 = 56$ True. ◆

E X A M P L E **3** When the largest of three consecutive odd integers is subtracted from three times the smallest, the result is 34. Find all three of the integers.

Equation Let n = 1st consecutive odd integer

$n + 2$ = 2nd consecutive odd integer

$n + 4$ = 3rd consecutive odd integer

Three times smallest minus largest result is 34.

$$3n \qquad\qquad - \quad (n + 4) \quad = \quad 34$$

Pay special attention to the parentheses around $n + 4$. Because the quantity being subtracted has more than one term in it, it has to be enclosed in parentheses so that *both* the n and the 4 will be subtracted.

Solution
$$3n - (n + 4) = 34$$
$$3n - 1(n + 4) = 34$$
$$3n - 1n - 4 = 34$$
$$2n - 4 = 34$$
$$\underline{\quad + 4 \qquad + 4\quad}$$
$$2n \quad = \quad 38$$
$$\frac{2n}{2} = \frac{38}{2}$$
$$n = 19$$
$$n + 2 = 21$$
$$n + 4 = 23$$

Check 19, 21, and 23 are consecutive odd numbers.

3 times the smallest minus the largest is 34.

$3(19) - 23 = 57 - 23 = 34$ True. ◆

14.3 ◆ Practice Problems

A. Write an equation for each problem. Then solve the equation and check the answer.

1. The sum of three consecutive integers is −84. Find the three integers.

2. The difference between the first and twice the third of three consecutive even integers is −16. Find the three integers.

3. The difference between the first and three times the second of two consecutive odd integers is −12. Find the integers.

4. The sum of three consecutive integers is −24. Find the three integers.

5. When you subtract the third of three consecutive even integers from 4 times the first, the difference is 50. Find the three integers.

6. The sum of an integer and 7 times the next consecutive integer is −49. Find the integers.

7. The sum of 4 times an integer and twice the next consecutive integer is −16. Find the integers.

8. Find two consecutive even integers such that twice the smaller decreased by the larger is 72.

9. Find two consecutive odd integers such that 3 times the smaller decreased by the larger is equal to 36.

B. **Write an equation for each of the following word problems. Then solve the equation and check the answer.**

10. The sum of the smallest and twice the largest of three consecutive odd integers is −13. Find the three integers.

11. The difference between the smallest and twice the largest of three consecutive odd integers is −13. Find the three integers.

12. The sum of twice the largest and the smallest of three consecutive integers is equal to −23. Find the three integers.

13. The difference between the smallest and twice the largest of three consecutive integers is −23. Find the largest integer.

14. When you subtract the third of three consecutive even integers from 4 times the first, the difference is −40. Find the largest integer.

15. Find three consecutive odd integers such that twice the first is equal to 7 more than the largest.

16. Find two consecutive odd numbers such that 7 times the first number is equal to 5 times the second number.

17. Find three consecutive even integers whose sum is −18.

18. Find three consecutive integers whose sum is −21.

19. Find two consecutive even integers such that twice the second equals three times the first.

20. Find two consecutive odd integers such that three times the second number equals five times the first.

21. When three times the first of three consecutive even integers is subtracted from twice the third, the result is −4. What are the three even integers?

22. When one-half of the first of two consecutive even integers is subtracted from three-fifths the next consecutive even integer, the result is 6. What are the integers?

23. The sum of two-thirds of the first of two consecutive odd integers and four-fifths of the next consecutive odd integer is −38. What are the integers?

24. The sum of the first of two consecutive integers and three times the second is −61. What are the integers?

14.4   Perimeter Problems

You will see a change in perimeter problems only if you use $P = 2L + 2W$ or $P = 2(L + W)$ as your formula instead of $P = L + W + L + W$. The first two versions of the formula involve using the Distributive Principle to solve the equation. Continue to draw a picture and to label the sides as you did in Chapter 11.

E X A M P L E   **1**   The length of a rectangle is 11 inches more than the width. Find the dimensions if the perimeter is 82 inches. Draw a sketch similar to Figure 1.

Equation

Let x = width

$x + 11$ = length

If $P = 2(L + W)$, then

$82 = 2(x + 11 + x)$

Solution

$82 = 2x + 22 + 2x$

$82 = 4x + 22$

$\underline{-22 = \qquad -22}$

$60 = 4x$

$\dfrac{60}{4} = \dfrac{4x}{4}$

$15 = x$

$26 = x + 11$

The width is 15 inches, and the length is 26 inches. Remember, *dimensions* means one length and one width.

FIGURE **1**

Check 26 is 11 more than 15.

15 + 26 + 15 + 26 = 82   True.   ◆

E X A M P L E 2 Find the width of a rectangle if the perimeter is 62 feet and the width is 4 feet more than one-half of the length. Draw a sketch like Figure 2.

Equation   Let x = length

$\dfrac{1}{2}x + 4$ = width

If $P = 2L + 2W$, then

$$62 = 2x + 2\left(\dfrac{1}{2}x + 4\right)$$

Solution $62 = 2x + x + 8$

$62 = 3x + 8$

$\underline{-\ 8 \qquad -\ 8}$

$54 = 3x$

$\dfrac{54}{3} = \dfrac{3x}{3}$

$18 = x$

Width is $\dfrac{1}{2}x + 4 = \dfrac{1}{2}(18) + 4 = 9 + 4 = 13$ feet.

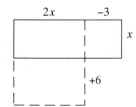

$\frac{1}{2}x + 4$ ▭ $\frac{1}{2}x + 4$

x

x

F I G U R E 2

Check   18 + 13 + 18 + 13 = 62   True.   ◆

E X A M P L E 3 Jo needs to change the dimensions of her rectangular garden. The original garden had a length that was twice the width. If she increases the width by 6 feet and decreases the length by 3 feet, the new perimeter will be 54 feet. What were the dimensions of the original garden? See Figure 3.

Equation   Let x = original width

$2x$ = original length

$x + 6$ = new width

$2x - 3$ = new length

If $P = 2L + 2W$, then

$54 = 2(2x - 3) + 2(x + 6)$

Solution $54 = 2(2x - 3) + 2(x + 6)$

$54 = 4x - 6 + 2x + 12$

$54 = 6x + 6$

$\underline{-\ 6 \qquad -\ 6}$

$48 = 6x$

$\dfrac{48}{6} = \dfrac{6x}{6}$

$8 = x$

$2x$ -3

x

$+6$

F I G U R E 3

The original width is $x = 8$ feet, and the original length is $2x = 2(8) = 16$ feet.

Check   Original length is twice the width: $16 = 2(8)$. New length $= 16 - 3 = 13$ feet; new width $= 8 + 6 = 14$ feet. $13 + 14 + 13 + 14 = 54$ feet.   True.   ◆

14.4 ◆ Practice Problems

Write an equation for each problem. Then solve the equation and check the answer.

1. Find the dimensions of a rectangle whose length is 8 feet more than its width. The perimeter is 76 feet.

2. Find the width of a rectangle if the perimeter is 138 inches, and the length is 3 inches more than the width.

3. The perimeter of a rectangle is 88 yards. Find the dimensions if the width is 6 yards less than the length.

4. Find the dimensions of a rectangle if the width is 7 meters less than the length, and the perimeter is 54 meters.

5. The length of a rectangle is twice the width. If the length is increased by 4 feet and the width is decreased by 1 foot, the new perimeter is 96 feet. Find the dimensions of the original rectangle.

6. The perimeter of a rectangle is 76 inches. The length is 2 inches more than 5 times the width. Find the dimensions.

7. A rectangular field is enclosed by 440 feet of fencing. Find the dimensions of the field if the length is 20 feet less than 3 times the width.

8. The length of Mac's wheat field is 40 yards more than the width. If he doubles the width of the field and increases the length by 30 yards, the perimeter will be 500 yards. What are the dimensions of the original field?

9. The second side of a triangle is 4 feet shorter than the first side. The third side is 2 times as long as the second side. Find the lengths of the sides if the perimeter is 28 feet.

10. The perimeter of a rectangular patio is 18 yards. Find the dimensions if the width is 1 yard less than the length.

11. Sharon has 56 feet of edging to use as a border for her flower bed. If she wants the bed to be 10 feet longer than it is wide, what should the dimensions be?

12. The length of a rectangular painting is 2 inches more than twice the width. If the perimeter is 52 inches, what are the dimensions of the painting?

13. The length of a rectangular garden is 3 times the width. If the width is increased by 5 feet and the length is decreased by 3 feet, the new perimeter will be 100 feet. What are the dimensions of the original garden?

14. The length of a rectangular bulletin board is 7 inches more than twice the width. If the perimeter is 224 inches, find the length of the bulletin board.

15. The width of a rectangular picture frame is 3 inches less than the length. If the perimeter is 38 inches, what are the dimensions of the frame?

16. Nancy needs 66 inches of bias tape to edge the sides of the rectangular place mat she is making. If the width of a mat is 3 inches less than the length, what are the dimensions of the mat?

17. The perimeter of a rectangle is 40 yards. The length is 5 yards more than twice the width. Find the length.

18. The width of a rectangle is one-half the length. If the width is increased by 4 feet and the length is decreased by 3 feet, the perimeter of the new rectangle will be 62 feet. What are the dimensions of the original rectangle?

19. Tom is having his family room enlarged. The original room had a width that was 6 feet less than the length. With remodeling, the length will not change but the width will double. The perimeter of the new room will be 66 feet. What were the dimensions of the original room?

20. The length of a rectangle is 3 yards less than twice the width. If the perimeter is 36 yards, what is the length of the rectangle?

14.5 ◆ Coin and Mixture Problems

The procedures for coin or mixture problems will be unchanged from the methods demonstrated in Chapter 11. Negative numbers will not appear in the answers to these problems (you cannot have a negative number of dimes, for instance), but the Distributive Principle will be used.

In these problems, you may be given a total number of coins or items rather than other information about the relative number of each kind. For example, you have previously been given information such as "there are 3 times as many dimes as quarters." You represented the number of coins as follows:

x = number of quarters (the one you know the least about)

$3x$ = number of dimes

In these new problems, you may be told only that the total number of dimes and quarters is 40. In this case you will let x equal the number of one of the coins, and $40 - x$ will equal the number of the other coin. It does not matter whether you say:

x = number of dimes	or	x = number of quarters
$40 - x$ = number of quarters		$40 - x$ = number of dimes

Both representations will give correct answers if used properly.

E X A M P L E 1　Simone has 56 coins in her bank. They are all either quarters or half dollars. The total amount in the bank is $23.00. How many quarters and how many half dollars does she have?

Draw a box like Chart 1 and fill in the given information. Then check your box with Chart 2.

CHART 1

Item	Value of Each	Number of Each	Value of Each Kind
			Total Value

CHART 2

Item	Value of Each	Number of Each	Value of Each Kind
Quarters	0.25	x	$0.25x$
Half dollars	0.50	$56 - x$	$0.50(56 - x)$
			Total Value
			23.00

Equation　$0.25x + 0.50(56 - x) = 23.00$

Solution

$$0.25x + 0.50(56 - x) = 23.00$$
$$0.25x + 28.00 - 0.50x = 23.00$$
$$-0.25x + 28.00 = 23.00$$
$$\underline{-28.00 \quad\quad -28.00}$$
$$-0.25x \quad\quad\quad = -5.00$$
$$\frac{-0.25x}{-0.25} = \frac{-5.00}{-0.25}$$
$$x = 20 \text{ quarters}$$
$$56 - x = 36 \text{ half dollars}$$

Check   20 + 36 does equal 56 coins.

$0.25(20) + 0.50(36) = 5.00 + 18.00 = \$23.00$ True.   ◆

Try solving the same problem with x = number of half dollars and $56 - x$ = number of quarters. Your equation should be $0.25(56 - x) + 0.50x = 23.00$. The solution is $x = 36$. This means that there are 36 half dollars and $56 - 36 = 20$ quarters. Does this agree with the previous solution?

E X A M P L E 2   The hot dog stand at the football game last Friday took in \$79.00. The money consisted of \$5 bills, \$1 bills, and quarters. There were twice as many \$1 bills as \$5 bills, and the number of quarters was six more than three times the number of \$5 bills. How many of each were there?

Draw a box and fill in the given information. Compare your box with Chart 3.

CHART 3

Item	Value of Each	Number of Each	Value of Each Kind
\$5 Bills	5	x	$5x$
\$1 Bills	1	$2x$	$1(2x)$
Quarters	0.25	$3x + 6$	$0.25(3x + 6)$
			Total Value
			79.00

Equation $5x + 1(2x) + 0.25(3x + 6) = 79.00$

Solution
$$5x + 2x + 0.25(3x + 6) =\quad 79.00$$
$$5x + 2x + 0.75x + 1.50 =\quad 79.00$$
$$7.75x + 1.50 =\quad 79.00$$
$$\underline{\; -\,1.50 \quad -\,1.50}$$
$$7.75x \qquad =\quad 77.50$$
$$\frac{7.75x}{7.75} = \frac{77.50}{7.75}$$
$$x = 10$$
$$2x = 2(10) = 20$$
$$3x + 6 = 3(10) + 6 = 36$$

There were ten \$5 bills, twenty \$1 bills, and thirty-six quarters.

Check   20 is 2 times 10, and 36 is 6 more than three times 10.

Ten \$5 bills = \$50.00

Twenty \$1 bills = \$20.00

Thirty-six \$0.25 = \$ 9.00

Total = \$79.00   True.   ◆

EXAMPLE 3   The Gourmet Coffee Shoppe blends two grades of coffee beans to produce its Deluxe Mix. How many pounds of Venetian beans, costing $10 per pound, should be blended with 40 pounds of Colombian beans, costing $4 per pound, to produce a mixture costing $6 per pound?

The procedure in this mixture problem is very similar to that used in coin problems. The unknown quantity is the number of pounds of Venetian beans. Let x = number of pounds of Venetian beans. Draw a box similar to the coin boxes, and fill in the information that is given (Chart 4). Note that the third row contains information about the final mixture.

CHART 4

Item	Cost per Pound	Number of Pounds of Each	Total Cost of Each
Venetian	10	x	$10x$
Colombian	4	40	$4(40)$
Deluxe	6	$x + 40$	$6(x + 40)$

Equation

Venetian beans	plus	Colombian beans	equals	Deluxe Mix.
$10x$	$+$	$4(40)$	$=$	$6(x + 40)$

Solution

$$10x + 4(40) = 6(x + 40)$$
$$10x + 160 = 6x + 240$$
$$\underline{-6x \qquad\quad -6x}$$
$$4x + 160 = 240$$
$$\underline{\quad -160 \qquad -160}$$
$$4x = 80$$
$$\frac{4x}{4} = \frac{80}{4}$$
$$x = 20 \text{ pounds of Venetian beans}$$

Check

$$10(20) + 4(40) = 6(20 + 40)$$
$$200 + 160 = 6(60)$$
$$360 = 360 \quad \text{True.}$$

◆

EXAMPLE 4   Craft and Company is preparing a new shredded cheese mixture. How many pounds of $4.30-per-pound romano cheese should be combined with 20 pounds of $2.25-per-pound mozzarella to produce a pizza cheese mix costing $3.80 per pound?

Record the information that is given in a box, and compare your box with Chart 5.

CHART 5

Item	Cost per Pound	Number of Pounds of Each	Total Cost of Each
Romano	4.30	x	$4.30x$
Mozzarella	2.25	20	$2.25(20)$
Mixture	3.80	$x + 20$	$3.80(x + 20)$

Equation $4.30x + 2.25(20) = 3.80(x + 20)$

Solution

$$
\begin{array}{rcl}
4.30x + & 45 = & 3.80x + 76 \\
-3.80x & & -3.80x \\
\hline
0.50x + & 45 = & 76 \\
& -45 & -45 \\
\hline
0.50x & = & 31 \\
\end{array}
$$

$$\frac{0.50x}{0.50} = \frac{31}{0.50}$$

$x = 62$ pounds of romano cheese

Check $4.30(62) + 2.25(20) = 3.80(62 + 20)$

$266.6 + 45.00 = 3.80(82)$

$\$311.60 = \$311.60$ True.   ◆

E X A M P L E 5   How much of a 13% salt solution should be added to 30 liters of an 18% salt solution to produce a solution that is 16% salt?

This is a mixture problem wherein percents, not costs, must be considered. Set up a chart similar to those you made in the previous problems, but change the Cost per Pound column to Percent of Each (see Chart 6). When placing values in that column, put them in as decimals so that you will remember to multiply by the decimal rather than by a percent.

CHART 6

Item	Percent of Each	Number of Liters of Each	Each Quantity
13% Salt	0.13	x	$0.13x$
18% Salt	0.18	30	$0.18(30)$
16% Salt	0.16	$x + 30$	$0.16(x + 30)$

Equation $0.13x + 0.18(30) = 0.16(x + 30)$

Solution

$$
\begin{array}{rcl}
0.13x + & 5.4 = & 0.16x + 4.8 \\
-0.13x & & -0.13x \\
\hline
& 5.4 = & 0.03x + 4.8 \\
& -4.8 & -4.8 \\
\hline
& 0.6 = & 0.03x \\
\end{array}
$$

$$\frac{0.6}{0.03} = \frac{0.03x}{0.03}$$

$20 = x$

Therefore, 20 liters of the 13% solution must be added.

Check   Substitute the answer into the expressions given in the last column to see if the sum of the two components does equal the final mixture.

$$0.13x + 0.18(30) = 0.16(x + 30)$$

$$0.13(20) + 0.18(30) = 0.16(20 + 30)$$

$$2.6 + 5.4 = 0.16(50)$$

$$8.0 = 8   \text{True.}   ◆$$

EXAMPLE 6   To meet consumer demand for a lower-fat ice cream that still tastes like ice cream, Elsie Dairy Company will blend ice cream that is 21% butterfat with a 15% butterfat product to make a 19% butterfat ice cream. How many gallons of each product need to be mixed to produce 120 gallons of the new product?

This is another mixture problem wherein percents, not costs, must be considered. Set up a chart and fill in the given information. See Chart 7 if you need help with this step.

CHART 7

Item	Percent of Each	Number of Gallons of Each	Each Quantity
15% Butterfat	0.15	x	$0.15x$
21% Butterfat	0.21	$120 - x$	$0.21(120 - x)$
19% Butterfat	0.19	120	$0.19(120)$

Equation $0.15x + 0.21(120 - x) = 0.19(120)$

Solution

$$0.15x + 25.2 - 0.21x = 22.8$$
$$+ 25.2 - 0.06x = 22.8$$
$$- 25.2 \qquad\qquad -25.2$$
$$- 0.06x = -2.4$$
$$\frac{-0.06x}{-0.06} = \frac{-2.4}{-0.06}$$
$$x = 40$$

Therefore, 40 gallons of the 15% butterfat product and $120 - x$, or 80, gallons of the 21% butterfat product will be used.

Check   Does $0.15(40) + 0.21(120 - 40) = 0.19(120)$?

$$6 + 16.8 = 22.8$$
$$22.8 = 22.8 \quad \text{True.}$$ ◆

14.5 ◆ Practice Problems

For each problem, draw a box and fill in the given information. Then write an equation, solve it, and check the answer.

1. A college student saves dimes and quarters so that he can go to the laundromat. The number of quarters he has is four more than twice the number of dimes. Altogether he has $3.40. How many of each coin does he have?

2. The Nut House blends $3.00-per-pound peanuts with $7.00-per-pound mixed nuts to produce an economy mix that sells for $6.00 per pound. How many pounds of peanuts and mixed nuts need to be blended to produce 20 pounds of economy mix?

3. Marge bought $0.25 and $0.15 stamps at the Post Office. She bought six fewer $0.25 stamps than $0.15 stamps and spent $8.90. How many of each stamp did she buy?

4. Tickets for the Cougar basketball games cost $2.00 for students and $3.50 for non-students. On Saturday, 120 tickets were sold and $360 was collected. How many of each kind of ticket was sold?

5. The Gamma Gamma Sorority is selling cakes to raise money for Children's Hospital. Pound cakes sell for $5 each, chocolate layer cakes for $7, and hazelnut tortes for $8. Group members sold six more tortes than pound cakes and twice as many chocolate cakes as tortes. They raised $321. How many chocolate cakes were sold?

6. A child's piggy bank contains five more quarters than dimes. If there is a total of $6.50, how many of each coin are there?

7. Lil's Card Shop sells two different styles of wrapping paper. One retails for $1.20 a roll and the other for $2.00 a roll. Last month the store sold a total of 40 rolls for $72.00. How many rolls of each kind were sold?

8. Marilyn bought ten stamps at the post office. Some were 15-cent stamps and the others were 25-cent stamps. If she spent $1.70, how many of each did she buy?

9. Ocean Splash Company blends cranberry and apple juices to make its fruit drink. If cranberry juice costs $3.25 per gallon, apple juice costs $2.25 per gallon, and the mixture must cost $2.50 per gallon, how much of each juice does the company need to mix to get 100 gallons of the blend?

10. How much $7.00-per-pound chocolate needs to be blended with 20 pounds of $3.50-per-pound chocolate to yield a mix that costs $4.50 per pound?

11. The office "coffee cash" box contains 32 coins, all either nickels or dimes. If the total is $2.00, how many dimes and nickels are in the box?

12. The student government has started a campaign to help the homeless. Collection boxes have been set out in the student lounge and in the cafeteria. One box contains twice as many nickels as dimes and has three more quarters than nickels. If there is $11.25 in the box, how many quarters are there?

13. Jamie's bank contains 24 more dimes than pennies and five times as many nickels as pennies. If the bank has $6.72, how many of each coin are there?

14. The petty cash box contains $50.00. There are 30 bills, all either $1 or $5 bills. How many $5 bills are there?

15. How much of an 8% alcohol solution and how much of a 5% alcohol solution must be combined to produce 300 milliliters of a 6% alcohol solution?

16. How many quarts of 20% salt solution must be added to 8 quarts of a 15% salt solution to get a solution that is 17% salt?

17. How many liters of a 30% sulfuric acid solution must be added to a 50% sulfuric acid solution to produce 10 liters of a 35% sulfuric acid solution?

18. When 30 liters of a 25% alcohol solution is added to 20 liters of a 20% alcohol solution, what is the percent alcohol in the resulting solution?

19. How many pounds of ground beef that is 10% fat and how many pounds of ground beef that is 15% fat must be combined to produce 50 pounds of ground beef that is 12% fat?

20. How many quarts of pure (100%) maple syrup and how many quarts of 20% maple syrup must be blended to produce 440 quarts of a 50% maple syrup product?

14.6 ◆ Motion Problems

One of the two new kinds of word problems you will solve in this chapter is *motion problems*. In solving these problems, you will use a concept you worked with earlier: The distance traveled is equal to the product of the rate times the time, or $D = rt$. You will be more successful with these problems if you organize the information in a box like Chart 8.

CHART **8**

	Rate	Time	Distance of Each
			Total Distance

Several different situations occur in the distance problems in this section. In some problems a total distance is given. And in some, each vehicle or person travels the same distance. The chart is completed in the same way in both cases, but the equations differ. Examples of both kinds of problems follow. Study them carefully so that you will be able to do similar problems on your own.

E X A M P L E 1

Two airplanes take off from Hayes Airport at the same time and travel in opposite directions. The northbound plane travels at an average rate of 400 miles per hour. The southbound plane travels at 350 miles per hour. After how many hours will they be 2,250 miles apart?

Making a sketch of the problem, such as Figure 4, may help you organize the information. You are looking for time. Because they start at the same time and travel until they are 2,250 miles apart, you can "read into" the problem the fact that the time is the same for both planes. The information has been placed in the box (Chart 9).

N

↑ 400 mph
 x hours

2,250 miles ● Airport

 350 mph
↓ *x* hours

S

F I G U R E 4

C H A R T 9

	Rate	Time	Distance of Each
Northbound (N) plane	400	*x*	
Southbound (S) plane	350	*x*	
			Total Distance 2,250

Multiply the information in the Rate box by the information in the Time box to get the Distance of Each, because $D = rt$ (see Chart 10).

C H A R T 10

	Rate	Time	Distance of Each
Northbound (N) plane	400	*x*	400*x*
Southbound (S) plane	350	*x*	350*x*
			Total Distance 2,250

Equation

If the total distance is given, as it is in this problem, *add* the two Distance of Each values and set them equal to the Total Distance value.

$$\underbrace{400x}_{N \text{ plane distance}} \text{ plus } \underbrace{350x}_{S \text{ plane distance}} = \underbrace{2{,}250}_{\text{total distance.}}$$

Solution

$$400x + 350x = 2{,}250$$
$$750x = 2{,}250$$
$$\frac{750x}{750} = \frac{2{,}250}{750}$$
$$x = 3$$

The planes will travel for 3 hours before they are 2,250 miles apart.

Check

$400x$ represents the distance that the northbound plane travels. If $x = 3$, $400x = 1{,}200$ miles. $350x$ for the southbound plane $= 350(3) = 1{,}050$ miles. Add the distances together to see if they equal the total distance.

$1{,}200 + 1{,}050 = 2{,}250$ True. ◆

E X A M P L E **2** Maria leaves for her evening walk traveling at 3 mph. One hour later, her husband Miguel leaves the house, following the same route and jogging at 5 mph. How long will it take him to overtake Maria?

Make a sketch representing the problem (Figure 5).

Maria •——— 3 mph ———→
Miguel •——— 5 mph ———→
        1 hour later

} Same distance

F I G U R E **5**

Draw a box and fill in the information that is given. You don't know anything about Maria's time, so call it x. Because Miguel leaves the house 1 hour later, his traveling time is $x - 1$ (he traveled 1 hour *less than* Maria). No total distance is given; in fact, they travel the same distance from their home to the point where Miguel catches up with Maria. Does your box look like Chart 11?

C H A R T **11**

	Rate	Time	Distance of Each
Maria	3	x	$3x$
Miguel	5	$x - 1$	$5(x - 1)$
			Total Distance same

Equation Because they travel the same distance, the equation is

$$\frac{\text{Maria's distance}}{3x} = \frac{\text{Miguel's distance.}}{5(x - 1)}$$

Solution
$$3x = 5x - 5$$
$$\underline{-5x \quad\quad -5x}$$
$$-2x = \quad\quad -5$$
$$\frac{-2x}{-2} = \frac{-5}{-2}$$
$$x = 2\frac{1}{2} \text{ or } 2.5$$

Because you are looking for Miguel's time,
$$x - 1 = 2.5 - 1 = 1.5 \text{ hours}$$

Check See if Maria's distance does equal Miguel's distance.

Maria's distance $= 3x = 3(2.5) = 7.5$ miles

Miguel's distance $= 5(x - 1) = 5(2.5 - 1)$
$$= 5(1.5) = 7.5 \text{ miles} \quad \text{True.}$$

◆

E X A M P L E   **3**   A motorboat, traveling 12 miles per hour faster than a sailboat, sets off from the same marina 2 hours after the sailboat left. The motorboat passes the sailboat 3 hours into the sailboat's trip. How fast is each traveling? Use a sketch like Figure 6 and a box like Chart 12.

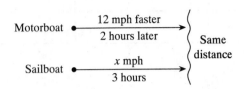

F I G U R E **6**

Filling in the Time column takes a little thought and careful reading of the problem. You need a traveling time for each boat. The sailboat had been traveling for 3 hours when the motorboat passed it, so the sailboat's travel time is 3 hours. The motorboat left 2 hours after the sailboat, so its time is 2 hours less, or $3 - 2 = 1$ hour.

C H A R T **12**

	Rate	Time	Distance of Each
Sailboat	x	3	$3x$
Motorboat	$x + 12$	1	$1(x + 12)$
			Total Distance same

Equation   They travel the same distance.

Solution
$$3x = x + 12$$
$$\underline{-x \quad -x}$$
$$2x = \quad 12$$
$$x = 6$$

Sailboat = 6 mph; motorboat = $3x$ = 18 mph.

Check   See if the sailboat's distance is equal to the motorboat's distance.

Sailboat: $3x = 3(6) = 18$ miles

Motorboat: $1(x + 12) = 1(6 + 12) = 18$ miles

Sailboat's distance = motorboat's distance   True.   ◆

E X A M P L E   **4**   Two cars start at the same time and drive toward each other from cities that are 570 miles apart. One car averages 55 mph, and the other averages 40 mph. In how many hours will they meet?

Make a sketch like Figure 7 and a box like Chart 13.

40 mph   55 mph

570 miles

F I G U R E **7**

CHART **13**

	Rate	Time	Distance of Each
Car 1	55	x	$55x$
Car 2	40	x	$40x$
			Total Distance 570

Equation $55x + 40x = 570$

Solution $95x = 570$

$$\frac{95x}{95} = \frac{570}{95}$$

$x = 6$ hours

Check See if the distances equal the total.

$55x = 55(6) = 330$ miles

$40x = 40(6) = \underline{240}$ miles

570 miles   True.

◆

14.6 ◆ Practice Problems

Use a sketch and a box to organize the information for each problem. Then write an equation and solve it.

1. Two motorcycles, leaving at the same time, are traveling in opposite directions from the same city. They are 658 miles apart after 7 hours. Find the rate of each motorcycle if one travels 10 miles per hour faster than the other.

2. Two trains start from the same station at the same time and travel in opposite directions for 6 hours, at which time they are 732 miles apart. One train travels 12 miles per hour faster than the other. How fast is each traveling?

3. Sara left home on her bicycle traveling at a rate of 6 mph. One hour later Becca left to catch up with her sister, traveling at 10 mph. How long will it take Becca to overtake Sara?

4. A freight train left Cleveland traveling at 40 mph. Two hours later a passenger train left the same station, following the same route and traveling at 60 mph. How long will it take the passenger train to overtake the freight train?

5. A truck and a car leave the city at the same time and travel in opposite directions. The car travels at 62 mph and the truck at 56 mph. After how many hours will they be 531 miles apart?

6. David left the campsite in his canoe and paddled upstream at a rate of 3 mph. Two hours later, Richard left camp in his motorboat and followed at 8 mph. How long will David be paddling before Richard catches him?

7. A jet and a private plane leave the same airport at the same time and travel in opposite directions. The jet travels 250 mph faster than the private plane. Find the rates of both planes if they are 2,550 miles apart after 3 hours.

8. Two planes leave the same airport at the same time and travel in opposite directions. One plane travels 50 mph faster than the other. If in 2 hours they are 860 miles apart, how fast is each plane flying?

9. At the same time, John and Nan start from two different cities that are 330 miles apart. They travel toward each other and meet for lunch 3 hours later. Find their rates if Nan travels 10 mph faster than John.

10. Charles and Don start at the same time from two different towns that are 120 miles apart and travel toward each other to meet for a picnic. Charles travels on the highway and can average 20 mph faster than Don. If they meet in 1.5 hours, how fast did each one travel?

11. Renée left her house on her bicycle traveling at 6 mph. One hour later Pierre set out to catch her and traveled at 8 mph. How long will it take Pierre to catch Renée?

12. Two airplanes take off from Scott Airport at the same time and travel in opposite directions. After three hours they are 1,272 miles apart. How fast is each flying if one plane averages 50 miles per hour faster than the other?

13. At the same time, Kay and Lisa leave their homes that are 24 miles apart and walk toward each other on the city bike path. If Kay travels at 5 miles per hour and Lisa at 3 miles per hour, how long will it take before they meet?

14. Two cars start at the same point at the same time and travel in opposite directions until they are 472 miles apart. How long will they travel if one averages 60 miles per hour and the other averages 58 miles per hour?

15. Two cars leave Berlin at 8:30 A.M. and travel in opposite directions. One of the cars averages 55 mph and the other averages 65 mph. How long will it take for the cars to be 480 miles apart?

16. Two bicyclists leave Meridan at 10:00 A.M. and travel in opposite directions. One averages 22 mph and the other averages 28 mph. How long will it take for them to be 200 miles apart?

17. Manuella and Marisol live in cities that are 440 miles apart. They leave their homes at the same time and drive toward each other until they meet 4 hours later. What is Marisol's average speed if she drives 20 miles per hour faster than Manuella?

18. Ingrid and Max leave their home at the same time and ride their bicycles in the same direction along the same road. Max averages 27 miles per hour and Ingrid averages 24 miles per hour. In how many hours will Max be seven miles ahead of Ingrid?

19. A bank robber leaves the scene of the crime and travels east on Route 6 at 60 miles per hour. One hour later, a police officer leaves the robbery site and travels along the same route at 75 miles per hour. In how many hours will the police officer overtake the robber?

20. A jet travels 200 mph faster than a propeller plane. The jet can fly in 3 hours the same distance that the propeller plane travels in 7 hours. What is the rate of travel of the jet?

14.7 Age Problems

These problems examine the relationship between two peoples' ages now and at some other time in the past or in the future. Have you ever looked at how such a relationship changes as you get older?

Suppose you are 30 years old and have a 10-year-old son. The ratio of your age to his is 30/10, or 3/1. Your age is 3 times his age. In 10 years, you will be 40 and he will be 20. That ratio is 40/20, or 2/1. Your age at that time will be twice his age. In another 10 years, the ratio will be 50/30, or 5/3. This is the type of thing that the word problems in this section consider.

In order to simplify age problems, try using a box to organize your information. It should look like Chart 14.

CHART 14

Name	Age Now	Other Time

Start by reading the problem carefully. Draw a box that is similar to Chart 14 and put in the same headings except for the last column. Replace Other Time with the time given, such as In 6 Years or 4 Years Ago. Follow the examples so that you will be able to do these problems yourself.

EXAMPLE 1 A mother is three times as old as her son. In 10 years she will be twice as old as he will be then. What are their ages now?

Draw a box like Chart 15 and label the Other Time column In 10 Years.

CHART 15

Name	Age Now	In 10 Years

The sentence "A mother is three times as old as her son" gives you the information for the Age Now column. His age is the one you know the least about, so call it x. Her age is three times his age, so call it $3x$ (Chart 16).

CHART 16

Name	Age Now	In 10 Years
Mother	$3x$	
Son	x	

You do not have to go back to the problem to get the rest of the information; it is already in the box. The last column should list the ages in 10 years. That means you simply *add 10* to whatever you wrote in the Age Now box. Do not include any other information in this column! (see Chart 17)

CHART 17

Name	Age Now	In 10 Years
Mother	$3x$	$3x + 10$
Son	x	$x + 10$

Now use the information in the last column and the middle sentence of the problem, "She will be twice as old as he is then," to write the equation.

Equation

She	will be twice as old as he.
$3x + 10$	$= \quad 2(x + 10)$

It makes sense that if you are going to set a larger and a smaller age equal to each other, then the smaller age will have to be multiplied by a quantity greater than 1 to make them equal.

Solution

$$3x + 10 = 2(x + 10)$$
$$3x + 10 = 2x + 20$$
$$\underline{-2x \qquad\quad -2x}$$
$$x + 10 = \qquad\quad 20$$
$$\underline{-10 \qquad\quad -10}$$
$$x = \qquad\quad 10$$

Now go back to the problem and the box to answer the question "What are their ages now?"

Son = x = 10 years old;   Mother = $3x$ = 30 years old.

Check   Use the In 10 Years column to check the answer.

Mother: $3x + 10 = 3(10) + 10 = 30 + 10 = 40$ years old

Son: $x + 10 = 10 + 10 = 20$ years old

Her age will be twice his age: $40 = 2(20)$ True.   ◆

E X A M P L E   2   Steve is 4 years older than his brother Marc. Two years ago, Steve's age was twice Marc's age. How old are they now?

Fill in Chart 18 and check your entries with Chart 19.

CHART 18

Name	Age Now	Other Time

CHART 19

Name	Age Now	2 Years Ago
Steve	$x + 4$	$x + 4 - 2$
Marc	x	$x - 2$

Equation   Use the last column of Chart 19 to write an equation that says:

$$\underline{\text{Steve}} \text{ was } \underline{\text{twice Marc's age}} .$$
$$x + 4 - 2 = \qquad 2(x - 2)$$

Solution $x + 4 - 2 = 2(x - 2)$
$$x + 2 = 2x - 4$$
$$\underline{-x \qquad\quad -x}$$
$$2 = x \quad - 4$$
$$\underline{+4 \qquad\quad +4}$$
$$6 = x$$

Using the Age Now column: Marc = x = 6 years old, and Steve = $x + 4 = 6 + 4 = 10$ years old.

Check Using the 2 Years Ago column:

Steve was $x + 4 - 2 = 6 + 4 - 2 = 8$ years old

Marc was $x - 2 = 6 - 2 = 4$ years old

Steve was twice Marc's age: $8 = 2(4)$ True.   ◆

E X A M P L E 3 Four years ago, Tony's age was four times his daughter's age. How old is each of them now if today her age is one-third of her father's age? Draw a box and list the given information. (Compare your result with Chart 20.)

CHART **20**

Name	Age Now	4 Years Ago
Tony	x	$x - 4$
Daughter	$\frac{1}{3}x$	$\frac{1}{3}x - 4$

Equation

$\underline{\text{Tony's age}}$ $\underline{\text{was}}$ $\underline{\text{four times his daughter's age.}}$

$$x - 4 \quad = \quad 4\left(\frac{1}{3}x - 4\right)$$

Solution $\quad x - 4 = 4\left(\frac{1}{3}x - 4\right)$

$$x - 4 = \frac{4}{3}x - 16$$

$$\underline{-x \qquad\quad -\frac{3}{3}x} \qquad\qquad \text{Use } \frac{3}{3}x \text{ for } 1x \text{ to have a common denominator.}$$

$$-4 = \frac{1}{3}x - 16$$

$$\underline{+16 = \qquad\qquad +16}$$

$$12 = \frac{1}{3}x \qquad\qquad \text{Multiply by the reciprocal of } \frac{1}{3}.$$

$$\frac{3}{1} \cdot \frac{12}{1} = \frac{3}{1} \cdot \frac{1x}{3}$$

$$36 = x$$

Ages now: Tony $= x = 36$ years old, and daughter $= \frac{1}{3}x = \frac{1}{3}(36) = 12$ years old

Check Four years ago: Tony $= x - 4 = 36 - 4$

$= 32$ years old

Daughter $= \frac{1}{3}x - 4 = \frac{1}{3}(36) - 4 = 8$ years old

Tony was four times his daughter's age:

$32 = 4(8)$ True.   ◆

E X A M P L E 4 Greg is 20 years old and Danny is 15. How many years ago was Greg's age twice Danny's age?

In this problem the unknown is the "other time," so label that column x Years Ago (Chart 21).

CHART 21

Name	Age Now	x Years Ago
Greg	20	$20 - x$
Danny	15	$15 - x$

Equation <u>Greg's age</u> <u>was</u> <u>twice Danny's age.</u>

$$20 - x \ = \ 2(15 - x)$$

Solution

$$20 - x = 2(15 - x)$$
$$20 - x = \ 30 - 2x$$
$$\underline{+ 2x \qquad + 2x}$$
$$20 + x = \ 30$$
$$\underline{- 20 \qquad - 20}$$
$$x = \ 10 \text{ years ago}$$

Check Greg $= 20 - x = 20 - 10 = 10$ years old

Danny $= 15 - x = 15 - 10 = 5$ years old

Greg was twice Danny's age: $10 = 2(5)$ True. ◆

14.7 ◆ Practice Problems

In each problem, draw a box to organize the given information. Then write an equation, solve it, and check the answer.

1. Mike is 4 years older than Steve. Twenty years ago, Mike was twice Steve's age. How old is each now?

2. Anne is 20 years older than her son Marc. Ten years ago, Anne's age was 5 times his age. How old is each one today?

3. Marie is 56 years old and her grandson is 16. In how many years will her age be three times his age?

4. Nancy's van is 3 years older than her husband's car. Next year her van will be twice as old as his car. How old is her van now?

5. Tony collects coins. He has a quarter that is 24 years older than a valuable dime. In 8 years, the quarter will be twice as old as the dime. How old is each coin now?

6. In an antique furniture store, an oak chair is 42 years old and a cherry end table is 14 years old. How many years ago was the age of the chair equal to five times the age of the table?

7. In Sue's jewelry collection, a diamond ring is 35 years old and a ruby ring is 7 years old. In how many years will the diamond be 3 times as old as the ruby?

8. Mike and Elena are cousins. Mike is 10 years older than Elena. Two years ago Mike was twice as old as Elena. How old is Mike now?

9. Sara's age is 1/5 of Becca's age. In three years, Becca will be twice as old as Sara. How old is each now?

10. Jennifer's age is 1/3 of her father's age. In 10 years, he will be twice as old as she is then. How old is Jennifer now?

11. A mother is 42 years old and her son is 14. How many years ago was she 5 times as old as he was?

12. Candace is 20 years old and her father is 44. In how many years will he be twice as old as Candace will be?

13. Brenda is five times as old as Jenna. In 5 years, Brenda's age will be three times Jenna's age. How old is Brenda now?

14. A first edition of a James Bond novel is 22 years old. A reprint is 2 years old. In how many years will the first edition be twice as old as the reprint?

15. Tony has a very old coin in his collection. He has a $5 gold piece that is 84 years older than a silver dollar. Twenty years ago, the gold piece was three times as old as the silver dollar. How old is each coin now?

16. Jim owns an antique auto that is 45 years older than his other car. In 13 years, the antique car will be four times as old as the other car. How old is the antique auto now?

17. The sum of the present ages of two collectible plates is 20 years. In 1 year the Hummel plate will be ten times as old as the Norman Rockwell plate. How old is each plate? (*Hint:* For the ages now, let one age be x and let the other be the total minus x.)

18. The sum of the present ages of Anya's and Uri's cars is 18 years. Two years ago, Anya's car was three times as old as Uri's car. How old is each car now? (See the hint in Problem 17.)

19. The sum of the ages today of an antique clock and an antique table is 160 years. If 65 years ago the clock was twice as old as the table, how old is each today? (See the hint in Problem 17.)

20. Valdez's and Monique's present ages add to 18 years. In 6 years, Valdez will be twice as old as Monique. What is Monique's age now? (See the hint in Problem 17.)

14.8   Inequality Word Problems

Some word problems require an answer that is a range of numbers rather than a single number. Others might be looking for a minimum or maximum value. Such problems translate into an inequality rather than an equation. We solve them the same way we solved word problems in Chapter 11 and earlier in this chapter. The examples that follow illustrate this type of problem.

E X A M P L E   1   Find the numbers that satisfy the following relationship: Six more than twice a number is greater than 32.

Inequality   Let n = number

<u>Six</u> <u>more than</u> <u>twice the number</u> <u>is greater than</u> <u>32.</u>
  6   + $2n$ >   32

Solution
$$6 + 2n > 32$$
$$\underline{-6 \qquad\quad -6}$$
$$2n > 26$$
$$n > 13$$

All numbers greater than 13

Check   Choose a value that is greater than 13, such as 15, and see if the inequality is satisfied.
$$6 + 2n > 32$$
$$6 + 2(15) > 32$$
$$6 + 30 > 32$$
$$36 > 32 \quad \text{True.}$$

◆

E X A M P L E   2   To earn a grade of "B" in Dr. Hocken's course, a student must have a grade average that is greater than or equal to 80 but less than 90. If Barry has test grades of 77, 89, and 68, what are the possible grades he can get on the next test so that he will earn a "B"?

Inequality   To find an *average*, add the four test grades together and divide by four. Let x = grade on the fourth test

$$80 \leq \frac{77 + 89 + 68 + x}{4} < 90$$

Solution Combine the like terms in the numerator, and then multiply all parts of the inequality by 4 to eliminate the denominator.

$$80 < \frac{234 + x}{4} < 90$$

$$4(80) < \overset{1}{\cancel{4}}\left(\frac{234 + x}{\underset{1}{\cancel{4}}}\right) < 4(90)$$

$$
\begin{array}{ccc}
320 < & 234 + x < & 360 \\
-234 & -234 & -234 \\
\hline
86 < & x < & 126
\end{array}
$$

Assuming that a grade greater than 100 is not possible, Barry must get a test grade greater than or equal to 86 or less than or equal to 100.

Check Use a value between 86 and 100, such as 88.

$$80 < \frac{77 + 89 + 68 + 88}{4} < 90$$

$$80 < \frac{322}{4} < 90$$

$$80 < 80.5 < 90 \quad \text{Both parts are true.} \qquad \blacklozenge$$

E X A M P L E 3 The Performing Arts Center has 500 seats. If admission for adults is \$5.00 and admission for children is \$3.00, what is the least number of adult tickets that can be sold and still cover the \$1,800 in production costs for a show?

Inequality Use a chart like Chart 22 to organize the information.

CHART **22**

Item	Cost of Each	Number of Each	Total Cost of Each
Adults' tickets	5.00	x	$5.00x$
Children's tickets	3.00	$500 - x$	$3.00(500 - x)$
			Total Needed 1,800

$$5.00x + 3.00(500 - x) \geq 1,800$$
$$\text{or } 5x + 3(500 - x) \geq 1,800$$

Solution
$$
\begin{array}{rcl}
5x + 1,500 - 3x & \geq & 1,800 \\
2x + 1,500 & \geq & 1,800 \\
-1,500 & & -1,500 \\
\hline
2x & \geq & 300 \\
x & \geq & 150
\end{array}
$$

At least 150 adult tickets must be sold.

Check   See that at least $1,800 is collected if 150 adults' tickets and 350 (500 − 150) children's tickets are sold.

$$5(150) = \quad \$750$$
$$3(250) = \underline{\$1,050}$$
$$\qquad\qquad \$1,800 \quad \text{True.}$$

◆

E X A M P L E   **4**   If six more than twice an integer is less than 18, what is the maximum value of the integer?

Solution   Let n = integer

$$2n + 6 < \quad 18$$
$$\underline{-6 \qquad -6}$$
$$2n \quad < \quad 12$$
$$n \quad < \quad 6$$

If the integer must be less than 6, then the maximum value for the integer is 5. The answer to the inequality may not be the answer to the word problem, just as answers to equations often weren't the answer to word problems. Be sure to take the solution back into the problem (words) to decide how to answer the question.

Check   Use the value 5 as a test.

$$2n + 6 < 18,$$
$$2(5) + 6 < 18,$$
$$10 + 6 < 18,$$
$$16 < 18 \quad \text{True.}$$

◆

14.8 ◆ Practice Problems

For each problem, write an inequality and then solve it.

1. If ten less than an integer is greater than eight, what is the minimum value of the integer?

2. Find the smallest integer that satisfies the condition that three more than five times a number is less than the difference between six times the number and four.

3. What is the smallest integer such that four less than three times the integer is greater than five plus twice the integer?

4. For what numbers is seven more than twice the number greater than three?

5. Juan has a bowling average of 175. So far tonight, he has bowled games of 210 and 160. What must he score in his third game if he is to exceed his average?

6. Katrina has received grades of 68 and 82 on her first two writing assignments. What grade must she get on the next assignment to have at least an 80 average?

7. In Dr. Garfield's Cultural Diversity course, a total of 400 points is needed on five exams to earn a grade of "B." So far, Marty has grades of 80, 72, 64, and 94. What grades on the fifth exam will result in her earning a "B" or better for the course?

8. Jack scored 69 and 73 during the first two rounds of a golf tournament. What is the minimum he can score in the third round to maintain his average of 70 on this course?

9. The length of a rectangle is three feet more than the width. What is the largest possible length if the perimeter cannot exceed 72 feet?

10. A Christmas tree farm needs to run fencing around its newly planted trees. The length of this new rectangular plot must be 4.5 times its width. What is the smallest that the length can be if the perimeter of the lot is at least 440 yards?

11. Find the smallest of three consecutive integers if the difference between three times the smallest and the largest is greater than 54.

12. Find the smallest of three consecutive even integers if the difference between three times the smallest and four times the largest is greater than zero.

13. Helen is going to buy fifteen stamps. Some will be 29-cent stamps and the others will be 20-cent stamps. What is the maximum number of 29-cent stamps she can buy if she spends no more than $3.90?

14. Carlos received grades of 83, 94, and 89 on the first three tests in his economics class. What is the lowest grade he can receive on the next test to have at least a 90 average?

15. Try Harder Car Rental Company charges customers $35 per day to rent a mid-size car. Number One Company charges $20 per day plus $0.15 per mile. For a five-day rental, what is the maximum number of miles a Number One car can be driven and still cost less than a Try Harder car that is rented for the same period?

16. A company rents trucks for $12 a day plus $0.12 per mile. What is the maximum number of miles the truck can be driven in one day if the rental fee is not to exceed $30?

14.9   Chapter Review

This chapter has brought together many of the concepts you have been working on and have mastered since you started this course. We hope that you have overcome your distaste for word problems by now and that you realize you now have the skills to determine what operation is called for, to organize the information, to write an equation, to solve that equation, and to check your answer(s). When you go on to other math courses and do more course work in your field of study, you will have to solve more application problems than simple calculation or equation problems. Also, when you must solve non-math problems, the skills you have been using in this chapter will help you to read (or listen) for details, determine what information is important and what is not, organize the facts into a usable form, decide what procedure(s) to follow to reach a solution, and check to be sure that the solution is reasonable. These skills go far beyond this algebra course!

As you review for the chapter test, be sure you are able to write and correctly solve equations for applications involving general problems, consecutive integers, perimeters, coins or mixtures, motion, and age. You will also need to be able to solve inequality word problems.

Review Problems

A. Write an equation for each word problem, and then solve it.

1. Three co-workers contributed toward a wedding gift for their secretary. One gave $10 more than the second. The third worker gave twice as much as the first. Altogether they donated $90. How much did each contribute?

2. Three numbers have a sum of 39. The second number is four less than the first, and the third number is equal to three times the second. What are the three numbers?

3. Michael goes to the drugstore to buy some snacks. Candy bars cost 45 cents each, and gum is 30 cents a pack. The number of candy bars that he buys is twice as large as the number of packs of gum, and he spends $2.40. How many candy bars does he buy?

4. The larger of two numbers is three more than twice the smaller. And when the larger number is subtracted from the smaller, the result is −15. What are the numbers?

5. Find three consecutive odd numbers such that the sum of three times the largest and twice the smallest is 97.

6. The total cost for padding and upholstery fabric to recover a small living room chair will be $73.50. The amount of fabric to be used is two-and-a-half times the amount of padding needed. Padding costs $3.25 per yard, and fabric costs $8.50 per yard. How much fabric will be needed?

7. The Birch and Maple Tree Farm needs to run fencing around its newly planted trees. The length of this new rectangular plot is 4.5 times its width. If the perimeter of the lot is 440 yards, what is its length?

8. Uncle Matt is two-and-a-half times as old as his niece, Cheryl. Six years ago he was four times as old as Cheryl was. How old is Matt now?

9. Find three consecutive numbers if the difference between three times the smallest and the largest is 50.

10. Maggie has agreed to take her daughter Carol's friends to Pizza King for her birthday party. Pizza King charges $1.75 per person and $2.50 for favors for the group. How many friends can Carol invite if her mom has set a $20 limit? (Remember that Maggie and Carol will eat too.)

11. Find three consecutive even integers such that the difference between three times the smallest and four times the largest is zero.

12. To mix a perfect blend of chocolate almond coffee, the clerk at the Roasted Bean needs to grind 1 ounce more of chocolate beans than the amount he uses of the almond-flavored beans. How many ounces of each will he mix if Mr. Harris asks for $4.00 worth of the blend? (The almond-flavored beans cost $0.95 an ounce, and the chocolate beans cost $0.70 an ounce.)

13. A propeller-driven plane leaves the Municipal Airport traveling at 200 mph. Two hours later, a jet leaves the same airport and follows the same flight path. It travels at 600 mph.

    a. In how many hours will the jet overtake the propeller plane?

    b. How far will the propeller plane have traveled in that time?

14. The width of a rectangular yard is one-fourth its length. Find the width if the perimeter is 250 feet.

15. Find three consecutive odd integers such that twice the smallest is equal to seven more than the largest.

16. A long-distance cyclist leaves Cleveland at 8 A.M. and travels toward Columbus at 12 mph. At 11 A.M. a support car carrying food and supplies sets out to follow the cyclist and travels at 48 mph.

    a. How long will it take the car to overtake the cyclist?

    b. How far will the cyclist have traveled?

17. An antique dealer has a 75-year-old ice box and an 85-year-old grandfather clock in stock. How many years ago was the clock's age twice the age of the ice box?

18. Karen has to drive to a resort town in North Carolina. Part of the 480-mile distance is on highways, and part is on winding mountain roads. She can average 20 miles per hour faster on highways. She drives for 4 hours on highways and 6 hours on mountain roads. What is her average speed on highways?

19. Sue is three-and-a-half times as old as her son David. In 15 years, her age will be twice his age. How old is each one now?

20. A stamp collector has a 20-year-old 2-cent stamp and a 16-year-old 5-cent stamp in his collection. How many years ago was the age of the 2-cent stamp twice the age of the 5-cent stamp?

21. A car and a bus set out from Rochester, New York, traveling in opposite directions. The car averaged 40 miles per hour less than twice the speed of the bus. After 2 hours, they were 220 miles apart. How fast was the car traveling?

22. Maria is putting a wallpaper border around her baby's nursery wall. She uses 46 feet of border paper. If the width of the room is 7 feet less than the length, what are the room's dimensions?

23. Twenty-four feet of weather-stripping material is needed to go around a door frame. If the width is 6 feet less than the length, what are the door's dimensions?

24. Four less than twice a number is the same as five times the sum of the number and 1. What is the number?

B. For each problem, write an inequality and then solve it.

25. Three times a number is greater than five times the sum of the number and four. Find the possible values for the number.

26. One-half a number is less than twice the difference between the number and 6. Find the possible values for the number.

27. Jo is preparing a new garden plot. She wants to use no more than one 72-foot package of edging to outline this rectangular plot. The width must be 20 feet. What is the maximum length she can make this garden?

28. Marian has bowled a 126 and a 147 in her league this evening. What score can she get in her third game to exceed her three-game average of 145? (A perfect game in bowling is 300.)

Chapter Test

A. For each word problem, write an equation and then solve it.

Find the dimensions of a rectangle if the length is 7 inches more than three times the width. The perimeter is 174 inches.

1. Equation _____

2. Solution _____

An antique table is 42 years older than a chair. Twenty years ago, the table was three times as old as the chair was then. How old is each item now?

3. Equation _____

4. Solution _____

A food company wants to improve its fruit punch product by adding more pure orange juice. How much orange juice (100% orange juice) should be added to 50 gallons of the fruit punch that is 20% orange juice to create a more healthful mixture that is 50% orange juice?

5. Equation _____

6. Solution _____

When Pat drove to Jamestown on Friday evening to visit her parents, it took her 6 hours. When she made the return trip on Sunday, with lighter traffic, the trip only took 5 hours. What was her average speed on the trip on Sunday if she averaged 10 miles per hour faster than on Friday?

7. Equation _____

8. Solution _____

An office has a jar to collect money to pay for coffee supplies. The jar contains $3.25 in nickels, dimes, and quarters. There are twice as many nickels as dimes and four more quarters than dimes. How many of each coin are there in the jar?

9. Equation _____

10. Solution _____

One number is five less than a second number. If three times the larger number is added to four times the smaller, the result is 204. What are the numbers?

11. Equation _____

12. Solution _____

Sue bought 28 stamps at the Post Office. How many 29-cent stamps and how many 20-cent stamps did she buy for $7.40?

13. Equation _____

14. Solution _____

The sum of twice the first and three times the second of two consecutive odd integers is −239. What are the integers?

15. Equation _____

16. Solution _____

A freight train leaves Baltimore for Boston at 7 A.M., traveling at an average of 35 miles per hour. One hour later, a passenger train leaves Boston for Baltimore, traveling the same route at 45 miles per hour. How long will it take before the passenger train meets the freight train if the cities are 315 miles apart?

17. Equation _____

18. Solution _____

B. Write an inequality and then solve it.

The length of a rectangle is 8 yards more than the width. What is the largest possible length (in whole yards) if the perimeter cannot exceed 68 yards?

19. Inequality _____

20. Solution _____

Thought-Provoking Problems

1. Describe a real-life situation (not necessarily related to math) wherein the problem-solving skills you have used in Chapters 11 and 14 could be applied. Explain each step of the process.

2. Write an inequality word problem. Be sure the wording is clear and complete. Then solve your problem.

3. The perimeter of a square is equal to the perimeter of a rectangle. The length of the rectangle is 10 inches less than twice the length of one side of the square. The width of the rectangle is 1 inch less than half the length of one side of the square. Find the dimensions of each figure.

4. a. How much pure salt should be added to 180 milliliters of a 20% salt solution to produce a 23% salt solution? [*Hint*: What is the concentration (percent) of a pure product?]

   b. How much distilled water should be added to 30 liters of a 15% sulfuric acid solution to dilute it to a 10% sulfuric acid solution? [*Hint*: What is the percent of sulfuric acid in water?]

5. Marco biked from Meridian to Jackson at 20 miles per hour. On the return trip the next day, he decreased his speed to 18 miles per hour, and the trip took 1 hour longer. How far apart are the two cities?

6. Sheena paddled her canoe 6 miles upstream against a 2-mph current and then returned to her starting point. If the whole trip took 4 hours, how fast can she travel in still water? (*Hint*: If r is the rate in still water, then $r + 2$ is the rate with the current, and $r - 2$ is the rate against the current.)

7. An airplane travels 700 miles against a steady head wind of 20 miles per hour. Then it returns to the same point with that same steady tail wind. How fast can the plane fly in still air? (*Hint*: The wind speed must be added to or subtracted from the plane's still-air rate.)

8. A motorboat takes 4 hours to travel a certain distance upstream, but it takes only 2 hours to go the same distance when traveling downstream. If the boat's rate of travel is 15 mph in still water, what is the speed of the current?

9. The sum of Tom's and Mike's present ages is 12 years. In 2 years, Tom's age will be equal to Mike's age 2 years ago. How old is each now?

10. Presently, Timmy's age is two-thirds of Natalie's age. In 6 years, his age will be seven-ninths of her age then. How old is Timmy now?

15

Polynomial Operations

Working with Polynomials

In the work you have been doing in Chapters 9 through 14, you have learned how to add and subtract like terms and how to multiply variable terms by numerical constants. You will extend those operations in this chapter.

◆ **Learning Objectives**

When you have completed this chapter, you should be able to:

1. Add and subtract monomials.
2. Multiply and divide monomials using the laws of exponents.
3. Simplify expressions that contain negative exponents.
4. Add and subtract polynomials.
5. Multiply polynomials.
   a. Multiply a monomial times a polynomial using the Distributive Principle.
   b. Multiply a polynomial times a polynomial using the Distributive Principle.
6. Divide polynomials:
   a. by monomials
   b. by polynomials
7. Work with scientific notation.

15.2 **Adding and Subtracting Monomials**

The term **polynomial** refers to algebraic expressions of one or more terms that have positive-integer exponents. A polynomial can have an infinite number of terms.

Polynomials	Not Polynomials
$8Y - 4$	$1/x$
$7b^2 + 4b - 15$	$A^{-3} + 2$
$18k$	$Y^{2/3} - 3$
$2c + 9d - 5e + 14f$	

Definition of Polynomial

An algebraic expression of one or more terms in which the terms are of the form ax^n, where n is a positive integer or zero.

Polynomials of one, two, or three terms have specific names. One-term polynomials are **monomials**, two-term polynomials are **binomials**, and polynomials of three terms are **trinomials**.

Monomials	Binomials	Trinomials
$3r^2s^3t$	$2x + 3$	$2m^2 - 3m + 7$
$4x$	$a^2 - b^2$	$c^2 - 7cd + 5d^2$
-19	$8f - g$	$-11 + 5t + 8s$

When you add or subtract monomials, you combine terms such as $7s$ and $-15s$. Such quantities can be added or subtracted only if they are *like terms*; that is, their variable and exponent parts are exactly the same. You have been combining like terms since Chapter 9. Combining these terms can be viewed as an application of the Distributive Principle, though it is seldom written this way when actually being calculated.

$$7s + (-15s) = [7 + (-15)]s = -8s$$

You can line up the terms vertically or horizontally. Remember to apply signed-number rules.

$$\begin{array}{r} 7s \\ + \underline{-15s} \\ -8s \end{array} \quad \text{or} \quad 7s + (-15s) = -8s$$

If you were to subtract $-15s$ from $7s$, your work would look like this:

$$\begin{array}{r} 7s \\ \text{S} \\ \underline{- (-15s)} \end{array} \quad \text{becomes} \quad \begin{array}{r} 7s \\ \text{A} \\ \underline{+ (+15s)} \\ 22s \end{array}$$

$$\begin{array}{cc} \text{S} & \text{A} \\ \text{or} \quad 7s - (-15s) & = 7s + (+15s) = 22s \end{array}$$

E X A M P L E **1** Find the difference between $-4a^2b^2c$ and $3a^2b^2c$.

Solution $-4a^2b^2c$ and $+3a^2b^2c$ are like terms because the variable part, a^2b^2c, **is the same in both** terms.

$$
\begin{array}{ll}
\quad -4a^2b^2c & \qquad\qquad -4a^2b^2c \\
\text{S} & \qquad\qquad \text{A} \\
\underline{-\ +3a^2b^2c} \quad \text{becomes} & \qquad \underline{+\ -3a^2b^2c} \\
& \qquad\qquad -7a^2b^2c
\end{array}
$$

or

$$
\begin{array}{l}
\qquad\qquad \text{S} \\
-4a^2b^2c \ - \ (+3a^2b^2c) \\
\qquad\quad \text{A} \\
= -4a^2b^2c + (-3a^2b^2c) \\
= -7a^2b^2c
\end{array}
$$

◆

E X A M P L E **2** Simplify: $11x^2 + (-2x^2) + (-7x^2) + 5x^2$

Solution There are two ways to simplify this problem. You can do it in the order given or do it by rearranging the terms (Associative and Commutative Properties of Addition). **Both will be** shown.

To do the problem in the order given, write

$$
\begin{array}{l}
\underline{11x^2 + (-2x^2)} + (-7x^2) + 5x^2 \\
= \underline{9x^2 \ + \qquad\qquad (-7x^2)} + 5x^2 \\
= \underline{2x^2 \ + \qquad\qquad\qquad\qquad 5x^2} \\
= 7x^2
\end{array}
$$

To do the problem by rearranging the terms, combine all the **positive terms and then all** the negative terms.

$$
\begin{array}{l}
11x^2 + (-2x^2) + (-7x^2) + 5x^2 \\
= \underline{11x^2 + 5x^2} + \underline{(-2x^2) + (-7x^2)} \\
= 16x^2 \qquad + \qquad (-9x^2) \\
= 7x^2
\end{array}
$$

◆

E X A M P L E **3** Simplify: $2a^2b - 3ab + 5ab^2$.

Solution There are no like terms, so there is nothing to add or subtract, **and the expression is al-** ready in simplest form.

◆

E X A M P L E **4** From the sum of $8c^2d^3$ and $(-11c^2d^3)$, subtract $(-5c^2d^3)$.

Solution
$$
\begin{array}{lll}
\text{Sum:} \quad 8c^2d^3 & \text{Difference:} \quad -3c^2d^3 = & -3c^2d^3 \\
\quad \underline{+\ -11c^2d^3} & \qquad\qquad \text{S} & \qquad \text{A} \\
\qquad -3c^2d^3 & \qquad \underline{-\ -5c^2d^3} = & \underline{+\ +5c^2d^3} \\
& & \qquad +2c^2d^3
\end{array}
$$

or

$$
\begin{array}{l}
\qquad\qquad\qquad \text{S} \\
\underline{8c^2d^3 + (-11c^2d^3)} - (-5c^2d^3) \\
\qquad\quad \text{A} \\
= -3c^2d^3 + (+5c^2d^3) \\
= 2c^2d^3
\end{array}
$$

◆

15.2 ◆ Practice Problems

A.   Add the monomial terms where possible.

1. $-3a$
$+4a$

2. $6x^2y$
$-2x^2y$

3. $-15g$
$-12g$

4. $+9abc^3$
$-9abc^3$

5. $12r^2s + (-7r^2s)$

6. $(-150x^2) + (-70x^2)$

7. $4b^2d^2 + 3b^2d$

8. $(-7yz) + (-3yz)$

9. $8m^2n + (-3m^2n)$

10. $5g^2h + 4gh^2$

B.   Subtract the polynomials where possible.

11. $-14k$
$- +11k$

12. $-5a^2b$
$- -7a^2b$

13. $6w^3$
$- -w^3$

14. $+ 8rs^2$
$- +12rs^2$

15. $6wxy^2 - (-5wxy^2)$

16. $(-3f^2) - (-5f^2)$

17. $8b^3c - 17b^3c$

18. $a^2b^2c^2 - 7a^2b^2c^2$

19. $44xyz - (-xyz)$

20. $-9m^2n - (-12m^2n)$

C.   Perform the indicated operation(s) in each of these problems.

21. $16gh + (-5gh) + 3gh$

22. $7q^3 + (-4q^3) + (-5q^3)$

23. $-7x^2 + 9x^2 - (-4x^2)$

24. $8m - 11m + 3m + (-m)$

25. $11a^2b^3 - 8a^2b^3 + a^2b^3 - 6a^2b^3$

26. $49x^2y - 11x^2y + 33x^2y - 16x^2y$

27. From the sum of $(-9mn)$ and $(17mn)$, subtract $(-3mn)$.

28. Subtract $(14fgh)$ from the sum of $(-6fgh)$ and $(11fgh)$.

29. Find the sum of $16r^2s$, $-3r^2s$, and $-6r^2s$.

30. From the sum of $(44z^2)$ and $(-56z^2)$, subtract $(20z^2)$.

31. Find the difference between $153p^3q^2$ and $(-119p^3q^2)$.

32. Find the sum of $75r^2s^3t^2$ and $(-87r^2s^3t^2)$.

33. Find the sum of $-86a^2b^2c^3$, $14a^2b^2c^3$, and $44a^2b^2c^3$.

34. Find the difference: $-128m^2n^3 - 116m^2n^2$.

35. From the sum of $(-8c^2d^3)$ and $(14c^2d^3)$, subtract $(-3c^2d^3)$.

15.3   Multiplying Monomials

When multiplying monomials, you do not need to work with only like terms as you do when adding and subtracting. Look at these monomial multiplications and see if you can discover a pattern and write a rule to describe that pattern.

In $2^3 \cdot 2^2$, the exponent (or power) tells you how many times the base is used as a factor. This means that

$$2^3 \cdot 2^2 = \underline{2 \cdot 2 \cdot 2} \cdot \underline{2 \cdot 2}.$$

But, by the definition of exponents, $2 \cdot 2 \cdot 2 \cdot 2 \cdot 2$ can be written 2^5. Therefore,

$$2^3 \cdot 2^2 = 2^5$$

Note: Variables and numbers with no exponent showing are understood to have an exponent of 1.

$$k = k^1 \qquad 2 = 2^1$$

E X A M P L E 1   Multiply: $k^2 \cdot k$

Solution $k^2 \cdot k$

$= k \cdot k \cdot k$

$= k^3$

◆

E X A M P L E **2** Multiply: $a^2 \cdot a^2$

Solution $a^2 \cdot a^2$

$= a \cdot a \cdot a \cdot a$

$= a^4$ ◆

Can you see a pattern in the expressions and answers? What relationship is there between the exponents in the original factors and the exponent in the answer? What happens to the base quantities?

Did you notice that the exponent in the answer is the *sum* of the exponents of the factors? And did you notice that the base quantity did *not* change?

$$2^3 \cdot 2^2 = 2^{3+2} = 2^5 \qquad k^2 \cdot k = k^{2+1} = k^3$$
$$a^2 \cdot a^2 = a^{2+2} = a^4$$

Multiplying Monomials with the Same Base

To multiply quantities with the same base, keep the base the same and add the exponents.

$$a^b \cdot a^c = a^{b+c}$$

If the monomials have numerical coefficients, then the Associative and Commutative Properties for Multiplication enable you to rearrange the factors as in Examples 3 through 6.

E X A M P L E **3** Multiply: $2b^3 \cdot 3b^2$

Solution $2b^3 \cdot 3b^2$

$= 2 \cdot b \cdot b \cdot b \cdot 3 \cdot b \cdot b$

$= 2 \cdot 3 \cdot b \cdot b \cdot b \cdot b \cdot b$

$= 6b^5$ ◆

E X A M P L E **4** Multiply: $(2x^3)(3x^4)$

Solution $(2x^3)(3x^4)$

$= 2 \cdot 3 \cdot x^3 \cdot x^4$

$= 6x^7$ ◆

E X A M P L E **5** Multiply: $11a(-2a^2b)$

Solution $11a(-2a^2b)$

$= 11 \cdot -2 \cdot a \cdot a^2 \cdot b$

$= -22a^3b$ ◆

E X A M P L E **6** Multiply: $(3r^2s)(2rs)(-5s^2)$

Solution $(3r^2s)(2rs)(-5s^2)$

$= 3 \cdot 2 \cdot -5 \cdot r^2 \cdot r \cdot s \cdot s \cdot s^2$

$= -30r^3s^4$ ◆

You do not need to write the intermediate steps in such problems. First multiply the numerical coefficients; then multiply the variables using the rule of exponents.

Exercises Try these exercises. If necessary, look at the examples.

1. $3ab(2a^2b)$ 2. $-2(-3x^2y)(2y^2)$ 3. $c^2d^2(cd^2)(c^2)$

Answers:

1. $6a^3b^2$ 2. $12x^2y^3$ 3. c^5d^4

There is another type of monomial multiplication. See if you can write a rule to describe the process you observe in Examples 7, 8, and 9.

E X A M P L E **7** Simplify: $(x^2)^3$

Solution $(x^2)^3$
$= (x \cdot x)^3$
$= \underline{x \cdot x} \cdot \underline{x \cdot x} \cdot \underline{x \cdot x}$
$= x^6$ ◆

E X A M P L E **8** Simplify: $(y^3)^3$

Solution $(y^3)^3$
$= (y \cdot y \cdot y)^3$
$= \underline{y \cdot y \cdot y} \cdot \underline{y \cdot y \cdot y} \cdot \underline{y \cdot y \cdot y}$
$= y^9$ ◆

E X A M P L E **9** Simplify: $(a^2)^4$

Solution $(a^2)^4$
$= (a \cdot a)^4$
$= \underline{a \cdot a} \cdot \underline{a \cdot a} \cdot \underline{a \cdot a} \cdot \underline{a \cdot a}$
$= a^8$ ◆

Can you see a pattern in $(x^2)^3 = x^6$, $(y^3)^3 = y^9$, and $(a^2)^4 = a^8$? In each case, when you raised an exponent form of the base to the power represented by an exponent, the base remained *the same* and the exponents were *multiplied* together to get the answer.

Raising a Base in Exponent Form to a Power

> To raise a base in exponent form to a power, keep the base the same and multiply the exponents.
>
> $(a^b)^c = a^{bc}$

Exercises Simplify each of these expressions, following the preceding examples if necessary.

1. $(m^3)^2$ 2. $(q^3)^4$ 3. $(y^6)^3$

Answers:

1. m^6 2. q^{12} 3. y^{18}

In problems where you are told to raise a product to the power represented by an exponent, raise each factor in the product to that power and apply the rules of exponents. For example, $(2b)^4$ is the same as $2b \cdot 2b \cdot 2b \cdot 2b = 2^4 b^4 = 16b^4$, but it is easier to apply the rule and get $(2b)^4 = 2^4 b^4 = 16b^4$.

E X A M P L E 10   Simplify: $(a^2 b)^3$

Solution $(a^2 b)^3$
$= (a^2)^3 (b)^3$
$= a^6 b^3$ ◆

E X A M P L E 11   Simplify: $(3x^2)^4$

Solution $(3x^2)^4$
$= (3)^4 (x^2)^4$
$= 81x^8$ ◆

Raising a Product to a Power

> To raise a product to a power, raise each factor to the power represented by that exponent.
>
> $$(ab)^c = a^c b^c$$

A similar rule applies to raising a fraction (quotient) to a power. You will raise each part (numerator and denominator) of the fraction to that power.

E X A M P L E 12   Simplify: $\left(\dfrac{a^2}{4}\right)^3$

Solution $\left(\dfrac{a^2}{4}\right)^3$
$= \dfrac{(a^2)^3}{4^3}$
$= \dfrac{a^6}{64}$ ◆

E X A M P L E 13   Simplify: $\left(\dfrac{2x}{3y^3}\right)^2$

Solution $\left(\dfrac{2x}{3y^3}\right)^2$
$= \dfrac{(2x)^2}{(3y^3)^2}$
$= \dfrac{2^2 x^2}{3^2 (y^3)^2}$
$= \dfrac{4x^2}{9y^6}$ ◆

Raising a Fraction to a Power

To raise a fraction to a power, raise both the numerator and the denominator to that power.

$$\left(\frac{a}{b}\right)^c = \frac{a^c}{b^c}$$

Exercises

Be sure that you understand these concepts by trying the sample problems that follow. Simplify each expression. The answers are provided so that you can check your work.

1. $(-2b^2)^2$ 2. $(5ax^2)^3$ 3. $\left(\dfrac{m^2}{n}\right)^3$ 4. $\left(\dfrac{0.4T^2}{R^3}\right)^2$ 5. $\left(\dfrac{2f^3}{3}\right)^2$

Answers:

1. $4b^4$ 2. $125a^3x^6$ 3. $\dfrac{m^6}{n^3}$ 4. $\dfrac{0.16T^4}{R^6}$ 5. $\dfrac{4f^6}{9}$

Pay close attention to how the problem is written! $(4x)^2$ and $4x^2$ are not equivalent because $(4x)^2 = 4^2x^2 = 16x^2$. In $4x^2$, only the x is raised to the second power. In the expression $4(x^2y^3)^5$, the exponent ^{5} is applied only to the factors inside the parentheses, not to the factor 4 that is outside. Thus, $4(x^2y^3)^5 = 4(x^2)^5(y^3)^5 = 4x^{10}y^{15}$.

Because an exponent applies only to the quantity that is immediately to its left, the expressions $(-2)^4$ and -2^4 are not equivalent. In $(-2)^4$, the quantity inside the parentheses is raised to the fourth power: $(-2)(-2)(-2)(-2) = 16$. In -2^4, only 2 is raised to the fourth power, and the negative sign means "the opposite of." Therefore, $-2^4 = -(2 \cdot 2 \cdot 2 \cdot 2) = -(16) = -16$, because the opposite of $+16$ is -16.

E X A M P L E   14   Simplify: -5^2

Solution $-5^2 = -(5 \cdot 5) = -(25) = -25$ ◆

E X A M P L E   15   Simplify: $(-5)^2$

Solution $(-5)^2 = (-5)(-5) = 25$ ◆

15.3 ◆ Practice Problems

A. Simplify each expression using the rules of exponents introduced in this section.

1. $y^2 \cdot y^3$ 2. $a^4 \cdot a^2$ 3. $b^3 \cdot b$

4. $m^2 \cdot m^2$ 5. $a^3 \cdot a^2$ 6. $m^3 \cdot m$

7. $z^3 \cdot z^4$ 8. $d \cdot d^2$ 9. $2^2 \cdot 2^3$

10. $3^2 \cdot 3^2$ 11. $2ab(3a^2b)$ 12. $3z(-2z^2)$

13. $A^2 \cdot A^3 \cdot A$ 14. $x \cdot x^2 \cdot x^4$ 15. $(-y)^3$

16. $(-x)^4$ 17. $y^2 \cdot y^4 \cdot y$ 18. $b \cdot b^2 \cdot b$

19. $(2x)^3$ 20. $(3y)^2$ 21. $(2m^2)^3$

22. $(6b^2)^2$ 23. $\left(\dfrac{-2g^2}{3}\right)^3$ 24. $\left(\dfrac{-x^3}{2y}\right)^2$

25. -2^6 26. -3^2 27. $(-2)^6$

28. $(-3)^2$ 29. -1^4 30. -1^6

B. Simplify each problem.

31. $(-3a^2)(2ab)(a^2b^2)$ 32. $x^2y(3xy^3)$

33. $(-4x^2)(-8x^3z^2)$ 34. $(7m^2n)(-2mn)(n^2)$

35. $(-a^3b^2)^5$ 36. $(-4x^3y^2)(-9x^5z^3)$

37. $y(x^2y^2)^3$ 38. $(-5m^2)^3$

39. $(-4abc)^3$ 40. $8a(a^2b)^3$

41. $\left(\dfrac{a^3b^2}{2}\right)^3$ 42. $\left(\dfrac{-x^2y}{2}\right)^3$

43. $(2m^2)(3mn^2)(-mn)$ 44. $\left(\dfrac{3}{4}c^3\right)^2$

45. $(2/5\,R^2S)^3$ 46. $6m(3m^4)^2$

47. $(5x^4)^2(-2x^5)^3$

49. $(8a^2b^3)^2$ 48. $(2y^2)^3(y^4)^2$

50. $(-7m^2n^2)(-2mn)(-3m^2n)$

15.4   Dividing Monomials

The rules for division will be developed as the rules for multiplication were developed. You will examine some step-by-step solutions to division problems and see if you can spot a pattern.

E X A M P L E   1   Divide: $\dfrac{x^3}{x}$

Solution $\dfrac{x^3}{x}$

$= \dfrac{x \cdot x \cdot x}{x}$

$= \dfrac{x \cdot x \cdot \overset{1}{\cancel{x}}}{\underset{1}{\cancel{x}}}$

$= x^2$ ◆

Because division by zero is undefined, a quotient such as $x^3 \div x$ has no value if $x = 0$. In this problem and in the rest of this chapter, assume that no denominator is zero.

E X A M P L E   2   Divide: $\dfrac{2^4}{2^3}$

Solution $\dfrac{2^4}{2^3}$

$= \dfrac{2 \cdot 2 \cdot 2 \cdot 2}{2 \cdot 2 \cdot 2}$

$= \dfrac{2 \cdot \overset{1}{\cancel{2}} \cdot \overset{1}{\cancel{2}} \cdot \overset{1}{\cancel{2}}}{\underset{1}{\cancel{2}} \cdot \underset{1}{\cancel{2}} \cdot \underset{1}{\cancel{2}}}$

$= 2^1$ or 2   ◆

E X A M P L E   3   Divide: $\dfrac{m^5}{m^2}$

Solution $\dfrac{m^5}{m^2}$

$= \dfrac{m \cdot m \cdot m \cdot m \cdot m}{m \cdot m}$

$= \dfrac{m \cdot m \cdot m \cdot \overset{1}{\cancel{m}} \cdot \overset{1}{\cancel{m}}}{\underset{1}{\cancel{m}} \cdot \underset{1}{\cancel{m}}}$

$= m^3$ ◆

Look at the problems and the answers again. What has happened to the base quantity? What about the exponents?

$$\frac{x^3}{x} = x^2 \qquad \frac{2^4}{2^3} = 2 \qquad \frac{m^5}{m^2} = m^3$$

In each example, the exponent of the answer was the *difference* between the exponents in the original fraction, and the base quantity was *unchanged*.

$$\frac{x^3}{x} = x^{3-1} = x^2 \qquad \frac{2^4}{2^3} = 2^{4-3} = 2^1 = 2 \qquad \frac{m^5}{m^2} = m^{5-2} = m^3$$

Dividing Monomials with the Same Base

To divide monomials with the same base, keep the same base and subtract the exponent in the divisor from the exponent in the dividend.

$$\frac{a^x}{a^y} = a^{x-y}$$

$$\frac{3^5}{3^2} = 3^{5-2} = 3^3 = 27 \qquad \frac{p^4}{p} = p^{4-1} = p^3$$

E X A M P L E 4   Divide: $\dfrac{x^{20}}{x^{11}}$

Solution $\dfrac{x^{20}}{x^{11}} = x^9$ ◆

E X A M P L E 5   Divide: $\dfrac{b^3}{b}$

Solution $\dfrac{b^3}{b} = b^2$ ◆

(Remember, b in the denominator is understood to have an exponent of 1.)

Exercises   Simplify these expressions and then check your answers with those given.

1. $\dfrac{z^3}{z}$ **2.** $\dfrac{h^5}{h^4}$ **3.** $\dfrac{a^6}{a^3}$

Answers:

1. z^2 **2.** h **3.** a^3

In all the division problems we have done so far, the difference between the exponents has been positive. That is, the exponent in the numerator was larger than the exponent in the denominator. In the next section you will learn to work with negative exponents, and later in this section you will learn about zero exponents.

When the monomials contain numerical coefficients as well as variables, you will divide or reduce those factors and then apply the exponent rule to the variables.

E X A M P L E **6** Divide: $\dfrac{10x^6}{2x^4}$

Solution $\dfrac{10x^6}{2x^4}$

$= 5x^2$ ◆

E X A M P L E **7** Divide: $\dfrac{-27a^8}{9a^3}$

Solution $\dfrac{-27a^8}{9a^3}$

$= -3a^5$ ◆

E X A M P L E **8** Divide: $\dfrac{12m^4}{16m}$

Solution $\dfrac{12m^4}{16m}$

$= \dfrac{3m^3}{4}$ ◆

When the monomials have more than one variable, you will repeat the procedures for each variable after reducing the numerical factors.

E X A M P L E **9** Divide: $\dfrac{36x^2y^4z^3}{12xy^2z^2}$

Solution $\dfrac{36x^2y^4z^3}{12xy^2z^2}$

$= 3xy^2z$ ◆

E X A M P L E **10** Divide: $\dfrac{-2a^5b^6}{8a^3b}$

Solution $\dfrac{-2a^5b^6}{8a^3b}$

$= -\dfrac{a^2b^5}{4}$ or $-\dfrac{1}{4}a^2b^5$

If either the numerator or the denominator is negative and the other part of the fraction is positive, the negative sign is written in front of the fraction (in dividing signed numbers, unlike signs give a negative answer). ◆

Now let's look at the meaning of an exponent of zero. This could occur in a division problem such as

$$\frac{y^5}{y^5} = y^{5-5} = y^0$$

Or you might see it in multiplication: $y^0 \cdot y^4 = y^{0+4} = y^4$. Both problems should suggest to you that $y^0 = 1$. In $y^5 \div y^5$, you divided a quantity by itself, and that always equals 1.

$$\frac{y^5}{y^5} = \frac{\overset{1}{\cancel{y}} \cdot \overset{1}{\cancel{y}} \cdot \overset{1}{\cancel{y}} \cdot \overset{1}{\cancel{y}} \cdot \overset{1}{\cancel{y}}}{\underset{1}{\cancel{y}} \cdot \underset{1}{\cancel{y}} \cdot \underset{1}{\cancel{y}} \cdot \underset{1}{\cancel{y}} \cdot \underset{1}{\cancel{y}}} = \frac{1}{1} = 1 \qquad \text{or} \qquad \frac{y^5}{y^5} = y^{5-5} = y^0 = 1$$

In the multiplication problem, $y^0 \cdot y^4 = y^4$. The expression y^4 is multiplied by y^0, but the answer is the same as the original quantity. The only number that can do this is 1. $y^0 \cdot y^4 = y^4$ and $1 \cdot y^4 = y^4$, so $y^0 = 1$.

Zero as an Exponent

Any nonzero quantity raised to a zero exponent is equal to 1.

$$a^0 = 1$$

You should always replace a quantity raised to a zero exponent with 1. Having a zero exponent in an answer is similar to not reducing a fraction to lowest terms. You have not finished simplifying the answer in either case.

E X A M P L E   11   Divide: $\dfrac{3x^5}{x^5}$

Solution $\dfrac{3x^5}{x^5}$

$= 3x^0$

$= 3 \cdot 1$

$= 3$ ◆

E X A M P L E   12   Divide: $\dfrac{16a^2b^3c}{24abc}$

Solution $\dfrac{16a^2b^3c}{24abc}$

$= \dfrac{2}{3} ab^2 c^0$

$= \dfrac{2}{3} ab^2 \cdot 1$

$= \dfrac{2}{3} ab^2$ ◆

Exercises   Simplify these expressions and check your answers.

1. $\dfrac{x^7 y^6}{x^7 y^2}$ 2. $\dfrac{12a^2 b^2 c^2}{4ab^2 c}$ 3. $\dfrac{144 m^2 n^3}{24 m^2 n}$

Answers:

1. y^4 2. $3ac$ 3. $6n^2$

In division problems such as Examples 13 and 14, you must simplify the numerator and denominator separately before doing the division.

E X A M P L E **13** Divide: $\dfrac{(2a^2)(-3ab^3)}{4a^2b}$

Solution $\dfrac{(2a^2)(-3ab^3)}{4a^2b}$

$= \dfrac{-6a^3b^3}{4a^2b}$

$= -\dfrac{3ab^2}{2}$ ◆

E X A M P L E **14** Divide: $\dfrac{(3m^2n)^2}{m^2(mn^2)}$

Solution $\dfrac{(3m^2n)^2}{m^2(mn^2)}$

$= \dfrac{(3)^2(m^2)^2(n)^2}{m^2(m)(n^2)}$

$= \dfrac{9m^4n^2}{m^3n^2}$

$= 9m$ ◆

15.4 ◆ Practice Problems

A. **Simplify each expression using the division rules.**

1. $\dfrac{x^7}{x^4}$ 2. $\dfrac{a^2}{a}$ 3. $\dfrac{c^5}{c^4}$ 4. $\dfrac{w^7}{w^3}$

5. $\dfrac{8d^4}{2d^2}$ 6. $\dfrac{2^8}{2^6}$ 7. $\dfrac{d^4}{d^4}$ 8. $\dfrac{64b^5}{72b^3}$

9. $\dfrac{-28g^3}{-7g}$ 10. $\dfrac{14c^2}{7c^2}$ 11. $\dfrac{3^4}{3^2}$ 12. $\dfrac{-81c^3}{-27c^2}$

13. $\dfrac{2^8}{2^8}$ 14. $\dfrac{-15z^4}{3z^2}$ 15. $\dfrac{12x^3}{27x}$ 16. $\dfrac{4^4}{4^4}$

B. **Simplify each expression.**

17. $(15x^2) \div (24x^2)$ 18. $(-11a^7) \div (44a^4)$

19. $(16x^2y^3) \div (4xy)$ 20. $(72a^2b^3) \div (12ab^2)$

21. $(-90b^2c^2d^2) \div (15cd)$

22. $(-144m^4n^3) \div (-24m^2n^2)$

23. $(4g^2h^4) \div (12g^2h^2)$ 24. $(15r^2s^2t^3) \div (3rst)$

25. $(-9a^4b^4) \div (3ab^4)$ 26. $(21x^3yz^2) \div (-6x^2z)$

27. $(-8m^2n) \div (16mn)$ 28. $(17c^3d^2) \div (51c^2d^2)$

29. $(72b^2c^2) \div (9b^2c)$

30. $(14x^2y^5) \div (-21x^2y^2)$

C. **Simplify completely.**

31. $\dfrac{2^8a^5b^4}{2^5a^3b^3}$ 32. $\dfrac{(3x^2)(-4y^3)}{2x^2y^2}$

33. $\dfrac{(-2m^2n)^2}{m^3n^2}$ 34. $\dfrac{3^3r^3s^3}{-3rs}$

35. $\dfrac{(3a^3)(3a^4)(-a)}{(9a)(-3a^2)}$ 36. $\dfrac{(-6x^4)(7x^6)}{7x}$

37. $\dfrac{(-2x^3y^3)(x^2y^2)}{(-2x^2y^3)(-2x^2y^2)}$ 38. $\dfrac{(9a^2b^3)(-7b^5)}{(7ab^2)(-11ab^3)}$

39. $\dfrac{(2ab^2)^3(3a^2b)}{(2a^2b^2)^2}$ 40. $\dfrac{(-3x^2y^3)^2}{(-4xy)(9x^2y)}$

41. $\dfrac{-(5m^2n)^2(3mn^2)}{(-5m)^2(mn^2)}$ 42. $\dfrac{-(a^2b)^2(-ab^3)^2}{(-a^2b)^3}$

43. $\dfrac{(-5x^2y)(2x^3)(-xy^2)}{3xy(-5x^2y)^2}$ 44. $\dfrac{-(pqr)^2(-p^2q^2r^3)}{(4pq^2)(-pr)^3}$

15.5   Negative Exponents

As we noted in Section 15.4, negative exponents can occur in monomial division. In $y^3 \div y^5$, the answer that results from subtracting exponents is y^{-2}. The meaning of y^{-2} is evident when we look at the problem in expanded form.

$$\frac{y^3}{y^5} = \frac{\overset{1}{\cancel{y}} \cdot \overset{1}{\cancel{y}} \cdot \overset{1}{\cancel{y}}}{\underset{1}{\cancel{y}} \cdot \underset{1}{\cancel{y}} \cdot \underset{1}{\cancel{y}} \cdot y \cdot y} = \frac{1}{y^2}$$

Therefore,

$$y^{-2} = \frac{1}{y^2}$$

Negative Exponents

If n is an integer,

$$a^{-n} = \frac{1}{a^n}$$

In simplifying the problems in this section, do not leave negative exponents in an answer. Replace them with the fraction form of the expression. This is usually the last step, after you have done the rest of the problem using the rules for exponents.

E X A M P L E 1   Divide m by m^4.

Solution

$\qquad m \div m^4$

$= m^{1-4}$ Subtract exponents in division.

$= m^{-3}$ Replace the negative exponent with its equivalent fraction form.

$= \dfrac{1}{m^3}$ ◆

E X A M P L E 2   Find the product of $-2a^{-2}b^3$ and $4ab^{-2}$.

Solution   Multiply the numerical coefficients, -2 and 4.

Multiply the same bases, adding exponents.

$\qquad (-2)(4) = -8$

$\qquad (a^{-2})(a) = a^{-2+1} = a^{-1}$

$\qquad (b^3)(b^{-2}) = b^{3+(-2)} = b^1 = b$

$\qquad (-2a^{-2}b^3)(4ab^{-2})$

$= -8a^{-1}b$ Replace a^{-1} with $\dfrac{1}{a}$.

$= -\dfrac{8b}{a}$ ◆

E X A M P L E 3 Simplify: $(2m^{-3}n^2)^{-2}$

Solution $(2m^{-3}n^2)^{-2}$

Raise each factor to the -2 power using the rules for exponents that appear in Section 15.3.

$= (2)^{-2}(m^{-3})^{-2}(n^2)^{-2}$

$= 2^{-2}m^{+6}n^{-4}$

Replace the negative exponents.

$= \dfrac{1}{2^2} \cdot \dfrac{m^6}{1} \cdot \dfrac{1}{n^4}$

$= \dfrac{m^6}{4n^4}$ ◆

E X A M P L E 4 Simplify: $\dfrac{X^{-8}}{X^{-5}}$

Solution $\dfrac{X^{-8}}{X^{-5}}$

$= X^{(-8)-(-5)}$

$= X^{(-8)+(+5)}$

$= X^{-3}$

$= \dfrac{1}{X^3}$ ◆

E X A M P L E 5 Simplify: $\dfrac{y^3 \cdot y^{-5}}{y^{-2}}$

Solution $\dfrac{y^3 \cdot y^{-5}}{y^{-2}}$ Simplify the numerator.

$= \dfrac{y^{3+(-5)}}{y^{-2}}$

$= \dfrac{y^{-2}}{y^{-2}}$

$= y^{(-2)-(-2)}$

$= y^{(-2)+(+2)}$

$= y^0 = 1$ ◆

E X A M P L E 6 Simplify: $\dfrac{1}{c^{-3}}$

Solution $\dfrac{1}{c^{-3}}$ Replace c^{-3} with $\dfrac{1}{c^3}$.

$= \dfrac{1}{\dfrac{1}{c^3}}$ Rewrite the complex fraction as division (Section 3.5).

$= 1 \div \dfrac{1}{c^3}$

$= \dfrac{1}{1} \cdot \dfrac{c^3}{1}$

$= c^3$ ◆

From Example 6, can you see that $1/c^{-3} = c^3$?

This relationship will be true all the time (if $c \neq 0$). It is stated in the following rule.

Negative Exponents

> Because $a^{-n} = \dfrac{1}{a^n}$, then $\dfrac{1}{a^{-n}} = a^n$ (where $c \neq 0$).

Factors that have negative exponents and are in the numerator will move to the denominator and the exponent will become positive. Factors that have negative exponents and are in the denominator will move to the numerator in the same way.

Example 4 could have been simplified this way:

$$\frac{X^{-8}}{X^{-5}} = \frac{X^5}{X^8} = X^{-3} = \frac{1}{X^3}$$

And Example 5 could be done as follows:

$$\frac{y^3 \cdot y^{-5}}{y^{-2}} = \frac{y^3 y^2}{y^5} = \frac{y^5}{y^5}$$

$$= y^0 = 1$$

EXAMPLE 7 Simplify: $\left(\dfrac{a^4}{b^5}\right)^{-3}$

Solution $\left(\dfrac{a^4}{b^5}\right)^{-3}$ Raise each exponential form to the -3 power.

$= \dfrac{a^{-12}}{b^{-15}}$ Move terms with negative exponents to the opposite part of the fraction.

$= \dfrac{b^{15}}{a^{12}}$ ◆

EXAMPLE 8 Simplify: $\left(\dfrac{4x^2}{3^{-1}y^3}\right)^{-2}$

Solution $\left(\dfrac{4x^2}{3^{-1}y^3}\right)^{-2}$ Apply the -2 exponent to both numerator and denominator.

$= \dfrac{(4x^2)^{-2}}{(3^{-1}y^3)^{-2}}$

$= \dfrac{4^{-2}x^{-4}}{3^2 y^{-6}}$ Move factors with negative powers and change to positive exponents.

$= \dfrac{y^6}{3^2 \cdot 4^2 x^4}$ Simplify.

$= \dfrac{y^6}{9 \cdot 16x^4}$

$= \dfrac{y^6}{144x^4}$ ◆

E X A M P L E **9**   Simplify: $\left(\dfrac{3a^2}{b^3}\right)^{-2}$

Solution $\left(\dfrac{3a^2}{b^3}\right)^{-2} = \dfrac{(3a^2)^{-2}}{(b^3)^{-2}}$

$$= \dfrac{(3)^{-2}(a^2)^{-2}}{(b^3)^{-2}}$$

$$= \dfrac{3^{-2}a^{-4}}{b^{-6}}$$

$$= \dfrac{b^6}{3^2 a^4}$$

$$= \dfrac{b^6}{9a^4}$$

◆

15.5 ◆ Practice Problems

Simplify each expression, applying exponent rules and eliminating negative exponents.

1. $a^2 \div a^5$

2. $x \div x^3$

3. $4R^2 \div R^3$

4. $b^5 \div 5b^7$

5. $9M^2 \div 6M^3$

6. $8n^4 \div 12n^7$

7. $(x^2)^{-2}$

8. $(a^3)^{-2}$

9. $(b^{-2})^{-2}$

10. $(c^{-1})^{-2}$

11. $(2a^{-1})^{-2}$

12. $(2c^2)^{-1}$

13. $(2x^2 y^{-1} z^3)^{-2}$

14. $(3a^{-2} b^2 c)^{-2}$

15. $\dfrac{3}{m^{-2}}$

16. $\dfrac{b^{-2}}{4}$

17. $\dfrac{x^{-2}}{x^{-5}}$

18. $\dfrac{n^{-4}}{n}$

19. $\left(\dfrac{a^2}{b^3}\right)^{-2}$

20. $\left(\dfrac{c^3}{d^{-1}}\right)^{-2}$

21. $\dfrac{3x^2 y^{-2} z}{2^{-1} x y^2 z^3}$

22. $\dfrac{5^{-1} m^{-3} n^2}{3m^2 n^{-3}}$

23. $\dfrac{(-2a^2 b^{-1} c^{-2})^2}{a^{-3} b c^{-1}}$

24. $\dfrac{(-5x^2 y^{-3} z)^{-2}}{3x^{-1} y^2 z}$

25. $\dfrac{(2^{-1} x^2 y^{-2})^{-2}}{(3x^{-1} y^2)(-2x^2 y^{-1})}$

26. $\left(\dfrac{-a^2 b^{-2}}{c^{-3}}\right)^{-2}$

27. $\left(\dfrac{2m^{-2} n}{-m^3 n^2}\right)^{-3}$

28. $\dfrac{(2p^{-1} q^{-2} r^2)^{-2}}{(5pq^2 r^{-1})(-2p^2)^{-1}}$

29. $[(-2a^2 b^{-2})(3a^{-1} b)]^{-2}$

30. $[(6c^3 d^{-2})(-b^{-2} cd^{-1})]^{-1}$

31. $(2ab^{-2})(4a^{-1} b)^{-3}(a^2 b^3)^{-2}$

32. $(-5xy^{-2})(2x^{-1} y)^{-2}(x^2 y^{-2})^{-3}$

33. $\dfrac{(2m^2 n^{-1})^{-2}}{(m^{-1} n^2)(m^3 n^{-1})}$

34. $\dfrac{(3a^{-1} b^{-2})^{-2}}{(a^2 b^{-1})(a^{-3} b^{-2})}$

35. $\dfrac{(-2a^2 b^{-3} c^{-1})^{-2}}{(3a^{-1} b^{-2} c^{-3})^3}$

36. $\dfrac{(m^2 n^{-3})^2 (m^{-1} n^{-2})}{(2m^3 n^{-2})(m^{-2} n^{-1})^{-1}}$

37. $\dfrac{(5x^2 y^{-3})(2x^{-1} y^{-2})^2}{(3x^{-1} y^2)^{-1}(x^2 y^{-2})}$

38. $[(x^{-3} y^{-2})^2]^{-1}$

39. $[(x^{-3} y^{-2})^{-2}]^{-1}$

40. $\left(\dfrac{(m^2 n^{-2})^{-1}}{(m^{-1} n^3)^{-1}}\right)^{-2}$

15.6 Adding and Subtracting Polynomials

The process of adding or subtracting polynomials simply involves combining like terms.

E X A M P L E **1**   Simplify: $\begin{aligned} -3x^2 + 4x - 6 \\ + \quad 6x^2 - 3x + 5 \end{aligned}$

Solution $\begin{aligned} -3x^2 + 4x - 6 \\ + \quad 6x^2 - 3x + 5 \\ \hline 3x^2 + \quad x - 1 \end{aligned}$

◆

E X A M P L E 2   Combine: $(13a^2b - 5ab + 4) + (7a^2b - 2ab + 5)$

Solution $(13a^2b - 5ab + 4) + (7a^2b - 2ab + 5)$
$= 13a^2b - 5ab + 4 + 7a^2b - 2ab + 5$
$= 13a^2b + 7a^2b - 5ab - 2ab + 4 + 5$
$= 20a^2b - 7ab + 9$ ◆

Note that the short cut presented in Chapters 12 and 13 was used. Terms were combined by using the operation sign as if it were a signed-number sign: $(-3x^2) + (+6x^2), (+4x) + (-3x), (-6) + (+5), (-5ab) + (-2ab)$, and so on. You can write the problem either vertically or horizontally, whichever you like better.

E X A M P L E 3   Simplify:

$$(-7a^2b^2 + 4ab - 8) + (15a^2b^2 + 3 - 4ab)$$

Solution $-7a^2b^2 + 4ab - 8$ Line up like terms.
$\underline{+\ 15a^2b^2 - 4ab + 3}$
$\ \ \ \ 8a^2b^2 + 0ab - 5 = 8a^2b^2 - 5$ ◆

E X A M P L E 4   Simplify completely:
$(9m^2 + 14m^2n - 5mn^2 + 3) + (2m^2 - 5m^2n + mn^2 - 1)$.

Solution $(9m^2 + 14m^2n - 5mn^2 + 3)$
$\underline{+\ (2m^2 -\ \ 5m^2n +\ \ mn^2 - 1)}$
$\ \ 11m^2 +\ \ 9m^2n - 4mn^2 + 2$ ◆

When subtracting polynomials, remember the signed-number rule (Chapter 12) about rewriting subtraction as addition of the opposite. You change the operation sign to addition and change the sign(s) of the number(s) you are taking away; then you proceed as in addition. In the problem $(11x^2 - 8x - 5) - (3x^2 + 9x + 1)$, the entire second polynomial is being subtracted. Therefore, the operation sign will be changed to addition, and the sign of *every* term in the second polynomial must be changed before you do the addition.

Vertically: $11x^2 - 8x - 5$
$\ \ \ \ \ \ \ \ \ \ \ \ S$
$\underline{-\ \ 3x^2 + 9x + 1}$

Change all the signs in the subtrahend (the number you are subtracting). Then add.

$\ \ \ \ \ 11x^2 -\ \ 8x - 5$
A
$\underline{+\ -3x^2 -\ \ 9x - 1}$
$\ \ \ \ \ \ \ 8x^2 - 17x - 6$

Horizontally: Multiply by -1 to do the sign changes, and then combine like terms.

$\ \ \ \ (11x^2 - 8x - 5) - (3x^2 + 9x + 1)$
$= (11x^2 - 8x - 5) - 1(3x^2 + 9x + 1)$
$= 11x^2 - 8x - 5 - 3x^2 - 9x - 1$
$= 8x^2 - 17x - 6$

Remember that multiplying the second polynomial by -1 accomplishes the sign changes required in subtraction.

E X A M P L E **5**   Simplify completely: $(3x + 6) - (2x - 3)$

Solution $(3x + 6) - (2x - 3)$
$= (3x + 6) - 1(2x - 3)$
$= 3x + 6 - 2x + 3$
$= x + 9$ ◆

E X A M P L E **6**   Simplify completely: $17a^2 - 4a + 6$
$\underline{- \ -5a^2 - 4a - 3}$

Solution

$17a^2 - 4a + 6$	$17a^2 - 4a + 6$
S	A
$\underline{- \ -5a^2 - 4a - 3}$	$\underline{+ \ +5a^2 + 4a + 3}$
	$22a^2 \qquad + 9$

◆

E X A M P L E **7**   From the sum of $(4a^2 - 3)$ and $(5a^2 + 7)$, subtract $(-3a^2 + 4)$.

Solution $(4a^2 - 3) + (5a^2 + 7) - (-3a^2 + 4)$
$= (4a^2 - 3) + (5a^2 + 7) - 1(-3a^2 + 4)$
$= 4a^2 - 3 + 5a^2 + 7 + 3a^2 - 4$
$= 12a^2$

or $4a^2 - 3$	$9a^2 + 4$	$9a^2 + 4$
	S	A
$\underline{+ \ 5a^2 + 7}$	$\underline{- \ -3a^2 + 4}$	$\underline{+ \ +3a^2 - 4}$
$9a^2 + 4$		$12a^2$

◆

Exercises   Simplify these expressions. If you have difficulty, review the examples.

1. $(2x^2 - 3x + 5) + (x^2 - 2x - 4)$
2. $(3m^4 - 4m + 1) - (8m^4 - m - 2)$
3. From the sum of $(-5a^2b - 13ab^2)$ and $(7a^2b - 3ab^2)$, subtract $(-4ab^2 + 12a^2b)$.

Answers:

1. $3x^2 - 5x + 1$ 2. $-5m^4 - 3m + 3$ 3. $-10a^2b - 12ab^2$

15.6 ◆ Practice Problems

A.   In each of the following problems, add.

1. $23x + 14y - 27$
   $\underline{15x - \ 8y - 12}$

2. $7q + 2r + 9s$
   $\underline{5q + 6r - 3s}$

3. $17m^2 + 5m - 14$
   $\underline{11m^2 - 7m + \ 5}$

4. $-4a + 12b - 6c$
   $\underline{\ 5a + \ 9b + 2c}$

5. $6R - 3S + 5T$
   $\underline{\ R + 4S - 2T}$

6. $6y^2 + 13y - 7$
   $\underline{-3y^2 + \ 6y - 1}$

B.   In each of the following problems, subtract. (Remember to use "S" and "A" if it helps.)

7. $11a + 4ab - \ 5b$
   $\underline{-7a + \ ab + 12b}$

8. $9mn + 4m - \ n$
   $\underline{\ mn + 2m - 3n}$

9. $15v + 2w - 7x$
   $\underline{23v - \ w + 5x}$

10. $-3c^2 - 4c + 8$
    $\underline{\ c^2 + 5c - 1}$

11. $15f^2 - 12f + \ h$
    $\underline{-3f^2 \qquad - 7h}$

12. $17xy - 2x + 9y$
    $\underline{-5xy - 4x}$

C.   Add or subtract as indicated.

13. $(3a + 4b) + (-7a + 6b)$

14. $(18 - 6x) + (15x - 3)$

15. $(-5m + 3) - (2m - 4)$

16. $(-3k + 2m) - (6m - k)$

17. $(9x^2 + 2x + 12) - (-7x + 9)$

18. $(-18y + 2z + 9) + (16 + 4z - 11y)$

19. $(2R - 4S + T) + (3S - 7T)$

20. $(5L - 2M + N) - (4N + 3L - 2M)$

21. $(6x^2y - 4xy + 2y^2) + (3y^2 - 5x^2y)$

22. $(9mn^2 - 2mn) + (3 + 4mn^2 - 2mn)$

23. $(3b^2 - 6b - 7) - (4b^2 + 5b + 6)$

24. $(5x^3 + 3x^2 - 7x) + (-x^2 + 5x - 7)$

25. $(4y^2 - 4y + 3) + (7y^3 + 3y^2 - 1)$

26. $(17c^3 - 4c^2) - (8c^2 + 2c + 3)$

27. $(m^2 - m^3 + 2 - 4m) + (3m - 2m^2 - 6)$

28. $(c^3 + 2c^2 - c + 1) + (5c^2 - 2c + 9) + (2c - 3c^3 + 2c^2 - 4)$

29. $(3x^2 + 2x^3 - 5) + (7x - 2x^2 + 3x^3 - 1) + (x^3 - 2x^2 - 3x)$

30. $(4w^3 - w - 1) - (2w^2 + 3w + 5)$

31. Subtract $(-p^2q - 2pq^2)$ from $(6p^2q + 2pq - pq^2)$.

32. Find the sum of $(2a - 9)$, $(3a + 4)$, and $(5a - 2)$.

33. Find the sum of $(4x^2 - 2x + 5)$, $(2x^2 - 3)$, and $(5x - 9)$.

34. From $(6c^2 - 7cd + d^3)$ subtract $(3d^3 - c^2 + 2cd)$.

35. From the sum of $(2x + 3y)$ and $(-x - 4y)$, subtract $(5x + 2y)$.

36. From the sum of $(4a - b)$ and $(-3a + 6b)$, subtract $(a - b)$.

37. From $(7y^2 + 5y - 3)$ subtract $(-3y^2 - 4y + 2)$.

38. Subtract $(-2R^2S^2 - 3R^2S)$ from $(-R^2S^2 + 2R^2S)$.

39. Find the difference between $(3m^2n^2 - 2m^2n + 5)$ and $(2m^2n^2 + mn^2 - 2)$.

40. Find the difference between $(5a^2b^2c^3 - 2ab^2c^2 + 7a^2b^2c^2)$ and $(4ab^2c^2 - 2a^2b^2c^2 + a^2b^2c)$.

15.7   Multiplying Polynomials

Multiplying polynomials by monomials combines the Distributive Principle and the rules for exponents. You will multiply every term inside the parentheses by the factor outside and, in doing so, will add exponents when the bases are the same.

E X A M P L E   1   Multiply: $3ab(2a^2b - b^2)$

Solution $(3ab)(2a^2b - b^2)$ Multiply $(3ab)(2a^2b)$ and $(3ab)(-b^2)$.
$= 6a^3b^2 - 3ab^3$

E X A M P L E   2   Simplify: $5x^2y(x^2y^2 + 2xy - y^2)$

Solution $5x^2y(x^2y^2 + 2xy - y^2)$
Multiply: $(5x^2y)(x^2y^2)$, $(5x^2y)(+2xy)$, and $(5x^2y)(-y^2)$.
$= 5x^4y^3 + 10x^3y^2 - 5x^2y^3$

E X A M P L E   3   Multiply: $(3mn^2)^2(4m^2n^2 - 5mn + 6)$

Solution   By the Order of Operations Agreement, simplify the first factor with the exponent before doing the multiplication.
$(3mn^2)^2 = (3)^2(m)^2(n^2)^2 = 9m^2n^4$

Then

$$(3mn^2)^2(4m^2n^2 - 5mn + 6)$$

$$= 9m^2n^4(4m^2n^2 - 5mn + 6)$$
$$= 36m^4n^6 - 45m^3n^5 + 54m^2n^4$$ ◆

When multiplying a polynomial by a polynomial, you will use a process similar to the one used in the previous examples. To multiply $(y + 3)$ by $(4y - 1)$, distribute each term of the first polynomial over every other term of the second polynomial.

$$(y + 3)(4y - 1)$$

$$y(4y - 1) + 3(4y - 1)$$

Then multiply through using the Distributive Principle. Be very careful with the signed numbers and exponents as you multiply and combine like terms.

$$= 4y^2 - y + 12y - 3$$
$$= 4y^2 + 11y - 3$$

This multiplication can also be arranged vertically, much like long multiplication in arithmetic. Multiply all the terms in the top binomial by -1; then multiply each term by $4y$ and line up the like terms.

$$
\begin{array}{r}
y + 3 \\
\underline{4y - 1} \\
-1y - 3 \\
\underline{4y^2 + 12y } \\
4y^2 + 11y - 3
\end{array}
$$

Add the partial products.

Let's look at another example of polynomial multiplication:

$$(2a + 3b)(a - 2b - 1)$$

$$2a(a - 2b - 1) + 3b(a - 2b - 1)$$
$$= 2a^2 - 4ab - 2a + 3ab - 6b^2 - 3b$$
$$= 2a^2 - 2a - ab - 3b - 6b^2$$

Combine like terms.

Doing this distributive multiplication vertically would look like this:

$$
\begin{array}{r}
a - 2b - 1 \\
\underline{2a + 3b} \\
3ab - 6b^2 - 3b \\
\underline{2a^2 - 2a - 4ab } \\
2a^2 - 2a - ab - 6b^2 - 3b
\end{array}
$$

Are the answers the two methods give the same? If you arranged the terms in the same order, they would be the same. Terms should be arranged in descending order of the exponents of the first (alphabetically first) variable and then in ascending order of the exponents of the next variable.

$$2a^2 - 2a - ab - 6b^2 - 3b = 2a^2 - 2a - ab - 3b - 6b^2$$

Watch the arrangement of answers in the next examples and see if you can follow the pattern as you do problems. If you happen to write the terms out of order, the answer will not be incorrect, but it is much easier for you—and everyone else—to check answers if they are all written in the same way.

E X A M P L E 4 Multiply: $(4x + 8)(x^2 + 2x - 9)$

Solution By the distributive method:

$$(4x + 8)(x^2 + 2x - 9)$$

Write each term from the first factor in front of the second polynomial.

$$= 4x(x^2 + 2x - 9) + 8(x^2 + 2x - 9)$$ Multiply using Distributive Principle.

$$= 4x^3 + 8x^2 - 36x + 8x^2 + 16x - 72$$ Combine like terms.

$$= 4x^3 + 16x^2 - 20x - 72$$ Notice the decreasing order of powers of x.

By long multiplication (which also uses the Distributive Principle):

$$
\begin{array}{r}
x^2 + 2x - 9 \\
4x + 8 \\
\hline
8x^2 + 16x - 72 \\
4x^3 + 8x^2 - 36x \\
\hline
4x^3 + 16x^2 - 20x - 72
\end{array}
$$

Multiply by $+8$.

Multiply by $+4x$, lining up like terms.

Add the partial products. ◆

E X A M P L E 5 Multiply: $(y^2 + 2y - 3)(y^2 - 4y - 8)$

Solution By the distributive method:

$$(y^2 + 2y - 3)(y^2 - 4y - 8)$$
$$= y^2(y^2 - 4y - 8) + 2y(y^2 - 4y - 8) - 3(y^2 - 4y - 8)$$
$$= y^4 - 4y^3 - 8y^2 + 2y^3 - 8y^2 - 16y - 3y^2 + 12y + 24$$
$$= y^4 - 2y^3 - 19y^2 - 4y + 24$$

By long multiplication:

$$
\begin{array}{r}
y^2 + 2y - 3 \\
y^2 - 4y - 8 \\
\hline
- 8y^2 - 16y + 24 \\
- 4y^3 - 8y^2 + 12y \\
y^4 + 2y^3 - 3y^2 \\
\hline
y^4 - 2y^3 - 19y^2 - 4y + 24
\end{array}
$$

◆

E X A M P L E 6 Find the product: $(k + 3)(2k - 7)$

Solution
$$(k + 3)(2k - 7)$$
$$= k(2k - 7) + 3(2k - 7)$$
$$= 2k^2 - 7k + 6k - 21$$
$$= 2k^2 - k - 21$$

◆

E X A M P L E **7**   Multiply: $(3m^2 - 2)(2m + 5)$

Solution $(3m^2 - 2)(2m + 5)$
$= 3m^2(2m + 5) - 2(2m + 5)$
$= 6m^3 + 15m^2 - 4m - 10$ ◆

E X A M P L E **8**   Simplify: $(3a - 4)^2$

Solution $(3a - 4)^2$
Rewrite as the product of the two binomials.
$= (3a - 4)(3a - 4)$
$= 3a(3a - 4) - 4(3a - 4)$
$= 9a^2 - 12a - 12a + 16$
$= 9a^2 - 24a + 16$ ◆

Students often want to do this problem by squaring the terms inside, $(3a)^2$ and $(-4)^2$. The rule that they are trying to apply to this difference of terms is the rule for raising a *product* to a power (Section 15.3). Note the difference between the expressions $(3ab^2)^2$ and $(3a - 4)^2$. In the product, all factors are to be raised to the second power: $3^2a^2(b^2)^2 = 9a^2b^4$. This is not the correct procedure for a binomial. The factor $(3a - 4)$ as a whole must be multiplied by itself as a whole: $(3a - 4)(3a - 4) = 9a^2 - 24a + 16$. Note that there is a middle term that would not have occurred if you only squared $(3a)$ and (-4). When you see a binomial in parentheses and the exponent outside, rewrite the problem as the product of the binomials.

$$(3a - 4)^2 = (3a - 4)(3a - 4)$$

Doing so should remind you to multiply it completely.

E X A M P L E **9**   Multiply: $(R - 5)^2$

Solution $(R - 5)^2$
$= (R - 5)(R - 5)$
$= R(R - 5) - 5(R - 5)$
$= R^2 - 5R - 5R + 25$
$= R^2 - 10R + 25$ ◆

E X A M P L E **10**   Simplify: $(x + 2)^3$

Solution $(x + 2)^3$
$= (x + 2)(x + 2)(x + 2)$
Multiply the first two binomials together.
$= [x(x + 2) + 2(x + 2)](x + 2)$
$= [x^2 + 2x + 2x + 4](x + 2)$
$= (x^2 + 4x + 4)(x + 2)$
Then multiply that product by the remaining binomial.
$= x(x^2 + 4x + 4) + 2(x^2 + 4x + 4)$
$= x^3 + 4x^2 + 4x + 2x^2 + 8x + 8$
Combine like terms.
$= x^3 + 6x^2 + 12x + 8$ ◆

E X A M P L E **11** Find the product: $(a^2 + b^2)(2a^2 + 3a)$

Solution $(a^2 + b^2)(2a^2 + 3a)$

$= a^2(2a^2 + 3a) + b^2(2a^2 + 3a)$

$= 2a^4 + 3a^3 + 2a^2b^2 + 3ab^2$

◆

Exercises Try these problems and check your answers before doing the Practice Problems. Review the examples if you need help.

1. $3xy(2x + 3y + 7)$ 2. $(a^2 - 2a + 9)(3a^2 - a - 7)$

3. $(2h - 5)(h + 2)$ 4. $(W + 4)^2$

Answers:

1. $6x^2y + 9xy^2 + 21xy$ 2. $3a^4 - 7a^3 + 22a^2 + 5a - 63$

3. $2h^2 - h - 10$ 4. $W^2 + 8W + 16$

15.7 ◆ Practice Problems

A. Multiply, and combine like terms where possible.

1. $3a(a^2 - 4)$

2. $5cd(2c^2d + 3cd + d^2)$

3. $(-8mn)(7m^2 + 4mn)$

4. $(-2x^2y)(3xy + 2y)$

5. $(5x + 2)(3x - 1)$

6. $(8a^2 + a + 3)(a^2 - 2a - 4)$

7. $(p^2 + 9p - 2)(-4p + 1)$

8. $(2g - 1)(g - 4)$

9. $(-x^2y)(3x^2y^2 + 2xy)$

10. $(m^2 - 2m)(2m - 3)$

11. $(c^2d - 4)(cd + 2d)$

12. $(-a^2)(2a^2 + 4a - 1)$

13. $-5x(x^2 + 2xy + 2y^2)$

14. $(4a - 5b)(4a - 5b)$

15. $(y + 2)(y^2 + 2y + 4)$

16. $(3a - 8)(2a - 1)$

B. Simplify completely.

17. $(a^2 - 7)(a^2 + 3)$ 18. $(3x - 5y)(x + 2y)$

19. $(0.3x - 0.4y)^2$ 20. $(4a - 5b)^2$

21. $(x - 3y)^2$ 22. $(h^2 + 0.4)^2$

23. $(g - 4)(g + 4)$ 24. $(h + 2)(h - 2)$

25. $(a - 2b)(3a^2 - 4ab + b)$

26. $(x + 3y)(x^2 - 2xy)$

27. $(-7x^2y)(4x^2y^2 - 2xy + 3y^2)$

28. $(5 + 4a)^2$

29. $(-2mn)(5m^3 - 4m^2n + 2n)$

30. $-5ab(a - b) + 2ab(3a - 2b)$

31. $(3 + 2w)^2$

32. $4x^2y(x + 2y) - 2xy(3x - y)$

33. $(y + 1)^3$ 34. $(k - 3)^3$

35. $(3y - 1)(3y + 1)$ 36. $(5w + 2)(5w - 2)$

37. $(x^2 - 3)(x^2 + 3)$ 38. $(m^3 - 2)(m^3 + 2)$

39. Compare the problems and your answers in Problems 21 and 31 with those in 35 and 37. What is the same in these problems and answers and what is different?

40. Compare the problems and your answers in Problems 20 and 28 with those in 36 and 38. What is the same in these problems and answers and what is different?

15.8 ◆ Dividing Polynomials

Dividing a polynomial by a monomial will be a very familiar operation if you first rewrite the problem as separate terms. This could be viewed as an application of the Distributive Principle:

$$\frac{9x^2y^2 + 24x^2y - 12xy^2}{3xy}$$

$$= \frac{1}{3xy} \cdot (9x^2y^2 + 24x^2y - 12xy^2)$$

$$= \frac{9x^2y^2}{3xy} + \frac{24x^2y}{3xy} - \frac{12xy^2}{3xy}$$

Now each term represents a monomial division like those you did earlier in this chapter. Divide the coefficients and subtract the exponents when the bases are the same.

$$= 3xy + 8x - 4y$$

In rewriting the problem, you are just putting each term of the numerator over the common denominator.

E X A M P L E 1 Divide: $\dfrac{a^2b^2 + 4a^2b - 8ab}{2ab}$

Solution $\dfrac{a^2b^2 + 4a^2b - 8ab}{2ab}$ Rewrite as separate terms.

$$= \frac{a^2b^2}{2ab} + \frac{4a^2b}{2ab} - \frac{8ab}{2ab}$$ Divide each term.

$$= \frac{1}{2}ab + 2a - 4$$ ◆

E X A M P L E 2 Divide: $\dfrac{36x^2y^2 - 18x^2y}{-6xy}$

Solution $\dfrac{36x^2y^2 - 18x^2y}{-6xy}$

$$= \frac{36x^2y^2}{-6xy} - \frac{18x^2y}{-6xy}$$

$$= -6xy - (-3x)$$

$$= -6xy + 3x$$ ◆

E X A M P L E   3   Divide: $\dfrac{14m^2 - 35m + 7}{7m}$

Solution

$$\frac{14m^2 - 35m + 7}{7m}$$

$$= \frac{14m^2}{7m} - \frac{35m}{7m} + \frac{7}{7m}$$

$$= 2m - 5m^0 + m^{-1}$$

Remember to simplify all negative and zero exponents.

$$= 2m - 5 + \frac{1}{m}$$ ◆

Exercises   Try the following problems to be sure that you understand this process. All negative and zero exponents should be simplified, and fractions should be reduced to lowest terms.

1. $\dfrac{20a^2b^2 - 16a^2b + 12ab^2}{-4ab}$ 2. $\dfrac{4x^2yz^2 - 18xy^2z}{6x^2yz}$ 3. $\dfrac{-20x^3y^4 + 10xy^2}{-5x^2y^3}$

Answers:

1. $-5ab + 4a - 3b$ 2. $\dfrac{2z}{3} - \dfrac{3y}{x}$ 3. $4xy - \dfrac{2}{xy}$

Dividing a polynomial by a polynomial is very similar to arithmetic long division. First arrange the terms of the polynomials in descending order of the exponents. In $(x^2 - x - 2) \div (x + 1)$, the terms are already in descending order, so just write them in the long division format.

$$x + 1 \overline{)x^2 - x - 2}$$

Determine how many times x will divide into x^2, and write that answer above the x^2 term.

$$\begin{array}{r} x \\ x + 1 \overline{)x^2 - x - 2} \end{array}$$

Multiply x times $x + 1$, and write the product under the like terms in the dividend.

$$\begin{array}{r} x \\ x + 1 \overline{)x^2 - x - 2} \\ x^2 + x \end{array}$$

Now subtract $x^2 + x$ from $x^2 - x$, remembering to change both signs in the polynomial being subtracted.

$$\begin{array}{r} x \\ x + 1 \overline{)x^2 - x - 2} \\ - x^2 + x \end{array} \qquad \begin{array}{r} x \\ x + 1 \overline{)\; x^2 - \; x - 2} \\ + -x^2 - \; x \\ \hline -2x \end{array}$$

Bring down the next term, divide x into $-2x$, and write that value above $-x$. Then multiply and subtract.

$$\begin{array}{r} x - 2 \\ x + 1 \overline{)\; x^2 - x - 2} \\ + -x^2 - x \\ \hline -2x - 2 \\ - \; -2x - 2 \end{array} \qquad \begin{array}{r} x - \; 2 \\ x + 1 \overline{)\; x^2 - \; x - 2} \\ + -x^2 - \; x \\ \hline -2x - 2 \\ + +2x + 2 \\ \hline 0 \end{array}$$

Because there is a zero remainder,

$$(x^2 - x - 2) \div (x + 1) = x - 2$$

You should check your answer by multiplying the quotient by the divisor and then adding any remainder to that product. The answer should be the dividend of the original problem.

$$(x - 2)(x + 1) = x^2 + x - 2x - 2 = x^2 - x - 2$$

After arranging the polynomial terms in descending order of the exponents, fill in any missing terms with a zero term. The following problem, which is worked in detail, illustrates this process and also involves an answer with a remainder.

E X A M P L E **4**   Divide: $(12y^4 + 3y - 4 - 6y^2)$ by $(2 + y)$

Solution   Rearrange the terms in both polynomials.

$$y + 2)\overline{12y^4 - 6y^2 + 3y - 4}$$

Fill in the missing y^3 term with $0y^3$.

$$y + 2)\overline{12y^4 + 0y^3 - 6y^2 + 3y - 4}$$

Divide y into $12y^4$ and place the answer above that term. Then multiply $12y^3$ times $y + 2$ and subtract the product (changing the sign of each term).

$$
\begin{array}{r}
12y^3 \\
y + 2)\overline{12y^4 + 0y^3 - 6y^2 + 3y - 4} \\
\underset{+}{\ominus}12y^4 \overset{\ominus}{+} 24y^3 \\
\hline
- 24y^3
\end{array}
$$

Bring down the next term, divide y into $-24y^3$, place the answer in the quotient, and then multiply and subtract.

$$
\begin{array}{r}
12y^3 - 24y^2 \\
y + 2)\overline{12y^4 + 0y^3 - 6y^2 + 3y - 4} \\
\underset{+}{\ominus}12y^4 \overset{\ominus}{+} 24y^3 \\
\hline
- 24y^3 - 6y^2 \\
\underset{-}{\oplus}24y^3 \overset{\oplus}{-} 48y^2 \\
\hline
+ 42y^2
\end{array}
$$

Repeat this sequence of steps until the last term in the dividend is brought down and used.

$$
\begin{array}{r}
12y^3 - 24y^2 + 42y - 81 \\
y + 2)\overline{12y^4 + 0y^3 - 6y^2 + 3y - 4} \\
\underset{+}{\ominus}12y^4 \overset{\ominus}{+} 24y^3 \\
\hline
- 24y^3 - 6y^2 \\
\underset{-}{\oplus}24y^3 \overset{\oplus}{-} 48y^2 \\
\hline
+ 42y^2 + 3y \\
\underset{+}{\ominus}42y^2 \overset{\ominus}{+} 84y \\
\hline
- 81y - 4 \\
\underset{-}{\oplus}81y \overset{\oplus}{-} 162 \\
\hline
+ 158
\end{array}
$$

Write the remainder as a fraction in the quotient, remainder over divisor. The quotient is

$$12y^3 - 24y^2 + 42y - 81 + \frac{158}{y + 2}$$

Check If you are not very confident with such an answer, check it by multiplying the answer times the divisor and then adding the remainder ($+158$) to the product.

$$
\begin{array}{r}
12y^3 - 24y^2 + 42y - 81 \\
\underline{y + 2} \\
+\, 24y^3 - 48y^2 + 84y - 162 \\
\underline{12y^4 - 24y^3 + 42y^2 - 81y} \\
12y^4 \phantom{{}- 24y^3} - 6y^2 + 3y - 162 \\
\underline{+ 158} \\
12y^4 \phantom{{}- 24y^3} - 6y^2 + 3y - 4
\end{array}
$$

Add the remainder.

◆

E X A M P L E 5 Find the quotient of $(k^2 - 4)$ and $(k + 2)$.

Solution
$$
\begin{array}{r}
k - 2 \\
k + 2 \overline{)\; k^2 + 0k - 4} \\
\ominus \quad \ominus \\
\underline{k^2 + 2k} \\
- 2k - 4 \\
\oplus \quad \oplus \\
\underline{- 2k - 4} \\
0
\end{array}
$$

Therefore, $(k^2 - 4) \div (k + 2) = k - 2$.

Check $(k - 2)(k + 2) = k^2 + 2k - 2k - 4 = k^2 - 4$

◆

E X A M P L E 6 Divide: $\dfrac{R^2 - 2RT + T^2}{R + T}$

Solution
$$
\begin{array}{r}
R - 3T + \dfrac{4T^2}{R + T} \\
R + T \overline{)\; R^2 - 2RT + T^2} \\
\ominus \quad \ominus \\
\underline{R^2 + RT} \\
- 3RT + T^2 \\
\oplus \qquad \oplus \\
\underline{- 3RT - 3T^2} \\
+ 4T^2
\end{array}
$$

Check $(R - 3T)(R + T) = R^2 + RT - 3RT - 3T^2 =$
$R^2 - 2RT - 3T^2$ Now add the remainder.
$$
\begin{array}{r}
\underline{+ 4T^2} \\
R^2 - 2RT + T^2
\end{array}
$$

◆

15.8 ◆ Practice Problems

A. Divide each polynomial by the monomial. Be sure that all answers are completely simplified.

1. $\dfrac{39ab - 13a}{13a}$

2. $\dfrac{12x^2 + 36x}{4x}$

3. $\dfrac{6x^3 - 12y^2}{-3}$

4. $\dfrac{8p^2 + 24q^2}{-4}$

5. $\dfrac{15m^3 - 10m^2 + 25m}{5m}$

6. $\dfrac{36a^2 - 24a + 12}{-6}$

7. $\dfrac{30x^2 + 15x + 20}{-10}$

8. $\dfrac{22d^4 - 33d^3 - 11d^2}{11d}$

9. $\dfrac{7r^3 - 5r}{r^2}$

10. $\dfrac{9t^3 - 5t}{t^2}$

11. $\dfrac{3p - 6q - 9r}{3}$

12. $\dfrac{7b + 14c - 28}{-7}$

13. $\dfrac{24v^3w^4x^2 - 18v^2w^2x^3}{6v^2wx}$

14. $\dfrac{8x^2yz^3 + 12x^2y^2z^2}{-6x^2yz}$

15. $\dfrac{9a^2b^3c + 15a^3b^2c^3}{6a^2b^2c}$

16. $\dfrac{27f^3g^2h^2 - 18f^2g^2h^2}{9f^2gh}$

17. $\dfrac{21a^2b - 14ab^2 - 7a^3b}{7a^2b}$

18. $\dfrac{p^3t - pt^2}{-p^2t}$

19. $\dfrac{x^4y^3 + x^2y^2}{-xy}$

20. $\dfrac{18r^2s^3 - 9r^2s^2 - 27r^3s^3}{9r^2s^2}$

B. Divide these polynomials as indicated.

21. $\dfrac{A^2 - 5A + 6}{A - 3}$

22. $\dfrac{T^2 - 7T + 12}{T - 4}$

23. $(n^2 - 4) \div (n + 2)$

24. $(c^2 - 36) \div (c + 6)$

25. $\dfrac{x^2 - x - 10}{x - 3}$

26. $\dfrac{y + y^2 - 5}{y + 2}$

27. $\dfrac{3t + 2t^2 + 1}{t + 1}$

28. $\dfrac{3P^2 - 2P + 1}{P - 1}$

29. $\dfrac{8z^4 - 2z^2 + 1}{z^2 + 1}$

30. $\dfrac{6m^4 + 3m^2 - 2}{m^2 - 1}$

31. $(6x - 2x^2 + 4 - 8x^3) \div (2 + x)$

32. $(7y^2 + 3y^3 + 4 + 6y + y^4) \div (y^2 + y + 1)$

33. $(2 - 8x^2 + 2x^4 - 2x + x^3) \div (x^2 - x - 2)$

34. $(1 - 3m^3 - m + 2m^2) \div (1 - m)$

35. $(a^4 - a^2 - 6) \div (a^2 + 2)$

36. $(c^4 + 3c^2 - 10) \div (c^2 - 2)$

15.9 ▶ Scientific Notation

One very useful application of the exponent rules occurs in scientific notation. This notation is used extensively in chemistry, physics, electronics, and other technical fields to simplify calculations with very large or very small numbers. Would you want to perform multiplication or division with numbers such as 5,880,000,000,000 (the approximate number of miles in 1 light year) or with 0.000000000000000000000602 (the number of molecules in a gram molecule)? Scientific notation will transform these quantities into manageable numbers.

From the short cut for multiplication of decimals (Section 5.5), when 2.5 is multiplied by 1,000 we find the answer by moving the decimal point three places to the right.

$$2.5 \times 1,000 = 2.500 \times 1,000 = 2,500$$

Because $1000 = 10 \cdot 10 \cdot 10$, it could be expressed as 10^3. Therefore, 2,500 could be written as 2.5×10^3.

2.5×10^3 is scientific notation for 2,500. It expresses the quantity as the product of a number between 1 and 10 and the power of 10 that would indicate the movement of the decimal point back to its original position.

$$10 = 10^1 \qquad 0.1 = 10^{-1}$$
$$100 = 10^2 \qquad 0.01 = 10^{-2}$$
$$1,000 = 10^3 \qquad 0.001 = 10^{-3}$$
$$10,000 = 10^4 \qquad 0.0001 = 10^{-4}$$

and so on   and so on

When 10 has a positive exponent, the decimal point is moved to the *right*, and when the exponent of 10 is negative, the decimal point is moved to the *left*.

To change a number to scientific notation, express the number as a value between 1 and 10 times the power of 10 that accounts for the number of places the decimal point would need to be moved—and the direction in which it would need to be moved—to return to the original number. For example,

5,880,000,000,000 would be 5.88×10^{12}

5.88 is between 1 and 10.

10^{12} indicates that the decimal point would have to be moved 12 places to the right to be returned to the original position.

5.880000000000

0.00000000000000000000000602 would be 6.02×10^{-23}

6.02 is between 1 and 10.

10^{-23} would indicate movement of the decimal point 23 places to the left.

Scientific Notation

> If N is a positive number, then $N = n \times 10^c$ where n is a number such that $1 \le n < 10$ and c is an integer.

To write a positive number in scientific notation:

1. Place the decimal point to the right of the first nonzero digit (make the number between 1 and 10).

2. Determine what power of 10 to use by counting the number of places the decimal point in this new number would have to be moved to return it to its original position.

   a. if the decimal point would have to be moved to the right, the exponent is positive.

   b. if the decimal point would have to be moved to the left, the exponent is negative.

   c. if the decimal point would not have to be moved, the exponent is zero.

E X A M P L E   **1**   Write 558 in scientific notation.

Solution $558 = 5.58 \times 10^2$ because the decimal point would have to move two places to the right to return to its original position.   ◆

E X A M P L E   **2**   Write 0.0065 in scientific notation.

Solution $0.0065 = 6.5 \times 10^{-3}$ because we would have to move the decimal point three places to the left to return to the original number.   ◆

E X A M P L E 3 Write 461,300 in scientific notation.

Solution $461,300 = 4.613 \times 10^5$ ◆

E X A M P L E 4 Change 29.8×10^3 to scientific notation.

Solution 29.8×10^3 is not in scientific notation because 29.8 is not between 1 and 10. 29.8 becomes 2.98×10^{-1}, so $29.8 \times 10^3 = 2.98 \times 10^1 \times 10^3$. The bases are the same, so $10^1 \times 10^3 = 10^4$. Thus

$$29.8 \times 10^3 = 2.98 \times 10^4$$ ◆

E X A M P L E 5 Change each of the following to *ordinary* notation.

    a. 1.57×10^{-5} **b.** 6.9×10^4 **c.** 4×10^{-3}

Solutions **a.** $1.57 \times 10^{-5} = 00001.57 \times 10^{-5}$
                $= 0.0000157$

      b. $6.9 \times 10^4 = 6.9000 \times 10^4$
                $= 69,000$

      c. $4 \times 10^{-3} = 0004. \times 10^{-3}$
                $= 0.004$ ◆

    When you are required to multiply or divide very large or small numbers, first change them to scientific notation. Then multiply or divide the numbers that are less than 10 and apply the rules of exponents to the powers of 10.

E X A M P L E 6 Multiply: $2,900,000 \times 13,000$

Solution $2,900,000 \times 13,000$

$= 2.9 \times 10^6 \times 1.3 \times 10^4$

$= 2.9 \times 1.3 \times 10^6 \times 10^4$

$= 3.77 \times 10^{10}$

It is permissible either to leave the answer in scientific notation or to write it in ordinary notation as 37,700,000,000. ◆

E X A M P L E 7 Find the quotient of 0.00384 and 3,000.

Solution $\dfrac{0.00384}{3,000}$

$= \dfrac{3.84 \times 10^{-3}}{3 \times 10^3}$

$= 1.28 \times 10^{-3-3} = 1.28 \times 10^{-3+(-3)}$

$= 1.28 \times 10^{-6}$ ◆

EXAMPLE 8 Simplify and express the answer in scientific notation: $\dfrac{0.0007 \times 0.042}{0.021}$

Solution $\dfrac{0.0007 \times 0.042}{0.021}$

$= \dfrac{7 \times 10^{-4} \times 4.2 \times 10^{-2}}{2.1 \times 10^{-2}}$ Simplify the numerator.

$= \dfrac{29.4 \times 10^{-6}}{2.1 \times 10^{-2}}$ Divide 29.4 by 2.1, and divide 10^{-6} by 10^{-2}.

$= 14 \times 10^{(-6)-(-2)}$

$= 14 \times 10^{-4}$

$= 1.4 \times 10^{1} \times 10^{-4}$

$= 1.4 \times 10^{-3}$ ◆

If you are using a calculator in your course work, try doing Examples 6, 7, and 8 with your calculator. Try each example first without using scientific notation, and then try it using the scientific notation keys. Compare the calculator displays and answers with the steps in each example.

15.9 ◆ Practice Problems

A. Write each number in scientific notation.

1. 2,780
2. 0.006
3. 0.00075
4. 18,000,000
5. 0.65
6. 3,000,000
7. 7,885
8. 43.8×10^{2}
9. 0.431
10. 675×10^{-3}
11. 0.0652
12. 8,841,000

B. Write each number in ordinary notation.

13. 6.5×10^{3}
14. 7.8×10^{2}
15. 1.57×10^{-2}
16. 9.8×10^{-3}
17. 9×10^{4}
18. 1×10^{-5}
19. 7×10^{-4}
20. 5×10^{5}
21. 3.144×10^{-2}
22. 6.8×10^{6}
23. 4.7×10^{5}
24. 4.157×10^{-3}

C. Change each number to scientific notation before doing the calculations. Express your answers in scientific notation.

25. $4,600,000 \times 9,000$
26. 0.0024×0.006
27. 0.009×0.00051
28. $27,000 \times 0.003$
29. 0.0064×800
30. $1,700,000 \times 40,000$
31. $\dfrac{6,800 \times 400}{40,000 \times 2,000}$
32. $\dfrac{0.06 \times 7,200}{0.008}$
33. $\dfrac{0.0024 \times 6,000}{120 \times 40,000}$
34. $\dfrac{480 \times 70,000}{8,000 \times 1,400}$
35. $0.135 \div 2,700$
36. $0.0185 \div 370$
37. $(0.008)^{2}$
38. $(20,000)^{2}$
39. $(12,000)^{2}$
40. $(0.0011)^{2}$

15.10   Chapter Review

In simplifying monomial and polynomial operations, you must follow these rules.

1. Add or subtract (combine) only *like* terms.

$$3x^{2} + 2x + 9x^{2} - 5x = 12x^{2} - 3x$$

2. To multiply quantities with the same base, keep the base and add the exponents.

$$a^{2} \cdot a^{3} = a^{5} \qquad (-2y^{2})(3y) = -6y^{3}$$

3. A number or variable with no exponent showing is understood to have an exponent of 1.

$$x^{2}y = x^{2}y^{1}$$

4. To raise a base in exponent form to a power, keep the base and multiply the exponents.

$$(x^3)^2 = x^6 \qquad (2^4)^2 = 2^8 \text{ or } 256$$

5. To divide quantities with the same base, keep the base and subtract the exponents.

$$\frac{m^5}{m^2} = m^3 \qquad \frac{3^4}{3} = 3^3 \text{ or } 27$$

6. Any nonzero number raised to an exponent of 0 is equal to 1 and must be simplified.

$$y^0 = 1 \qquad a^2 b^0 c = a^2 \cdot 1 \cdot c = a^2 c$$

7. When simplifying polynomials, replace negative exponents with their positive exponent equivalents.

$$2^{-x} = \frac{1}{2^x} \qquad a^{-2} b^3 = \frac{b^3}{a^2}$$

8. Use either the Distributive Principle itself or its long multiplication format to multiply polynomials.

$$a^2 b(3a^2 - 4ab + 2b^2) = 3a^4 b - 4a^3 b^2 + 2a^2 b^3$$

$$(x^2 + 2x + 4)(2x^2 - x - 6)$$
$$= x^2(2x^2 - x - 6) + 2x(2x^2 - x - 6) + 4(2x^2 - x - 6)$$
$$= 2x^4 - x^3 - 6x^2 + 4x^3 - 2x^2 - 12x + 8x^2 - 4x - 24$$
$$= 2x^4 + 3x^3 - 16x - 24$$

or

$$\begin{array}{r} x^2 + 2x + 4 \\ 2x^2 - x - 6 \\ \hline -6x^2 - 12x - 24 \\ -x^3 - 2x^2 - 4x \\ 2x^4 + 4x^3 + 8x^2 \\ \hline 2x^4 + 3x^3 + 0x^2 - 16x - 24 = 2x^4 + 3x^3 - 16x - 24 \end{array}$$

$$(3d + 4)(d - 6)$$
$$= 3d(d - 6) + 4(d - 6)$$
$$= 3d^2 - 18d + 4d - 24$$
$$= 3d^2 - 14d - 24$$

9. In dividing a polynomial by a monomial, write each term over the denominator and then follow the procedure for dividing monomials.

$$\frac{15x^3 - 10x^2 + 20x}{5x}$$
$$= \frac{15x^3}{5x} - \frac{10x^2}{5x} + \frac{20x}{5x}$$
$$= 3x^2 - 2x + 4$$

10. In dividing a polynomial by a polynomial, use the same procedures as in arithmetic long division after arranging the terms in descending order and filling in any missing terms.

11. Very large or very small numbers can be written in scientific notation, with powers of 10 used to represent the decimal places. Once such numbers have been expressed in scientific notation, follow the arithmetic rules for multiplication and division and the exponent rules for the powers of 10.

12. In all operations, the correct signed-number rules must be applied.

Review Problems

A. Perform the indicated operation.

1. $24x - 5x$ 2. $14b^2 + 11b^2$

3. $a - 7a$ 4. $y - 3y$

5. $8c^2d + 3c^2$ 6. $8a - 3a$

7. $p^7 \div p^5$ 8. $q^3 \div q^2$

9. $x^4 \div x$ 10. $(-2a^2)(4ab^2)$

11. $6mn(-3mn^2)$ 12. $b^5 \div b^2$

13. $n^6 \div n^2$ 14. $k^5 \div k$

15. $(t^3)^5$ 16. $(m^2)^4$

17. $(3b)^2$ 18. $(-2a^2)^3$

19. $(5a^3)(-3a^7)$ 20. $(-2r^2)(3r^4)$

21. $8m^2n - 4n$ 22. $12c^2d^2 - 4cd^2$

23. $3(2a + b)$ 24. $2(3x - 2y)$

25. $2b^{-2}$ 26. $a^{-1}b^{-2}c^3$

27. $b^2 \div b^5$ 28. $y^3 \div y^4$

29. $(2b)^{-2}$ 30. $(ab^2)^{-1}$

B. Simplify each expression by performing the indicated operations and eliminating zero and negative exponents from the answers.

31. $\dfrac{9t^9}{12t^4}$ 32. $\dfrac{45n^8}{30n^5}$

33. $(2x^2y^4)(7x^5y^2)$ 34. $(-6a^8b^3)(-4ab^8)$

35. $\dfrac{-56a^5b^3}{7a^2b^2}$ 36. $\dfrac{72x^7y^5}{12x^3y^3}$

37. $4pq^4 + p^4q - 7pq^4$ 38. $18r^2s - 3rs + 5rs$

39. $(5f + 2g)(4f - g)$ 40. $(2z - 3)(5z + 6)$

41. $5x^5 - 8x^5 + 6x^5$ 42. $-3y^7 + 9y^7 - y^7$

43. $2rs - 3r^2s$

44. $5p^2q^3 - 2p^2q^2 - 7p^2q^3$

45. $-2a^3b^6 + 8a^7b^3 - 9a^3b^6$

46. $8wz - 11wz^2$

47. $\dfrac{39t^4}{13t^6}$ 48. $\dfrac{24x^2}{16x^5}$

49. $(3a^{-2}bc^{-1})^{-2}$ 50. $(2x^2y^{-3}z^{-1})^{-3}$

C. Simplify completely, eliminating zero and negative exponents from the answers.

51. $2(5a - 7) - 3(4a + 2)$

52. $(-3x^3y)^2(2xy)$

53. $(3c^2d^2)^2(-2cd)$

54. $8(t - 4) - 7(3t - 1)$

55. $(y^2 + 2y - 1)(3y^2 - 5y - 2)$

56. $(2a^2 - a + 7)(a^2 - 4)$

57. $\dfrac{9w^6 + 4w^4 + 2w^3}{w^2}$

58. $\dfrac{7g^5 - 5g^3 + 3g^2}{g^2}$

59. $(5p + 2)^2$

60. $(5x^2 - 6x + 2) - (3x^2 - 4x + 9)$

61. $(y^2 - 6y - 4) - (3y^2 - 8y + 1)$

62. $(7x - 4y)^2$

63. $(5 - 3p)(p + 6)$

64. $(3k - 2) - (3 + 4k)$

65. $(4y + 3) - (9 - 7y)$

66. $(5 - 3z)(7z + 6)$

67. $\dfrac{12b^3 + 16b^2 - 9b}{-6b + 1}$

68. $\dfrac{9d^3 - 15d^2 + 4d}{-3d + 1}$

69. $\dfrac{15x^3y^2z^2 - 25x^2y^2z^2 + 10xy^3z^2}{5xy^2z}$

70. $\dfrac{24a^2b^3c^3 - 18a^2b^2c^2 - 6a^2b^3c^4}{-6ab^2c}$

71. $\dfrac{a^2 + 7a + 10}{a + 5}$

72. $\dfrac{x^2 - 5x + 6}{x - 3}$

73. $(2m^{-2}n^{-3})(-3m^3n^{-2})$

74. $(3m^2 - 1) \div (m - 3)$

75. $(4x^3 - 2x + 1) \div (x - 1)$

76. $(-5r^{-2}s^{-1}t^3)(2rs^{-3}t^{-2})$

77. $\dfrac{(2x^{-1}y^2)^2(x^2y^{-3})^{-1}}{(2^{-3}xy^3)^2}$

78. $(1 + 2t)^3$

79. $(2 + a)^3$

80. $\dfrac{(5m^2n^{-1})^{-1}(2m^2n^2)^{-1}}{(4m^{-3}n^2)^{-1}}$

D.   Complete the following chart.

Scientific Notation	Ordinary Notation
81.	0.0087
82. 3.44×10^{-3}	
83. 9×10^7	
84.	0.00005
85.	62.3×10^3
86.	0.8×10^{-3}
87. 6.5×10^{-4}	
88. 3.89×10^4	
89.	6,400,000
90. 1.5×10^{-1}	
91. 9.9×10^3	
92.	2,960,000

E.   Perform the indicated operations using scientific notation and the properties of exponents. Express each answer in scientific notation.

93. $0.045 \div 9{,}000$ 94. $1{,}114 \times 0.003$

95. 0.0093×0.006 96. $4{,}200{,}000 \div 0.21$

97. $0.007 \times 690 \times 3{,}000$

98. $910 \times 0.008 \times 40{,}000$

99. $\dfrac{64{,}000 \times 0.004}{0.0008 \times 0.02}$

100. $\dfrac{850 \times 0.00001 \times 5{,}000}{17{,}000 \times 0.000005}$

Chapter Test

Perform the indicated operation(s) and simplify completely, eliminating zero and negative exponents from the answers.

1. $(3a^2b)(-2a^3b^2)$

2. $(y - 4)(2y + 1)$

3. $-8m^2n \div 16mn$

4. $4a^2c + (-6a^2c)$

5. $c^4 \div c^8$

6. $(-2r^2st^3)^2$

7. $(3a^{-2}bc^3)^{-1}$

8. $2c^{-2}$

9. $-2(5x - 3y)$

10. $F^2 \cdot F \cdot F^3$

11. $(3a^{-2}b^{-3})(-2a^{-3}b^2)$

12. $(x + 3)^2$

13. $\left(\dfrac{4x^2}{3y}\right)^{-2}$

14. $\dfrac{56c^2d^2 - 14cd^2 + 21cd^3}{7cd}$

15. From the sum of $(5x)$ and $(-11x)$, subtract $(-3x)$.

16. Subtract: $(5m^2n - 2mn + 7)$
    $- (9mn - 11 - 2m^2n)$

17. Multiply: $(a + 2)(3a^2 - 2a + 5)$

18. Divide: $(4y^3 + 3y^2 + 2) \div (y + 1)$

19. Change 0.00067 to scientific notation.

20. Use scientific notation to solve $\dfrac{97{,}000 \times 600}{3{,}000{,}000}$.

    Show your work.

Thought-Provoking Problems

1. Explain why, in the definition of zero as an exponent that appears on page 408, the condition that it applies to "any nonzero quantity" is included. In other words, is $0^0 \neq 1$? Why?

2. Simplify completely:

   a. $\left(\dfrac{1}{2}a^2b\right)^3\left(\dfrac{2}{3}ab^2\right)^2$

   b. $\left(\dfrac{3}{8}x + \dfrac{1}{2}\right)^2$

   c. $\left(\dfrac{5}{6}b^2 + \dfrac{1}{3}b + 4\right) + \left(\dfrac{1}{8}b^2 - \dfrac{3}{4}b + \dfrac{2}{3}\right)$

   d. $\left(\dfrac{2}{3}m^3 - \dfrac{3}{4}m^2 + \dfrac{5}{8}m\right) \div \left(\dfrac{1}{2}m\right)$

3. Simplify completely, using a calculator if you wish.

   a. $(0.4x + 1.5)^2$

   b. $(4.36y^2 + 1.75y - 3.2) - (2y^2 - 0.4y + 0.16)$

   c. $\dfrac{12.4y^8 + 7.6y^6 - 5.2y^5}{0.02y^3}$

   d. $(0.4a^2b^{-3})^{-2}(1.2a^{-1}b)^2$

4. Simplify completely, eliminating negative exponents.

   a. $(c - x)^{-2}$ b. $\left(\dfrac{4a^2b^{-3}}{6a^{-1}b}\right)^{-1}$

   c. $\dfrac{2m^{-3}}{5m^{-5}}$ d. $\left(\dfrac{c^3}{d}\right)^{-4}$

   e. $\left(\dfrac{6p^{-3}r^2}{pr^{-1}}\right)^0$

5. Simplify: $(y^2 + 2y + 3)^3$

6. The mass of an electron of a neutral atom is 9.11×10^{-28} gram. The mass of the neutron of such an atom is 4.67×10^{-24} gram. About how many times heavier is the neutron than the electron?

7. Simplify, and state the answers in scientific notation.

   a. $6,850,000 \times 2,450 \times 0.0072$

   b. $\dfrac{8,795 \times 0.0725}{6,050,000}$

8. In this chapter you have multiplied and divided quantities in scientific notation. As with multiplication and division of fractions, wherein like denominators were not necessary, like exponents on the power of 10 were not needed to do these operations. However, in the addition and subtraction of quantities in scientific notation, the powers of 10 must be the same.

   Example: $(3.26 + 10^4) + (1.8 \times 10^5)$ must be changed to either

   $$(0.326 \times 10^5) + (1.8 \times 10^5) = 2.126 \times 10^5$$

   or

   $$(3.26 \times 10^4) + (18 \times 10^4) = 21.26 \times 10^4$$
   $$= 2.126 \times 10^5$$

   Try these problems:

   a. $(1.6 \times 10^3) + (2 \times 10^2)$

   b. $(9.2 \times 10^{-2}) + (1.1 \times 10^{-3})$

   c. $(7 \times 10^6) - (8.2 \times 10^5)$

9. Find the product of each of the following.

   $(x + 2)^2$ $(y + 3)^2$ $(2a + 1)^2$ $(x + a)^2$

   Examine the middle term in each answer and look for a pattern that exists. Make a statement in words about the middle terms of squared binomials.

10. Find the product of each of the following, and then look for a pattern for the middle terms. State your findings in words.

    $(x + 2)(x - 2)$ $(y - 3)(y + 3)$

    $(2a + 1)(2a - 1)$ $(x + a)(x - a)$

16

Introduction to Factoring

16.1   Introduction

In the previous chapters you learned how to multiply monomial and/or polynomial factors together. In this chapter, you will reverse the multiplication process. You will begin with products containing two, three, or four terms and will determine what prime factors were multiplied together to give that answer. This process is called **factoring**. Some of the factoring techniques may seem complicated at first, but you will come to understand them with practice.

♦ **Learning Objectives**

When you have completed this chapter, you should be able to:

1. Completely factor expressions containing:
   a. greatest common factors
   b. the difference of two squares
   c. trinomials
   d. perfect-square trinomials
   e. the sum or difference of perfect cubes
2. Check factoring answers by multiplying.
3. Solve quadratic equations by factoring.

16.2   Greatest Common Factor

Comparing the kinds of problems presented in Chapters 15 and 16 reveals that factoring is the reverse of multiplication.

	Chapter 15	Chapter 16
Problem:	$3x^2y(-2xy + 5)$	$14a^2b + 21a^2b^2$
Answer:	$-6x^3y^2 + 15x^2y$	$7a^2b(2 + 3b)$

In the Chapter 16 example, each term in the original expression was examined to see if it contained a factor common to the others. If so, as was the case here, the **greatest common factor** (GCF) was removed from each term. (You may find it helpful to review the arithmetic procedure for finding the GCF in Section 2.7.) The greatest common factor was written in front of a set of parentheses, and the remaining factors of each term were written inside the parentheses. In fact, the Distributive Principle was used to write the factorization. Let's look closely to see how the answer was found.

First, each term was written in prime factored form. (See prime factorization in Section 2.5.)

$$14a^2b + 21a^2b^2 = 2 \cdot 7 \cdot a \cdot a \cdot b + 3 \cdot 7 \cdot a \cdot a \cdot b \cdot b$$

Note that 7, a, a, and b are in both lists of factors. Consequently, their product, $7a^2b$, was removed from each term.

$$= 2 \cdot \not{7} \cdot \not{a} \cdot \not{a} \cdot \not{b} + 3 \cdot \not{7} \cdot \not{a} \cdot \not{a} \cdot \not{b} \cdot b$$

Then $2 + 3b$ remained, so the factorization of $14a^2b + 21a^2b^2$ was written as $7a^2b(2 + 3b)$.

To check your answer, multiply the factors together to see if the answer agrees exactly with the original expression. If the answer and the original expression do agree, your answer is correct, as long as you have done all the factoring that was possible.

Check $7a^2b(2 + 3b) = 14a^2b + 21a^2b^2$, which is the same as the original problem.

E X A M P L E   1   Factor completely: $15m^2n^2 - 20m^2n + 35mn$

Solution $15m^2n^2 - 20m^2n + 35mn$

$5mn$ is in every term and therefore is the GCF. Take it out of every term (divide each term by $5mn$).

$$= 3 \cdot \not{5} \cdot \not{m} \cdot m \cdot \not{n} \cdot n - 2 \cdot 2 \cdot \not{5} \cdot \not{m} \cdot m \cdot \not{n} + \not{5} \cdot 7 \cdot \not{m} \cdot \not{n}$$

$3mn - 4m + 7$ remains as the quotient. Write the GCF outside the parentheses and the quotient inside.

$$= 5mn(3mn - 4m + 7)$$

Check   Multiply the answer.

$$5mn(3mn - 4m + 7) = 15m^2n^2 - 20m^2n + 35mn$$

◆

E X A M P L E **2** Factor completely: $26x^2 + 13x + 39$

Solution $26x^2 + 13x + 39$

13 is the GCF, so factor it out.

$= 13(2x^2 + x + 3)$

Check Multiply the answer.

$13(2x^2 + x + 3) = 26x^2 + 13x + 39$ ◆

E X A M P L E **3** Factor completely: $28y^2z^2 + 81y^2z + 15$

Solution $28y^2z^2 + 81y^2z + 15$

$= 2^2 \cdot 7 \cdot y^2z^2 + 3^4y^2z + 3 \cdot 5$

There is no factor common to all of the terms. Therefore, this expression cannot be factored by this method. ◆

E X A M P L E **4** Factor completely: $55a^2 + 33a + 121$

Solution You do not need to write a list of factors of each term to see that there is no common literal factor. However, 55, 33, and 121 are all multiples of 11.

$= 11(5a^2 + 3a + 11)$

Check Multiply the answer.

$11(5a^2 + 3a + 11) = 55a^2 + 33a + 121$ ◆

E X A M P L E **5** Factor completely: $-2x^2 + 28$

Solution When the first term of a polynomial is negative, factor out the common negative factor.

$-2x^2 + 28 = -2(x^2 - 14)$

Check $-2(x^2 - 14) = -2x^2 + 28$ ◆

Looking for the greatest common factor in a polynomial should be the first step in any factoring problem you do. The resulting expression will be much easier to factor further.

Greatest Common Factor

> If all terms in an expression contain a common factor or factors, remove the greatest common factor from each term and multiply that GCF by the sum of the remaining terms.

Thus

$$a^2b^2 + a^2b + ab^2 = ab(ab + a + b)$$

and

$$14x^2 + 7x + 28 = 7(2x^2 + x + 4)$$

E X A M P L E **6**   Factor completely: $a(x + 2) + b(x + 2)$

Solution   Although this doesn't look like the previous problems, there is a common factor in both terms of the expression; it is $x + 2$.

$$a(x + 2) + b(x + 2) = (x + 2)(a + b)$$

Check   Multiply the answer.

$$(x + 2)(a + b) = ax + bx + 2a + 2b$$

Is that the same as $a(x + 2) + b(x + 2)$?

Multiply the original expression.

$$a(x + 2) + b(x + 2) = ax + 2a + bx + 2b$$

The answers would be the same if the terms were rearranged.   ◆

E X A M P L E **7**   Factor completely: $y(y + 5) - 4(y + 5)$

Solution $(y + 5)$ is the common factor.

$$y(y + 5) - 4(y + 5)$$
$$= (y + 5)(y - 4)$$

Check $$(y + 5)(y - 4)$$
$$= y^2 - 4y + 5y - 20$$
$$= y^2 + y - 20$$

And $y(y + 5) - 4(y + 5)$
$$= y^2 + 5y - 4y - 20$$
$$= y^2 + y - 20$$ ◆

E X A M P L E **8**   Factor completely: $y^2 + 2y + yz + 2z$

Solution   There isn't a common factor in all four terms, but you can factor by taking a y out of the first two terms and a z out of the last two terms.

$$y(y + 2) + z(y + 2)$$

Now can you see the common factor, $(y + 2)$? Remove it from both terms and multiply it by what remains.

$$y(y + 2) + z(y + 2) = (y + 2)(y + z)$$

Check $(y + 2)(y + z) = y^2 + yz + 2y + 2z$ ◆

Do you think you would have come up with the same answer if you had grouped the terms differently? Let's see:

$$y^2 + yz + 2y + 2z$$
$$= y(y + z) + 2(y + z)$$
$$= (y + z)(y + 2)$$ The answers are the same.

The process used in Example 8 is called **factoring by grouping**. If you are to factor an expression with four terms and there is no factor common to all of them, try this method of working with two factors at a time.

E X A M P L E **9**   Factor completely: $mn + kn + 2m + 2k$

Solution   Use *factoring by grouping* because there are four terms that have no factor in common.

$$mn + kn + 2m + 2k$$
$$= n(m + k) + 2(m + k)$$
$$= (m + k)(n + 2)$$

Check $(m + k)(n + 2) = mn + 2m + kn + 2k$ ◆

E X A M P L E **10**   Factor completely: $3x^2 + 6y^2 - 2x^2y - 4y$

Solution   Try factoring by grouping.

$$3x^2 + 6y^2 - 2x^2y - 4y$$
$$= 3(x^2 + 2y^2) - 2y(x^2 + 2)$$

You cannot factor any further because $(x^2 + 2y^2)$ and $(x^2 + 2)$ are not like factors.   ◆

E X A M P L E **11**   Factor completely: $4m + 8 - 12my - 24y$

Solution $4m + 8 - 12my - 24y$ Remove the GCF of 4 from all the terms.
$= 4[m + 2 - 3my - 6y]$ Factor by grouping.
$= 4[1(m + 2) - 3y(m + 2)]$
$= 4(m + 2)(1 - 3y)$

Check   Multiply: $4(m + 2)(1 - 3y)$
$= (4m + 8)(1 - 3y)$
$= 4m - 12my + 8 - 24y$ Rearrange terms.
$= 4m + 8 - 12my - 24y$ ◆

16.2 ◆ Practice Problems

A. Multiply each set of factors to see if the product that is given is correct.

1. $4(2x + 7)$ and $8x + 28$
2. $-5(7y^2 - 1)$ and $-35y^2 - 5$
3. $-8(a^2 + 4b^2)$ and $-8a^2 + 32b^2$
4. $7(m^2n + 2mn)$ and $7m^2n + 14mn$
5. $c^2b(3b + 5c^2)$ and $3c^2b + c^4b$
6. $ax^2(-4x + 2a^2)$ and $-4ax^2 + 2a^3x^2$
7. $9(w^2 + 3w - 2)$ and $9w^2 + 27w - 18$
8. $2y^3(4y^4 - y^3 + 5y^2 - 1)$ and
   $8y^7 - 2y^6 + 10y^5 - 1$
9. $m^5n^6(m^2n^2 - 2mn + 1)$ and
   $m^7n^8 - 2m^6n^7 + m^5n^6$
10. $-x^2(x^2 - 2x + 10)$ and $-x^4 + 2x^3 - 10x^2$

B. Factor out the greatest common factor from each expression. Check answers by multiplying.

11. $2x^2 - 4x + 16$ 12. $8m^2 - 16m + 24$
13. $5a^2b^2 \div 12a^2b + 4ab$
14. $2.8b^3 + 1.4b^2 + 4.2b$
15. $-12r^2 - 12r$ 16. $-3y^2 + 9$
17. $1.6m^3 + 6.4m^2 + 2.4m$ 18. $-4x^2 + 20$
19. $-a^3 + 2a^2 + 6a$
20. $15d^2 - 10d - 50$

C.   Factor out the GCF. If there is none, write "none" for the answer. Check answers by multiplying.

21. $11f^3 + 44g^3$

22. $13c^3 - 39d^3$

23. $-a^3 - a^2 + 8a$

24. $-b^5 - 4b^3 + 2b^2$

25. $64b^4 + 16b^3$

26. $16b^2 - 27$

27. $5a^3 - 15a^2 + 4$

28. $-12m^6 + 36m^5 - 48m^4$

29. $x^5y^4z^2 + 7x^4y^2z^2 - x^2z^2$

30. $2x^2y^2z^3 - 8x^2y^3z^2 + 4x^3y^2z^2$

31. $a(a + 4) - 4(a + 4)$

32. $m(n - 1) + 6(n - 1)$

D.   Factor by grouping. Check answers by multiplying.

33. $ax^2 - 7ax + 2x - 14$

34. $8xy - 16x + y^2 - 2y$

35. $18wx + 6x + w^2 + w$

36. $a^2 - 2a - 8ab + 16b$

37. $acx + acy - bcx - bcy$

38. $3rt + 3st + 6r + 6s$

39. $p^2 + 5p + 12pq + 60q$

40. $m^2 - 4m - 9mn + 36n$

16.3   Difference of Two Perfect Squares

The numbers 1, 4, 9, 16, 25, 36, 49, 64, 81, 100, 121, 144, and so on are called **perfect squares**. This means that a rational number was multiplied by itself to produce this product (see Section 2.2).

$1 \times 1 = 1$	$2 \times 2 = 4$	$3 \times 3 = 9$	$4 \times 4 = 16$
$5 \times 5 = 25$	$6 \times 6 = 36$	$7 \times 7 = 49$	$8 \times 8 = 64$
$9 \times 9 = 81$	$10 \times 10 = 100$	$11 \times 11 = 121$	$12 \times 12 = 144$

The quantities x^2, y^2, b^6, $4a^2$, and so on are also perfect squares. They are the product of a factor times itself.

$$x \cdot x = x^2 \qquad y \cdot y = y^2$$
$$b^3 \cdot b^3 = b^6 \qquad 2a \cdot 2a = 4a^2$$

Five is said to be the square root of 25 ($\sqrt{25} = 5$). Eleven is the square root of 121 ($\sqrt{121} = 11$), and $2a$ is the square root of $4a^2$ ($\sqrt{4a^2} = 2a$, $a \geq 0$). The **square root** of a quantity is the number that, when multiplied by itself, gives the original quantity. Quantities such as 4, y^2, $121b^2$, and $16y^2$ appear in a special type of factoring problem. When the *difference* of two perfect squares is factored, the result is always the product of the sum of their square roots and the difference of their square roots. This definitely sounds more complicated than it is!

If you are asked to factor $x^2 - 4$, first check for a GCF. There is none. Note that the expression contains only *two terms* and that x^2 and 4 are *perfect squares*. The factors of their *difference* will always be the sum and the difference of their square roots:

$$x^2 - 4 = (x)^2 - (2)^2 = (x + 2)(x - 2)$$

Check the answer by multiplying the factors together.

$$(x + 2)(x - 2)$$
$$= x^2 - 2x + 2x - 4$$

When you combine the like terms, the result is $x^2 + 0x - 4$, which equals $x^2 - 4$. Because this is the original expression, the factors are correct.

E X A M P L E　**1**　Factor completely: $4b^2 - 9$

Solution　There is no GCF, so look to see if this might be the difference of two perfect squares.

$\sqrt{4b^2} = 2b$　and　$\sqrt{9} = 3$

Therefore, $4b^2 - 9$ fits the description of the difference of two perfect squares.

$4b^2 - 9 = (2b)^2 - (3)^2 = (2b + 3)(2b - 3)$

Check　　$(2b + 3)(2b - 3)$
$= 4b^2 - 6b + 6b - 9$
$= 4b^2 - 9$　　　　　　　　　　　◆

Multiply the following sets of factors:

$(y + 6)(y - 6) =$
$(2a + 7)(2a - 7) =$
$(b^2 + 5)(b^2 - 5) =$

Did you see the pattern in these products? In each case, the middle terms add to zero. This occurs every time you multiply the sum and the difference of the same two terms.

Would you expect that to occur if you were multiplying the sum of the same terms? How about the difference of the same terms? Let's find out.

$(x + 2)(x + 2)$
$= x^2 + 2x + 2x + 4$
$= x^2 + 4x + 4$

$(2a - 7)(2a - 7)$
$= 4a^2 - 14a - 14a + 49$
$= 4a^2 - 28a + 49$

Apparently, with any combination other than the sum and difference, the middle terms do *not* add to zero and do not disappear.

Check it further by doing these multiplications:

$(y + 6)(y + 6)$　　　$(3z + 2)(3z - 2)$　　　$(b^2 - 5)(b^2 - 5)$

Your products should have been $y^2 + 12y + 36$, $9z^2 - 4$, and $b^4 - 10b^2 + 25$. In each case, the *same signs* in the factors *produced a middle term* in the product. But when the *signs in the factors were different*, the *middle terms added to zero*.

Difference of Two Perfect Squares

> The factors of the difference of two perfect squares are always the sum and difference of the square roots of the terms.

Thus,

$a^2 - b^2 = (a + b)(a - b)$
$16 - x^2 = (4 + x)(4 - x)$
$64m^4 - 9 = (8m^2 + 3)(8m^2 - 3)$

The sum of two perfect squares cannot be factored this way. If you tried to factor $x^2 + y^2$, you might guess $(x + y)(x - y)$, $(x - y)(x - y)$, or $(x + y)(x + y)$ as possible solutions. When those sets of factors are multiplied, they give the products $x^2 - y^2$, $x^2 - 2xy + y^2$, and $x^2 + 2xy + y^2$. None agrees with the original expression. In fact, $x^2 + y^2$ is *prime*; it has *no factors but itself and 1*.

When you begin to factor an expression that you recognize as the difference of two squares, always look first to see if there is a common factor (GCF) that can be removed.

$$64x^2 - 16 = 16(4x^2 - 1) = 16(2x + 1)(2x - 1)$$

$(8x + 4)(8x - 4)$ would not have been a correct answer, because each factor could be further simplified by removing a common factor of 4. When you are asked to factor an expression, your answer must have each part factored *completely*. Except for the GCF, all other factors must be prime.

Factor Completely

> After completing each step of a factoring problem, see if any more factoring can be done.

E X A M P L E 2   Factor completely: $64x^2 - 4$

Solution

$$64x^2 - 4$$ First remove the GCF.

$$= 4(16x^2 - 1)$$ Factor the difference of two perfect squares.

$$= 4(4x + 1)(4x - 1)$$

Check

$$4(4x + 1)(4x - 1)$$

$$= (16x + 4)(4x - 1)$$

$$= 64x^2 - 16x + 16x - 4$$

$$= 64x^2 - 4$$ ◆

E X A M P L E 3   Factor completely: $162r^4 - 2$

Solution

$$162r^4 - 2$$

$$= 2(81r^4 - 1)$$

$$= 2(9r^2 + 1)(9r^2 - 1)$$ $9r^2 - 1$ is the difference of two perfect squares.

$$= 2(9r^2 + 1)(3r + 1)(3r - 1)$$

Check

$$2(9r^2 + 1)(3r + 1)(3r - 1)$$

$$= (18r^2 + 2)(3r + 1)(3r - 1)$$

$$= (54r^3 + 18r^2 + 6r + 2)(3r - 1)$$

$$= 162r^4 + 54r^3 + 18r^2 + 6r - 54r^3 - 18r^2 - 6r - 2$$

$$= 162r^4 - 2$$ ◆

Exercises   Try these problems. If you have difficulty, review the steps in the previous examples.

1. $4 - a^2$ 2. $36K^2 - 49$ 3. $3m^3 - 27m$

Answers:

1. $(2 + a)(2 - a)$ 2. $(6k + 7)(6k - 7)$

3. $3m(m^2 - 9) = 3m(m + 3)(m - 3)$

16.3 ◆ Practice Problems

A. Find the product.

1. $(x + 2)(x - 2)$
2. $(b + 6)(b - 6)$
3. $(m + 3)(m + 3)$
4. $(z - 2)(z - 2)$
5. $(2b + 1)(2b - 1)$
6. $(3x + 5)(3x - 5)$
7. $(x + 2z)(x - 2z)$
8. $(m + n)(m - n)$
9. $(1 - 3a)(1 + 3a)$
10. $(6 - 5y)(6 + 5y)$
11. $(3y - 2z)(3y + 2z)$
12. $(7c - 2b)(7c + 2b)$

B. Factor completely. If the expression cannot be factored, write "prime" as your answer. Check your answers by multiplying.

13. $x^2 - 49$
14. $b^2 - 36$
15. $a^2 - 1$
16. $m^2 - 16$
17. $k^2 - 121$
18. $r^2 - 1$
19. $z^2 - 81$
20. $s^2 - 49$
21. $c^2 - 4$
22. $w^2 - 25$
23. $4f^2 - 9$
24. $144d^2 - 1$
25. $25T^2 - 36$
26. $4G^2 - 9$
27. $49J^2 - 64$
28. $121v^2 - 36$
29. $x^2 + y^2$
30. $a^2 - 4b^2$

31. $36 - c^2$
32. $B^2 + 4$
33. $a^2 - 64b^2$
34. $49 - w^2$
35. $16a^2 - 1$
36. $4c^4 - 25c^2$
37. $9b^2c^2 - 16a^2$
38. $36m^2 - 1$
39. $1 - 100x^2$
40. $64x^2y^2 - 49z^2$

C. Factor completely. Check your answers by multiplying. Write "prime" if the expression cannot be factored.

41. $6b^3 - 150b$
42. $2a^2 - 32$
43. $x^6 - 81a^2$
44. $c^6 - 16c^2$
45. $3d^4 - 243$
46. $11x^2 - 44y^2$
47. $5m^2 - 45n^2$
48. $9a^2 - 2$
49. $49a^2b^2 - 6$
50. $5z^4 - 80$
51. $a^3 - 9a$
52. $64 - b^4$
53. $4m^2 - 64n^2$
54. $c^4 + c^2d^2$
55. $25x^3 - x$
56. $9m^3 - 36m$
57. $98 + 2a^2$
58. $3a^2 - 27b^2$
59. $27b^2 - 75$
60. $147 - 3y^2$

16.4 ◆ Factoring Trinomials

When you are asked to factor a polynomial completely, first look to see if either GCF or difference of two squares applies. Neither apply for the trinomial $x^2 + 13x + 12$, so we need to explore other methods of factoring.

Some students and instructors use a trial-and-error approach to finding the factors of trinomials. Such a strategy for the foregoing problem would follow this reasoning:

$x^2 + 13x + 12$

Factors of x^2 are $(x)(x)$.

Possible factors of $+12$ are $(3)(4)$, $(-3)(-4)$, $(2)(6)$, $(-2)(-6)$, $(1)(12)$, and $(-1)(-12)$.

Try these combinations to see which produces a middle term of $+13$.

$(x + 3)(x + 4) = x^2 + 4x + 3x + 12 = x^2 + 7x + 12$
$(x - 3)(x - 4) = x^2 - 4x - 3x + 12 = x^2 - 7x + 12$
$(x + 2)(x + 6) = x^2 + 6x + 2x + 12 = x^2 + 8x + 12$
$(x - 2)(x - 6) = x^2 - 6x - 2x + 12 = x^2 - 8x + 12$
$(x + 1)(x + 12) = x^2 + 12x + x + 12 = x^2 + 13x + 12$

That's the one! $(x + 1)(x + 12)$ has a middle term of $+13x$.

Therefore, $x^2 + 13x + 12 = (x + 1)(x + 12)$.

As students use this process, they begin to see clues that enable them to eliminate from consideration some of the factors of the last term without going through the multiplication. For instance, in trinomials where the squared term has a coefficient of 1, if the last term is positive, then both factors will have the same sign. That sign will be the sign of the middle term. And if the last term is negative, one term will have a positive sign and the other a negative sign. The sum of these two terms must equal the middle term.

Apply these clues to the last problem, $x^2 + 13x + 12$. Because the last term is positive and the middle term is positive, both of the factors will have $+$ signs:

$$(x + \quad)(x + \quad)$$

This eliminates the need to look at $(-3)(-4)$, $(-2)(-6)$, and $(-1)(-12)$. And the only factors left that can add to $+13$ are $(+1)(+12)$.

Let's look at another trinomial and apply the trial-and-error method to factor it. Factor completely: $m^2 - 12m - 45$

1. The last term is negative, so the factors will have different signs; and their sum must be -12 (the coefficient of the middle term).

2. Possible factors of -45 are $(-1)(45)$, $(1)(-45)$, $(-3)(15)$, $(3)(-15)$, $(-5)(9)$, and $(5)(-9)$. The only pair that would combine to give -12 are (3) and (-15).

3. Try that combination and see if it checks.

$$(m + 3)(m - 15) = m^2 - 15m + 3m - 45$$
$$= m^2 - 12m - 45$$

This method is more tedious when the coefficient of the squared term is not 1. Factor completely: $12R^2 - 17R - 5$

1. Factors of $12R^2$ must be tested in combination with factors of -5 to see which products combine to give a middle term of $-17R$.

    Factors of $12R^2$: $(12R)(R)$, $(-12R)(-R)$, $(6R)(2R)$, $(-6R)(-2R)$, $(4R)(3R)$, $(-4R)(-3R)$

    Factors of -5: $(1)(-5)$, $(-1)(5)$

2. All possible combinations of factors are

$(12R - 5)(R + 1)$	$(6R - 5)(2R + 1)$
$(12R + 5)(R - 1)$	$(6R + 5)(2R - 1)$
$(12R - 1)(R + 5)$	$(6R - 1)(2R + 5)$
$(12R + 1)(R - 5)$	$(6R + 1)(2R - 5)$
$(-12R - 5)(-R + 1)$	$(-6R - 5)(-2R + 1)$
$(-12R + 5)(-R - 1)$	$(-6R + 5)(-2R - 1)$
$(-12R - 1)(-R + 5)$	$(-6R - 1)(-2R + 5)$
$(-12R + 1)(-R - 5)$	$(-6R + 1)(-2R - 5)$
$(4R - 5)(3R + 1)$	$(-4R - 5)(-3R + 1)$
$(4R + 5)(3R - 1)$	$(-4R + 5)(-3R - 1)$
$(4R - 1)(3R + 5)$	$(-4R - 1)(-3R + 5)$
$(4R + 1)(3R - 5)$	$(-4R + 1)(-3R - 5)$

3. After multiplying, look at the sum of the second and third terms to see which one gives $-17R$.

$$(4R + 1)(3R - 5)$$
$$= 12R^2 - 20R + 3R - 5$$
$$-17R$$

4. Therefore, $12R^2 - 17R - 5 = (4R + 1)(3R - 5)$.

5. Check: $(4R + 1)(3R - 5) = 12R^2 - 20R + 3R - 5$
$$= 12R^2 - 17R - 5$$

For students who prefer a more structured approach to factoring trinomials, the **product–sum method** can be used. *Product* means the product of the coefficient of the first term and the coefficient of the last term. *Sum* means the sum of the factors of the products that add to give the coefficient of the middle term.

In $x^2 + 8x + 12$, the product of the coefficient of the first term and the last term is $+12$ from $(+1)(+12)$. Remember that a variable with no coefficient showing is understood to have a coefficient of $+1$. Your job is to find the factors of that product, $+12$, that add together to equal the coefficient of the middle term, $+8$.

$$x^2 + 8x + 12$$

$$\text{Product} = (+1)(+12) = +12$$
$$x^2 + 8x + 12$$

$$\text{Sum} = +8$$

Factors of $+12$: $\quad (+1)(+12) \quad (-2)(-6)$
$$(-1)(-12) \quad (+3)(+4)$$
$$(+2)(+6) \quad (-3)(-4)$$

The only factors of $+12$ that add to $+8$ are $(+2)$ and $(+6)$. Next, write the original expression in an expanded form, with $+8x$ written as the sum of $+2x$ and $+6x$.

$$x^2 + 8x + 12 = x^2 + 2x + 6x + 12$$

(Think back to the process of multiplying two binomials, such as $(y + 2)(y - 1)$. The distributive multiplication produces: $y^2 - y + 2y - 2$, which was then simplified by combining like terms. The expanded form in this factoring problem is similar to this expression, before the terms are combined.)

From the first two terms, factor their GCF, and then factor a GCF from the third and fourth terms. This is the same kind of factoring you studied in Examples 6 through 10 in Section 16.2.

$$x^2 + 2x + 6x + 12$$
$$= x(x + 2) + 6(x + 2)$$

Look at the two groups that were formed. Is there anything similar between them? If so, factor it out of both, and write what remains. Because $(x + 2)$ is the common factor, multiply it by the remaining terms.

$$x(x + 2) + 6(x + 2)$$
$$= (x + 2)(x + 6)$$

Therefore, the factorization of the trinomial is

$$x^2 + 8x + 12 = (x + 2)(x + 6)$$

Check the answer by multiplying the factors together.

$$(x + 2)(x + 6) = x^2 + 6x + 2x + 12 = x^2 + 8x + 12$$

This whole process seems difficult, but once you see the pattern of steps, you will be able to apply them when factoring any trinomial. Follow the detailed steps in the examples, and then try some on your own.

E X A M P L E **1**   Factor completely: $y^2 + 6y + 9$

Solution   1.   Product $= (+1)(+9) = +9$ Sum $= +6$

Factors of $+9$: $(+1)(+9)$
$(-1)(-9)$
$(+3)(+3)$
$(-3)(-3)$

$(+3)$ and $(+3)$ are the only factors of $+9$ that add to $+6$.

2.   Write the original expression in expanded form:

$y^2 + 3y + 3y + 9$

3.   Factor from the first and second terms their GCF, and from the third and fourth terms their GCF. This is the process you used in factoring by grouping in Section 16.2.

$$y^2 + 3y + 3y + 9$$

$$= y(y + 3) + 3(y + 3)$$

4.   Take out the common factor from both groups and multiply it by the remaining terms.

$(y + 3)(y + 3)$

This is factored completely. Therefore,

$y^2 + 6y + 9 = (y + 3)(y + 3)$

5.   Check by multiplication.

$$(y + 3)(y + 3) = y^2 + 3y + 3y + 9$$
$$= y^2 + 6y + 9$$ ◆

E X A M P L E **2**   Factor completely: $m^2 - 12m - 45$

Solution   1.   Product $= (+1)(-45) = -45$ Sum $= -12$

Factors of -45: $(+1)(-45)$
$(-1)(+45)$
$(+3)(-15)$
$(-3)(+15)$
$(+5)(-9)$
$(-5)(+9)$

2.   Write in expanded form.

$m^2 + 3m - 15m - 45$

3.   Factor the GCFs from pairs of terms.

$$m^2 + 3m - 15m - 45$$

$$= m(m + 3) - 15(m + 3)$$

Note that (-15) was factored from the third and fourth terms. This choice was made so that the terms inside the second parentheses would have the same signs as those inside the first parentheses.

4.   Write the common factor times the remaining terms.

$(m + 3)(m - 15)$

Therefore, the complete factoring is

$m^2 - 12m - 45 = (m + 3)(m - 15)$

In step 2, the order of the middle terms could be reversed and still produce the same answer.

2. Write in expanded form.

$$m^2 - 15m + 3m - 45$$

3. Factor the GCFs from the pairs of terms.

$$m(m - 15) + 3(m - 15)$$

4. Remove the common factor.

$$(m - 15)(m + 3)$$

5. Check:

$$(m + 3)(m - 15)$$
$$= m^2 - 15m + 3m - 45$$
$$= m^2 - 12m - 45$$

E X A M P L E **3** Factor completely: $2T^2 - 8T + 8$

Solution Remove the GCF first; then factor the trinomial.

$$2(T^2 - 4T + 4)$$

1. Product = $(+1)(+4) = +4$ Sum = -4

   Factors of $+4$: $(+1)(+4)$
                            $(-1)(-4)$
                            $(+2)(+2)$
                            $(-2)(-2)$

   (-2) and (-2) are the only factors of $+4$ that add to -4.

2. Write the original expression in expanded form.

$$2(T^2 - 2T - 2T + 4)$$

3. Factor GCFs from the first and second terms and from the third and fourth terms. The brackets are needed here to keep the first greatest common factor (2) outside the rest of the expression that is now being factored. If the 2 weren't carried along throughout the problem, it might get forgotten in the answer.

$$2[T(T - 2) - 2(T - 2)]$$

4. Take out the common factor from both groups and multiply it by what remains.

$$2[(T - 2)(T - 2)]$$

   Check to be sure that no more factoring can be done, as is the case here. Therefore,

$$2T^2 - 8T + 8 = 2(T - 2)(T - 2)$$

5. Check by multiplication:

$$2(T - 2)(T - 2)$$
$$= (2T - 4)(T - 2)$$
$$= 2T^2 - 4T - 4T + 8$$
$$= 2T^2 - 8T + 8$$

E X A M P L E **4** Factor completely: $x^2 - 6xy + 9y^2$

Solution This is the first problem you've seen where the last term also contains a variable. Because y^2 will factor like x^2 does, be sure to include the y with the numerical factors you obtain from the product–sum procedure.

1. Product $= (+1)(+9) = +9$ Sum $= -6$

   Factors of $+9$: $(+1)(+9)$
                           $(-1)(-9)$
                           $(+3)(+3)$
                           $(-3)(-3)$

2. Expanded form:

   $x^2 - 6x + 9y^2$
   $= x^2 - 3xy - 3xy + 9y^2$

3. Factor GCFs from the pairs of terms.

   $x^2 - 3xy - 3xy + 9y^2$

   $= x(x - 3y) - 3y(x - 9y)$

4. Take out the common factor from both groups:

   $(x - 3y)(x - 3y)$

   This is factored completely. Therefore,

   $x^2 - 6xy + 9y^2$
   $= (x - 3y)(x - 3y)$ or $(x - 3y)^2$

5. Check: $(x - 3y)(x - 3y)$
   $= x^2 - 3xy - 3xy + 9y^2$
   $= x^2 - 6xy + 9y^2$ ◆

E X A M P L E   **5**   Factor completely: $12R^2 - 17R - 5$

Solution   This is the first example we have done using the product–sum method wherein the squared term has a coefficient other than 1. You worked this problem earlier in this section using trial-and-error method. Try it now by following the same steps used in Examples 1 through 4.

1. Product $= (12)(-5) = -60$ Sum $= -17$

   Factors of -60: $(-1)(+60)$
                        $(+1)(-60)$
                        $(-2)(+30)$
                        $(+2)(-30)$
                        $(-3)(20)$
                        $(3)(-20)$

   You can stop here, because $(3)(-20)$ are the factors of -60 that have a sum of -17.

2. Expanded form: $12R^2 + 3R - 20R - 5$

3. Factor the GCFs from the pairs of terms.

   $3R(4R + 1) - 5(4R + 1)$

   Have you noticed by now that you get the same factors in both sets of parentheses each time? If that doesn't happen at this step in a problem, then either you have made a mistake somewhere and should start over, or the expression is not factorable.

4. Remove the common factor from both groups.

   $(4R + 1)(3R - 5)$

   This is completely factored. Therefore,

   $12R^2 - 17R - 5 = (4R + 1)(3R - 5)$

5. Check: $(4R + 1)(3R - 5)$
   $= 12R^2 - 20R + 3R - 5$
   $= 12R^2 - 17R - 5$ ◆

E X A M P L E 6   Factor completely: $22X^3 + 77X^2 - 44X$

Solution   First take out the greatest common factor.

$11X(2X^2 + 7X - 4)$

1. Product $= (+2)(-4) = -8$ Sum $= +7$

    Factors of -8: $(+1)(-8)$
    $(-1)(+8)$

    No need to list any more factors, because (-1) and $(+8)$ add to $+7$.

2. Expanded form: $11X(2X^2 - 1X + 8X - 4)$

3. Factor GCFs from the pairs of terms.

    $11X[X(2X - 1) + 4(2X - 1)]$

    (There's that matching pair again!)

4. Remove the common factor from the groups.

    $11X[(2X - 1)(X + 4)]$

    This is completely factored. Therefore,

    $22X^3 + 77X^2 - 44X = 11X(2X - 1)(X + 4)$

5. Check: $11X(2X - 1)(X + 4)$

    $= (22X^2 - 11X)(X + 4)$

    $= 22X^3 + 88X^2 - 11X^2 - 44X$

    $= 22X^3 + 77X^2 - 44X$ ◆

E X A M P L E 7   Factor completely: $3A^2 - 20A - 4$

Solution   Product $= (+3)(-4) = -12$ Sum $= -20$

Factors of -12: $(-1)(+12)$
$(-12)(+1)$
$(-2)(+6)$
$(+2)(-6)$
$(-3)(+4)$
$(+3)(-4)$

No factors of -12 add to -20, so this trinomial cannot be factored. Therefore,

$3A^2 - 20A - 4$ is *prime*.   ◆

Product–Sum Method

> In factoring trinomials of the type $ax^2 + bx + c$, the product–sum method can be used, where *product* means the product of a and c. *Sum* means the sum of the factors a and c, which must equal b.

E X A M P L E 8   Factor completely: $X^2 - 5XY + 6Y^2$

Solution   Product $= (+1)(+6) = +6$ Sum $= -5$

Factors of $+6$: $(+1)(+6)$
$(-1)(-6)$
$(+2)(+3)$
$(-2)(-3)$

$X^2 - 2XY - 3XY + 6Y^2$

$= X(X - 2Y) - 3Y(X - 2Y)$

$= (X - 2Y)(X - 3Y)$

Therefore,

$X^2 - 5XY + 6Y^2 = (X - 2Y)(X - 3Y)$

Check $(X - 2Y)(X - 3Y)$

$= X^2 - 3XY - 2XY + 6Y^2$

$= X^2 - 5XY + 6Y^2$ ◆

Exercises   Try these problems using the trial-and-error method or product–sum. Review the steps in the examples if you have difficulty. The answers follow, but you should check your answers by multiplying.

1. $a^2 + 7a + 12$ 2. $2r^2 - 5r + 2$ 3. $2x^2 - 18x + 36$
4. $6x^2 + 11x - 10$

Answers:

1. $(a + 4)(a + 3)$ 2. $(2r - 1)(r - 2)$ 3. $2(x - 3)(x - 6)$
4. $(3x - 2)(2x + 5)$

16.4 ◆ Practice Problems

A. Factor completely. Check by multiplying.

1. $y^2 + 8y + 15$ 2. $x^2 + 4x + 4$
3. $t^2 - 16t + 15$ 4. $r^2 - 5r + 4$
5. $a^2 - 7a + 10$ 6. $b^2 - 2b - 35$
7. $c^2 + c - 20$ 8. $y^2 + 13y + 12$
9. $w^2 + 10w + 25$ 10. $d^2 + 11d + 18$
11. $m^2 - 7m + 6$ 12. $n^2 - 7n + 12$
13. $x^2 - 10x + 21$ 14. $c^2 - c - 12$
15. $r^2 - 2r + 1$ 16. $t^2 - 11t + 10$
17. $b^2 + 9b + 20$ 18. $m^2 - m - 30$
19. $z^2 - 8z - 33$ 20. $x^2 + x - 72$
21. $2a^2 + 3a + 1$ 22. $3x^2 + 4x + 1$
23. $3x^2 + 7x + 2$ 24. $4y^2 + 8y + 3$

B. Factor completely. Check by multiplying.

25. $r^2 + 7r - 8$ 26. $x^2 - 2x - 8$
27. $3v^2 - 3v - 60$ 28. $3R^2 - 21R + 36$

29. $5X^2 + 40X + 75$ 30. $2x^2 - 2x - 24$
31. $7t^2 - 42t + 63$ 32. $5a^2 + 25a + 30$
33. $m^2 + 6mn + 9n^2$ 34. $R^2 - 4RS + 4S^2$
35. $w^2 - 2wz + z^2$ 36. $b^2 + 2bc + c^2$
37. $3a^2 + 11ab + 10b^2$ 38. $7x^2 + 8xy + y^2$
39. $2x^2 - 11xy + 5y^2$ 40. $3m^2 + 11mn + 6n^2$
41. $4a^3 + 8a^2 - 60a$ 42. $-3y^3 - 24y^2 + 27y$
43. $-12Y^3 - 22Y^2 + 20Y$ 44. $5m^2 + 39m + 72$
45. $4c^2 - 4c - 3$ 46. $8t^2 + 14t + 3$
47. $3r^2 + 11r - 20$ 48. $2x^2 + 7x + 5$
49. $18x^2 + 3x - 6$ 50. $6n^2 + 33n - 18$
51. $7m^2n + 13mn - 2n$ 52. $4x^2 + 20x + 25$
53. $16y^2 - 24y + 9$ 54. $20 + 3z - 2z^2$
55. $3t^2 - 30t + 63$ 56. $21a^2 - 13ab + 2b^2$
57. $20 - 19m + 3m^2$ 58. $4x^2 + 4ax - 8a^2$
59. $9a^2b - 12ab + 4b$ 60. $m^4 - 2m^2n - 8n^2$

16.5 ◆ Special Products

In Section 16.3 you worked with the difference of two perfect squares. This is just one of several special types of products that are easy to factor if you memorize their patterns.

Factoring Patterns for Special Products

$$x^2 - y^2 = (x + y)(x - y)$$
$$x^2 + 2xy + y^2 = (x + y)(x + y) \text{ or } (x + y)^2$$
$$x^2 - 2xy + y^2 = (x - y)(x - y) \text{ or } (x - y)^2$$
$$x^3 + y^3 = (x + y)(x^2 - xy + y^2)$$
$$x^3 - y^3 = (x - y)(x^2 + xy + y^2)$$

The trinomials $x^2 + 2xy + y^2$ and $x^2 - 2xy + y^2$ are called **perfect-square trinomials**. In both, the first and last terms are perfect squares and the middle term is twice the product of the square roots of the first and last terms. Look at $x^2 + 2xy + y^2$ more closely, for example. You can see that x^2 and y^2 are perfect squares and that twice the product of their square roots, $2(x \cdot y)$, gives the middle term, $2xy$.

Some trinomial factoring problems are easier to do if you apply special-product techniques instead of the product–sum method or trial and error.

E X A M P L E 1   Factor completely: $25x^2 + 30x + 9$

Solution   See if this is a perfect-square trinomial.

$$25x^2 = (5x)^2 \qquad 9 = (3)^2$$

$$2(5x)(3) = 30x, \text{ the middle term}$$

Yes, it is a perfect-square trinomial, so it will factor like the pattern.

$$25x^2 + 30x + 9 = (5x + 3)(5x + 3)$$
$$= (5x + 3)^2$$

Check   Multiply: $(5x + 3)^2$
$$= (5x + 3)(5x + 3)$$
$$= 25x^2 + 15x + 15x + 9$$
$$= 25x^2 + 30x + 9$$ ◆

E X A M P L E 2   Factor completely: $a^2 - 12a + 36$

Solution   This is a perfect-square trinomial because $a^2 = (a)^2$, $36 = (6)^2$, and $2 \cdot a \cdot 6 = 12a$.

$$a^2 - 12a + 36 = (a - 6)^2$$

Check $(a - 6)^2$
$$= (a - 6)(a - 6)$$
$$= a^2 - 12a + 36$$ ◆

The last two patterns that you need to memorize apply to the sum of perfect cubes and to the difference of perfect cubes. Pay special attention to the sign of each term in the factors. Use your checking process to catch any errors in sign.

E X A M P L E   **3**   Factor completely: $a^3 + 27$

Solution $a^3 + 27 = (a)^3 + (3)^3$

Factor these cubes by following the pattern shown earlier in this section. For this problem, x is a and y is 3.

$$a^3 + 27 = (a + 3)(a^2 - 3a + 3^2)$$
$$= (a + 3)(a^2 - 3a + 9)$$

Check $(a + 3)(a^2 - 3a + 9)$
$$= a^3 - 3a^2 + 9a + 3a^2 - 9a + 27$$
$$= a^3 + 27$$

E X A M P L E   **4**   Simplify completely: $125y^3 - 1$

Solution   This is the difference of two terms, but they are perfect cubes, not perfect squares.

$$125y^3 - 1 = (5y)^3 - (1)^3$$
$$= (5y - 1)[(5y)^2 + 5y + (1)^2]$$
$$125y^3 - 1 = (5y - 1)(25y^2 + 5y + 1)$$

Check $(5y - 1)(25y^2 + 5y + 1)$
$$= 125y^3 + 25y^2 + 5y - 25y^2 - 5y - 1$$
$$= 125y^3 - 1$$

Factoring techniques have been presented one at a time in this chapter. That made it pretty easy to do the homework, because you had only one or two types of factoring to worry about at a time. Now that you have practiced all the techniques, however, you should be able to handle any type of problem that you are asked to factor. Any one or more of the following skills will be needed to completely factor a problem: Remove the greatest common factor, factor by grouping, factor the difference of two perfect squares, factor perfect-square trinomials, factor perfect cubes, and factor trinomials by the product–sum method or the trial-and-error method.

Each time you approach a problem, examine it to see which of these skills must be used. Follow the order shown in the list below as you work through each problem. The examples that follow will require you to use two or three of the skills before the problem is *factored completely*.

1.   Look to see if there is a common factor in every term. If there is, *remove the greatest common factor* (GCF) from every term.

2.   **a.**   If there are only *two* terms in the expression, look to see if it is the *difference of two perfect squares*. If so, factor as the product of the sum and difference of the square roots.

    b.   If it is the sum of two perfect squares, $a^2 + b^2$, then the expression is *prime* and cannot be factored.

    c.   If it is the sum or difference of perfect cubes, then factor according to the appropriate pattern (which you have memorized).

3.   If the expression (with the GCF removed) is a trinomial, check to see if it is a perfect-square trinomial. If it is, then factor according to the following pattern: The expression $x^2 + 2cx + c^2$, where x^2 and c^2 are perfect squares and twice the product of their square roots is equal to the middle term, factors as $(x + c)^2$.

4.   If the trinomial is not a perfect square, use the *product–sum method* or the *trial-and-error method* to factor.

5.   Check all factoring answers by multiplication.

E X A M P L E 5   Factor completely: $3x^2 + 18x + 15$

Solution   Take out the GCF.

$$3x^2 + 18x + 15 = 3(x^2 + 6x + 5)$$

The trinomial is not a perfect-square trinomial, so use trial and error or product–sum. Product–sum is shown. Product $= +5$. Sum $= +6$. The factors are $(+1)(+5)$.

$$3(x^2 + 6x + 5) = 3(x^2 + 1x + 5x + 5)$$
$$= 3[x(x + 1) + 5(x + 1)] = 3(x + 1)(x + 5)$$

Check $3(x + 1)(x + 5)$
$$= (3x + 3)(x + 5)$$
$$= 3x^2 + 15x + 3x + 15$$
$$= 3x^2 + 18x + 15$$ ◆

E X A M P L E 6   Factor completely: $64t^4 - 4$

Solution $64t^4 - 4$ Take out the GCF.
$$= 4(16t^4 - 1)$$ Now factor the difference of two perfect squares.
$$= 4[(4t^2 + 1)(4t^2 - 1)]$$

Because $4t^2 - 1$ is also the difference of two perfect squares, there is still more factoring to be done.

$$= 4(4t^2 + 1)(2t + 1)(2t - 1)$$

Check $4(4t^2 + 1)(2t + 1)(2t - 1)$
$$= (16t^2 + 4)(2t + 1)(2t - 1)$$
$$= (16t^2 + 4)(4t^2 - 1)$$
$$= 64t^4 - 16t^2 + 16t^2 - 4$$
$$= 64t^4 - 4$$ ◆

E X A M P L E 7   Factor completely: $4a^2 + 28a + 49$

Solution $4a^2 + 28a + 49$
There is no GCF, so check to see if this is a perfect-square trinomial.
$4a^2 = (2a)^2$, $49 = (7)^2$, and $2 \cdot 2a \cdot 7 = 28a$. Therefore,
$$4a^2 + 28a + 49 = (2a + 7)^2$$

Check $(2a + 7)^2$
$$= (2a + 7)(2a + 7)$$
$$= 4a^2 + 14a + 14a + 49$$
$$= 4a^2 + 28a + 49$$ ◆

E X A M P L E 8   Factor completely: $AB^2 + 3AB + 6B + 18$

Solution   There are four terms, so try factoring by grouping.
$$AB^2 + 3AB + 6B + 18 = AB(B + 3) + 6(B + 3)$$
$$= (B + 3)(AB + 6)$$

Check $(B + 3)(AB + 6) = AB^2 + 6B + 3AB + 18$ ◆

E X A M P L E   **9**   Simplify completely: $64x^{12} - 1$

Solution   There are only two terms, so this might be the difference of two perfect squares or cubes. $64x^{12} = (8x^6)^2$ or $(4x^4)^3$, and 1 is both a perfect square and a perfect cube. So this expression will factor as a perfect cube.

$$64x^{12} - 1 = (4x^4 - 1)[(4x^4)^2 + 4x^4 + 1]$$
$$= (4x^4 - 1)(16x^8 + 4x^4 + 1)$$

You need to check to see if all factors are prime. $4x^4 - 1$ will factor as the difference of two perfect squares, but the trinomial will not factor further.

$$64x^{12} - 1 = (2x^2 + 1)(2x^2 - 1)(16x^8 + 4x^4 + 1)$$

Check $(2x^2 + 1)(2x^2 - 1)(16x^8 + 4x^4 + 1)$
$$= (4x^4 - 2x^2 + 2x^2 + 1)(16x^8 + 4x^4 + 1)$$
$$= (4x^4 - 1)(16x^8 + 4x^4 + 1)$$
$$= 64x^{12} + 16x^8 + 4x^4 - 16x^8 - 4x^4 - 1$$
$$= 64x^{12} - 1$$

◆

16.5   ◆   Practice Problems

A. Factor completely and check the answer by multiplying.

1. $x^2 + 8x + 16$

2. $y^2 - 14y + 49$

3. $a^2 + 16a + 64$

4. $m^2 - 12m + 36$

5. $4y^2 + 4y + 1$

6. $25x^2 - 90x + 81$

7. $16 - 8c + c^2$

8. $4 - 20p + 25p^2$

9. $125a^3 + 64$

10. $27t^3 - 8$

11. $36t^2 + 60tq + 25q^2$

12. $25 - 10a + a^2$

13. $16 + 8m + m^2$

14. $49m^2 + 42mn + 9n^2$

15. $4a^2 - 12ab + 9b^2$

16. $9b^2 + 6b + 1$

B. Factor completely.

17. $3m + 3n - am - an$

18. $2a^3 - 10a^2 - 4a + 20$

19. $y^3 - 4y^2 - 3y + 12$

20. $2ac + 10a + c + 5$

21. $x^4 + 2x^3 - 80x^2$

22. $2z^2 - 28z + 98$

23. $5n^2 - 30n + 45$

24. $b^4 - 12b^3 + 35b^2$

25. $5b^3 - 40$

26. $2x^3 + 3x^2 - 2x - 3$

27. $a^3 + 3a^2 - 4a - 12$

28. $3a^3 - 375$

29. $5ab + bc - 20a - 4c$

30. $2rt + 24s - 16r - 3st$

31. $4t^2 + 14t + 6$

32. $m^3 + 8m^2 + 16m$

33. $3m^4 - 48$

34. $ax - 3x - 2ay + 6y$

35. $ab - ac - 2c + 2b$

36. $2p^4 - 162$

37. $3y^3 - 30y^2 + 75y$

38. $8m^2 + 56m + 98$

39. $20x^4y^2 - 7x^3y^3 - 3x^2y^4$

40. $15a^4b^2 + 2a^3b^3 - 8a^2b^4$

16.6   Solving Quadratic Equations by Factoring

Until now, when asked to solve equations, you have been dealing with first-degree equations (the exponent of the variable was 1). In this section, you will learn to solve second-degree equations—that is, equations with variables that have an exponent of 2. For example, $x^2 - x = 0$ and $y^2 - y = 20$ are second-degree equations.

Before you learn how to solve such equations, you must understand a basic property of the real number system.

Products Equal to Zero

If $ab = 0$, then either $a = 0$ or $b = 0$.

In other words, if you multiply quantities together and their product is zero, then one or more of the quantities must be zero. This situation will arise in the solutions to second-degree equations, because your first step is to rearrange the terms so that they are all on one side of the equals sign and there is a zero on the other side. Returning to our earlier examples, $x^2 - x = 0$ is already in correct form but $y^2 - y = 20$ needs to be rearranged.

$$y^2 - y \quad = \quad 20$$
$$\underline{\quad -20 = -20}$$
$$y^2 - y - 20 = \quad 0$$

Once the equation is set equal to zero, factor the polynomial, if possible.

$$x^2 - x = 0 \qquad y^2 - y - 20 = 0$$
$$x(x - 1) = 0 \qquad (y - 5)(y + 4) = 0$$

The products equal zero, so one of the factors must equal zero. Because you don't know which one, set each factor equal to zero and solve the resulting first-degree equation.

$$x(x - 1) = 0$$

If the x factor is zero, then $x = 0$.

If the $x - 1$ factor is zero, then $x - 1 = \quad 0$
$$\underline{\quad +1 \quad +1}$$
$$x = \quad 1$$

The equation has two *possible* answers, so both have to be listed. The solution is given as $x = 0$, $x = 1$. We check the solutions by substituting them into the original equation to see if they produce true statements.

$$
\begin{array}{ll}
x^2 - x = 0 & x^2 - x = 0 \\
0^2 - 0 = 0 & 1^2 - 1 = 0 \\
0 - 0 = 0 & 1 - 1 = 0 \\
\quad 0 = 0 \;\; \text{True.} & \quad 0 = 0 \;\; \text{True.}
\end{array}
$$

Now solve the other equation in the same fashion.

$$y^2 - y = 20$$
$$y^2 - y - 20 = 0$$
$$(y - 5)(y + 4) = 0$$

$$
\begin{array}{ll}
\text{If } y - 5 = \quad 0 & \text{If } y + 4 = \quad 0 \\
\underline{\quad +5 \quad +5} & \underline{\quad -4 \quad -4} \\
\quad y = \quad 5 & \quad y = -4
\end{array}
$$

To check the solutions, $y = 5$ and $y = -4$, substitute each into the original equation.

$$
\begin{array}{ll}
y^2 - y = 20 & y^2 - y = 20 \\
5^2 - 5 = 20 & (-4)^2 - (-4) = 20 \\
25 - 5 = 20 & 16 + (+4) = 20 \\
\quad 20 = 20 \;\; \text{True.} & \quad 20 = 20 \;\; \text{True.}
\end{array}
$$

In this section, all of the second-degree equations, which are called **quadratic equations,** can be solved by factoring. In Chapter 21 you will learn techniques for solving quadratics that cannot be factored.

Quadratic Equation

A quadratic equation is an equation of the form $ax^2 + bx + c = 0$, where a, b, and c are real numbers and $a \neq 0$. The $ax^2 + bx + c = 0$ form is called the standard form of a quadratic equation.

Study the detailed solutions and checks that follow so that you will be able to do similar steps when you solve the homework problems.

E X A M P L E **1** Solve and check: $x^2 + 6x + 8 = 0$

Solution The equation is already set equal to zero.

$$x^2 + 6x + 8 = 0 \qquad \text{Factor.}$$

$$(x + 4)(x + 2) = 0$$

Set each factor equal to zero and solve the simple equations.

$$
\begin{array}{ll}
x + 4 = 0 & x + 2 = 0 \\
\underline{-4 \quad -4} & \underline{-2 \quad -2} \\
x = -4 & x = -2
\end{array}
$$

The solutions are $x = -4$ and $x = -2$.

Check

$$
\begin{array}{ll}
x^2 + 6x + 8 = 0 & x^2 + 6x + 8 = 0 \\
(-4)^2 + 6(-4) + 8 = 0 & (-2)^2 + 6(-2) + 8 = 0 \\
(16) + (-24) + 8 = 0 & (4) + (-12) + 8 = 0 \\
-8 + 8 = 0 & -8 + 8 = 0 \\
 0 = 0 \quad \text{True.} & 0 = 0 \quad \text{True.}
\end{array}
$$

◆

E X A M P L E **2** Solve and check: $c^2 - 4c - 45 = 0$

Solution

$$c^2 - 4c - 45 = 0 \qquad \text{Factor.}$$

$$(c - 9)(c + 5) = 0 \qquad \text{Set each factor equal to zero.}$$

$$
\begin{array}{ll}
c - 9 = 0 \qquad c + 5 = 0 & \text{Solve each equation.} \\
\underline{+9 \quad +9} \qquad \underline{-5 \quad -5} \\
c = 9 \qquad c = -5
\end{array}
$$

The solutions are $c = 9$ and $c = -5$.

Check

$$
\begin{array}{ll}
c^2 - 4c - 45 = 0 & c^2 - 4c - 45 = 0 \\
9^2 - 4(9) - 45 = 0 & (-5)^2 - 4(-5) - 45 = 0 \\
81 - 36 - 45 = 0 & 25 + 20 - 45 = 0 \\
45 - 45 = 0 & 45 - 45 = 0 \\
 0 = 0 \quad \text{True.} & 0 = 0 \quad \text{True.}
\end{array}
$$

◆

E X A M P L E **3**   Solve and check: $f^2 + 12f + 36 = 0$

Solution $f^2 + 12f + 36 = 0$ This is a perfect-square trinomial.

$(f + 6)^2 = 0$

Both factors are the same, so there will be only one solution.

$f + 6 = 0$

$f = -6$

When this happens, -6 is called a **double root** or a **double solution**, because every solv- . able second-degree equation has *two* solutions.

Check $f + 12f + 36 = 0$

$(-6)^2 + 12(-6) + 36 = 0$

$36 + (-72) + 36 = 0$

$-36 + 36 = 0$

$0 = 0$ True.   ◆

E X A M P L E **4**   Solve and check: $z^2 = 10z - 24$

Solution $z^2 = 10z - 24$ Rearrange by subtracting $10z$ and adding 24 to both sides.

$z^2 - 10z + 24 = 0$ Factor the trinomial.

$(z - 6)(z - 4) = 0$ Set each factor equal to zero and solve.

$z - 6 = 0$ $z - 4 = 0$

$z = 6$ $z = 4$

The solutions are $z = 6$ and $z = 4$.

Check $z^2 = 10z - 24$ $z^2 = 10z - 24$

$6^2 = 10(6) - 24$ $4^2 = 10(4) - 24$

$36 = 60 - 24$ $16 = 40 - 24$

$36 = 36$ True. $16 = 16$ True.   ◆

E X A M P L E **5**   Solve and check: $3w^2 - 6w + 3 = 0$

Solution $3w^2 - 6w + 3 = 0$

$3(w^2 - 2w + 1) = 0$

$3(w - 1)^2 = 0$

The constant term, 3, can never be zero and will not affect the solution.

$w - 1 = 0$

$w = 1$

Check $3w^2 - 6w + 3 = 0$

$3(1)^2 - 6(1) + 3 = 0$

$3 - 6 + 3 = 0$

$-3 + 3 = 0$

$0 = 0$ True.   ◆

E X A M P L E   **6**   Solve and check: $\dfrac{y^2}{2} + 4y - 10 = 0$

Solution   Because you need to have integer coefficients before you can factor, multiply through any fractional equation by the least common denominator (LCD) to eliminate the denominator(s).

$$\frac{y^2}{2} + 4y - 10 = 0$$

$$\overset{1}{(2)}\frac{y^2}{\underset{1}{2}} + (2)4y - (2)10 = (2)0$$

$$y^2 + 8y - 20 = 0$$

$$(y + 10)(y - 2) = 0$$

$$y + 10 = \quad 10 \qquad y - 2 = 0$$

$$y = -10 \qquad\qquad y = 2$$

The solutions are $y = -10$, $y = 2$.

Checks

$$\frac{y^2}{2} + 4y - 10 = 0 \qquad\qquad \frac{y^2}{2} + 4y - 10 = 0$$

$$\frac{(-10)^2}{2} + 4(-10) - 10 = 0 \qquad\qquad \frac{(2)^2}{2} + 4(2) - 10 = 0$$

$$\frac{100}{2} + (-40) - 10 = 0 \qquad\qquad \frac{4}{2} + 8 - 10 = 0$$

$$50 + (-40) - 10 = 0 \qquad\qquad 2 + 8 - 10 = 0$$

$$10 - 10 = 0 \qquad\qquad 10 - 10 = 0$$

$$0 = 0 \quad \text{True.} \qquad\qquad 0 = 0 \quad \text{True.} \qquad \blacklozenge$$

As you see from these examples, the process of solving quadratic equations by factoring is always the same.

Solving Quadratic Equations by Factoring

1. Rearrange the terms so that zero is alone on one side of the equals sign (standard form).
2. If there are fractional coefficients, multiply every term of the equation by the LCD value.
3. Factor the equation, if possible.
4. Set each non-constant factor equal to zero.
5. Solve each equation that has been formed.
6. Check each answer by substituting it into the original equation.

E X A M P L E   **7**   Solve and check: $(x - 3)(x + 4) = 8$

Solution   Do not set the factors $(x - 3)$ and $(x + 4)$ equal to zero, because the equation must first be arranged in standard form. Multiply $(x - 3)(x + 4)$ to remove the parentheses.

$$(x - 3)(x + 4) = 8$$
$$x^2 + x - 12 = 8$$
$$\underline{\quad -\ 8 \quad -8 \quad}$$
$$x^2 + x - 20 = 0$$

Then factor the new equation.

$$(x + 5)(x - 4) = 0$$
$$x + 5 = 0 \qquad x - 4 = 0$$
$$x = -5 \qquad\quad x = 4$$

Check

$(x - 3)(x + 4) = 8$	$(x - 3)(x + 4) = 8$
$(-5 - 3)(-5 + 4) = 8$	$(4 - 3)(4 + 4) = 8$
$(-8)(-1) = 8$	$(1)(8) = 8$
$8 = 8$ True.	$8 = 8$ True.

◆

E X A M P L E 8 Solve and check: $4a^2 = 3 - 11a$

Solution $4a^2 = 3 - 11a$ Rearrange to standard form.

$4a^2 + 11a - 3 = 0$ Factor.

Product $= -12$ Sum $= +11$

$(+12)(-1)$ add to $+11$.

$$4a^2 + 12a - a - 3 = 0$$
$$4a(a + 3) - 1(a + 3) = 0$$
$$(a + 3)(4a - 1) = 0$$
$$a + 3 = 0 \qquad 4a - 1 = 0$$
$$a = -3 \qquad\quad 4a = 1$$
$$a = \frac{1}{4}$$

Check

$4a^2 = 3 - 11a$	$4a^2 = 3 - 11a$
$4(-3)^2 = 3 - 11(-3)$	$4(1/4)^2 = 3 - 11(1/4)$
$4(9) = 3 - (-33)$	$4(1/16) = 3 - 11/4$
$36 = 3 + 33$	$1/4 = 12/4 - 11/4$
$36 = 36$ True.	$1/4 = 1/4$ True.

◆

16.6 ◆ Practice Problems

A. Solve these equations by factoring.

1. $(y + 2)(y - 5) = 0$
2. $(b + 4)(b - 3) = 0$
3. $(a + 8)(a - 9) = 0$
4. $x(x + 5) = 0$
5. $m(3m - 5) = 0$
6. $4a(2a - 3) = 0$
7. $x^2 - 5x = 0$
8. $a^2 - 12a = 0$
9. $b^2 - 49 = 0$
10. $7c^2 = -3c$
11. $4y^2 = 8y$
12. $9y = 36y^2$
13. $3c^3 - 27c = 0$
14. $r^2 - 121 = 0$
15. $5m = -7m^2$
16. $6 - 54t^2 = 0$
17. $t^3 - t = 0$
18. $19x = -x^2$
19. $x^2 + 5x + 6 = 0$
20. $m^2 + 6m + 8 = 0$

B. Solve these quadratic equations.

21. $n^2 - 3n - 10 = 0$
22. $x^2 - 6x - 40 = 0$
23. $a^2 + a - 72 = 0$
24. $t^2 - t - 56 = 0$
25. $x^2 + 10x = -21$
26. $w^2 = 3w + 40$

27. $4s^2 + 12s = -9$

28. $m^2 - 24m - 25 = 0$

29. $3y^2 - 7y - 6 = 0$

30. $h^2 = 22h - 121$

31. $x^3 - 6x^2 + 9x = 0$

32. $6x^2 + x - 15 = 0$

33. $\dfrac{R^2}{16} = 1$

34. $V^2 = \dfrac{1}{16}$

35. $32 - 12x - 2x^2 = 0$

36. $\dfrac{y^2}{3} - 2y + 3 = 0$

37. $y(y + 3) = 28$

38. $(x + 4)(x - 1) = 6$

39. $\dfrac{x^2}{3} - 2x - 9 = 0$

40. $\dfrac{x^2}{2} + 4 = 3x$

41. $(a - 5)(a + 4) = 52$

42. $2m^2 - 5m - 12 = 0$

43. $5a^2 + 13a + 6 = 0$

44. $6y^2 - 13y + 6 = 0$

45. $(m - 4)(m + 6) = -9$

46. $21 - 4x - x^2 = 0$

47. $m^2 - 3m = 10$

48. $a(a - 1) = 20$

49. $3k^2 - 13k + 4 = 0$

50. $(n + 1)(n - 6) = -10$

16.7 Chapter Review

When you are asked to factor an expression completely, follow these steps:

1. First look to see if there is a common factor in every term. If there is, *remove the greatest common factor* (GCF) from every term.

$$ax^2 + ax + ac = a(x^2 + x + c)$$

2. If there are only *two* terms in the expression, look to see if it is the *difference of two perfect squares*. If so, factor as the product of the sum and difference of the square roots.

$$a^2 - b^2 = (a + b)(a - b)$$

If it is the sum of two perfect squares, $a^2 + b^2$, then the expression is *prime* and cannot be factored.

If it is the sum or difference of two perfect cubes, factor according to the following patterns.

$$x^3 + y^3 = (x + y)(x^2 - xy + y^2)$$
$$x^3 - y^3 = (x - y)(x^2 + xy + y^2)$$

3. If the expression (with the GCF removed) is a trinomial, check to see if it is a perfect-square trinomial. If it is, factor according to the following pattern:

$x^2 + 2cx + c^2$, where x^2 and c^2 are perfect squares and twice the product of their square roots is equal to the middle term, factors as $(x + c)^2$.

4. If the trinomial is not a perfect square, use the *product–sum* method or the *trial-and-error* method to factor it.

5. If the polynomial has four terms, see if there are common factors in pairs of terms and factor them out. Look again for GCFs.

6. Check all factoring answers by multiplication.

When solving quadratic equations such as $ax^2 + bx + c = 0$, apply the steps listed on page 456. Check the answers in the original equation.

Review Problems

A. Multiply each of the following expressions. Simplify answers by combining like terms.

1. $3xy(x^2y - 2xy)$

2. $(x + 2)(x - 2)$

3. $(y - 3)(y + 6)$

4. $(8 + b)(8 + b)$

5. $(k - 3)(k + 3)$

6. $5a^2b(3ab + 4)$

7. $3(2c - 1)(c + 3)$

8. $5(m + 4)(m - 2)$

9. $(x + 2)(x + 2)$

10. $(y + 7)(y + 7)$

11. $(a + 3)^2$

12. $(x - 4)^2$

B. Factor these expressions completely. If an expression cannot be factored, write "prime" for the answer. Check all answers by multiplying.

13. $3y^4 + 3y$

14. $m^5 - 3m^3 + 2m$

15. $b^2 - 16$

16. $-6x^2 + 12$

17. $c^2 - 8c + 15$

18. $b^2 - 2b - 35$

19. $x^6 + 2x^4 + x^2$

20. $c^2 + 9$

21. $-t^3 + 2t^2$

22. $10ab - 2a$

23. $4d^2 - 1$

24. $x^2 + 7x - 18$

25. $3m^2n^2 - 5m^2n$

26. $b^2 - 4b - 21$

27. $39 - 16m + m^2$

28. $3y^4 - 81y$

29. $a^2 + 12a + 27$

30. $m^2 + mn - 6n^2$

31. $x^2 + 14x + 49$

32. $72 - 17p + p^2$

C. Factor completely and check answers by multiplication. If prime, write "prime."

33. $35 - 21b$

34. $a^2 + 3ab - 10b^2$

35. $m^2 - 4mn - 45n^2$

36. $10a^3b - 24a^2b^2 + 16ab^3$

37. $4a^3 + 8a^2 - 60a$

38. $-2n^3 - 6n^2 + 20n$

39. $-3y^3 - 24y^2 + 27y$

40. $7 + 14k$

41. $3m^2 - 48$

42. $18x^4 - 8x^2y^2$

43. $9x^3y^3 - 27x^2y^2 - 18xy$

44. $64 - 3x^2$

45. $4a^2 - 64b^2$

46. $25 - 4y^2$

47. $w^4 - 16z^4$

48. $c^4 - 81d^4$

49. $6m^4 - 24m^2n^2 - 30n^4$

50. $a^3bc - a^2b^2c - 30ab^3c$

51. $x^3yz - 2x^2y^2z - 24xy^3z$

52. $4x^4 + 4x^2y^2 - 24y^4$

D. Factor completely.

53. $-8m^4 + 216m$

54. $15a^2 - 12a$

55. $-12a^2 + 6a + 6$

56. $-6x^3 + 54x$

57. $x^2 - 49$

58. $a^3 - 64b^3$

59. $10n^2 + 130n + 220$

60. $y^4 - 14y^2 + 45$

61. $a^2 + 4a + 3$

62. $2x^2 + 9x + 7$

63. $x^2 + 11x + 24$

64. $x^2 + 15x + 14$

65. $-4b^2 + 64$

66. $2b^2 - 13b - 24$

67. $2m^2 - 18mn - 20n^2$

68. $4c^2 - 20cd + 16d^2$

69. $3x^2 - 8x - 16$

70. $x^2 - 2x - 15$

71. $x^4yz + x^3yz^2 + x^2y^3z$

72. $a^5b^3c + 2a^3b^3c^3 - 3a^3b^4c^2$

73. $x^2y + xy - 2x - 2$

74. $4ab^2 + 8ab + 8b + 16$

75. $y^3 - 4y^2 - y + 4$

76. $10a + 2ac + 5 + c$

77. $32g^2 - 160gh + 200h^2$

78. $12m^3 + 6m^2 - 30m$

79. $20p^2 + 17p - 24$

80. $5x^3 + 10x^2 - 7x$

81. $18w^2 + 27w - 35$

82. $15b^2 - 65bc + 20c^2$

E. Solve and check.

83. $c^2 - 6c = 0$

84. $\dfrac{w^2}{4} - 100 = 0$

85. $14x = 7x^2$

86. $a^2 + a - 30 = 0$

87. $6m^2 - 19m + 15 = 0$

88. $y^2 + 21 = -10y$

89. $20 + 3n = \dfrac{n^2}{2}$

90. $3p^2 - 9p = -6$

91. $3 - 48t^2 = 0$

92. $4x^2 - 12x = 40$

93. $4m^2 - 63 = 4m$

94. $y^2 + 20 = 9y$

95. $(x - 5)(x + 2) = 18$

96. $(m + 4)(m - 7) = 60$

Chapter Test

1. Multiply the expression: $3xy(4x^2y - 7xy + 2)$
2. Multiply: $(4a - 3c)(a + 2c)$

Completely factor each of the following expressions, if possible.

3. $4R^2 - 6R - 40$
4. $c^3 - cd^2$
5. $ax + ay + 5x + 5y$
6. $25 + 40y + 16y^2$
7. $15t^2 - 7t - 2$
8. $2m^2 + 3mn - 2n^2$
9. $9p^2 + 63p + 90$
10. $4w^2 - 16$

11. $x^2 - 5x - 14$
12. $F^2 + G^2$
13. $m^2 - 4m + 6mn - 24n$
14. $2x^2 + 3x + 1$
15. $3a^3 + 21a^2 - 9a$
16. $c^2 - 4c - 12$
17. $25 - 9z^2$
18. $w^2 - 10w + 21$

Solve each quadratic equation by factoring.

19. $s^2 = s + 30$
20. $v^2 + 3v - 28 = 0$

Thought-Provoking Problems

1. Factor completely.

    a. $\dfrac{9}{4}y^2 - \dfrac{1}{4}$ b. $0.16m^2 - 1.44$

    c. $0.5a^2b^2 + 2.5a^2b + 3ab - 6.5ab^2$

2. Factor completely.

    a. $P^8 - 256$ b. $12R^3 - 56R^2 + 40R$

3. Describe in your own words how to factor an expression that is the difference of two perfect squares.

4. How can you test to see if an expression is a perfect-square trinomial?

5. Which of the following are perfect-square trinomials?

    a. $121 - 44b + 4b^2$

    b. $8a^2 + 4a + 1$

    c. $0.04d^2 - 0.1d + 0.25$

    d. $\dfrac{c^2}{4} + c + 1$

6. Compare the method used on page 442 to factor $12R^2 - 17R - 5$ with that used in Example 5 on page 446. Which do you prefer and why?

7. Solve and check.

    a. $2a^2 = 8a - a$ b. $(x - 3)(x + 4) = 0$

    c. $y^3 = -8y^2$ d. $m = m^3$

8. Find two consecutive odd integers such that 1 less than the sum of the two numbers is equal to their product.

9. The length of a rectangle is 2 feet less than three times the width. The area $(A = LW)$ is 21 square feet. What are the dimensions of the rectangle? Only one of your solutions can be used as the answer to this problem. Why is this?

10. One number is 3 more than twice another number. Their product is -1. What are the numbers? (*Hint*: Two pairs of numbers satisfy this word problem. Find both pairs of numbers.)

Introduction to Graphing

Introduction

In Chapter 12, on signed numbers, we first looked at the idea of a number line. A number line is a graph that shows what the picture of a set of signed numbers might look like. It can also be an aid in doing calculations with signed numbers. If we expand that notion to a two-dimensional graph, similar to a map, we graph points whose location is indicated by a *pair* of numbers. In this chapter, we will talk about how to identify the location of points on that graph and about the need for an ordered pair of numbers to indicate up–down position and left–right position. We will also show a picture of the solution set for a first-degree linear equation and a first-degree inequality.

Learning Objectives

When you have completed this chapter, you should be able to:

1. Identify the components of a Cartesian coordinate system (graph).
2. Locate a point on a Cartesian coordinate system and identify that location by an ordered pair of numbers.
3. Write a first-degree linear equation in the form $y = mx + b$.
4. Produce a table of at least three pairs of values that satisfy the given first-degree equation.
5. Determine the slope and y intercept of a line from its equation in slope–intercept form.
6. Given two points on a line, find the slope of that line from the graph or by using the two-point formula for slope.
7. Graph the solution of a first-degree linear equation.
8. Write the equation of a line given its slope and y intercept.
9. Write the equation of a line given two points on the line.
10. Write the equation of a line given one point on the line and its slope.
11. Graph the solution of a first-degree inequality.

You will need some graph paper of your own to do the work in this chapter. We recommend 1/4-inch graph paper, but a small quantity of any type will do.

17.2 Points on a Graph

Take a few minutes to remember what you learned in Chapter 12 about a number line. It is a picture—a **graph**—of signed numbers that extends indefinitely in both directions. A thermometer is the same type of graph, but it normally is read up and down instead of from side to side like a number line. When two graphs (visual representations) such as a number line and a thermometer are combined, the resulting figure is called a **Cartesian coordinate system**. Figure 1 is such a Cartesian coordinate system, a two-dimensional graph.

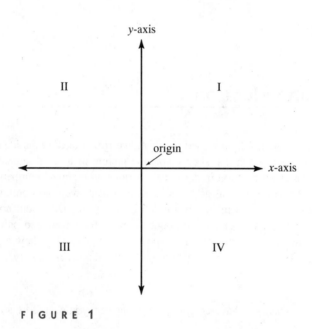

FIGURE 1

The vertical scale is called the **y-axis**, and the horizontal scale is the **x-axis**. The point at which the two scales, or axes, cross is called the **origin**. We measure distance to the *right* of the y-axis as *positive* (+) and distance to the *left* as *negative* (−). Distance *up* from the x-axis is *positive* (+), and distance *below* the x-axis is *negative* (−). We use ordered pairs of numbers, (x, y), to indicate location on the graph. In an ordered pair, the first coordinate always measures the distance left or right, and the second coordinate always measures the distance up or down.

The points that do not lie on the axes are separated into four sections called **quadrants**. The quadrants are designated by roman numerals (I, II, III, and IV) and are numbered counterclockwise as shown in Figure 1. Each point on the graph is either right (+) or left (−) of zero and is also either up (+) or down (−) from zero. Pairs of numbers are used to indicate a particular location on the graph.

Look at Figure 2. This graph shows the location of several points. For instance, (2,5) is two units to the right of zero and five units up. (−3,4) is three units to the left of zero and four units up. (−1,−3) is one unit to the left of zero and three units down. (3,−2 1/2) is three units to the right of zero and 2 1/2 units down. Are the point (−3,5) and the point (5,−3) in the same place on the graph? Find them. They are different. Even though the integers are the same, −3 and 5, their order in the ordered pair is switched. (−3,5) is not the same point as (5,−3). The right–left location is always given first in the pair of numbers, and the up–down location is always given second.

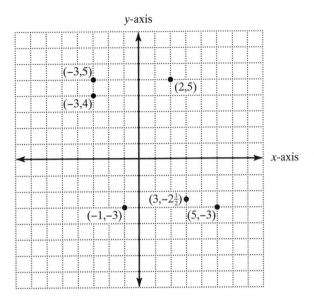

FIGURE 2

Look at Figure 3. What can you say about the signs of the numbers in the ordered pairs in Quadrant I? All points there are to the right of the origin, (+), and above the *x* axis, (+). The ordered pairs would therefore look like (+,+). What about points in Quadrant II? in Quadrant III? in Quadrant IV? It will be helpful later on to have figured this out.

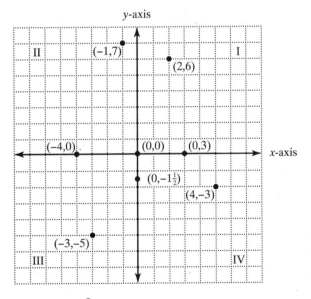

FIGURE 3

(+3,0) is three units to the right of zero but zero units up or down. It lies on the horizontal axis, the *x*-axis. Points that fall *on* one of the axes have a special feature. Points on the *x*-axis haven't moved up or down, so they have a *y* value of 0, and their ordered pairs have the form (left or right value,0). Points on the *y*-axis haven't moved left or right, so they have an *x* value of 0 and take the form (0,up or down value). (0,−1.5) lies on the *y*-axis, 1.5 units down. (0,0) is neither right nor left and neither up nor down, so (0,0) is the location of the origin. (−4,0) lies on the *x*-axis. (0,+3) lies on the *y*-axis. Each of these points is graphed on the coordinate system shown in Figure 3.

Exercises   See if you can give the ordered pairs that identify the location of the points A through J in Figure 4. Then check your answers to see if you have a good understanding of these ideas. Estimate the coordinates of points that do not fall on two intersecting lines. If the point is between (1,3) and (2,3) but is closer to (2,3) than to (1,3), you might use (1 3/4,3). If the point appears to be halfway between (1,3) and (2,3), you could use (1 1/2,3) or (1.5,3).

Answers:

A.   (2,3)   **B.**   (0,−4)   **C.**   (2,0)   **D.**   (−2,2 1/2)   **E.**   (−3,−5)
F.   (0,2 1/2)   **G.**   (3 1/2,−1 1/2)   **H.**   (−3,5)   **I.**   (0,0)
J.   (−3,0)

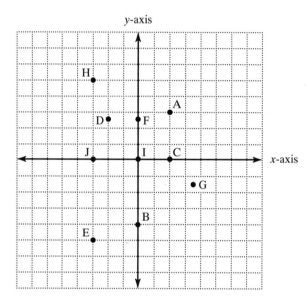

F I G U R E **4**

17.2 ◆ Practice Problems

A.   Complete the following sentences.

1. Points to the right of the *y*-axis have a _____ first coordinate.

2. Points to the left of the *y*-axis have a _____ first coordinate.

3. Points above the *x*-axis have a _____ second coordinate.

4. Points below the *x*-axis have a _____ second coordinate.

5. Ordered pairs of the form (−,+) lie in Quadrant _____ .

6. Ordered pairs of the form (+,−) lie in Quadrant _____ .

7. Ordered pairs of the form (0,*y*) lie on the _____ .

8. Ordered pairs of the form (*x*,0) lie on the _____ .

9. Ordered pairs with two _____ coordinates lie in Quadrant I.

10. Ordered pairs with two _____ coordinates lie in Quadrant III.

11. Two intersecting axes form a _____ .

12. The point where the two axes meet is called the

_____ .

B.   Give the coordinates of each point labeled in Figure 5.

13. A	14. B	15. C	16. D
17. E	18. F	19. G	20. H
21. I	22. J	23. K	24. L

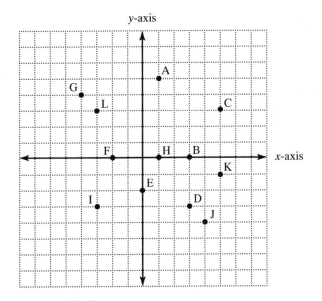

y-axis

x-axis

FIGURE 5

C. Construct a graph on a sheet of graph paper. Label each of the following items on the graph. Plot each point listed, and label it using its capital-letter designation.

25. x-axis
26. y-axis
27. Quadrant I
28. Quadrant II
29. Quadrant III
30. Quadrant IV
31. origin
32. point A $(0,-4)$
33. point B $(-2,5)$
34. point C $(3,0)$
35. point D $(3/4,6)$
36. point E $(-3,-3)$
37. point F $(-4,1)$
38. point G $(1,-4)$
39. point H $(5,-2)$
40. point I $(3/2,3)$

17.3 Graphing Linear Equations by Using Three Points

It is time to discover why graphs can be so helpful and why they have a place in the study of algebra. You have already learned how to solve an equation such as $x + 5 = 7$. Your job was to find the value for the letter x that would make $x + 5 = 7$ a true statement. Just by inspection, you could tell that when x has the value 2, the statement is true. $2 + 5$ *is* equal to 7.

In the equation $x + y = 7$, you have to replace both x *and* y with values that would give a true statement.

If you try $x = 2$ and $y = 5$, you get $x + y = 7 \rightarrow 2 + 5 = 7$.   True.

If you try $x = 4$, then y must be 3, because $4 + 3 = 7$ is true.

If you try $x = 6$ and $y = 1$, you get $x + y = 7 \rightarrow 6 + 1 = 7$.   True.

Obviously, there are many pairs of numbers that would work. Next, put your findings into an orderly display by using a *table of values* like this:

$x + y = 7$

x	y
2	5
4	3
6	1

Now see if you can find two *more* pairs of numbers that would work. How about $(0, 7)$? and $(3\ 1/2, 3\ 1/2)$?

There was no special reason for choosing any one of these values; that is, the choice was by chance. Generally, you just pull any value for x out of the air and then see what y must be to give a true statement. If you choose different values from those the book shows, and your values give true statements, then they are also correct.

In the table of values you just created, the first set of numbers ($x = 2$ and $y = 5$) can be represented as an ordered pair (x, y), or $(2,5)$. Every other point on the table can also be written in that form.

See what happens when you put (or, as we say, *plot*) these ordered pairs of numbers on a graph (Figure 6).

E X A M P L E 1   Graph the solution of the equation $x + y = 7$.

x	y
2	5
4	3
6	1
0	7
3 1/2	3 1/2

FIGURE 6

If the pairs of numbers that have been graphed on a Cartesian coordinate system were connected, they would form a straight line. This line also appears to go through the points $(7, 0)$ and $(-1, 8)$, which are not values on the table but which certainly satisfy the equation $x + y = 7$. In fact, the coordinates of every point on the line satisfy the equation. This line is the picture (graph) of the **solution set** of the equation. The solutions for equations such as $2x + 3y = 6$, $y = 8x - 4$, and $4y - 2x = 11$ all graph as straight lines. Equations of the form $ax + by = c$, where a, b, and c represent real numbers, are called **linear equations** because the graphs of their solutions form a *straight line*.

Here are the steps you should follow to graph the solution set for a simple linear equation.

1. Find three pairs of values that satisfy the equation. (Two points determine a line, and the third point is for insurance. When all three points fall on the same straight line, you know you have not made any errors.) Don't let the values get too large for your graph.

2. List that information in a table.

3. Transfer those ordered pairs to a graph.

4. Connect the points to see if they do form a straight line. If they do not fall in the same straight line, check for an error in your table of values.

E X A M P L E 2   Graph the solution set for the equation $x - y = 5$.

Solution   Begin by picking three pairs of numbers that satisfy the equation. Because this is a simple equation, you should be able to find these pairs just by looking at the equation. Figure 7 shows three possible choices. After completing a table of values, plot the ordered pairs on the graph and connect the points.

$x - y = 5$

x	y
6	1
8	3
10	5

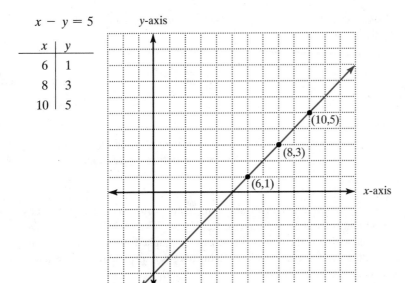

FIGURE 7

It would appear that the line also goes through the point (5,0). See if this is true.

$$x - y = 0$$
$$5 - 0 = 5 \quad \text{True.}$$

Therefore, (5,0) is also a solution of $x - y = 5$. What about $(3,-2)$? See if it gives a true statement in the equation.

$$x - y = 5$$
$$3 - (-2) = 5$$
$$3 + (+2) = 5$$
$$5 = 5 \quad \text{True.}$$

From Examples 1 and 2, you should see that the coordinates of every point on the line form a solution set for the equation. Also, any ordered pair of values that is a solution of a linear equation falls on the line that is its graph. Thus, unlike the equations you studied earlier (such as $x + 5 = 7$ and $3M + 4 = 5M - 2$), linear equations such as $y + 2x = 5$ don't have just one solution. They have an infinite number of solutions, or values that give a true statement. The graph displays these solutions.

Sometimes, if the equation is not a simple one, you will want to rearrange its terms so that you can more easily find three solutions. It is very helpful to put the equation into the same form each time. The form that is easiest to use has y all alone on one side of the equals sign. You will want the equation to follow the pattern $y = mx + b$, where m is the numerical coefficient of x, and b is the numerical term. This arrangement is called **slope–intercept form** (we'll see why in a moment). Use the same steps to rearrange the equation that you used to solve equations in Chapter 10. To eliminate a term from one side of an equation, add its opposite to both sides of the equation. To eliminate a coefficient from the y term, divide all the terms by that coefficient.

Remember, you want to arrive at the form $y = mx + b$, where y always has a coefficient of $+1$. The coefficient of x, represented by m in the equation $y = mx + b$, is called the **slope** of the line. *Slope* means the way the line slants. The constant term (numerical term) is b. The value b is called the **y intercept**; it tells where the graph intercepts, or cuts through, the y-axis.

For example, let's rearrange $x + y = 7$ into slope–intercept form.

$$x + y = 7$$

$$\underline{-x} \qquad \underline{\quad -x \quad}$$ To isolate y, you need to subtract the x term from each side.

$$y = 7 - x$$

$$y = -x + 7$$ Rearrange the right side of the equation into the $y = mx + b$ form. The slope of the line is -1 and the y intercept is $+7$.

Once the equation is in slope–intercept form, you can give x any value that you choose and solve the resulting equation for y. When you find y, you have the ordered pair (x,y) that satisfies the equation and can be graphed. Pick three values for x for the equation, and solve for each y value. (Remember not to pick an x value that is too large to fit onto your graphing surface.)

Consider the following examples, which have been worked out completely for you.

E X A M P L E 3   Graph the solution set for $x + y = 7$.

Solution   1. You want the $y = mx + b$ form. We rearranged this equation before and got $y = -x + 7$, or $y = -1x + 7$, where m (the numerical coefficient of x) is -1, and b (the constant term) is $+7$.

    2. Pick three x values and solve for the corresponding y values.

Value for x	Substitute into Slope–Intercept Form	Value for y	Point
x	$y = -x + 7$	y	(x,y)
0	$y = -0 + 7$	7	$(0,7)$
2	$y = -2 + 7$	5	$(2,5)$
4	$y = -4 + 7$	3	$(4,3)$

It is always a good idea to take the three points you have found and substitute them into the original equation. This enables you to make sure that you have not made an error in writing the equation in slope–intercept form or in solving for the y values by using the x values you chose.

If $x + y = 7$, then $0 + 7 = 7$.   True.

If $x + y = 7$, then $2 + 5 = 7$.   True.

If $x + y = 7$, then $4 + 3 = 7$.   True.

    3. Graph the solution of $x + y = 7$, or $y = -x + 7$, using $(0,7)$, $(2,5)$, and $(4,3)$. (See Figure 8.) Label the points.

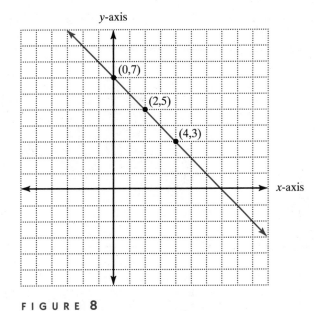

F I G U R E 8   ◆

Note that when $x = 0$, the line cuts through, or intercepts, the y axis at the point $(0,7)$. Therefore, 7 is indeed the y intercept. Also note that a negative slope means that the line slants down from left to right.

E X A M P L E 4   Graph the solution set for $y - x = 5$.

Solution   1.   Rearrange $y - x = 5$ into slope–intercept form.

$y - x = 5$

| $+ x$ | $+ x$ | To isolate y, you must add x to both sides of the equals sign. |

$y\quad\quad = 5 + x$

$y = x + 5$ Rearrange the terms on the right-hand side of the equals sign and you have the $y = mx + b$ form.

$y = +1x + 5$ Remember that you may always write the coefficient of x as $+1$.

2.   Pick three x values and solve for the corresponding y values.

Value for x	Substitute into Slope–Intercept Form	Value for y	Point
x	$y = \quad x + 5$	y	(x,y)
0	$y = \quad 0 + 5$	5	$(0,5)$
3	$y = \quad 3 + 5$	8	$(3,8)$
-2	$y = -2 + 5$	3	$(-2,3)$

Check: (Be sure to substitute correctly.)

$y - x = 5$	$y - x = 5$	$y - x = 5$
$5 - 0 = 5$	$8 - 3 = 5$	$3 - (-2) = 5$
$5 = 5$	$5 = 5$	$3 + 2 = 5$
		$5 = 5$

3. Graph the solution of $y - x = 5$, or $y = x + 5$, using (0,5), (3,8), and (−2,3). (See Figure 9.)

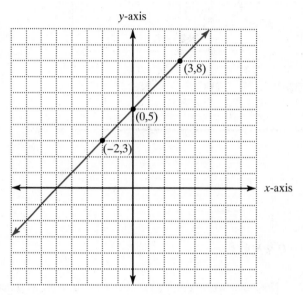

FIGURE 9 ◆

Look again closely at the graphs in Examples 3 and 4. The equation in Example 3 is $y = -x + 7$ and the m value, or slope, is −1. In Example 4, $y = x + 5$, the m value, or slope, is +1. Now examine the graphs in Figures 8 and 9. What might slope indicate about the graph of the line? Which way do the lines slant?

E X A M P L E 5   Graph the solution set for $3y = 6x - 15$.

Solution   **1.** To isolate y, you would divide *every* term in the equation $3y = 6x - 15$ by 3. Then $y = 2x - 5$ in $y = mx + b$ form, where $m = +2$ and $b = -5$.

2.

x	$y = 2x - 5$	y	(x,y)
0	$y = 2(0) - 5$	−5	(0,−5)
2	$y = 2(2) - 5$	−1	(2,−1)
−1	$y = 2(-1) - 5$	−7	(−1,−7)

Substitute the three points into the *original* equation to check your work so far.

$$3y = 6x - 15 \qquad\qquad 3y = 6x - 15 \qquad\qquad 3y = 6x - 15$$
$$3(-5) = 6(0) - 15 \qquad 3(-1) = 6(2) - 15 \qquad 3(-7) = 6(-1) - 15$$
$$-15 = 0 - 15 \qquad\qquad -3 = 12 - 15 \qquad\qquad -21 = -6 - 15$$
$$-15 = -15 \qquad\qquad -3 = -3 \qquad\qquad -21 = -21$$

3. Graph the solution of $3y = 6x - 15$, or $y = 2x - 5$, using (0,−5), (2,−1), and (−1,−7). (See Figure 10.)

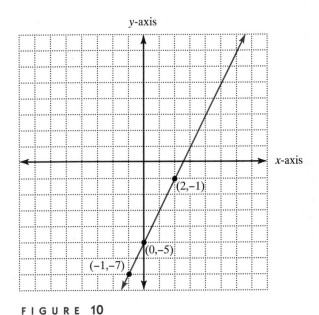

FIGURE 10

E X A M P L E   6   Graph the solution set for $x - y = 4$.

Solution   **1.** You could first subtract the x term from both sides of the equation. The changed form of the equation would be $-y = -x + 4$. The coefficient of y has to be a positive 1, so you must divide each term of the equation by -1.

$$\frac{-y}{-1} = \frac{-x}{-1} + \frac{4}{-1}$$

Now you have $y = x - 4$, where $m = +1$ and $b = -4$. Remember that you always want the coefficient of y to be a positive 1 (which is not written).

2.

x	$y = x - 4$	y	(x,y)
0	$y = 0 - 4$	-4	$(0,-4)$
2	$y = 2 - 4$	-2	$(2,-2)$
-1	$y = -1 - 4$	-5	$(-1,-5)$

Check

$x - y = 4$	$x - y = 4$	$x - y = 4$
$0 - (-4) = 4$	$2 - (-2) = 4$	$(-1) - (-5) = 4$
$0 + 4 = 4$	$2 + 2 = 4$	$-1 + 5 = 4$
$4 = 4$	$4 = 4$	$4 = 4$

3. Graph the solution of $x - y = 4$, or $y = x - 4$, using $(0,-4)$, $(2,-2)$, and $(-1,-5)$. (See Figure 11.)

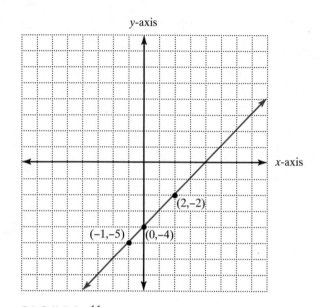

E X A M P L E   7   Graph the solution set for $2y - 4 = 3x$.

Solution   1.   Rearrange $2y - 4 = 3x$ into slope–intercept form.

$$2y - 4 = 3x$$

$$\underline{+\,4} \qquad \underline{+\,4} \qquad\qquad \text{To isolate the } y \text{ term, add 4 on } both\ sides \text{ of the equals}$$
sign.

$$2y \quad = 3x + 4$$

$$\frac{2y}{2} \quad = \frac{3}{2}x + \frac{4}{2} \qquad\qquad \text{To change the coefficient of } y \text{ to a positive 1, divide } \\ each \text{ term of the equation by } +2.$$

$$y = \frac{3}{2}x + 2$$

If $y = 3/2\,x + 2$, $m = 3/2$ and $b = +2$.

2.   Pick three x values and solve for the corresponding y values. Because the slope is 3/2, you should pick even x values so that you will not have a lot of fraction calculations to do.

x	$y = 3/2x + 2$	y	(x,y)
0	$y = 3/2(0) + 2$	2	(0,2)
2	$y = 3/2(2) + 2$	5	(2,5)
-2	$y = 3/2(-2) + 2$	-1	$(-2,-1)$

Check

$2y - 4 = 3x$	$2y - 4 = 3x$	$2y - 4 = 3x$
$2(2) - 4 = 3(0)$	$2(5) - 4 = 3(2)$	$2(-1) - 4 = 3(-2)$
$4 - 4 = 0$	$10 - 4 = 6$	$-2 - 4 = -6$
$0 = 0$	$6 = 6$	$-6 = -6$

3.   Graph the solution of $2y - 4 = 3x$, or $y = 3/2x + 2$, using (0,2), (2,5), and $(-2,-1)$. (See Figure 12.)

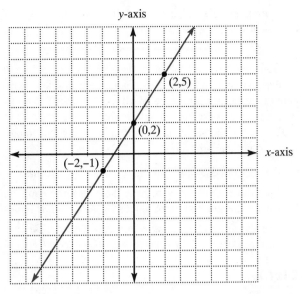

FIGURE 12

E X A M P L E **8**   Graph the solution set for $y = -2x$.

Solution   1. To write the equation in slope–intercept form, all you need to do is add a constant term, zero. $y = -2x + 0$. The slope is -2 and the y intercept is 0.

2. Pick three x values and solve for the corresponding y values.

x	$y = -2x + 0$	y	(x,y)
0	$y = -2(0) + 0$	0	$(0,0)$
2	$y = -2(2) + 0$	-4	$(2,-4)$
-2	$y = -2(-2) + 0$	$+4$	$(-2,4)$

Check $y = -2x$ $y = -2x$ $y = -2x$

$0 = -2(0)$ $-4 = -2(2)$ $4 = -2(-2)$

$0 = 0$ $-4 = -4$ $4 = 4$

3. Graph the solution of $y = -2x$, using $(0,0)$, $(2,-4)$, and $(-2,4)$. (See Figure 13.)

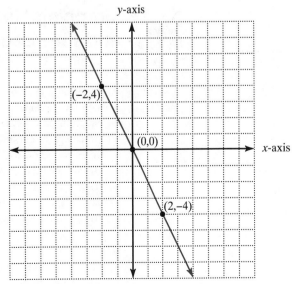

FIGURE 13

Note from this graph that an equation with a *b*-value of 0—that is, a *y* intercept of 0—passes through the origin. ◆

You now have all the parts necessary to graph the solutions of these equations. Remember that although only two points are necessary to determine a straight line, the third point is needed for insurance. If all three do not fall on the same line, then you know that something is wrong and can try to find your error.

These are the steps you should now follow to graph the solution of a linear equation using slope–intercept form:

1. Change the equation into the $y = mx + b$ form.

2. Pick three x values. Substitute them into the equation and solve for the y values. Check the three pairs of values in the *original equation*.

3. Use the three ordered pairs to graph the line. Be sure that it passes through all three points. Label the three points you used so that someone else can follow your work.

17.3 ◆ Practice Problems

For each of the following equations:

a. **Change the equation into the $y = mx + b$ form.**

b. **Find three ordered pairs of numbers that satisfy the equation, and check those values in the original equation.**

c. **Graph the solution set for the equation. Be sure to label the three points that you used. (When you check your answers against the answer key, your three points may be different from those labeled on the graph shown in the answer key, but the line should look the same.)**

1. $x + y = -4$
2. $2x - 4 = y$
3. $-3x + 5 = y$
4. $x + y = 7$

5. $x - y = 3$
6. $6x = 3y + 12$
7. $2y = -x$
8. $y = 4x$
9. $2y = 8x - 4$
10. $x - y = 5$
11. $7x - 2y = 6$
12. $y + 4x = -2$
13. $-2x = 6y$
14. $3x - 2y = 0$
15. $x - 2y = 8$
16. $3y = x + 9$
17. $y + 2 = 3x$
18. $3x - y = 4$
19. $3y = x + 12$
20. $-2x = 3y - 6$
21. $3x = -4y + 6$
22. $5x + 2y = 7$
23. $2y - 3x = -4$
24. $-3y + 2x = -12$

17.4 ◆ Graphing Linear Equations by Using Slope and Intercept

An equation that is in **slope–intercept form** follows the pattern $y = mx + b$. If the *y* is isolated on one side of the equation, you can tell a lot about the graph just by looking at the equation. In $y = mx + b$, the *b* value is called the *y* intercept; it tells you where the line will intercept (or cut through) the *y*-axis. Because that point on the line will lie on the *y*-axis, its first coordinate will be zero. The ordered pair associated with that point is $(0,b)$.

In an equation in slope–intercept form ($y = mx + b$), the letter *m* is the numerical coefficient of *x* and has a special name, the **slope**. That *m* value indicates which way the line will slant. If *m* is positive, the line might be described as slanting upward or increasing to the right. For example, a slope of $+2/3$ means that the line increases, because the slope is positive. If *m* is negative, the line is decreasing to the right; that is, it slants downward.

The slope value, *m*, is also the ratio of the vertical change (up or down) to the horizontal change (right or left) between two points on the line. As a ratio, the slope is written in fraction form. A slope of $+2/3$, 2 over 3, means that if you begin at any point that lies on the line and move up two units (+2) and to the right three units (+3), you will come to another point that lies on the line. From that point, you can move up 2 units and over to the right 3 units and come to yet another point on the line. Because $-2/-3$ is the same as $+2/+3$, you could also move down two units (-2) and to the left three units (-3) and arrive at another point that lies on the line. A slope of 4 is written as $4/1$ in fraction form. From any point on the line, you could move up four units (+4) and to the right one unit (+1) and come to another point on the line. Or you could move down four units (-4) and to the left one unit (-1) and come to still another point.

   The ratio of the vertical change to the horizontal change is often called "the rise over the run," which means up–down change over right–left change. Consider the graph of the line $y = 2/3\,x - 3$ shown in Figure 14, and study "the rise" and "the run."

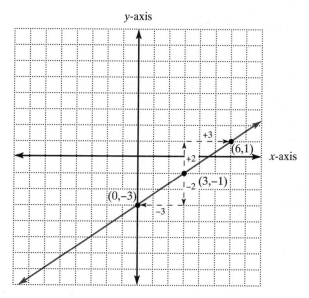

F I G U R E 14

   Because the slope of the graph of $y = 2/3\,x - 3$ is 2 over 3, the rise is +2 and the run is +3. From any point on the graph you can rise +2 units and run +3 units and come to another point that lies on the line. Beginning at $(3,-1)$ you would go up two units (rise +2) and go over three units (run +3) and come to $(6,1)$, another point on the line. Because $-2/-3$ has the same value as $+2/+3$, you could also "rise" -2 and "run" -3 to find another point. Again beginning at $(3,-1)$, you would go down two units (rise -2) and back to the left three units (run -3) and come to $(0,-3)$, which is also a point on the line. Both points that you found using rise and run—$(6,1)$ and $(0,-3)$—satisfy the equation and therefore lie on the line.

$$y = \frac{2}{3}x - 3 \qquad\qquad y = \frac{2}{3}x - 3$$

$$1 = \frac{2}{3}(6) - 3 \qquad\qquad -3 = \frac{2}{3}(0) - 3$$

$$1 = 4 - 3 \qquad\qquad\qquad -3 = 0 - 3$$

$$1 = 1 \quad \text{True.} \qquad\qquad -3 = -3 \quad \text{True.}$$

   Look at the following examples of linear equations in slope–intercept form and at their graphs, which have been drawn for you. Note the slopes of the lines, the rise over the run, and the coordinates of the point at which each line intercepts the y axis $(0,b)$.

E X A M P L E **1**   Use the slope and y intercept to graph the solution set for the equation $y = -x + 7$.

Solution $b = +7$, so the line intercepts the axis at $(0,7)$.

$m = -1$, so the line should slant down toward the right.

$m = -1$ could be $-4/+4$ (down 4/right 4), for example, or $+1/-1$ (up 1/left 1).

Begin at $(0,7)$ and find two more points on the line by using the rise over the run. Because the slope is -1, you could use $-1/+1$ (down 1/right 1) to find another point. That takes you to $(1,6)$. You could use $+2/-2$ (up 2/left 2) to find $(-2,9)$, a third point on the line. (See Figure 15.) You should check all three points in the original equation to be sure that they do, in fact, satisfy the equation.

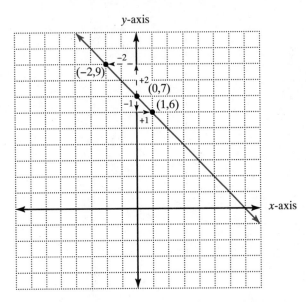

F I G U R E **15**

$y = -x + 7$	$y = -x + 7$	$y = -x + 7$
$9 = -(-2) + 7$	$7 = -(0) + 7$	$6 = -(1) + 7$
$9 = 2 + 7$	$7 = 0 + 7$	$6 = -1 + 7$
$9 = 9$	$7 = 7$	$6 = 6$

E X A M P L E **2**   Graph the solution set for $y = x + 5$ by using the slope and y intercept.

Solution   Because $b = +5$, the y intercept is $+5$ and the line goes through the point $(0,5)$.

$m = +1$ could be $+2/+2$ (up 2/right 2) or $-1/-1$ (down 1/left 1).

From the graph in Figure 16 you can see that $+2/+2$ takes you to $(2,7)$ and that, beginning again at $(0,5)$, $-1/-1$ takes you to $(-1,4)$. To check the accuracy of the graphing process, substitute all three points back into the original equation and see if they all result in a true statement.

$y = x + 5$	$y = x + 5$	$y = x + 5$
$7 = 2 + 5$	$5 = 0 + 5$	$4 = -1 + 5$
$7 = 7$	$5 = 5$	$4 = 4$

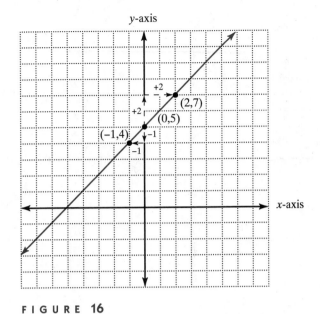

FIGURE 16

EXAMPLE 3 Graph the solution set for $3x + 2y = 4$ by using the slope and y intercept.

Solution In this problem, you will first need to change $3x + 2y = 4$ into $y = mx + b$ form.

$$3x + 2y = 4$$

$$\underline{-3x} \qquad \underline{\quad -3x}$$

$$2y = 4 - 3x$$

$$\frac{2y}{2} = \frac{-3x}{2} + \frac{4}{2}$$

$$y = -\frac{3}{2}x + 2$$

The slope is $-3/2$, so the line slants down.

The y intercept is 2, so the line goes through the point $(0,2)$.

To graph the line, begin at $(0,2)$ and use the rise/run ratio to find two more points. $-3/2$ could mean $-3/2$ (down 3/right 2) or $3/-2$ (up 3/left 2). Studying Figure 17 reveals that two other points on the line are $(-2,5)$ and $(2,-1)$. Check the graph by substituting all three points into the original equation to see if they all result in a true statement.

$3x + 2y = 4$	$3x + 2y = 4$	$3x + 2y = 4$
$3(-2) + 2(5) = 4$	$3(0) + 2(2) = 4$	$3(2) + 2(-1) = 4$
$-6 + 10 = 4$	$0 + 4 = 4$	$6 + (-2) = 4$
$4 = 4$	$4 = 4$	$4 = 4$

F I G U R E **17**

◆

To review: You can graph the solution of any first-degree equation using the y intercept and slope by following these steps:

1. Write the equation in slope–intercept form, $y = mx + b$.

2. Identify the slope and the y intercept.

3. Begin by graphing the y intercept and using the slope (the ratio of rise to run) to find two more points on the line. Draw the straight line through all three points.

4. Check that the graph is correct by substituting all three points back into the original equation to make sure that they all produce true statements.

Here are two more examples worked out completely for you to follow.

E X A M P L E 4 Graph the solution set for $2x + y = 6$.

Solution 1. $2x + y = 6 \rightarrow y = -2x + 6$

2. $b = +6$, so the line goes through $(0,6)$.

$m = -2$, so the line decreases to the right.

$m = -2$ could be $-2/+1$ (down 2/right 1), for example, or $+2/-1$ (up 2/left 1).

3. Begin at $(0,6)$. Go down 2 and right 1; you reach $(1,4)$. Begin again at $(0,6)$. Go up 2 and left 1; you reach $(-1,8)$. (See Figure 18.)

4. Check the three points in the original equation to make sure that they satisfy it.

$2x + y = 6$	$2x + y = 6$	$2x + y = 6$
$2(-1) + 8 = 6$	$2(0) + 6 = 6$	$2(1) + 4 = 6$
$-2 + 8 = 6$	$0 + 6 = 6$	$2 + 4 = 6$
$6 = 6$	$6 = 6$	$6 = 6$

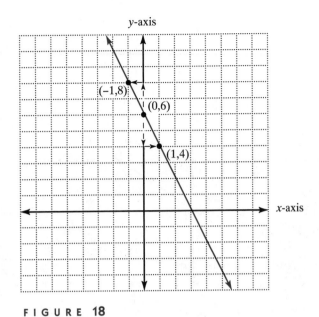

FIGURE **18**

E X A M P L E 5   Graph the solution set for $2x - 3y = 9$.

Solution   **1.** Transform the equation into $y = mx + b$ form.

$$2x - 3y = 9$$

$$\underline{-2x} \qquad\qquad \underline{-2x}$$

$$-3y = 9 - 2x$$

$$\frac{-3y}{-3} = \frac{9}{-3} - \frac{2x}{-3}$$

$$y = -3 + \frac{2}{3}x$$

$$y = \frac{2}{3}x - 3$$

2. The y intercept is -3, so the line goes through the point $(0,-3)$. The slope is $+2/3$, so the line increases; that is, it slants upward.

3. Begin at $(0,-3)$. Use $+2/+3$ (up 2/right 3) and/or $-2/-3$ (down 2/left 3) to locate two other points on the line. (See Figure 19.)

4. Check all three points in the original equation.

$2x - 3y = 9$	$2x - 3y = 9$	$2x - 3y = 9$
$2(3) - 3(-1) = 9$	$2(0) - 3(-3) = 9$	$2(-3) - 3(-5) = 9$
$6 + 3 = 9$	$0 + 9 = 9$	$-6 + 15 = 9$
$9 = 9$	$9 = 9$	$9 = 9$

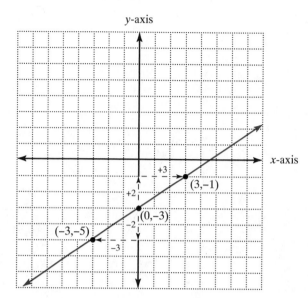

FIGURE **19**

17.4 ◆ Practice Problems

For each of the following equations:

a. Change the equation into the $y = mx + b$ form.

b. Identify the slope and the y intercept.

c. Graph the solution set for the equation by using the y intercept and the slope. Be sure to label the three points that you found to graph the equation.

1. $x + y = -2$
2. $3x + 2 = y$
3. $y = 2x$
4. $-3x = y$

5. $x - y = 1$
6. $x - y = 5$
7. $3y + x = 0$
8. $2y - 4x = 0$
9. $3y = -9x - 6$
10. $8x = 4y + 4$
11. $3x - y = -1$
12. $2x - 3 = -y$
13. $x - 2y = 6$
14. $2y = x - 8$
15. $y + 4 = -2x$
16. $x - 2y = 3$
17. $3x - 4y = -4$
18. $3y + 6 = 4x$
19. $2y + x = -8$
20. $2x + 5y = -15$

17.5 More About Slope

Now that you can picture slope as the slant of a line and know that it equals the ratio rise/run, you need to look at slope in more detail. Slope can be defined in several ways, but the definition that is the most helpful in calculating the slope of a line is

Definition of Slope

$$\text{Slope} = \frac{\text{change in } y \text{ values}}{\text{change in } x \text{ values}}$$

This definition can also be expressed as a formula.

Formula for Slope

Given two points (x_1, y_1) and (x_2, y_2), we can find the slope of the line through those two points by using the formula

$$m = \frac{y_2 - y_1}{x_2 - x_1}$$

This formula may look complicated, but it's very easy to use. The subscript notation is used to indicate that you are working with two different points but that each of them has two coordinates, x and y. It does not matter which point you call (x_1, y_1), but you must be consistent. For safety's sake, always refer to the first point mentioned as (x_1, y_1) and to the second point as (x_2, y_2). Then you will not get them confused.

Consider two points, $(2,1)$ and $(4,7)$. Let $(2,1)$ be (x_1, y_1) and let $(4,7)$ be (x_2, y_2). Then, by the definition and formula,

$$m = \frac{y_2 - y_1}{x_2 - x_1} = \frac{7 - 1}{4 - 2} = \frac{6}{2} = \frac{3}{1} = 3$$

Therefore, the slope of the line through $(2,1)$ and $(4,7)$ is positive 3, so the line increases to the right. Graph the line through $(2,1)$ and $(4,7)$ and you will see that this is correct. Remember also that slope indicates the ratio of the vertical change to the horizontal change (Section 17.4). A slope of 3/1 tells you that if you begin at $(2,1)$ and move up three units $(+3)$ and to the right one unit $(+1)$, you will find another point on the line. Begin at $(2,1)$ and move up three units and over one unit to $(3,4)$. Then from $(3,4)$ move up three units and over one unit, and you come to $(4,7)$. Check this in Figure 20.

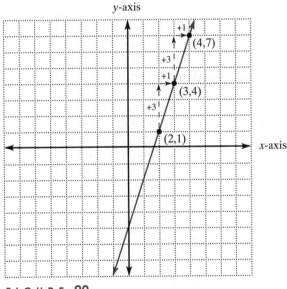

The line is increasing to the right.
Rise/run is $+3/+1$.

F I G U R E **20**

Let's try another example. Let (x_1, y_1) be $(3,1)$, and let (x_2, y_2) be $(-2,2)$. Then

$$m = \frac{y_2 - y_1}{x_2 - x_1} = \frac{2 - 1}{-2 - 3} = \frac{1}{-5} = -\frac{1}{5}$$

Because the slope is negative, the line should decrease; that is, it should slant downward to the right. Look at the graph through $(3,1)$ and $(-2,2)$. (See Figure 21.) It does, in fact, decrease to the right.

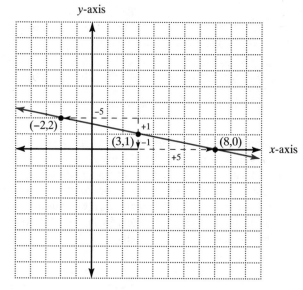

The line is decreasing to the right.
Rise/run is $-5/+1$.
or $+5/-1$.

FIGURE **21**

A slope of $-1/5$ could mean -1 over $+5$ ($-1/+5$) or $+1$ over -5 ($+1/-5$). If you begin at $(3,1)$ and move down one unit and over five units, you come to $(8,0)$. If you begin at $(3,1)$ and move up one unit and back to the left five units, you come to $(-2,2)$. As you can see from the graph in Figure 21, all three of those points are part of the line.

To examine the case where both y values are the same, let (x_1,y_1) be $(4,3)$ and let (x_2,y_2) be $(7,3)$. Then

$$m = \frac{y_2 - y_1}{x_2 - x_1} = \frac{3 - 3}{7 - 4} = \frac{0}{3} = 0$$

Look at the graph in Figure 22.

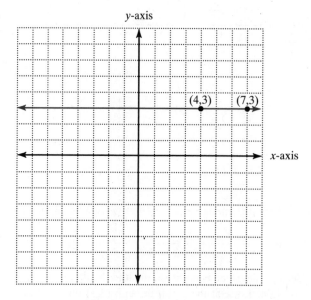

FIGURE **22**

The line has no slant—no tilt either way. A ratio of 0/3 means that there is no movement up or down. When you begin at (4,3), you move (three spaces) to the right or left only. Therefore, *a line parallel to the x-axis has slope = 0.* That means that the line will never intersect the *x*-axis.

What happens when both *x* values are the same? Well, if (x_1,y_1) is (4,3) and (x_2,y_2) is (4,9), then

$$m = \frac{y_2 - y_1}{x_2 - x_1} = \frac{9 - 3}{4 - 4} = \frac{6}{0} \qquad \text{Undefined.}$$

Look at Figure 23.

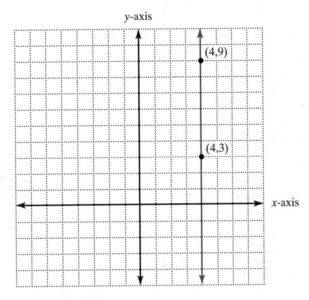

F I G U R E **23**

Because division by zero is impossible in our number system, the slope is said to be *undefined*. A ratio of 6/0 (or a ratio of any other number to zero) means there is no left–right movement between points. Looking at the graph, you can see that the line is straight up and down. Therefore, *a line parallel to the y-axis has an undefined slope.*

When you are given two points, it does not make any difference which point you call (x_1,y_1) and which you call (x_2,y_2). The most important thing to remember in doing these problems is the need to calculate carefully. The two-point formula is used to calculate slope in the following three examples.

E X A M P L E 1   Calculate the slope of the line that passes through the points (9,4) and (7,−2).

Solution $m = \dfrac{y_2 - y_1}{x_2 - x_1} = \dfrac{-2 - 4}{7 - 9} = \dfrac{-6}{-2} = +3$

The slope is +3, and the line increases to the right. The ratio could be +3/+1 or −3/−1, for example, because both are equivalent to +3.   ◆

E X A M P L E **2**   Calculate the slope of the line that passes through the points $(-2,-3)$ and $(4,-1)$.

Solution $m = \dfrac{y_2 - y_1}{x_2 - x_1} = \dfrac{-1 - (-3)}{4 - (-2)} = \dfrac{-1 + 3}{4 + 2} = \dfrac{2}{6} = \dfrac{1}{3}$

The slope is $+\dfrac{1}{3}$, and the line increases to the right. The ratio could be $+1/+3$ or $-1/-3$; both are equivalent to $1/3$.   ◆

E X A M P L E **3**   Calculate the slope of the line that passes through the points $(-2,4)$ and $(5,0)$.

Solution $m = \dfrac{y_2 - y_1}{x_2 - x_1} = \dfrac{0 - (+4)}{5 - (-2)} = \dfrac{0 - 4}{5 + 2} = \dfrac{-4}{7}$ or $-4/7$

The slope is $-\dfrac{4}{7}$ and the line decreases to the right. The ratio could be $-4/+7$ or $+4/-7$; both are equivalent to $-\dfrac{4}{7}$.   ◆

Look at the graph in Figure 24. Calculate the slope by using the two points labeled on the graph.

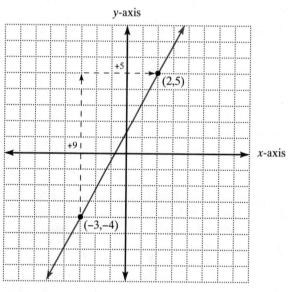

F I G U R E **24**

To move from $(-3,-4)$ to $(2,5)$, you need to go up nine units and over five units. Therefore, the slope of the line is $+9/+5$, or $9/5$. The slope is positive and the line is increasing to the right. Check the value for the slope by using the formula. Let (x_1, y_1) be $(-3,-4)$ and let (x_2, y_2) be $(2,5)$. Then

$$m = \frac{y_2 - y_1}{x_2 - x_1} = \frac{5 - (-4)}{2 - (-3)} = \frac{5 + 4}{2 + 3} = \frac{9}{5}$$

The formula value agrees with the value you found for the slope by moving from one point to the other on the graph.

To summarize:

1. Slope can be found by putting the equation into slope–intercept form, $y = mx + b$. The numerical coefficient of x, the m value, is the slope.

2. Slope can be found by using two points (x_1, y_1) and (x_2, y_2), and applying the formula

$$m = \frac{y_2 - y_1}{x_2 - x_1}$$

3. Slope can be found from a graph by moving from one point on the line to another and counting the rise and the run.

4. A line with a positive slope increases, or slants upward to the right.

5. A line with a negative slope decreases, or slants downward to the right.

6. A line *parallel to the x-axis has slope = 0.*

7. A line *parallel to the y-axis has an undefined slope.*

8. The slope written in ratio (fraction) form describes movement from one point to another on the line. The ratio is movement up or down (rise) compared to movement right or left (run).

17.5 ◆ Practice Problems

Study the slope of the line through each pair of points by following the steps listed below.

A. Graph each pair of points and draw the line passing through them.

1. (2,3) and (4,6)
2. (1,4) and (−2,−2)
3. (2,6) and (3,5)
4. (7,4) and (3,5)
5. (0,4) and (3,8)
6. (6,2) and (4,0)
7. (2,4) and (2,8)
8. (3,2) and (−2,2)
9. (5,4) and (2,6)
10. (0,3) and (5,2)
11. (3,4) and (−4,4)
12. (6,3) and (6,−2)
13. (4,−3) and (2,−5)
14. (−3,5) and (−2,−3)

B. Using the graphs, calculate each slope by studying movement from one point to the other and calculating rise/run.

C. Using the two-point formula for slope, calculate the slope of the line through each pair of points.

D. Make sure that your number answers in parts B and C agree.

E. Compare the numerical result you got in parts B and C with the way the line actually slants by inspecting the graphs you drew in part A.

17.6   Writing the Equation of a Line

In some situations it is helpful to be able to look at the graph of a line and work in reverse, so to speak, to find the equation of that line.

Consider the graph in Figure 25. The line passes through two points, (2,2) and (0,−2). You know from looking at the graph that its y intercept is −2. You can calculate the slope of the line by using those points. To move from (0,−2) to (2,2), you would go up four units (+4) and then over two units (+2). The rise over the run—the slope—is +4/+2, or +2.

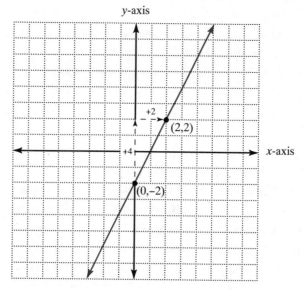

F I G U R E 25

To write the equation of this line, you will use the slope–intercept form for the equation of a line, $y = mx + b$. You know that the slope is $+2$, so $m = +2$. You can see from the graph that the y intercept is -2, so $b = -2$. By substituting into $y = mx + b$, you can write the equation of the line $y = +2x - 2$. This is the equation of the line that passes through the points $(2,2)$ and $(0,-2)$. Whenever you write the equation of a line, you should check to make sure that, in fact, both points do make the equation true.

$$y = 2x - 2 \qquad\qquad y = 2x - 2$$

Substitute $(2,2)$.   Substitute $(0,-2)$.

$$2 = 2(2) - 2 \qquad\qquad -2 = 2(0) - 2$$

$$2 = 4 - 2 \qquad\qquad\qquad -2 = 0 - 2$$

$$2 = 2 \quad \text{True.} \qquad\qquad -2 = -2 \quad \text{True.}$$

Whenever you are given the slope and y intercept, it is quick to write the equation of the line directly by substituting into the general form, $y = mx + b$. Study Example 1.

E X A M P L E 1   Write the equation of a line whose slope is -5 and whose y intercept is $+1$.

Solution   You know that $m = -5$ and $b = +1$, so substitute these values into the slope–intercept form for the equation of a line, $y = mx + b$.

$$y = mx + b$$

$$y = -5x + 1$$ ◆

Whenever you are given two points and are asked to write the equation, you can graph the line through the two points and see where the line cuts through the y axis to identify the y intercept. You can use the graph or the formula to determine the slope and then write the equation as in Example 1. Study Example 2 to see how you might find the equation by using the graph.

E X A M P L E   **2**   Write the equation of the line that passes through the points $(-2,8)$ and $(0,2)$ by using its graph to determine the slope and y intercept.

Solution   **1.**   Sketch the graph (see Figure 26).

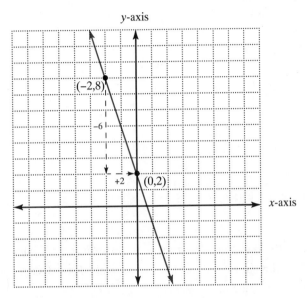

F I G U R E   **26**

    2.   You can see that the y intercept, b, is $+2$.

    3.   Calculate the slope.

$$m = \text{rise/run} = -6/+2 = -3$$

    4.   Write the slope–intercept form and substitute.

$$y = mx + b$$
$$y = -3x + 2$$

Therefore, the equation of the line is $y = -3x + 2$.

    5.   Check by substituting the x and y values for both points into the equation.

$y = -3x + 2$	$y = -3x + 2$
Substitute $(-2,8)$.	Substitute $(0,2)$.
$8 = -3(-2) + 2$	$2 = -3(0) + 2$
$8 = 6 + 2$	$2 = 0 + 2$
$8 = 8$ True.	$2 = 2$ True.

◆

E X A M P L E   **3**   Write the equation of the line that passes through the points $(-5,4)$ and $(3,4)$ by using the graph of the line to determine the slope and y intercept.

Solution   **1.**   Draw the line through $(-5,4)$ and $(3,4)$. (See Figure 27.)

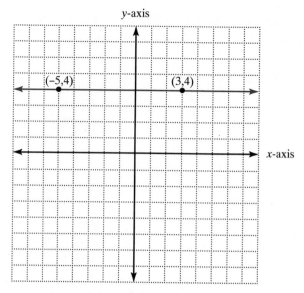

F I G U R E **27**

2. By looking at the graph you can see that the y intercept, b, is $+4$.
3. Moving from $(-5,4)$ to $(3,4)$ reveals that the slope, m, is rise/run $= 0/+8 = 0$.
4. Substitute into $y = mx + b$.

$$y = mx + b$$
$$y = 0x + 4$$

Therefore, the equation of the line is $y = 0x + 4$, or $y = 4$.

5. Check your answer by substituting the x and y values for both points into the equation.

$y = 0x + 4$	$y = 0x + 4$
Substitute $(-5,4)$.	Substitute $(3,4)$.
$4 = 0(-5) + 4$	$4 = 0(3) + 4$
$4 = 0 + 4$	$4 = 0 + 4$
$4 = 4$ True.	$4 = 4$ True.

◆

E X A M P L E   4   Write the equation of the line that passes through the points $(3,3)$ and $(-3,7)$ by using its graph to determine the slope and y intercept.

Solution

1. Draw the line through the two points. (See Figure 28.)
2. Looking at the graph, you can see that the y intercept is $+5$.
3. Moving from $(-3,7)$ to $(3,3)$ reveals that the slope, m, is rise/run $= -4/+6 = -\dfrac{2}{3}$.
4. Substitute into $y = mx + b$.

$$y = mx + b$$
$$y = -\frac{2}{3}x + 5$$

Therefore, the equation is $y = -\dfrac{2}{3}x + 5$. Or multiplying through by 3 gives the equation $3y = -2x + 15$. Either form is acceptable.

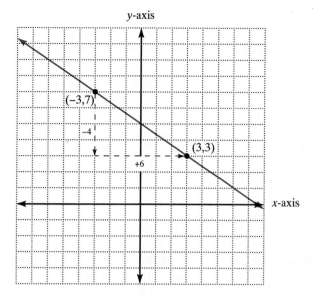

y-axis

(−3,7)

−4

(3,3)

+6

x-axis

FIGURE **28**

5. Check the equation by substituting both points into either form of the equation.

$$y = -\frac{2}{3}x + 5 \qquad\qquad 3y = -2x + 15$$

Substitute (−3,7).   Substitute (3,3).

7 = −2/3(−3) + 5   3(3) = −2(3) + 15

7 = 2 + 5   9 = −6 + 15

7 = 7 True.   9 = 9 True.

As you can see from these examples, it is very helpful to be given the *y* intercept, as in Example 1, or to be able to read it directly from the graph, as in Examples 2, 3, and 4. When this is not possible because the *y* intercept is not an integer that can readily be determined from the graph, you will need to use another general form of a linear equation. This *two-point formula* can be used when you are given any two points, but it *must* be used when the *y* intercept (*b* value) is not readily available.

When you are given two points and are asked to write the equation of the line through those points, use this two-point formula:

Two-Point Formula

The equation of the line passing through points (x_1, y_1) and (x_2, y_2), is

$$y - y_1 = \left(\frac{y_2 - y_1}{x_2 - x_1}\right)(x - x_1)$$

or

$$y - y_1 = m(x - x_1).$$

Because the second form looks so much easier to use and to remember, it is generally a good idea to calculate the slope (the *m* value) first and then substitute it into $y - y_1 = m(x - x_1)$.

If you want to write the equation of the line through the points $(-3,2)$ and $(2,-3)$, use $y - y_1 = m(x - x_1)$. First find the value for m, letting $(-3,2)$ be (x_1, y_1) and letting $(2,-3)$ be (x_2, y_2).

$$m = \frac{y_2 - y_1}{x_2 - x_1} = \frac{-3 - 2}{2 - (-3)} = \frac{-3 - 2}{2 + 3} = \frac{-5}{+5} = -1$$

Then $y - y_1 = m(x - x_1)$.   Substitute $(-3,2)$ for (x_1, y_1) and -1 for m.

$\quad y - 2 = -1[x - (-3)]$ Use the subtraction rule.

$\quad y - 2 = -1[x + 3]$ Distribute the -1.

$\quad y - 2 = -1x - 3$ Solve for y.

$\quad\underline{\quad + 2 \qquad\quad + 2\quad}$

$\quad y \qquad = -1x - 1 \quad$ or $\quad y = -x - 1$

Check both points in the equation to make sure that it is, in fact, the equation of the line that passes through the points $(-3,2)$ and $(2,-3)$.

$y = -x - 1$	$y = -x - 1$
Substitute $(-3,2)$.	Substitute $(2,-3)$.
$2 = -1(-3) - 1$	$-3 = -1(2) - 1$
$2 = 3 - 1$	$-3 = -2 - 1$
$2 = 2$ True.	$-3 = -3$ True.

You would have arrived at the same equation if you had used the longer form. The only difference between the two is that here the slope is calculated separately first and then the simpler formula is used. Try it for yourself.

The only real problem that most students have in writing equations like these is in keeping the points straight. Do not use x_2 when you need x_1, and don't use y_1 when you need y_2! Be sure to label the points when you begin and to substitute carefully.

E X A M P L E   **5**   Write the equation of the line that passes through the points $(-2,5)$ and $(0,-3)$.

Solution   First calculate the slope. Let $(-2,5)$ be (x_1, y_1), and let $(0,-3)$ be (x_2, y_2).

$$m = \frac{y_2 - y_1}{x_2 - x_1} = \frac{-3 - 5}{0 - (-2)} = \frac{-3 - 5}{0 + 2} = \frac{-8}{+2} = -4$$

Then $y - y_1 = m(x - x_1)$.

$\quad y - 5 = -4[x - (-2)]$

$\quad y - 5 = -4[x + 2]$

$\quad y - 5 = -4x - 8$

$\quad\underline{\quad + 5 \qquad\quad + 5\quad}$

$\quad y \qquad = -4x - 3$

Check

$y = -4x - 3$	$y = -4x - 3$
Substitute $(-2,5)$.	Substitute $(0,-3)$.
$5 = -4(-2) - 3$	$-3 = -4(0) - 3$
$5 = 8 - 3$	$-3 = 0 - 3$
$5 = 5$ True.	$-3 = -3$ True.

◆

E X A M P L E 6   Write the equation of the line that passes through the points (2,4) and (2,−3).

Solution   Let (2,4) be (x_1, y_1), and let (2,−3) be (x_2, y_2).

Then $m = \dfrac{y_2 - y_1}{x_2 - x_1} = \dfrac{-3 - 4}{2 - 2} = \dfrac{-7}{0}$, undefined.

The slope is undefined, so you know that the line is straight up and down, and because the line passes through (2,4) and (2,−3), you know that x has the value 2. The first coordinate for every point on that line is 2. Therefore, the equation of the line is $x = 2$. ◆

E X A M P L E 7   Write the equation of the line that passes through the points (−1,−3) and (4,−3).

Solution $m = \dfrac{y_2 - y_1}{x_2 - x_1} = \dfrac{-3 - (-3)}{4 - (-1)} = \dfrac{-3 + 3}{4 + 1} = \dfrac{0}{5} = 0$

Substitute into the general formula for finding an equation given two points.

$$y - y_1 = m(x - x_1)$$
$$y - (-3) = 0[x - (-1)]$$
$$y + 3 = 0[x + 1]$$
$$y + 3 = 0x + 0$$
$$y + 3 = 0x + 0$$
$$\underline{ -3 \qquad\quad -3}$$
$$y \qquad\quad = 0x - 3 \quad \text{or} \quad y = -3$$

The graph of a line with zero slope is horizontal (flat) and is of the form $y = b$. This line goes through two points whose y coordinates are −3. Therefore, $y = -3$ is the equation of this line.

Check

$y = 0x - 3$	$y = 0x - 3$
Substitute (−1,−3).	Substitute (4,−3).
$-3 = 0(-1) - 3$	$-3 = 0(4) - 3$
$-3 = 0 - 3$	$-3 = 0 - 3$
$-3 = -3$ True.	$-3 = -3$ True.

◆

If you are given the slope of a line and only one point, you can still write the equation of the line by using $y = mx + b$. You are given x, y, and m and are missing only b. Therefore, substitute the values you have for x, y, and m into $y = mx + b$ and solve the resulting equation for b. Then go back to the general form $y = mx + b$ and substitute the given m value and the b value that you found. Be sure to check the original point, (x,y), in the equation you have written.

E X A M P L E 8   Write the equation of the line that has a slope of −2 and passes through the point (3,−5). Use $y = mx + b$.

Solution   Begin with the slope–intercept form of an equation, $y = mx + b$. You know that $m = -2$, and because the point you are given is (3,−5), you also know that $x = 3$ and $y = -5$.

$$y = mx + b$$

$$-5 = -2(3) + b$$

$$-5 = -6 + b$$

$$\underline{+6 \quad +6}$$

$$+1 = \qquad b$$

Go back to $y = mx + b$, and use $m = -2$ and $b = +1$ to write the equation.

$$y = mx + b$$

$$y = -2x + 1$$

Check   You need to be sure that the given point, $(3, -5)$ does in fact lie on the graph of $y = -2x + 1$.

$$y = -2x + 1$$

Substitute $(3, -5)$.

$$-5 = -2(3) + 1$$

$$-5 = -6 + 1$$

$$-5 = -5 \quad \text{True.}$$ ◆

E X A M P L E   9   Write the equation of the line that passes through the point $(-4, 7)$ and has a slope of $-1/2$. Use $y = mx + b$.

Solution   You know that $m = -1/2$, $x = -4$, and $y = 7$.

$$y = mx + b$$

$$7 = -1/2(-4) + b$$

$$7 = \quad 2 + b$$

$$\underline{-2 \quad -2}$$

$$5 = \qquad b$$

Now you know that $m = -1/2$ and $b = +5$.

$$y = mx + b$$

$$y = -\frac{1}{2}x + 5$$

You may choose to multiply through by the common denominator, 2, to clear the fraction and produce an equation with integral coefficients.

$$y = -\frac{1}{2}x + 5$$

$$(2)y = (2)(-1/2)x + (2)(5)$$

$$2y = -x + 10$$

Check   Substitute $(-4, 7)$ to see if it satisfies the equation you have written.

$$2y = -x + 10$$

$$2(7) = -(-4) + 10$$

$$14 = +4 + 10$$

$$14 = 14 \quad \text{True.}$$ ◆

E X A M P L E   **10**   Write the equation of the line that passes through the point $(-5, -4)$ and has a slope of $2/3$.

Solution   You know that $x = -5$, $y = -4$, and $m = 2/3$.

$$y = mx + b$$
$$-4 = (2/3)(-5) + b$$
$$-4 = -10/3 + b$$
$$-12/3 = -10/3 + b$$
$$\underline{+10/3 \quad +10/3}$$
$$-2/3 = \qquad b$$

Now you know that $m = 2/3$ and $b = -2/3$.

$$y = mx + b$$
$$y = \frac{2}{3}x - 2/3$$

Multiply through by the LCD, 3, if you choose to write the equation with integral coefficients instead of fractions. Either form of the equation is correct, but it may be easier to check with fewer fractions involved.

$$(3)y = (3)(2/3)x - (3)(2/3)$$
$$3y = 2x - 2$$

Check   Substitute the original point, $(-5, -4)$, into this new equation to make sure that the point does, in fact, lie on the graph of the line.

$$3y = 2x - 2$$

Substitute $(-5, -4)$.

$$3(-4) = 2(-5) - 2$$
$$-12 = -10 - 2$$
$$-12 = -12 \quad \text{True.}$$ ◆

Writing the equation of a line when you are given the slope and the y intercept is just a matter of substituting into the slope–intercept form of an equation, $y = mx + b$. Writing the equation from two points is easiest to do by first calculating the slope and then using $y - y_1 = m(x - x_1)$ and simplifying the resulting equation. To write the equation given the slope and one point, (x, y), you must first substitute the values x, y, and m into $y = mx + b$ to solve for b. Then use $y = mx + b$ again and substitute the original slope value, m, and the y-intercept value that you found. In equations that have fractions, you may wish to multiply through by the least common denominator to clear the fractions and therefore to be able to work with an equation that has all integral coefficients.

17.6 ◆ Practice Problems

A. Write the equation of a line that has the given slope and y intercept.

  1. $m = 3$, $b = -2$

  2. $m = -4$, $b = +1$

  3. $m = 1/2$, $b = 3$

  4. $m = 1/3$, $b = 2/3$

  5. slope $= 0$, y intercept $= 2$

  6. $m = 2$, $b = -3$

  7. $m = -5$, $b = 6$

  8. slope $= 0$, y intercept $= -4$

  9. slope $= -3/4$, y intercept $= 2$

  10. slope $= -1$, y intercept $= -1$

B. Write the equation of the line that passes through the given points.

11. (0,3) and (2,6)

12. (−3,4) and (3,−5)

13. (−5,2) and (−2,8)

14. (3,0) and (−1,4)

15. (3,5) and (0,0)

16. (4,2) and (4,−3)

17. (6,1) and (6,−2)

18. (−2,−3) and (0,0)

19. (−2,3) and (3,3)

20. (5,−2) and (−1,−2)

C. Write the equation of the line that passes through the given point and has the given slope. Write your final equation with integral coefficients.

21. (6,−1); $m = 3$

22. (−1,−6); $m = 5$

23. (−2,−7); $m = −2$

24. (5,−3); $m = −4$

25. (−3,0); $m = 1/2$

26. (−2,−4); $m = 1/3$

27. (5,−2); $m = −1/2$

28. (2,0); $m = −1/4$

29. (2,−1); $m = 3/4$

30. (3,−2); $m = −2/5$

◆ 17.7   Graphing Inequalities

To find the graphical solution for a first-degree inequality, you follow all of the same steps that you used to graph a line and then add to that graph the inequality portion of the solution. The solution of an inequality may or may not include its boundary line, but it includes all the points on one side of that line.

Remember that when you were working with a number line (Section 17.5), you could graph individual points or whole sections of the number line. When you graphed an inequality, the end point of the graph was either an open circle or a closed (filled-in) circle, depending on whether the inequality included $>$ or $<$ (an open circle), or $\geq$ or $\leq$ (a closed circle). Study these simple graphs to refresh your memory.

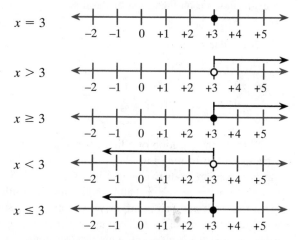

When you want to graph the solution set for an inequality that involves two variables, you need to take the steps used in the following sample problem:

Graph the solution set for $x + y > 3$.

First you need to graph the boundary line. You will graph the equation $x + y = 3$. Isolate y, obtaining $y = −x + 3$. Then the table of values is completed.

x	$y = −x + 3$	y	(x,y)
2	$y = −2 + 3$	1	(2,1)
0	$y = −0 + 3$	3	(0,3)
−2	$y = −(−2) + 3$	5	(−2,5)

Next, graph the three points and connect them with a *dashed line* because the inequality is $x + y > 3$, not $x + y \geq 3$. The dashed line shows that the boundary line is not included in the solution set. (See Figure 29.)

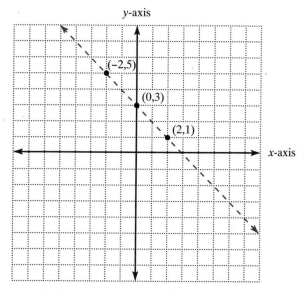

FIGURE 29

Finally, the only question is which side of the line to shade. Looking at Figure 30, pick a point that is obviously on one side of the boundary line. Don't get too close to the line. (0,0) is a convenient choice.

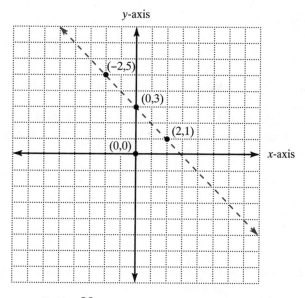

FIGURE 30

Substitute that point into the inequality and see if it makes a true statement. If the resulting inequality is true, shade the side of the boundary line that includes (0,0). If the inequality is not true, shade the other side of the boundary line.

$$x + y > 3$$
$$0 + 0 > 3$$
$$0 > 3 \quad \text{False.}$$

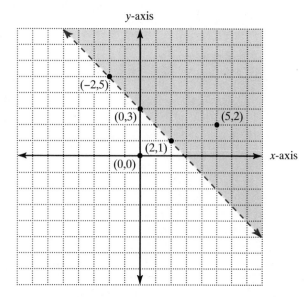

F I G U R E **31**

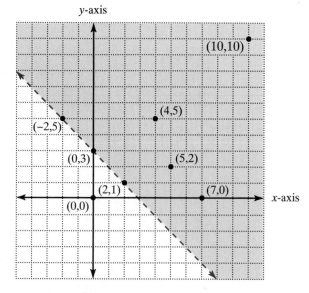

F I G U R E **32**

Therefore, $(0,0)$ is not part of the solution set, and that side cannot be shaded. If you try the point $(5,2)$, which is clearly on the other side of the boundary line, the resulting inequality is true.

$$x + y > 3$$

$$5 + 2 > 3$$

$$7 > 3 \quad \text{True.}$$

Therefore, you shade the side that includes $(5,2)$, and the solution set looks like the graph in Figure 31.

You could have picked any point on the "top side" of the line and it would have satisfied the inequality. The point $(5,2)$ was chosen at random. Pick $(7,0)$, $(4,5)$, and $(10,10)$, and you will see that they also satisfy the inequality.

$x + y > 3$	$x + y > 3$	$x + y > 3$
$7 + 0 > 3$	$4 + 5 > 3$	$10 + 10 > 3$
$7 > 3 \quad \text{True.}$	$9 > 3 \quad \text{True.}$	$20 > 3 \quad \text{True.}$

Figure 32 shows the location of all four points, and you can see that all four are on the shaded part of the graph.

The reason for shading one whole side of the line is to show that any point in that part of the coordinate system is a solution for the inequality. It also shows that no point in the unshaded part is in the solution set for $x + y > 3$.

Study the following examples and then try the practice problems.

E X A M P L E 1   Graph the solution set for $x + 2y \le 4$.

Solution   First isolate y by writing the inequality as an equation in slope–intercept form.

$$x + 2y = 4$$

$$\underline{-x \qquad\qquad -x}$$

$$2y = 4 - x$$

$$2y = -x + 4$$

$$\frac{2y}{2} = \frac{-1x}{2} + \frac{4}{2}$$

$$y = -\frac{1}{2}x + 2$$

Therefore, the equation of the boundary line is $y = -\dfrac{1}{2}x + 2$.

Graph the boundary line by using the slope and y intercept. The boundary line will be solid because the points that are on the line are part of the solution set for $x + 2y \le 4$. (See Figure 33.)

$$b = 2$$

$$m = -1/2 \text{ (down 1/right 2)}$$

Pick a point—say (0,0)—and see if it is part of the solution set for the inequality.

$$x + 2y \le 4$$

$$0 + 2(0) \le 4$$

$$0 + 0 \le 4$$

$$0 \le 4 \quad \text{True.}$$

Therefore, shade the side of the boundary line that includes (0,0). The graph is shown in Figure 34.

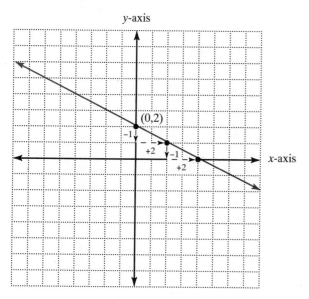

FIGURE 33

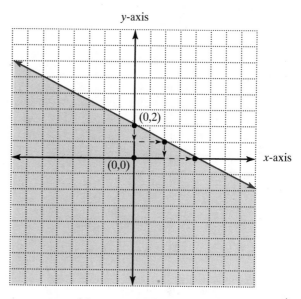

FIGURE 34

EXAMPLE **2**   Graph the solution set for the inequality $2x - y < 3$.

Solution   First isolate the y in the inequality, and then write the equation for the boundary line in $y = mx + b$ form.

$$2x - y < 3$$

$$\underline{-2x \qquad\qquad -2x}$$

$$-y < 3 - 2x$$

$$-y < -2x + 3$$

$$\frac{-1y}{-1} > \frac{-2x}{-1} + \frac{3}{-1} \qquad \text{Remember to reverse the direction of the inequality when you multiply or divide by a negative value.}$$

$$y > 2x - 3$$

Therefore, the equation for the boundary line is $y = 2x - 3$. The boundary line will be dashed because the points on the line are not part of the solution set (the inequality is $2x - y < 3$, not $2x - y \le 3$).

The slope is $+2$ ($+2/+1$, up 2/right 1).

The y-intercept is -3.

The boundary line for $2x - y < 3$ is graphed in Figure 35.

To determine the shading, try $(0,0)$ in the inequality to see if $(0,0)$ is part of the solution set.

$$2x - y < 3$$

$$2(0) - 0 < 3$$

$$0 - 0 < 3$$

$$0 < 3 \quad \text{True.}$$

Shade the portion of the graph that includes $(0,0)$. (See Figure 36.)

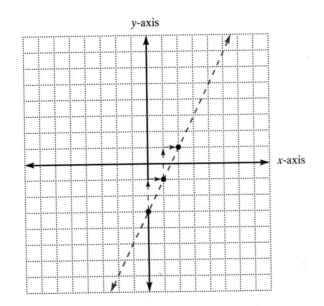

FIGURE **35**

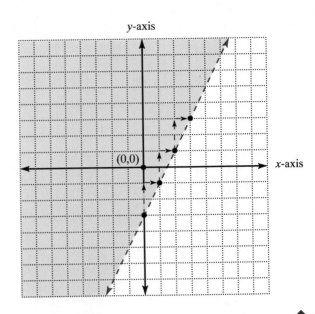

FIGURE **36**

EXAMPLE 3   Graph the solution set for the inequality $y < x - 4$.

Solution   Write the boundary line in $y = mx + b$ form.

$$y = x - 4$$

Find three points that lie on that line.

x	$y = x - 4$	y	(x,y)
2	$y = 2 - 4$	-2	$(2,-2)$
0	$y = 0 - 4$	-4	$(0,-4)$
-2	$y = -2 - 4$	-6	$(-2,-6)$

The boundary line will be dashed. Because the inequality is $y < x - 4$, the line is not included in the solution set.

Check the point $(0,0)$ to see which side of the line should be shaded.

$$y < x - 4$$

$$0 < 0 - 4$$

$$0 < -4 \quad \text{False.}$$

Because $(0,0)$ does not satisfy the inequality, shade the other side of the boundary line. (See Figure 37.)

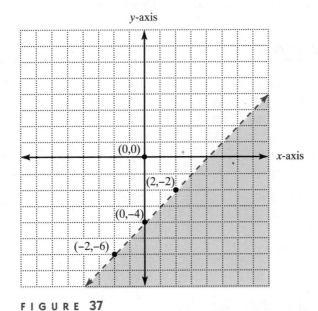

FIGURE 37

EXAMPLE 4   Graph the solution set for $x \leq 5$.

Solution   There is no y value, so the boundary line is made up of all points whose first coordinate, the x coordinate, is 5. Therefore, $x = 5$ is the equation for the boundary line. The line will be solid because the inequality states that x is less than *or equal to* 5. When we check the point $(0,0)$, we find that the inequality is true because $0 \leq 5$. The graph is shown in Figure 38.

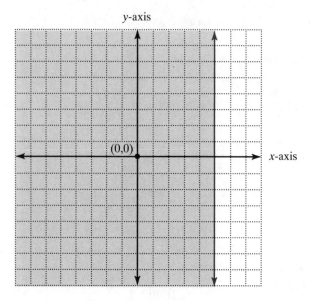

F I G U R E 38   ◆

As you can see from these examples, there are only two new things you need to be careful about when graphing an inequality. Is the boundary line included (solid) or not included (dashed)? And which side of the boundary line shows the solution set? To answer this question, you pick a point that is obviously on one side of the line, and if that point satisfies the inequality, you shade that side of the boundary line. If the point does not satisfy the inequality, you shade the other side of the boundary line.

17.7 ◆ Practice Problems

Graph the solution set for each of the following first-degree inequalities. Make sure that your graphs are neatly done and that the boundary line is clearly marked.

1. $x + y < 5$ **2.** $x + y \geq 4$

3. $x - y \geq 3$ **4.** $x - y < 4$

5. $2y > x$ **6.** $3x \leq y$

7. $2x + y > 3$ **8.** $3x + y > 1$

9. $x \leq -3y$ **10.** $-2y > x$

11. $y - x \leq 4$ **12.** $y - 2x \leq 3$

13. $y > -2$ **14.** $2y + x > 6$

15. $2y - x < 4$ **16.** $x < -2$

17. $3x - 2y \geq 12$ **18.** $2x + 3y \geq 6$

19. $x \leq 3$ **20.** $y \geq -3$

21. $-2y < x + 4$ **22.** $-y < 2x + 3$

23. $x \geq y - 5$ **24.** $y - 4 \leq 3x$

17.8 Chapter Review

In this chapter, you have learned about the Cartesian coordinate system, more commonly called a graph. There will be many other times when you will work with graphs in math, science, and business classes. It is crucial that you have a clear understanding of what it means to graph the solution of a linear equation. Most often, you will be asked just to "graph the line," but what you are really doing is drawing a picture of all the pairs of numbers that make your equation or inequality true.

Now that you have completed this chapter, you should be able to:

1. Discuss the following terms and concepts: graph, Cartesian coordinate system, ordered pair, quadrant, origin, axes, table of values, linear equation, slope–intercept form, slope, y intercept, rise/run, slope = 0, undefined slope, parallel, undefined, inequality, boundary line, dashed or solid line, and shading.

2. Plot ordered pairs of numbers on a graph.

3. Isolate y in order to write any equation or inequality in slope–intercept form.

4. Graph the solution set of a linear equation by following these steps:

   a. Write the equation in $y = mx + b$ form.

   b. Pick three x values and solve for the corresponding y values.

   c. Graph the three points and the straight line that passes through them.

5. Graph the solution of a linear equation using the y intercept and the slope by following these steps:

   a. Write the equation in $y = mx + b$ form.

   b. Identify the slope and y intercept.

   c. Begin with the y intercept and use the ratio rise/run to find two more points on the line.

   d. Check that the three points satisfy the *original* equation.

6. Find the slope from a graph by using rise/run.

7. Find the slope of a line that passes through two given points by using the formula

$$m = \frac{y_2 - y_1}{x_2 - x_1}$$

8. Write the equation of a line given the slope and y intercept by substituting into $y = mx + b$.

9. Write the equation of a line given two points by first calculating the slope and then substituting into $y - y_1 = m(x - x_1)$.

10. Write the equation of a line given one point and the slope by substituting into $y = mx + b$ and solving for b. Then substitute again into $y = mx + b$, using the given slope and the newly found b value.

11. Graph the solution of a first-degree inequality by following these steps:

    a. Isolate y in the inequality.

    b. Write the equation of the boundary line by using $y = mx + b$ and determining whether the line is solid (included, $\geq$ or $\leq$) or dashed (not included, $>$ or $<$).

    c. Graph the boundary line.

    d. Pick a point on one side of the boundary line and substitute its coordinates into the inequality to determine which side of the boundary line is to be shaded.

Review Problems

A. Give the location (which quadrant or which axis) of each of the following points.

1. $(-3,-1)$ **2.** $(5,-2)$ **3.** $(0,-4)$

4. $(-2,4)$ **5.** $(3,-8)$ **6.** $(7,0)$

7. $(2,6)$ **8.** $(-7,-2)$ **9.** $(-3,0)$

10. $(0,-11)$ **11.** $(-1,8)$ **12.** $(5,4)$

B. In Problems 13 through 22:

   a. Change the equation into $y = mx + b$ form.

   b. Identify the slope and y intercept.

   c. Find three ordered pairs of numbers that satisfy the equation.

   d. Graph the line that is the solution for each equation. Be sure to label the three points that you used.

13. $2y = -6x - 4$ **14.** $5 - y = 3x$

15. $2x - y = 5$ **16.** $y + 4 = -x$

17. $x - y = 3$ **18.** $3y + 9 = 6x$

19. $6 = x$ **20.** $y = -1$

21. $2y + 3x = 6$ **22.** $3y + 2x = 6$

C. Find the slope of the line that passes through each pair of points.

23. $(4,5)$ and $(2,3)$ **24.** $(5,4)$ and $(2,3)$

25. $(1,4)$ and $(3,5)$ **26.** $(3,2)$ and $(6,7)$

27. $(4,5)$ and $(4,8)$ **28.** $(6,3)$ and $(6,7)$

29. $(-1,-3)$ and $(5,2)$ **30.** $(-2,4)$ and $(5,-3)$

D. Write the equation of a line given the following information. Write the final equations with integral coefficients.

31. slope $= -2$, y intercept $= +3$

32. passes through $(2,6)$ and $(-3,2)$

33. passes through $(2,5)$; $m = -3$

34. passes through $(-3,4)$; $m = -2$

35. passes through $(4,5)$ and $(0,-3)$

36. $m = -4$, $b = -2$

37. passes through $(1,3)$ and $(2,0)$

38. passes through $(1,4)$ and $(-2,-5)$

39. $m = 2/3$, $b = -1$

40. slope $= 3$, y intercept $= +2$

41. passes through $(-1,-4)$; $m = -1/2$

42. passes through $(5,6)$; $m = 1$

43. $m = 0$, $b = -2$

44. $m = -1/4$, $b = 3$

45. passes through $(7,0)$; $m = 3$

46. passes through $(-2,-3)$; $m = 1/4$

47. slope $= -1/2$, y intercept $= 3/4$

48. slope $= 0$, y intercept $= -2/3$

49. passes through $(5,1)$ and $(-2,-3)$

50. passes through $(3,-1)$ and $(-2,2)$

51. passes through $(3,5)$; $m = 2/3$

52. passes through $(6,2)$; $m = 1/3$

53. passes through $(-2,3)$ and $(0,-1)$

54. passes through $(-1,2)$ and $(3,0)$

55. passes through $(-4,1)$; $m = -2/3$

56. passes through $(-3,2)$; $m = -1/2$

57. $m = -3$, $b = 1/2$

58. $m = -1$, $b = 3/4$

59. passes through $(3,-2)$ and $(-4,3)$

60. passes through $(2,2)$ and $(-5,-5)$

E. Graph the solution set for each of the following inequalities.

61. $x + y > 4$ **62.** $x + y \leq 3$

63. $2x - y \leq 3$ **64.** $-3x + y > 4$

65. $x - 2y < 4$ **66.** $x + 2y < 6$

67. $2x + 3y \geq 3$ **68.** $3y - 2x \leq 6$

69. $-2y + x < -4$ **70.** $2x - y < 5$

Chapter Test

A. Complete the following sentences.

1. Points of the form $(0,y)$ lie on the _____ axis.

2. The point $(-3,2)$ is in Quadrant _____ .

3. The point $(0,0)$ is called the _____ .

4. In the equation $2y + x = 5$, _____ is the slope.

5. In the equation $-3y + 2x = 6$, _____ is the y intercept.

6. The equation $2x - y = 4$ would be _____ in slope–intercept form.

7. The slope of the line that passes through the points $(-2,5)$ and $(3,-7)$ is _____ .

8. The slope of the line that passes through the points $(-5,7)$ and $(-5,-4)$ is _____ .

B. Write the equation for a line given the following information.

9. slope $= -2$, y intercept $= +4$

10. slope $= 2/3$, y intercept $= -3$

11. passes through $(-2,4)$ and $(3,-1)$

12. passes through $(4,0)$ and $(4,-2)$

13. passes through $(-3,-2)$ and $(4,-2)$

14. passes through $(2,-2)$ and has a slope of -3

C. Graph the solution set for each of the following equations and inequalities.

15. $2x + y = 5$

16. $-3y = x + 6$

17. $x - y = 4$

18. $-2y + x \leq 4$

19. $2y > 6$

20. $x \geq y - 1$

Thought-Provoking Problems

1. Describe the relationship between the equation of a line and the graph of that line.

2. Write four facts that you know about slope.

3. How does "slope $= 0$" differ from "undefined slope"?

4. Why are coordinates of the form (a,b) called ordered pairs? Use a numerical example to show why ordering is important.

5. On a single set of coordinate axes, draw the graphs for these four equations.

$$y = 2x$$
$$y = 2x + 1$$
$$y = 2x - 3$$
$$y = 2x + 3$$

What do you notice about all four equations?

What does this composite graph look like?

What conclusion can you draw?

6. On a single set of coordinate axes, draw the graphs for these four lines.

$$y = 2$$
$$y = -3x + 2$$
$$y = 2x + 2$$
$$y = 4x + 2$$

What do you notice about all four equations?

What do you notice about all four graphs?

What conclusion can you draw?

7. a. On a single set of coordinate axes, draw the graph for $y = \dfrac{2}{3}x + 1$ and the graph for

$$y = -\dfrac{3}{2}x + 3.$$

What does this graph look like?

How might you describe the relationship between the lines?

b. On a single set of coordinate axes, draw the graph for $y = -2x - 1$ and the graph for

$$y = \dfrac{1}{2}x + 2.$$

What does this graph look like?

c. What do you notice about the slopes of the lines in Part a and in Part b?

What is the product of the slopes in Part a?

What is the product of the slopes in Part b?

Can you draw any conclusion from these graphs and from the slopes of the lines?

8. In what ways is graphing the solution set for an equation the same as graphing the solution set for an inequality? In what ways are the two processes different?

9. List three different sets of information from which you can write the equation of a line.

10. Graph the solution set for $x + y > 4$ and the solution set for $2x - y < 3$ on the same Cartesian coordinate system. Give three ordered pairs of values that satisfy *both* inequalities.

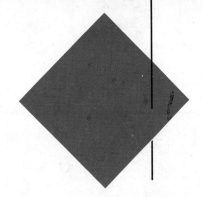

Cumulative Review, Chapters 1–17

A. Answer each of the following questions. Reduce all fraction answers to lowest terms.
In Problems 1 through 5, add.

1. $1\ 3/8 + 2\ 5/9$

2. $3a + 4a + a + 11a$

3. $(7y^2 - 2y + 3) + (8y - 2y^2 + 5)$

4. $(-69) + (14) + (-8)$

5. $7.6 + 0.09 + 44$

B. In Problems 6 through 10, subtract.

6. $9 - [6 - 3(3 - 5)]$

7. $0 - (-12) - (3)$

8. $(5a - 2ab + 3b) - (12a + 5ab + 6b)$

9. $11xyz - (-3xyz)$

10. $\dfrac{5}{9} - \dfrac{1}{4}$

C. In Problems 11 through 15, multiply.

11. Find 3/8 of 19.

12. $(-6)(-3)(2)(-1)$

13. $3x^2y(2x^2y^2z)$

14. 0.654×100

15. $(a + 2b)^2$

D. In Problems 16 through 20, divide.

16. $\dfrac{4a^2bc^3}{-2abc}$

17. $4\ 7/8 \div 13$

18. $0.06\overline{)44.04}$

19. $\dfrac{15m^2n^2 - 10m^2n + 5mn^2}{-5mn}$

20. $x + 2\overline{)x^2 - 4x - 12}$

E. In Problems 21 through 30, answer each question.

21. What is the slope of a line that passes through $(-1,4)$ and $(-5,-6)$?

22. Change 21% to a decimal.

23. Identify the slope and y-intercept of $7y - 3 = 4x$.

24. Reduce $\dfrac{48}{90}$ to lowest terms.

25. Factor completely: $4x^2 - 20xy + 25y^2$.

26. Perform the indicated operations and express the answer in scientific notation: $\dfrac{0.006 \times 51.6}{0.008}$

27. Evaluate $-4a^2 + 2a - 3$ when $a = -2$.

F. In Problems 28 through 30, use Figure 1.

FIGURE 1

28. During what year did Company B have its greatest increase in sales?

29. How much more sales did Company A have than Company B in 1990?

30. What was the average amount of sales for Company B during the 4-year period?

G. In Problems 31 through 41, solve each equation or inequality. If necessary, round answers to the nearest tenth.

31. $4x - 3 = 15$

32. $1 \le 3x + 4 \le 7$

33. 15% of what number is 16?

34. $x^2 + 6x + 8 = 0$

35. $5(y + 2) = 3y - 4$

36. $\dfrac{8}{N} = \dfrac{45}{10}$

37. $Y^2 = 9$

38. 15 is what percent of 12?

39. $3(2x - 1) \ge 9$

40. $5T - 5(6 - T) = -12$

41. $7Y - \dfrac{2}{3}(Y - 5) = 2Y - 1$

42. Solve $A = 1/2h\,(b_1 + b_2)$ for h if $b_1 = 9$ feet, $b_2 = 3$ feet, and $A = 42$ square feet.

H. Solve each of the following problems. You may find it helpful to write an equation or a proportion or to use a formula. When necessary, round answers to the nearest tenth.

43. Six less than twice a number is equal to the product of the number and eight. Find the number.

44. Hani has a total of $2.20 in his pocket, all nickels and quarters. If he has twelve coins, how many are quarters?

45. The daily rainfall during a certain week in Columbus, Ohio, was 0.7 inch on Monday, 0.2 inch on Tuesday, 1.2 inches on Wednesday, 0.9 inch on Thursday, and 1.3 inches on Saturday. On Friday and Sunday, it did not rain. What was the average daily rainfall for the seven days of that week?

46. Using the information from Problem 45, give the median and range for rainfall during the seven-day period.

47. It took Greg two hours to drive from his home to campus during the morning rush hour. It took him only one hour to go the same distance in the afternoon. If he averaged twenty miles per hour faster on the afternoon trip, how fast did he drive in the morning?

48. Forty-eight percent of the students attending this college are male. If there are 7,872 male students, what is the total enrollment for this term?

49. If 8 more than 5 times an integer is less than 38, what is the maximum value of the integer?

50. The perimeter of a rectangular field is 224 meters. The length is 8 meters less than 3 times the width. Find the dimensions of the field.

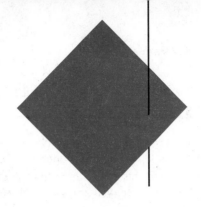

Appendix A: Overcoming Math Anxiety

Math anxiety! You probably know it as that awful feeling you get in the pit of your stomach when a teacher announces a math test, or when fellow students start talking about all the math they know and you can't remember a thing! Rest assured, you are not alone. More than one-fourth of all students suffer from math anxiety. And that amount is higher in some classrooms than in others.

Is this disease terminal? How did you get it? What can you do about it? If you didn't have it, would you do a lot better in your math courses? All of these are the kinds of questions and comments that might run through your mind when you think about math anxiety. Let's take a look at this horrible phenomenon and see if perhaps there isn't something that can be done about it.

First, you need to understand that math anxiety is a *learned response*. You were not born with it. You developed it as a result of something that happened in your life. And *because it is a learned response, you can unlearn it*. Second, math anxiety is just another type of stress. In these hectic times, life can be full of stress. You probably don't need another kind to add to your list. There are several things that you can and should do to overcome it.

You should also know that a little stress is good for you. When it is used in the right way, it can help you to perform at your very best. It can keep you on your toes during a test so that you don't become too complacent or too relaxed and allow yourself to make silly mistakes. It is a problem only when it becomes so powerful that it overtakes your whole person and makes you unable to function normally.

In order to unlearn math anxiety, it is helpful to try to figure out when in your own life the problem originated. In fact, there are "math inventories" and "math autobiographies" that you can complete to help you pinpoint when you first developed the problem and what situation originally caused you to experience math anxiety. Most often, students find out that it was not something that they themselves did or didn't do, but rather someone else who caused them to have a bad experience with math. Perhaps it was a teacher who embarrassed you or a parent who pressured you to bring home "good math grades." Maybe a particular math topic has always caused you trouble, and now you are sure you will never be able to learn it—or any other math. Ask your instructor or the counseling center on campus for information about math inventories and math autobiographies. Once you identify where the problem started, you can begin to break its hold on you.

In the meantime, let's look at some other things that might help. When you find yourself going in to take a test—an anxiety producer for sure—there are some things you can do to help you deal with your anxiety. Study the following suggestions and see which ones you might be able to use.

1. Don't study right up to the time of the test. Do your studying far enough in advance so that the night before the test, all you are doing is reviewing what you already know. This is not the time to be learning new material.

2. Don't box your time in. Try to plan not to do anything major just before or just after a test. Otherwise, your mind will be on those things and not on the test.

3. Just before the test, don't listen to other students cram. If they remember something you have forgotten, you will only feel worse and get more anxious.

4. Go into the testing room a few minutes early, and practice some relaxation techniques to get yourself prepared physically for the task ahead. Take a few deep breaths; think of something that you find very peaceful or soothing; concentrate on individual limbs of your body and try to relax each one. A specific relaxation technique that you might find helpful is outlined later in this section.

5. Once the teacher has asked for your attention, listen carefully to the last-minute directions that are given. Be sure to read *all* the directions on the test before you begin to work any problems.

6. Once you have the test in your hands, put a smiley face or a phrase like "I can do this!" on the top of the test to proclaim your ability to do well on the test. Positive thinking never hurts.

7. When you get the test, turn it over and begin to write down any formulas or facts that you think you may need but are afraid of forgetting. Jotting down just a few will make you feel very confident.

8. Preview the whole test and write down anything else that you want to remember.

9. Work through the problems carefully, doing the easiest ones first. Concentrate on being careful. You don't want to rush through the easy problems and make careless mistakes.

10. If you come to a problem that you know you cannot do, read through it carefully and go on. Sometimes something will jog your memory and you will be able to return to this problem and complete it.

11. Finally, allow enough time to go back over the whole test when you have finished. There is no prize for turning in your test first. Use all the time you have been given to be sure that you have not forgotten to label answers or to complete problems.

Here is a simple relaxation technique that you can employ once you are in the classroom, in the last few minutes just before any test.

1. Take a few deep breaths. Breathe in through your mouth and out through your nose, counting to three during each phase so that you breathe in and out slowly.

2. Put your feet flat on the floor in front of you and grab the bottom of the chair with your hands. Press down against the floor and pull up on the chair at the same time. Concentrate on the way your muscles feel when they are tense, and then let them slowly relax. Do this rather quickly the first time, and then repeat the process more slowly for a second, third, and fourth time. When you can feel your muscles tense, you will be more aware of how they feel when they are relaxed. Concentrate on relaxing the muscles in the back of your neck and in your shoulders. They won't be used much in taking the test, but tension tends to build there. You can repeat this process during the test when you feel yourself becoming tense. No one around you will even be aware that you are doing it.

Of course, the best way to overcome a serious problem with math anxiety is to study carefully and steadily all along. Keeping up with the course work and not cramming at the last minute are the strongest defense against anxiety. Read through the suggestions in the next section on math study skills. Apply the ideas offered there. Believe in yourself and in your ability to be successful. The best way to develop that self-confidence is to study your math faithfully. A very learned college math professor once said that you should never have to study for a math test. He said that doing math is like riding a bike. Once you have practiced it, you can do it over and over again. The key is practice.

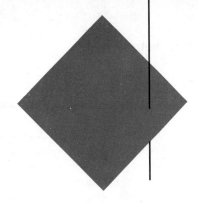

Appendix B: How to Study Math

One of the biggest changes you will notice between high school work and college work is in mathematics. In high school and elementary school, you did most of your drill and practice work in class and turned in homework to be corrected and graded. In college courses, you do the practice work outside of class and often have to correct those assignments yourself. Much of the homework won't even be collected. Often students think that this arrangement enables them to skip the work, but that kind of thinking only gets them into serious trouble!

Math assignments in college are not given as "busy work." They are given because the instructor knows that you can learn math only by practicing it. The drill and practice are necessary for you to master the subject. This means that you must find time or make time to do such work, and it should be done as soon after the lecture as possible. You must discipline yourself to do some homework following each class and not to "save" it all to do the night before the exam. If you fall behind in this work, it will be very hard to catch up.

Taking enough notes during class and study times will help you with the homework assignments. The key is to take down enough information—but not too much. Notes should be taken during math lectures, but they should be kept to a minimum. You should record the main ideas on each topic, but don't try to copy down every example. Rather, listen to the explanation of the problem and try to follow the instructor's reasoning. If the lectures closely follow the presentation in the textbook, then reading the material in advance will be helpful. You can then limit note-taking to those lecture topics or approaches that vary from the text.

You should ask questions during a lecture if you can't follow the reasoning being presented. Get immediate clarification so that difficulties do not occur later, when you are on your own doing the homework. Of course, it is only fair that you do your part first. Don't interrupt or ask questions if you haven't been paying attention to the presentation or if you haven't kept up with previously assigned work.

For maximum benefit, study your notes and try some problems as soon after the lecture as you can—within a few hours if possible. If you wait too long, you will have forgotten much of the material presented. Don't attempt to work problems without reading the text explanation and your class notes. Doing problems without understanding the material is a waste of time, and it often results in your learning incorrect procedures that are hard to correct later.

Talk to other students in your class and try to organize a study group. You can meet regularly to discuss problems, go over notes, and study for tests. If you are absent, someone in the group can be responsible for taking notes for you. In study groups or "study buddy" sessions, if you don't know how to do a problem or don't understand a concept, someone else may be able to explain it to you. You will be able to return the favor on some other topic that you understand but that someone else doesn't.

Read your math textbook. Most of what is written in it is very significant. It gives you background information and history that are enlightening, shows relationships between principles, and demonstrates applications of the concepts. Definitions and rules are very important; they are usually displayed clearly in your text. Use the explanations that accompany examples and sample problems to be sure you understand the steps in the solution. Use the table of contents or index to refer to previous topics that are essential to the current material. Read the introduction to each chapter in order to survey the objectives and the topics to be presented, and use the chapter review and review exercises to see if you have accomplished those objectives.

The best way to prepare for a math test is to do your work daily, to clear up small difficulties as they arise, and constantly to review earlier material that still applies. In this way, all you will need to do is refresh yourself before the exam. You have already learned the material and have been practicing it. If you had some difficulty when doing parts of the assignment, work similar problems to be sure you understand them now. The anxiety of test-taking is greatly reduced when you have only to review material the night before the exam, not learn new material.

If the test is to include definitions, formulas, or principles that have to be memorized, write each on a separate index card and study them for several days before the test. If you carry the cards around with you, you can study them any time in a spare moment.

Learn from your mistakes. When you get a test back, even if you have done well on it, rework the problems that you missed and find out what you did wrong. If you follow this suggestion, you probably won't make that same mistake again.

You shouldn't rely too much on memory in mathematics. It is more important to understand why certain steps are taken than to memorize the steps. Although we usually memorize the multiplication facts, it is more important to understand that multiplication is just repeated addition. If you forget what 6×7 equals but remember that $6 \times 5 = 30$, you can build to the product that you need.

$$
\begin{array}{rl}
6 \times 5 = 30 & \\
\underline{+\ 6} & \\
36 & \text{which is } 6 \times 6 \\
\underline{+\ 6} & \\
42 & \text{which must be } 6 \times 7
\end{array}
$$

You can often fill in memory lapses if the understanding of the concept is there.

If you find that you are experiencing difficulties, talk to your instructor and find out where you can get additional help. Do this as soon as you sense trouble. Don't wait! The instructor may be able to work with you during office hours, or your college may have support services such as a learning center or tutors. If math anxiety or test anxiety interferes with your efforts, then talk to your instructor and seek help for those problems through the counseling center or student services department. Be sure also to read Appendix A on overcoming math anxiety.

If you follow these suggestions, and if you have a good background in the prerequisite skills needed for the course, you should succeed with any math course you attempt. The fact that you are reading this material indicates that you are willing to put a little extra effort into your studies and that you want to do well in this course. Those two factors alone should help you achieve success.

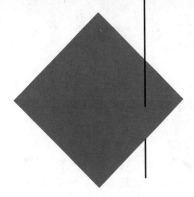

Appendix C: Skills Checks

There are four parts to Appendix C. Turn to the one that is appropriate to your needs, and use the Skills Checks and materials as directed.

1. The section entitled "Basic Facts" provides a brief test of your knowledge of addition, subtraction, multiplication, and division facts. Activities are given to help you overcome any diagnosed weakness(es).

2. "Addition Facts" contains a chart and activities to help you learn the addition and subtraction facts.

3. "Multiplication Facts" contains a chart and activities to use in memorizing the multiplication and division facts.

4. "Fractions and Decimals" allows you to test your basic abilities in working with these kinds of numbers. References to the textbook sections where these topics are covered enable you to review any topic that you have trouble with.

Basic Facts

Much of the math that you will do in this course is an extension or a review of work that began in elementary school and continued through later grades into high school. This is one of the few subjects that has such a long, on-going development in your education and builds the new material on the material you have previously learned and used. If you have had difficulty with mathematics, it may be that you never learned early concepts or principles that were essential for success with later topics.

If that is the case, it is important that you stop right now, identify those weaknesses, and take steps to correct them. If you master the basics now, you will be able to move on and succeed with future math courses that are necessary to your educational or career goals.

Many difficulties encountered with arithmetic can be attributed to weakness in the most fundamental step: operations with whole numbers. Many people never learned (memorized) the addition or multiplication facts. Test yourself right now to see if that might be a problem for you.

Find a quiet place where you can work uninterrupted for 3 minutes, and have a watch or clock available. Then do the Arithmetic Facts Check that follows. Allow yourself only 3 minutes. Go to the test now, cover up the answers, and write your answers on a sheet of paper. Stop working at the end of 3 minutes, whether you are finished or not, and check your answers.

Arithmetic Facts Check

1. $2 + 3$	**2.** $7 - 4$	**3.** 6×9	**4.** $12 \div 4$
5. $45 \div 5$	**6.** 3×0	**7.** 7×3	**8.** $6 + 2$
9. $27 \div 3$	**10.** $8 + 7$	**11.** 4×6	**12.** $14 - 8$
13. $2 + 0$	**14.** 5×1	**15.** $72 \div 8$	**16.** $9 + 9$
17. 4×9	**18.** $32 \div 8$	**19.** $13 - 5$	**20.** 6×3
21. $49 \div 7$	**22.** $13 - 8$	**23.** $21 \div 3$	**24.** 0×5
25. $6 + 7$	**26.** $42 \div 7$	**27.** $12 - 3$	**28.** 2×7
29. $9 + 1$	**30.** $17 - 9$	**31.** 5×6	**32.** 8×3
33. $24 \div 6$	**34.** $24 \div 8$	**35.** $12 - 7$	**36.** $4 - 0$
37. $0 \div 4$	**38.** $48 \div 8$	**39.** $3 + 8$	**40.** $15 - 7$

Answers

1. 5	**2.** 3	**3.** 54	**4.** 3	**5.** 9	**6.** 0
7. 21	**8.** 8	**9.** 9	**10.** 15	**11.** 24	**12.** 6
13. 2	**14.** 5	**15.** 9	**16.** 18	**17.** 36	**18.** 4
19. 8	**20.** 18	**21.** 7	**22.** 5	**23.** 7	**24.** 0
25. 13	**26.** 6	**27.** 9	**28.** 14	**29.** 10	**30.** 8
31. 30	**32.** 24	**33.** 4	**34.** 3	**35.** 5	**36.** 4
37. 0	**38.** 6	**39.** 11	**40.** 8		

After checking your answers, answer the following questions.

1. How many problems didn't you have time to do? _____

2. How many errors did you make in addition? _____

3. How many errors did you make in subtraction? _____

4. How many errors did you make in multiplication? _____

5. How many errors did you make in division? _____

If you had fewer than five problems not finished, and had very few errors, then you should be all right in this course. You know your basic facts, and the practice you get doing homework will help you to increase your speed.

If you had more than five problems not completed, then you need to study all the facts so that you really memorize them and don't have to calculate each time, as must be happening now.

If you missed more than three problems in addition *and* subtraction, then you need to study the addition facts that follow, using the suggestions for learning them.

If you missed more than three problems in multiplication *and* division, then you need to study and learn the multiplication facts that follow.

After you complete the practice indicated, retake the Arithmetic Facts Test and see what progress you have made. When you can make fewer than three errors in the three minutes, you will know the facts very well. If you apply the suggested study methods and still show little progress, then discuss the problem with your instructor.

Addition Facts

If your weakness is in adding and subtracting, work with Table 1. To use the fact table to find the answer to an addition problem such as $8 + 5$, follow the 8 row until you are under the 5 column. The number that appears where that row and column intersect is the answer to the problem: $8 + 5 = 13$.

T A B L E 1   Addition Facts

	1	2	3	4	5	6	7	8	9
1	2	3	4	5	6	7	8	9	10
2	3	4	5	6	7	8	9	10	11
3	4	5	6	7	8	9	10	11	12
4	5	6	7	8	9	10	11	12	13
5	6	7	8	9	10	11	12	13	14
6	7	8	9	10	11	12	13	14	15
7	8	9	10	11	12	13	14	15	16
8	9	10	11	12	13	14	15	16	17
9	10	11	12	13	14	15	16	17	18

To use the table to practice subtraction, start at a number inside the chart, and trace back and up to the row and column headings. If you start at 11 and trace back and up to the headings 7 and 4, then you know that $11 - 7 = 4$ and $11 - 4 = 7$.

Besides the facts in the table, remember that zero added to any number does not change the number: $0 + 5 = 5$. Similarly, when zero is subtracted from a number, the number does not change: $8 - 0 = 8$.

Here are some other ways of learning the basic addition and subtraction facts.

1. Make flash cards with the "fact" (such as $5 + 3 = $) on one side and the answer (8) on the other. Writing the facts with a bright-colored marker may help you "see" them in your memory. Drill with the facts, separating the ones you get correct from the ones you miss. Then work with the incorrect ones until you know them.

2. Punch a fact into a calculator, but guess the answer before the answer is displayed. Check your guess against the display. Record the ones you get wrong so that you can work on them.

3. Record an audio tape, saying the fact and the answer. Listen to the tape during spare moments.

4. Write the facts several times, saying each one to yourself as you write it.

5. If you want some more strategies, see your instructor.

Multiplication Facts

If one of your weaknesses is in multiplying or dividing, use Table 2 to work on these problems. To use the fact table to find the answer to a multiplication problem such as 8×5, follow the 8 row until you are under the 5 column. The number that appears where that row and column intersect is the answer to the problem: $8 \times 5 = 40$.

To use the table to practice division, start at a number inside the chart, and trace back and up to the row and column headings. If you start at 28 and trace back and up to the headings 7 and 4, then you know that $28 \div 7 = 4$ and $28 \div 4 = 7$.

T A B L E **2**   Multiplication Facts

	1	2	3	4	5	6	7	8	9
1	1	2	3	4	5	6	7	8	9
2	2	4	6	8	10	12	14	16	18
3	3	6	9	12	15	18	21	24	27
4	4	8	12	16	20	24	28	32	36
5	5	10	15	20	25	30	35	40	45
6	6	12	18	24	30	36	42	48	54
7	7	14	21	28	35	42	49	56	63
8	8	16	24	32	40	48	56	64	72
9	9	18	27	36	45	54	63	72	81

In addition to these facts, you need to learn that zero times any number is equal to zero: $7 \times 0 = 0$. Also, any number divided *into* zero is equal to zero: $0 \div 4 = 0$. But if any number is divided *by* zero, there is no answer; division by zero is said to be *undefined*.

Here are some other ways of learning the basic multiplication and division facts.

1. Make flash cards with the "fact" (such as $7 \times 9 =$) on one side and the answer (63) on the other. Drill with the facts, separating the ones you get correct from the ones you miss. Then work with the incorrect ones until you know them.

2. Punch a fact into a calculator, but guess the answer before the solution is displayed. Check your guess against the display. Record the ones you get wrong so that you can work on them.

3. Record an audio tape, saying the fact and the answer. Listen to the tape during spare moments.

4. Write the facts several times, saying it to yourself as you write it.

5. If you want some more strategies, see your instructor.

Fractions and Decimals

If you want to "diagnose" weaknesses in fraction and/or decimal operations, then try the Skills Checks that follow. These are not timed tests, but you should find a quiet spot to work so that you can concentrate. When you have finished, check your answers and follow the recommendations given.

Fraction Skills Check   Reduce all answers to lowest terms.

1. $\dfrac{3}{8} + \dfrac{5}{6}$ 2. $3\dfrac{11}{12} + 2\dfrac{3}{4}$ 3. $\dfrac{7}{9} - \dfrac{2}{3}$ 4. $5\dfrac{1}{8} - 2\dfrac{3}{4}$

5. $13 - 4\dfrac{5}{8}$ 6. $\dfrac{3}{4} \times \dfrac{8}{15}$ 7. $1\dfrac{3}{5} \times 2\dfrac{1}{2}$ 8. $\dfrac{5}{8} \div \dfrac{4}{5}$

9. $\dfrac{4}{9} \div 8$ 10. $\dfrac{\dfrac{3}{4}}{\dfrac{2}{3}}$

Check your answers with those that follow. If you made an error, go back to the section in the book that is indicated beside the correct answer. Study that material and try some Practice Problems. When you can do those correctly, go on to other sections that you need to study.

Answers

1. $\dfrac{29}{24}$ or $1\dfrac{5}{24}$ 2. $5\dfrac{20}{12} = 6\dfrac{2}{3}$ Sections 4.2 and 4.3

3. $\dfrac{1}{9}$ 4. $2\dfrac{3}{8}$ 5. $8\dfrac{3}{8}$ Sections 4.2 and 4.3

6. $\dfrac{24}{60} = \dfrac{2}{5}$ 7. $\dfrac{40}{10} = 4$ Section 3.4

8. $\dfrac{25}{32}$ 9. $\dfrac{1}{18}$ 10. $\dfrac{9}{8}$ or $1\dfrac{1}{8}$ Section 3.5

If answers are correct but not reduced, then study Section 3.3.

Decimal Skills Check   **1.** $0.6 + 2.14$ **2.** $32.657 + 2.43$ **3.** $12.009 - 3.597$

4. $9 - 2.48$ **5.** 9.46×0.38 **6.** 0.493×1000

7. $3.68 \div 0.04$ **8.** $\dfrac{12}{0.8}$ **9.** $\dfrac{10.65}{6}$

10. Round 5.693 to the nearest hundredth.

Check your answers with those that follow. If you made an error, go back to the section in the book that is indicated beside the correct answer. Study that material and try some Practice Problems. When you can do those correctly, go on to other sections that you need to study.

Answers

1. 2.74   **2.** 34.847   Section 5.3

3. 8.412   **4.** 6.52   Section 5.3

5. 3.5948   **6.** 493   Section 5.4

7. 92   **8.** 15   **9.** 1.775   Section 5.5

10. 5.69   Section 5.2

You may find that you need to go back periodically to these sections to review the facts. That's a good thing to do, because getting a lot of practice is the best way to learn math.

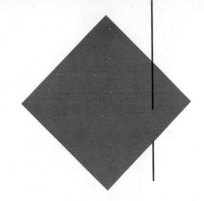

Answer Key

Practice Problems—Chapter 1

Section 1.2

A.
  1. 8,265

  5. 48,002,085

  9. 29,460

  13. 72,480,000

  17. 2,000,045,152

  3. 6,014,385

  7. 8,000,004,007

  11. 46,980,002

  15. 8,006,043

  19. 200,085,206

B.
  21. sixteen thousand, two hundred five

  23. eight thousand, nine hundred fourteen

  25. two thousand, nine

  27. seventy thousand, seventy

  29. one million, two hundred thirty-four thousand, five hundred sixty-seven

  31. fourteen thousand, eight

  33. twelve billion, fourteen million, two hundred thirty-five thousand, six

  35. thirty-five billion, eight

  37. nine million, eighty-one thousand, sixty

  39. seventy-six thousand, eight

C.
  41. 14,300

  45. 18,600

  49. 42,400

  53. 20,000

  57. 37,070,000

  43. 2,000

  47. 33,000

  51. 24,000

  55. 110,000

  59. 200,000

D.
  61. $4,400   **63.** $36,000   **65.** $28,000,000

Section 1.3

A.
  1. 811

  7. 651

  13. 637

  3. 272

  9. 31,672

  15. 1,192

  5. 2,869

  11. 174,064

  17. 11,443

  19. 11,576   **21.** 3,128   **23.** 2,110

  25. 6,644   **27.** 1,203   **29.** 738

  31. 5,281   **33.** 34,510   **35.** 26,489

  37. 1,535   **39.** 30,186

B.
  41. 1,131,391   **43.** 46,009,775

  45. 10,001,277   **47.** 226,579

  49. 16,041,675

Section 1.4

A.
  1. 267

  7. 217

  13. 1,033 R2

  19. 405 R13

  25. 2,064

  3. 672

  9. 609

  15. 2,670 R28

  21. 39 R26

  27. 308 R12

  5. 6,104

  11. 526 R1

  17. 691 R8

  23. 124 R8

B.
  29. 3,615

  35. 3,658

  31. 17,602

  37. 368

  33. 2,003

  39. 1,028

Section 1.5

A. **1.** D   **3.** F   **5.** H   **7.** C

B. **9.** 112 = 112   **11.** 42 = 42

  13. 48 = 48   **15.** 56 = 56

  17. 154 = 154   **19.** 2,970 = 2,970

Section 1.6

  1. $8690

  5. $20,363

  9. $101

  13. 86 miles

  17. 304 miles

  3. 28 acres

  7. $42

  11. 464 miles

  15. $2,697,367

  19. $132

Section 1.7

A. **1. a.** 18   **b.** 19   **c.** 19   **d.** 16

     3. a. 193   **b.** 204   **c.** 204   **d.** 100

B. **5.** mean: 84   median: 85   mode: none   range: 16

     7. Maria and Chris   **9.** Igor

     11. 60°   **13.** 30°

     15. August   **17.** Game 3

     19. 27 more points   **21.** 27 points

     23. 21 points

Section 1.8 Review Problems

A. **1. a.** 9   **b.** 8   **c.** 2   **3. a.** 2   **b.** 1   **c.** 5

B. **5.** 18,004,017   **7.** 9,042

     9. 17,000,009,055   **11.** 846,000

C. **13.** Thirty-two thousand, one hundred eighty-nine

     15. eight hundred ninety-one thousand, four hundred seven

     17. one million, four hundred twenty-four thousand, eight hundred fifty-six

     19. seventy-seven thousand, nine hundred twenty-three

     21. four hundred ninety-two thousand, one hundred eight

     23. twenty-three thousand, eleven

D. **25.** 185,000   **27.** 358,900   **29.** 120,000

     31. $92,300   **33.** $390,000

E. **35.** 45,115   **37.** 2,367   **39.** 317

     41. 479,136   **43.** 44,158   **45.** 171,342

F. **47.** G   **49.** A   **51.** B   **53.** H

G. **55.** $72 = 72$ True   **57.** $22 \neq 110$ False

     59. $140 = 140$ True   **61.** $102 \neq 84$ False

     63. $286 = 286$ True

H. **65.** 92 laps   **67.** $14 each

     69. 535,300 people   **71.** $25,872

     73. 24 pounds   **75.** 6 pounds per month

     77. 18,500 miles

I. **79.** 50 minutes   **81.** 225 minutes

J. **83.** 3,235 points   **85.** 3,225 and 3,240

     87. Monday   **89.** up 10 points

Test—Chapter 1

1. 47,006,804

2. thirty million, two hundred six thousand, eleven

3. 176,900   **4.** 300,000   **5.** 10,063

6. 4,028   **7.** 12,087   **8.** 2,088

9. 273   **10.** $83 + 29$ **11.** $(3 \times 7) \times 6$

12. Addition Property of Zero   **13.** 0

14. Undefined, no solution   **15.** no

16. 65 quarters   **17.** $70

18. $7,871   **19.** $9,900

20. mean: 85   median: 84   mode: 76 and 84   range: 24

Practice Problems—Chapter 2

Section 2.2

A. **1.** 27   **3.** 25   **5.** 1

     7. 256   **9.** 0

B. **11.** 5.916   **13.** 4.123   **15.** 6

     17. 7.211   **19.** 4

C. **21.** 4   **23.** 2   **25.** 1

     27. 10   **29.** 11

D. **31.** 14   **33.** 8

     35. yes, $17 \times 17 = 289$

     37. yes, $7 \times 7 \times 7 = 343$ **39.** 4

E. **41.** 4,096   **43.** 2   **45.** 9

     47. 1,048,576   **49.** 100

Section 2.3

A. **1.** 22   **3.** 6   **5.** 7   **7.** 6

     9. 28   **11.** 0   **13.** 19   **15.** 6

     17. 26   **19.** 23

B. **21.** 44   **23.** 27   **25.** 2   **27.** 10

     29. 4   **31.** 10   **33.** 29   **35.** 9

C. **37.** 14   **39.** 12   **41.** 6   **43.** 12

     45. 12   **47.** 28   **49.** 16   **51.** 0

     53. 12   **55.** 45   **57.** 154   **59.** 10

D. **61.** 18.866   **63.** 225.036   **65.** 4,915.277

     67. 319.77   **69.** 29.25

Section 2.4

A. 1. 62 yards   3. 480 square feet

B. 5. 46 inches

C. 7. 384 miles   9. 18,000 feet

D. 11. 22   13. 18   15. 11

    17. 34   19. 13

E. 21. 7,812 feet   23. $693, 960

    25. 320 yards   27. 252   29. 3,517

Section 2.5

A. 1. 1, 2, 4, 8   3. 1, 3, 7, 21

    5. 1, 7   7. 1, 2, 4, 8, 16, 32

    9. 1, 2, 3, 6, 9, 18   11. 1, 3, 5, 9, 15, 45

    13. 1, 2, 5, 7, 10, 14, 35, 70

    15. 1, 2, 3, 4, 5, 6, 8, 10, 12, 15, 20, 24, 30, 40, 60, 120

    17. 1, 2, 3, 5, 6, 10, 15, 25, 30, 50, 75, 150

    19. 1, 2, 4, 5, 8, 10, 20, 25, 40, 50, 100, 200

B. 21. No, $1 + 4 = 5$ 23. Yes, ends in 6

    25. No, ends in 6   27. Yes, sum is 12

    29. Yes, ends in 5   31. No, ends in 5

    33. Yes, ends in 2   35. No, ends in 2

    37. Yes, ends in 8   39. Yes, sum is 9

C. 41. $2 \cdot 2 \cdot 3$ 43. $2 \cdot 2 \cdot 5$

    45. $2 \cdot 2 \cdot 3 \cdot 3$ 47. $2 \cdot 3 \cdot 3 \cdot 3$

    49. $2 \cdot 2 \cdot 3 \cdot 7$ 51. $5 \cdot 17$

    53. $2 \cdot 2 \cdot 2 \cdot 3 \cdot 5$ 55. $2 \cdot 7 \cdot 11$

    57. $2 \cdot 2 \cdot 2 \cdot 2 \cdot 3 \cdot 3$ 59. $2 \cdot 11 \cdot 11$

D. 61. $3 \cdot 5 \cdot 11 \cdot 17$ 63. $3 \cdot 7 \cdot 11 \cdot 17$

    65. $3 \cdot 3 \cdot 3 \cdot 7 \cdot 7 \cdot 11$ 67. $3 \cdot 3 \cdot 5 \cdot 7 \cdot 13$

    69. $5 \cdot 11 \cdot 13 \cdot 23$ 71. $2 \cdot 2 \cdot 5 \cdot 7 \cdot 7 \cdot 11$

Section 2.6

A. 1. 24   3. 84   5. 60   7. 40

    9. 60   11. 180   13. 90   15. 60

    17. 120   19. 84

B. 21. 180   23. 7,560   25. 1,575

    27. 300   29. 1,080

Section 2.7

A. 1. 2   3. 4

    5. 8   7. 1, relatively prime

    9. 15   11. 2

    13. 7   15. 1, relatively prime

    17. 5   19. 6

B. 21. 36   23. 42   25. 72

    27. 63   29. 24

Section 2.8 Review Problems

A. 1. 125   3. 4   5. 12   7. 2

    9. 1

B. 11. 12   13. 4   15. 26   17. 28

    19. 7   21. 147   23. 10   25. 2

    27. 34   29. 22   31. 36   33. 50

C. 35. 1,232 miles   37. 108 square feet

    39. 19 inches   41. 5.385

    43. 28   45. 28

    47. 66

D. 49. 1, 2, 3, 6, 9, 18

    51. 1, 2, 3, 5, 6, 10, 15, 30

    53. 1, 2, 3, 4, 5, 6, 8, 10, 12, 15, 20, 24, 30, 40, 60, 120

    55. Yes, ends in 6   57. Yes, ends in 0

    59. No, sum is 10   61. Yes, sum is 12

E. 63. $2 \cdot 2 \cdot 2 \cdot 2 \cdot 5$ 65. $2 \cdot 3 \cdot 5 \cdot 7$

    67. $2 \cdot 3 \cdot 19$ 69. $3 \cdot 5 \cdot 5$

    71. $2 \cdot 3 \cdot 11$ 73. $2 \cdot 2 \cdot 3 \cdot 11$

F. 75. 60   77. 120   79. 84   81. 180

    83. 210

G. 85. 14   87. 1   89. 6

Test—Chapter 2

1. 2, 3, 5, 7, 11, 13, 17, 19, 23

2. yes, $2 + 7 + 6 = 15$ and 15 is divisible by 3

3. $2 \cdot 3 \cdot 3 \cdot 5 \cdot 7$ or $2 \cdot 3^2 \cdot 5 \cdot 7$

4. $5 \cdot 5 \cdot 7$ or $5^2 \cdot 7$ 5. 210

6. 120   7. 180

8. 6   9. relatively prime

10. 20   11. 9

12. 47   13. 55

14. 1, 2, 5, 10, 25, 50   15. $R = 20$

16. $K = 20$ 17. $M = 34$

18. $A = 126$ square feet   19. $P = 50$ feet

20. $D = 301$ miles

Practice Problems—Chapter 3

Section 3.2

A. **1. a.** 13/35   **b.** 22/35
      3. a. 13/24   **b.** 11/24
      5. a. 11/20   **b.** 9/20

B. **7.** 4 1/5   **9.** 3 1/2   **11.** 1 2/3
      13. 2 7/11   **15.** 2 3/32   **17.** 1

C. **19.** 44/7   **21.** 58/5   **23.** 11/8
      25. 68/9   **27.** 18/7

D. **29.** 19/4, 8/8, 6/5
      31. 1 1/2, 19/4, 6/5, 4 4/9, 25, 3
      33. 8/8
      35. 1 1/2, 19/4, 6/5, 4 4/9

Section 3.3

A. **1.** Yes, 5/8 = 5/8   **3.** Yes, 4/5 = 4/5
      5. No, 2/3 ≠ 1/2   **7.** Yes, 12/20 = 12/20
      9. No, 5/8 ≠ 15/32

B. **11.** 9   **13.** 15   **15.** 16
      17. 5   **19.** 6   **21.** 28
      23. 3 14/16 or 62/16

C. **25.** 5/6   **27.** 3/5   **29.** 1/4
      31. 3/4   **33.** 2/3   **35.** 5/8
      37. 8/15   **39.** 13/7   **41.** 9/20
      43. 49/72   **45.** 24/31   **47.** 16/21
      49. 3/4

Section 3.4

A. **1.** 10/27   **3.** 12/35   **5.** 5/24
      7. 1/2   **9.** 1/5   **11.** 15
      13. 95   **15.** 22   **17.** 3 17/27
      19. 26   **21.** 497/12 or 41 5/12
      23. 290

B. **25.** 13 1/2   **27.** 4 1/2   **29.** 7 1/2
      31. 285 5/7   **33.** 72
      35. 315/4 or 78 3/4

C. **37.** 18 students   **39.** 10 bars
      41. 29,037 paperbacks   **43.** 110 students
      45. 1 3/4 points   **47.** 24 points
      49. 1 11/16 boxes

Section 3.5

A. **1.** 3/2   **3.** 1/4   **5.** 1/12
      7. 7/9   **9.** 3/13

B. **11.** 8/9   **13.** 5/2 or 2 1/2
      15. 48/5 or 9 3/5   **17.** 5/2 or 2 1/2
      19. 7/12   **21.** 117   **23.** 32/35
      25. 7/9   **27.** 2/483

C. **29.** 1/6   **31.** 1/6   **33.** 2/3
      35. 7/8   **37.** 3/2 or 1 1/2

D. **39.** 5/16 pounds   **41.** 36 cars
      43. 8 34/45   **45.** 5 1/2 dollars
      47. 10 scarves   **49.** 54 packages
      51. 56 plots

Section 3.6 Review Problems

A. **1.** 2 14/15   **3.** 6 1/3
      5. 1 5/6   **7.** 12 1/5

B. **9.** 27/8   **11.** 38/3
      13. 272/5   **15.** 81/8

C. **17.** 3/4   **19.** 2/5
      21. 16/21   **23.** 2/5
      25. 8/13   **27.** 2/3
      29. 48/63   **31.** 7/10

D. **33.** 8   **35.** 9
      37. 12   **39.** 18
      41. 1   **43.** 4

E. **45.** 5/9   **47.** 3 1/2
      49. 6 2/3   **51.** 10
      53. 4 1/2   **55.** 8/9
      57. 1 3/5   **59.** 4 11/16
      61. 1 7/9   **63.** 5/6

F. **65.** 64 patties   **67.** 18 days
      69. 800 disks   **71.** 46 1/2 mph

Test—Chapter 3

1. 2 5/7   **2.** 37/8   **3.** 9/20

4. $\dfrac{2 \times 2 \times 2 \times 2 \times 2 \times 5}{2 \times 2 \times 3 \times 3 \times 7} = \dfrac{40}{63}$

5. 28   **6.** 5/3   **7.** 10/21

8. 25/9 or 2 7/9   **9.** 22/9 or 2 4/9

10. 46/3 or 15 1/3   **11.** $\dfrac{128}{3}$ or 42 2/3

12. 35   **13.** 13/5 or 2 3/5

14. 3/2 or 1 1/2   **15.** 15/8 or 1 7/8

16. 32/3 or 10 2/3   **17.** $1,980

18. 35 packages   **19.** 28 miles per gallon

20. 40 1/2 feet

Practice Problems—Chapter 4

Section 4.2

A.   **1.** 12   **3.** 8   **5.** 12   **7.** 12

   9. 120   **11.** 210   **13.** 288   **15.** 720

   17. 180   **19.** 72   **21.** 288

B.   **23.** 7/9   **25.** 3/5

   27. 16/15 or 1 1/15   **29.** 1/2

   31. 29/18 or 1 11/18   **33.** 19/15 or 1 4/15

   35. 11/80   **37.** 23/12 or 1 11/12

   39. 5/4 or 1 1/4   **41.** 9/20

   43. 11/36   **45.** 167/180

   47. 353/180 or 1 173/180

   49. 137/72 or 1 65/72

C.   **51.** 2 1/2 pounds   **53.** 3/8 pound

   55. 2 1/8 hours   **57.** 1/8 hour

   59. 1 11/12 hours

Section 4.3

A.   **1.** 8 1/8   **3.** 2 1/6   **5.** 10 1/12

   7. 12 7/36   **9.** 8 1/6   **11.** 5 4/15

   13. 15 19/28   **15.** 32 1/40   **17.** 6 5/9

   19. 5/6   **21.** 143 13/16   **23.** 4 11/12

   25. 11 25/36   **27.** 42 3/5   **29.** 29 23/30

B.   **31.** 20 7/8   **33.** 38 17/24

   35. 16 5/6 pounds   **37.** 131 3/4 pounds

   39. 11 1/4 pounds   **41.** 5/6 cup

   43. 12 7/12 hours   **45.** 5/6 hour

Section 4.4

1. 13/12 or 1 1/12   **3.** 23/30

5. 3/8   **7.** 1/6

9. 1 5/9   **11.** 1/96

13. 9/16   **15.** 9/8 or 1 1/8

17. 47/48   **19.** 59/48 or 1 11/48

21. 2 11/12   **23.** 1 5/12

25. 1/18   **27.** 23/40

29. 35/54

Section 4.5

A.   **1.** >   **3.** >   **5.** =   **7.** >

   9. <   **11.** <   **13.** =   **15.** >

B.   **17.** 3/8, 2/3, 5/6   **19.** 3/4, 7/8, 15/16

   21. 2/9, 2/3, 4/5   **23.** 2/3, 3/4, 15/16

   25. 1/2, 5/8, 2/3, 3/4   **27.** 3/8, 1/2, 4/7, 2/3

   29. 7/16, 11/24, 17/32, 5/8

C.   **31.** 1/12   **33.** 1/48   **35.** 1/16

   37. 1/48   **39.** 1/32   **41.** 1/196

Section 4.6

1. 8 cups   **3.** 1 1/3 cups

5. 19 3/4 laps   **7.** 3/4 lap

9. 1 5/6 miles   **11.** 34 8/11 miles

13. 214 bottles   **15.** 5/8 hour

17. 11/16 of the class   **19.** 6 players

21. 7/8 yard   **23.** 18 1/8 yards

25. 1 13/18 hours   **27.** 15 volumes

Section 4.7 Review Problems

A.   **1.** 19/12 or 1 7/12   **3.** 1/12

   5. 4 11/36   **7.** 4 27/56

   9. 3/4   **11.** 13 33/40

   13. 31/28 or 1 3/28   **15.** 1 7/8

   17. 8 1/3   **19.** 34 23/24

   21. 31 4/9   **23.** 2 5/8

   25. 20 37/45   **27.** 1 71/160

   29. 1 7/8

B. **31.** 7/8   **33.** 1/2
     35. 0   **37.** 9/2 or 4 1/2
     39. 35/4 or 8 3/4

C. **41.** 4/5, 4/9, 4/15   **43.** 5/12, 1/4, 3/16
     45. 2/3, 11/18, 5/9   **47.** 5/12, 3/8, 5/16
     49. 11/12, 9/10, 5/6

D. **51.** 8 3/4 hours   **53.** 14 7/8 hours
     55. 10 25/44 miles   **57.** 5/8 hour
     59. 20 lots

Test—Chapter 4

1. 120   **2.** 3/2 or 1 1/2
3. 3/10   **4.** 21/16 or 1 5/16
5. 6 5/12   **6.** 1 4/15
7. 3 4/7   **8.** 47/24 or 1 23/24
9. 14/45   **10.** 1 7/12
11. 13 2/5   **12.** 6 2/5
13. 3 2/3   **14.** 29/18 or 1 11/18
15. 1/2   **16.** 3/8, 1/4, 3/16
17. 18 17/24 hours   **18.** 1 7/12 hours
19. 7/12 acre remains **20.** 24 miles on one gallon

Cumulative Review 1–4

1. 376   **2.** 75,330
3. 3/2 or 1 1/2   **4.** 36/25 or 1 11/25
5. 16 1/18   **6.** 274
7. 27,039   **8.** 2/5
9. 2/3   **10.** 71 5/8
11. 42,042   **12.** 4,500
13. 27/64   **14.** 4/9
15. 315/4 or 78 3/4   **16.** 143
17. 180 R 34 or 180 17/21 **18.** 5/2 or 2 1/2
19. 175/16 or 10 15/16 **20.** 30
21. no; yes; no   **22.** 28
23. 56   **24.** 3 · 3 · 7
25. 2 · 2 · 2 · 2 · 3 · 5 **26.** 32
27. 9
28. 1, 2, 3, 4, 6, 8, 12, 16, 24, 48
29. 16   **30.** 5 1/6
31. c   **32.** a   **33.** d   **34.** e
35. b   **36.** 7   **37.** 18   **38.** 13
39. 11   **40.** 11/12 **41.** 41/45 **42.** 13
43. 32   **44.** 4,800 women
45. 575 miles   **46.** $32
47. 19 miles   **48.** 12 pieces
49. 12 7/24   **50.** 69

Practice Problems—Chapter 5

Section 5.2

A. **1. a.** 4   **b.** 3   **c.** 6   **d.** 1
     3. a. 7   **b.** 4   **c.** 0   **d.** 3

B. **5.** one hundred sixty and six tenths
     7. one hundred eighty-two and thirty-four thousandths
     9. four hundred fifty-six ten-thousandths
     11. three and seventeen thousandths
     13. forty-five and forty-five hundredths

C. **15.** 6.074   **17.** 0.107   **19.** 116,000

D. **21.** 16.05   **23.** 170   **25.** 0.08
     27. 14.264 **29.** 358.0 **31.** $286
     33. $24.86 **35.** 268

E. **37.** 2.03, 2.12, 2.2   **39.** 0.635, 0.67, 0.7
     41. 4.003, 4.03, 4.304

F. **43.** 0.401 > 0.4 > 0.392 > 0.04
     45. 2.015 < 2.123 < 2.146 < 2.15
     47. 2.402 > 2.24 > 2.204 > 2.004
     49. 0.053 < 0.305 < 0.31 < 0.351

Section 5.3

A. **1.** 7.91   **3.** 21.235   **5.** 3.432
     7. 30.14   **9.** 5.778   **11.** 154.95
     13. 2,866   **15.** 1.64   **17.** $193.89
     19. $37.27

B. **21.** 17.735 **23.** 147.09 **25.** 32.66
     27. $12.41 **29.** 57.68

C. **31.** 963.165　　**33.** 3,047.57　　**35.** 509.8165
　　37. 986.046　　**39.** 165,205.43　　**41.** 854.418
　　43. 120,790.33　　**45.** 1,220.916　　**47.** 0.01827
　　49. 0.007809

Section 5.4

A. **1.** 48.776　　**3.** 0.0414　　**5.** 101.4
　　7. 0.035　　**9.** 34.8　　**11.** 2,876.8
　　13. $1.01　　**15.** 2,300　　**17.** $404.20
　　19. 8,848　　**21.** 0.1337　　**23.** 1,417.83
　　25. 59　　**27.** 0.00189　　**29.** 0.000008096
　　31. 472.211　　**33.** 154.224

B. **35.** 4.390864　　**37.** 19,309.101
　　39. 173.55775　　**41.** 1,650.627
　　43. 0.0010602　　**45.** 30,638.1
　　47. 867.33258　　**49.** 0.0292

Section 5.5

A. **1.** 20　　**3.** 320　　**5.** 0.25
　　7. 0.0245　　**9.** 400　　**11.** 20
　　13. 0.265　　**15.** 0.09　　**17.** 0.64
　　19. 600　　**21.** 0.45　　**23.** 0.1834

B. **25.** 21.667　　**27.** 0.867　　**29.** 0.329
　　31. 5.217　　**33.** 0.026　　**35.** 0.007
　　37. 0.571　　**39.** 0.271　　**41.** 0.214
　　43. 28.696　　**45.** 0.059　　**47.** 0.001
　　49. 0.001

C. **51.** 43,104.17　　**53.** 1.4223　　**55.** 36.541
　　57. 9.7　　**59.** 344.1

Section 5.6

A. **1.** 7/50　　**3.** 2 3/1000　　**5.** 17 4/5
　　7. 9/125　　**9.** 36 1/8

B. **11.** 0.273　　**13.** 2.833　　**15.** 0.083
　　17. 10.5　　**19.** 16.211

C. **21.** 2.175 or 87/40 or 2 7/40
　　23. 4.1 or 41/10 or 4 1/10
　　25. 38.8 or 194/5 or 38 4/5
　　27. 3.45 or 69/20 or 3 9/20
　　29. 5.25 or 21/4 or 5 1/4

D. **31.** 0.461　　**33.** 3.85　　**35.** 0.280
　　37. 9.84　　**39.** 0.0350

E. **41.** 59.415　　**43.** 2,815.2　　**45.** 3.6291
　　47. 186.4　　**49.** 2,391.45

Section 5.7

A. **1.** $47.77　　**3.** $49.05　　**5.** $10.12
　　7. $1.39　　**9.** $17.22　　**11.** 10.8 gals.
　　13. 14 vials　　**15.** $311.25　　**17.** $77.78
　　19. $119.26

B. **21.** 25.6　　**23.** 3.927
　　25. 315.99　　**27.** 18,622 people
　　29. $2,084,394.60

Section 5.8 Review Problems

A. **1.** twenty-four and sixty-three thousandths
　　3. six and forty-seven thousandths
　　5. two hundred three and twenty-three thousandths
　　7. one thousand four hundred thirteen and three thousandths

B. **9.** 4.016　　**11.** 409.09　　**13.** 0.0408
　　15. 12,000.012

C. **17.** 4.7　　**19.** 50　　**21.** 92.5
　　23. 0.101

D. **25.** 0.602, 0.609, 0.61
　　27. 2.002, 2.02, 2.2
　　29. 0.08, 0.801, 0.81

E. **31.** 17.32 > 17.3 > 17.2 > 17.03
　　33. 0.006 < 0.06 < 0.066 < 0.606

F. **35.** 90.767　　**37.** 147.886　　**39.** 57.818
　　41. 2,772.04　　**43.** 1,897.2　　**45.** 53.06
　　47. 135.48　　**49.** 75.76　　**51.** $45.17
　　53. 214.93

G. **55.** 0.07　　**57.** 2.09　　**59.** 89.47
　　61. 0.00　　**63.** 80　　**65.** 0.03
　　67. 0.05　　**69.** 4.97　　**71.** 4.82
　　73. 0.07

H. **75.** 7/25   **77.** 22 9/20   **79.** 1/125
81. 45 43/200
I. **83.** 0.625   **85.** 0.778   **87.** 3.833
89. 0.004
J. **91.** $0.57   **93.** 3.8 yards   **95.** 0.35 quart
97. $675   **99.** $7.94

Test—Chapter 5

1. thirty-two thousandths   **2.** 16.0406
3. 315.3   **4.** 27.89

5. $37.70   **6.** 0.023, 0.203, 0.23
7. 23.775   **8.** 7.785
9. 37.166   **10.** 0.0051
11. $1.06   **12.** 321.7
13. 7,200   **14.** 0.8
15. 0.000215   **16.** 23/125
17. 0.917   **18.** 15.9 gallons
19. $113.68   **20.** $8.29

Practice Problems—Chapter 6

Section 6.2

A. **1.** 1/3   **3.** 5/9   **5.** 40/3
7. 8/1   **9.** 3/10   **11.** 4/5
13. 7/27   **15.** 145/63   **17.** 24/1
B. **19.** 3/2   **21.** 1/8   **23.** 2/7
25. 3/5   **27.** 3/1   **29.** 4/5
31. 1/4   **33.** 16/15   **35.** 3/4
37. 1/12   **39.** 1/2
C. **41. a.** 32/3   **b.** 15/32   **c.** 3/32
d. 3/10   **e.** 16/25
43. a. 21/38   **b.** 21/17   **c.** 1/28
d. 3/152
D. **45.** $0.15 per ounce   **47.** 40/1
49. 2/1   **51.** 24 miles per gallon
53. 110 miles per hour   **55.** 35 miles per gallon
57. $0.995 = $1.00   **59.** $2.50 per bag
E. **61.** A, $0.163 each   **63.** A, $0.233 each
65. A, $0.74 each   **67.** B, $0.044 per ounce
69. B, $0.156 per ounce
F. **71.** 16.14 miles per gallon, $15.03
73. $0.083 per ounce

Section 6.3

A. **1.** Yes, 36 = 36   **3.** No, 225 ≠ 315
5. Yes, 1,372 = 1,372   **7.** Yes, 1.728 = 1.728
9. Yes, 17.55 = 17.55   **11.** Yes, 15/8 = 15/8
B. **13.** N = 81   **15.** R = 3   **17.** A = 40
19. C = 5   **21.** N = 4.2   **23.** R = 6.67
25. A = 16.63   **27.** B = 189.96   **29.** T = 16

31. A = 160   **33.** W = 2 5/8   **35.** N = 3/5
37. A = 10   **39.** R = 3.17
C. **41.** Yes, 132.066 = 132.066   **43.** N = 4.23
45. N = 5,563.18

Section 6.4

A. **1.** 12.5 defective VCRs   **3.** 265 miles
5. 1.6 inches   **7.** 2 2/3 houses
9. $0.80 per pound   **11.** $0.59
13. 250 cans
15. 5,500 pounds of gravel
17. 12 hours   **19.** $75 interest
B. **21.** $94.45   **23.** $1,039.31
25. 214.4 or 215 absentees

Section 6.5

A. **1.** 4.5 quarts   **3.** 4,320 minutes
5. 28,800 seconds   **7.** 0.7 miles per minute
9. 146.7 feet per second
B. **11.** 30 decimeters   **13.** 15 centimeters
15. 71.1 centimeters   **17.** 6.4 kilometers
19. 2.1 feet   per mile
C. **21.** 1.65 decimeters   **23.** 1,653.5 inches
25. 61.4 meters   **27.** 22.4 miles per hour
29. 1,234.4 centimeters

Section 6.6

1. $\dfrac{12}{20} = \dfrac{3}{5}$ **3.** $\dfrac{20}{28} = \dfrac{5}{7}$

5. 23 grams   **7.** $\dfrac{10}{6} = \dfrac{5}{3}$

9. 1/16   **11.** 11.5 wins

13. 15/22   **15.** $15,000

17. $45,000   **19.** $\dfrac{30}{55} = \dfrac{6}{11}$

21. $\dfrac{55}{60} = \dfrac{11}{12}$ **23.** 39°

25. $\dfrac{1}{48} = \dfrac{3}{M}$, 144 miles

Section 6.7 Review Problems

A.   **1.** 2/9   **3.** 7:2   **5.** 160:3

     7. 2:3   **9.** 1:9   **11.** 16 to 5

     13. 36 to 5   **15.** 2:5   **17.** 16 to 7

     19. a. 6:23   **b.** 6:29   **c.** 23:29

B.   **21.** $0.13   **23.** $0.98   **25.** $0.43

     27. $16.59   **29.** $0.367

C.   **31.** Yes, 140 = 140   **33.** No, 156 ≠ 144

     35. Yes, 0.144 = 0.144   **37.** No, 880 ≠ 636

     39. No, 75/4 ≠ 90/4

D.   **41.** $C = 3$ **43.** $M = 2.67$ **45.** $C = 7.2$

     47. $F = 0.47$ **49.** $K = 48$ **51.** $T = 16/15$

E.   **53.** $3.30   **55.** 140 miles   **57.** $7.35

     59. 3.75 pizzas, so order 4

     61. 3.77 inches

F.   **63.** 1.35 feet   **65.** 18,000 seconds

     67. 64 cups   **69.** 15 mph

     71. 152.4 centimeters

G.   **73.** 5 to 9   **75.** 58 cars   **77.** 50 cars

     79. No, ratio is still 2 to 3. Explanations will vary.

Test—Chapter 6

1. 4/9   **2.** 60/1

3. 6/5   **4.** 4/15

5. 60/17   **6.** $0.205

7. 12 ounce, $0.274   **8.** 1.3 hours

9. 393.7 inches   **10.** Yes, 105 = 105

11. No, 0.128 ≠ 0.0144   **12.** $W = 63$

13. $C = 25.6$ **14.** $X = 72$

15. $A = 34.8$ **16.** $T = 5/4$

17. $X = \$31.25$ **18.** $X = 191.25$ miles

19. $X = 25$ feet   **20.** $X = \$5.85$

Practice Problems—Chapter 7

Section 7.2

1. a. 62   **b.** 12   **c.** 5

3. a. 89   **b.** 11   **c.** 11%

5. 27.4%   **7.** 8%   **9.** 18%

Section 7.3

A.   **1.** 32%   **3.** 69.8%   **5.** 347%

     7. 46.4%   **9.** 1200%   **11.** 60%

     13. 71.4%   **15.** 225%   **17.** 350%

     19. 6.3%

B.   **21.** 0.12   **23.** 1.26   **25.** 0.009

     27. 8.0 or 8   **29.** 0.105   **31.** 0.253

C.   **33.** 43/100   **35.** 13/25   **37.** 8/1

     39. 1/125

D.

	Fraction	Decimal	Percent
41.	143/500		28.6%
43.		0.8	80%
45.	1/40	0.025	
47.	3/400	0.008	
49.		0.583	58.3%

Section 7.4

A.   **1.** $R = 23\%$ $B = 80$ $A = ?$
        $A = 18.4$

     3. $R = 7\%$ $B = 95$ $A = ?$
        $A = 6.65$

     5. $R = 124\%$ $B = 70$ $A = ?$
        $A = 86.8$

7. $R = 12\ 1/2\%$ $B = 100$ $A = ?$
$A = 12.5$

9. $R = 7\%$ $B = ?$ $A = 35$
$B = 500$

11. $R = 0.5\%$ $B = ?$ $A = 15$
$B = 3,000$

13. $R = 32\%$ $B = ?$ $A = 8$
$B = 25$

15. $R = 12\%$ $B = ?$ $A = 43$
$B = 358.33$

17. $R = ?$ $B = 10$ $A = 12$
$R = 120\%$

19. $R = ?$ $B = 40$ $A = 4.8$
$R = 12\%$

21. $R = ?$ $B = 30$ $A = 16$
$R = 53.3\%$

23. $R = ?$ $B = 8$ $A = 7$
$R = 87.5\%$

25. $R = 16.5\%$ $B = 42$ $A = ?$
$A = 6.93$

27. $R = 24\%$ $B = ?$ $A = 12$
$B = 50$

29. $R = ?$ $B = 65$ $A = 21$
$R = 32.3\%$

B. **31.** $R = 77.3\%$ $B = 92,476$ $A = ?$
$A = 71,483.95$

33. $R = ?$ $B = 146$ $A = 91.4$
$R = 62.6\%$

35. $R = 29.4\%$ $B = 37,645$ $A = ?$
$A = 11,067.63$

37. $R = ?$ $B = 8,276$ $A = 189.4$
$R = 2.3\%$

39. $R = 38.4\%$ $B = ?$ $A = 127$
$B = 330.73$

41. $R = 82\%$ $B = ?$ $A = 67.4$
$B = 82.20$

43. $R = 46.5\%$ $B = ?$ $A = 35.4$
$B = 76.13$

45. $R = ?$ $B = 8,375$ $A = 1,276$
$R = 15.2\%$

Section 7.5

A. **1.** $24/B = 15/100$ $B = 160$
3. $12/40 = P/100$ $P = 30\%$
5. $A/30 = 94/100$ $A = 28.2$
7. $6\ 1/2/B = 18/100$ $B = 36.11$
9. $30/40 = P/100$ $P = 75\%$

11. $A/40 = 18/100$ $A = 7.2$
13. $48/B = 8/100$ $B = 600$
15. $45/80 = P/100$ $P = 56.3\%$
17. $54/B = 1.8/100$ $B = 3,000$
19. $A/300 = 14.8/100$ $A = 44.4$
21. $36/B = 8\ 1/2/100$ $B = 423.53$
23. $52/B = 120/100$ $B = 43.33$
25. $60/32 = P/100$ $P = 187.5\%$
27. $45/B = 150/100$ $B = 30$
29. $21.5/38 = P/100$ $P = 56.6\%$
B. **31.** $18.4/2,006 = P/100$ $P = 0.9\%$
33. $21/B = 43.2/100$ $B = 48.61$
35. $A/13,465 = 67.5/100$ $A = 9,088.88$
37. $623/18,540 = P/100$ $P = 3.4\%$
39. $1,765/B = 46.8/100$ $B = 3,771.37$

Section 7.6

A. **1.** \$288 **3.** \$108 **5.** \$67.20
7. \$625 **9.** \$1,625
B. **11.** \$62.30 **13.** \$115.70 **15.** 12%
17. 400 people **19.** \$944.20 **21.** \$300
23. 12% **25.** 8% **27.** \$38.68
29. \$365.88

Section 7.7

A. **1.** television **3.** \$9,000 **5.** \$7,200
B. **7.** 33% **9.** \$1,056 **11.** \$350
C. **13.** \$20.50 **15.** 2.5% **17.** \$19.80
D. **19.** \$1,000 **21.** 32.5% **23.** 17.5%
25. See Figure 9 Answer

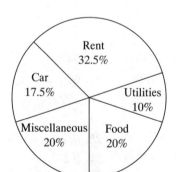

FIGURE **9** **Answer**

Section 7.8

A. **1.** 38%   **3.** 15%

B.

	Fraction	Decimal	Percent
5.	4/25		16%
7.		0.25	25%
9.		0.125	12.5%
11.	1 27/50		154%
13.	1 3/25	1.12	
15.		0.083	8.3%

C. **17.** $360   **19.** $462.50   **21.** $9,975

D. **23.** 14.63   **25.** 10.08   **27.** 75

   29. 168   **31.** 20%   **33.** 40%

   35. 42.9%   **37.** 705.88   **39.** 2.5%

   41. 15.36   **43.** 718.37   **45.** 909.09

   47. 15%   **49.** 13.12

E. **51.** 40 people   **53.** 86.4%   **55.** $5,000

   57. $70.50   **59.** $1,500   **61.** 7.0%

   63. $80   **65.** $96.60

F. **67.** 20%   **69.** $175

Test—Chapter 7

1. $R = 16\%, B = 40, A = 6.4$

2. 11%   **3.** 0.7%   **4.** 0.1675

5. 762.5%   **6.** 29/200   **7.** $324

8. $17.60 = R \times 320$

9. $17.60/320 = P/100$

10. 35   **11.** 2%   **12.** 20.3%

13. 70   **14.** 1,800   **15.** $137

16. $52.50   **17.** 12%   **18.** $16,500

19. $20,000   **20.** $1,166,160

Cumulative Review 1–7

1. 48.9   **2.** 5/4 or 1 1/4

3. 3,979   **4.** 10 7/24

5. 26.885   **6.** 9 1/9

7. 529   **8.** 55.39

9. 147.7   **10.** 30 1/5

11. 48   **12.** 1.1

13. $11,293   **14.** 1,541/32 or 48 5/32

15. 7.6   **16.** 1,273.33

17. 372   **18.** 19/6 or 3 1/6

19. 41.43   **20.** 33/16 or 2 1/16

21. 45/4   **22.** 12

23. 13/42   **24.** 45%

25. 1/8   **26.** 65.2

27. 0.005   **28.** 2 3/8 or 19/8

29. 0.42   **30.** 16/1

31. 4/3   **32.** 8.556

33. 13.3%   **34.** 145

35. 8   **36.** 8/15

37. 0.7476   **38.** 182

39. 1/2   **40.** $N = 120$

41. $N = 7 1/3$ or $N = 7.3$ **42.** $N = 1.87$

43. $A = 14.4$ **44.** $9.54

45. 24 pieces   **46.** $96.11

47. 60%   **48.** 3/5

49. $2.00   **50.** 41 1/4 miles

Practice Problems—Chapter 8

Section 8.2

A. **1.** supplementary   **3.** an infinite number

   5. complementary   **7.** obtuse

   9. vertical   **11.** protractor

   13. degrees   **15.** intersecting

   17. 61°   **19.** 73°

B. **21.** 127°   **23.** 118°

C. **25.** 52°   **27.** 180°

   29. vertical, congruent

Section 8.3

A. **1.** equilateral or equiangular   **3.** vertices

   5. isosceles   **7.** altitude

   9. 180°   **11.** congruent

B. **13.** 64°   **15.** 75°

   17. 60°

C. **19.** 24 square feet   **21.** 60 inches

D. **23.** 60 meters   **25.** 9.4 inches

E. **27.** 384 square feet   **29.** 96 feet
F. **31.** 480 square feet   **33.** 18.5 feet

Section 8.4

A. **1.** congruent   **3.** 1 to 1
    5. proportional   **7.** equiangular, similar
B. **9.** $BC = 12$ yards   **11.** $ML = 9$ yards
    13. $XY = 18$ meters   **15.** $YZ = 36$ meters
    17. $AE = 8$ yards   **19.** $CE = 8$ yards
C. **21.** $KN = 25$ yards   **23.** $LM = 24$ yards
    25. $JL = 12$ yards   **27.** 384 square yards
    29. 96 to 60 or 8 to 5

Section 8.5

A. **1.** B, C   **3.** B, C, D, E
    5. B, C, D, E   **7.** none
B. **9.** 140 square feet   **11.** 324 square inches
    13. 100 feet   **15.** 92 inches
    17. 7,070 square feet   **19.** 36 square feet
    21. 106.5 square yards   **23.** 52 yards
C. **25.** $4,462.50   **27.** $9,825.00
    29. 28 passes

Section 8.6

A. **1.** 4.65 feet   **3.** 314 square meters
    5. 25.12 inches   **7.** 21.98 feet
    9. 254.34 square meters
B. **11.** 6.8π meters   **13.** 11.56π square meters
    15. $7\,1/4\pi$ feet   **17.** $13\,9/64\pi$ square feet
C. **19.** 69.08 inches   **21.** 628 square inches
    23. 81.12 feet   **25.** 104.52 square inches
    27. 150.72 square feet   **29.** 395.16 square inches

Section 8.7

A. **1.** 216 cubic centimeters
    3. 328 square inches
    5. 48π square inches

    7. $166\,2/3\pi$ cubic inches
    9. 120 cubic inches
    11. 160π cubic inches
B. **13.** 872 square inches
    15. 6.01 gallons
    17. 15.87 ounces
    19. 3,260,000 square kilometers
    21. 1.79 centimeters
    23. 2.54 square centimeters
    25. 13.74 square meters

Section 8.8

A. **1.** acute   **3.** straight
    5. 180°   **7.** 32°
    9. equilateral   **11.** scalene
    13. similar   **15.** cubic
B. **17.** 122°
    19. adjacent or supplementary
C. **21.** 15 inches   **23.** 54 square inches
D. **25.** 3 inches   **27.** 12 inches
E. **29.** $DE = 4.5$ inches
F. **31.** 68 square meters   **33.** 126 square yards
    35. 12.56 inches   **37.** 59 square feet
G. **39.** 5.4 cubic feet   **41.** 121.5 cubic inches

Test—Chapter 8

1. 36 square yards   **2.** 512 cubic centimeters
3. 27 square feet   **4.** 43.96 inches
5. $PQ = 7.2$ feet   **6.** $\angle$MKJ
7. 142°   **8.** 45°
9. 26 feet   **10.** $AB = 9$ centimeters
11. 104°   **12.** 30 feet
13. 42 square feet   **14.** 10.24π square inches
15. 432 square inches   **16.** 576 cubic inches
17. 138.16 square inches   **18.** 20.8 pounds
19. 22.5π cubic inches   **20.** 7,234.56 cubic inches

Practice Problems—Chapter 9

Section 9.2

A. **1.** $q = 4g$ **3.** $M = m/g$
    5. $C = 25s$ **7.** $t = m + f$
    9. $f = i/12$

B. **11.** $q - 3 = p$ or $p + 3 = q$
    13. $b/2 = a$ or $2a = b$
    15. $x - 5 = w$ or $w + 5 = x$

C.   **17.** $A = R - S$; $A = 8$
   19. $L = W/Y$; $L = 9$
   21. $d = 52h$; $h = 364$ miles
   23. $c = rs$; $c = 1,944$ people

Section 9.3

A.   **1. a.** numerical coefficients   **b.** variables
   3. a. 6   **b.** m and n **c.** 4, 3, and 8
   5. expression
   7. numerical coefficient
   9. exponent or power

B.   **11.** $5w$ **13.** $12X$ **15.** $2a^3$
   17. $13P$ **19.** $9c^2$

C.   **21.** $11y + 7z$ **23.** $15f + 9g$
   25. $13a + 4b$ **27.** $7a^2 + 4b^2$
   29. $19b + 5c$

Section 9.4

A.   **1.** 31   **3.** 11   **5.** 21   **7.** 8
   9. 14   **11.** 77   **13.** 18   **15.** 75
   17. 18   **19.** 31

B.   **21.** 72   **23.** 23   **25.** 108   **27.** 18
   29. 79   **31.** 11   **33.** 48   **35.** 3
   37. 31   **39.** 44

Section 9.5

   1. b **3.** c **5.** g **7.** a **9.** i
   11. k **13.** m **15.** e **17.** d **19.** k

Section 9.6

A.   **1.** equation   **3.** solution
   5. 9   **7.** false

B.   **9.** $y = 9$ **11.** $M = 15$
   13. $T = 17$ **15.** $N = 12$
   17. $K = 10$ **19.** $C = 4$

C.   **21.** $x + 3 < 9$ whenever $x < 6$
   23. $y - 2 > 10$ whenever $y > 12$
   25. $4a \leq 20$ whenever $a \leq 5$
   27. $30 \geq a + 10$ whenever $20 \geq a$
   29. $m \div 2 > 5$ whenever $m > 10$

Section 9.7 Review Problems

A.   **1.** $i = 36Y$ **3.** $c = 9D$
   5. $s = p + a$ **7.** $r = d/t$
   9. $t + f = V$

B.   **11.** $K = L + 6$ or $L = K - 6$
   13. $A = 3B$ or $B = A/3$
   15. $P = R \div 4$ or $R = 4P$
   17. $C = D - 2$ or $D = C + 2$
   19. $T = V - 5$ or $V = T + 5$

C.   **21.** x and y **23.** 7 and 3
   25. p and r **27.** 6
   29. 5, 3, and 1

D.   **31.** $2W + 4Y$ **33.** $11c$
   35. $5a^2$ **37.** $5B$
   39. $6w - 3x + 2y$ **41.** $8y^2$
   43. $4a^2$ **45.** $4t^3 + 11t^2$

E.   **47.** 30   **49.** 12   **51.** 48   **53.** 248
   55. 8   **57.** 32   **59.** 20

F.   **61.** 28   **63.** 60   **65.** 14   **67.** 96
   69. 13

G.   **71.** $X = 13$ **73.** $f = 6$
   75. $B = 8$ **77.** $M = 15$
   79. $y = 4$

H.   **81.** $g - 7 > 2$ whenever $g > 9$
   83. $y \div 6 < 3$ whenever $y < 18$
   85. $2d \leq 12$ whenever $d \leq 6$
   87. $40 > m + 10$ whenever $30 > m$
   89. $x + 7 \geq 11$ whenever $x \geq 4$

Test—Chapter 9

   1. $p = s - a$
   2. $P = c + t$
   3. $s = 60m$
   4. $A = P/S$
   5. $X = Y + 4$ or $Y = X - 4$
   6. $G = H \div 6$ or $G = H/6$ or $H = 6G$
   7. $5a$
   8. $11X + 7Y$
   9. 78   **10.** 17   **11.** 11   **12.** 6
   13. 33   **14.** 90   **15.** 5   **16.** $R = 96$
   17. $M = 11$ **18.** $K = 48$
   19. $Y < 8$ **20.** $P \geq 5$

Practice Problems—Chapter 10

Section 10.2

1. $B = 5$
ck: $9 = 9$

3. $W = 4$
ck: $12 = 12$

5. $m = 10$
ck: $7 = 7$

7. $t = 9$
ck: $6 = 6$

9. $4 = N$
ck: $12 = 12$

11. $s = 6$
ck: $30 = 30$

13. $C = 18$
ck: $6 = 6$

15. $P = 27$
ck: $9 = 9$

17. $y = 5/4$
ck: $5/6 = 5/6$

19. $X = 4$
ck: $7 = 7$

21. $W = 7$
ck: $21 = 21$

23. $y = 12$
ck: $6 = 6$

25. $S = 20$
ck: $12 = 12$

27. $B = 17$
ck: $3 = 3$

29. $m = 35/9$
ck: $5/9 = 5/9$

31. $X = 12$
ck: $22 = 22$

33. $10/3 = t$
ck: $10 = 10$

35. $W = 80$
ck: $20 = 20$

37. $A = 3$
ck: $14 = 14$

39. $R = 0$
ck: $8 = 8$

Section 10.3

1. $T = 9$
ck: $11 = 11$

3. $F = 6$
ck: $25 = 25$

5. $Y = 6$
ck: $38 = 38$

7. $P = 8$
ck: $1.6 = 1.6$

9. $T = 7$
ck: $41 = 41$

11. $C = 4$
ck: $48 = 48$

13. $X = 0$
ck: $3 = 3$

15. $W = 20$
ck: $2.4 = 2.4$

17. $F = 180$
ck: $206 = 206$

19. $N = +16$
ck: $44 = 44$

21. $T = 3$
ck: $0 = 0$

23. $K = 9/8$
ck: $13 = 13$

25. $F = 48$
ck: $17 = 17$

27. $A = 11/15$
ck: $3 = 3$

29. $W = 0.16$
ck: $0.8 = 0.8$

Section 10.4

A. **1.** $T = 2$
ck: $16 = 16$

3. $F = 5$
ck: $13 = 13$

5. $R = 3$
ck: $43 = 43$

7. $40 = n$
ck: $240 = 240$

9. $T = 2$
ck: $22 = 22$

11. $M = 9$
ck: $18 = 18$

13. $r = 7$
ck: $21 = 21$

15. $M = 5$
ck: $25 = 25$

B. **17.** $4 = Y$
ck: $32 = 32$

19. $T = 3$
ck: $30 = 30$

21. $M = 4$
ck: $52 = 52$

23. $R = 4$
ck: $24 = 24$

25. $12 = C$
ck: $240 = 240$

27. $x = 4$
ck: $28 = 28$

29. $t = 1$
ck: $16 = 16$

31. $C = 1$
ck: $11 = 11$

33. $R = 12$
ck: $36 = 36$

35. $T = 3/2$
ck: $15 = 15$

37. $M = 0$
ck: $6 = 6$

39. $N = 0$
ck: $9 = 9$

Section 10.5

A. **1.** $R = 3$
ck: $27 = 27$

3. $t = 1/3$
ck: $2\ 2/3 = 2\ 2/3$

5. $W = 2$
ck: $22 = 22$

7. $1 = k$
ck: $12 = 12$

9. $m = 4$
ck: $24 = 24$

11. $T = 3$
ck: $20 = 20$

13. $w = 3$
ck: $11 = 11$

15. $Y = 3$
ck: $34 = 34$

B. **17.** $x = 3$
ck: $23 = 23$

19. $W = 2$
ck: $7 = 7$

21. $x = 3$
ck: $3 = 3$

23. $a = 0$
ck: $7 = 7$

25. $x = 2$
ck: $21 = 21$

27. $C = 3$
ck: $29 = 29$

29. $k = 4$
ck: $24 = 24$

C. **31.** $b = 33$
ck: $27 = 27$

33. $y = 30$
ck: $2 = 2$

35. $a = 7$
ck: $28 = 28$

37. $T = 11$
ck: $8 = 8$

39. $y = 8$
ck: $35 = 35$

41. $C = 5$
ck: $22 = 22$

43. $a = 7$
ck: $7 = 7$

45. $X = 30$
ck: $6 = 6$

47. $Z = 9$
ck: $37 = 37$

49. $R = 2$
ck: $73 = 73$

Section 10.6

A.   **1.** yes   **3.** no   **5.** yes

B.   **7.** $K = 10/9$
ck: $1/2 = 1/2$

9. $N = 3$
ck: $7.12 = 7.12$

11. $W = 20$
ck: $2.4 = 2.4$

13. $F = 180$
ck: $206 = 206$

C.   **15.** $X = 15/2$
ck: $7/2 = 7/2$

17. $K = 3.3$
ck: $1.12 = 1.12$

19. $N = 2/3$
ck: $15/2 = 15/2$

21. $2.45 = X$
ck: $17.64 = 17.64$

23. $M = 2$
ck: $13/4 = 13/4$

25. $X = 5$
ck: $22.5 = 22.5$

27. $N = 0.25$
ck: $1.35 = 1.35$

29. $A = 4/3$
ck: $6 = 6$

D.   **31.** $Y \doteq 0.31$
ck: $6.12 \doteq 6.35$

33. $t \doteq 0.25$
ck: $0.23 \doteq 0.24$

Section 10.7

A.   **1.** 24 feet

3. 17 cm.

5. 50 inches

7. 119.32 feet

9. 17 1/2 square cm.

11. 36 square feet

13. 1,133.5 square feet

B.   **15.** $3,000

17. 4 years

19. 13%

21. $500

23. $17,187.50

C.   **25.** $157.30

27. 65 square inches

29. 12 centimeters

31. 7 inches

D.   **33.** $\dfrac{3V}{b} = h$

35. $\dfrac{I}{rt} = p$

37. $\dfrac{y - b}{m} = x$

39. $\dfrac{A - P}{Pr} = t$

41. $\dfrac{E}{I} = R$

43. $\dfrac{3V}{\pi r^2} = h$

45. $r^3 = \dfrac{3v}{4\pi}$

Section 10.8 Review Problems

A.   **1.** $X = 4$
ck: $12 = 12$

3. $F = 42$
ck: $23 = 23$

5. $5 = H$
ck: $9 = 9$

7. $A = 6$
ck: $16 = 16$

9. $4 = W$
ck: $11 = 11$

11. $c = 10.81$
ck: $9.44 = 9.44$

13. $Y = 36$
ck: $24 = 24$

15. $2 = M$
ck: $14 = 14$

17. $4 = S$
ck: $32.1 = 32.1$

19. $Y = 24$
ck: $24 = 24$

21. $P = 8/3$
ck: $24 = 24$

23. $Y = 3$
ck: $3.6 = 3.6$

25. $C = 4$
ck: $64 = 64$

27. $A = 24/7$
ck: $6 = 6$

29. $S = 11/2$
ck: $18 = 18$

31. $F = 4$
ck: $7 = 7$

33. $Y = 4$
ck: $16 = 16$

35. $1 = R$
ck: $24 = 24$

37. $y = 54/11$
ck: $45/11 = 45/11$

39. $M = 2$
ck: $2 = 2$

41. $y = 16/15$
ck: $2/3 = 2/3$

43. $W = 2$
ck: $12 = 12$

45. $R = 4$
ck: $8 = 8$

47. $R = 3$
ck: $39 = 39$

49. $y = 10$
ck: $14 = 14$

B.   **51.** 7 1/2 hours

53. 70 feet

55. 6%

57. $2.34

59. 8 inches

61. 24 cm.

63. 254.34 meters

C.   **65.** $\dfrac{P - 2L}{2} = W$

67. $\dfrac{V}{LW} = H$

69. $x = \dfrac{36 - 9y}{2}$

Test—Chapter 10

1. $R = 27$
ck: $81 = 81$

2. $R = 7$
ck: $38 = 38$

3. $C = 14$
ck: $35 = 35$

4. $W = 48$
ck: $12 = 12$

5. $A = 7$
ck: $35 = 35$

6. $M = 5.8$
ck: $14.32 = 14.32$

7. $B = 15/2$
ck: $3/4 = 3/4$

8. $R = 2$
ck: $6 = 6$

9. $W = 1$
ck: $3 = 3$

10. $T = 20/9$
ck: $1 = 1$

11. $A = 15$
ck: $25 = 25$

12. $Y = 3$
ck: $11 = 11$

13. $C = 3$
ck: $20 = 20$

14. $N = 2$
ck: $21 = 21$

15. $T = 32/3$
ck: $4 = 4$

16. $M = 20$
ck: $14.2 = 14.2$

17. $P = 3$
ck: $34 = 34$

18. $Y = 1$
ck: $8 = 8$

19. $L = 32$ yards

20. $C = 25.12$ inches

Practice Problems—Chapter 11

Section 11.2

A. **1.** $68x$ **3.** $M - 5$ **5.** $R/4$
  7. $P + 11$

B. **9.** D **11.** A **13.** J **15.** G **17.** H

C. **19.** $T + 11$ **21.** $T - 10$ **23.** $2T$
  25. $17/T$ **27.** $2/3\,T + 12$ **29.** $T - 8$
  31. $T + 17$ **33.** $T - 20$ **35.** $8 - T$
  37. $2T - 4$ **39.** $T + 18$

Section 11.3

A. **1.** $N + 17 = 39$ **3.** $11N = 99$
    number is 22   number is 9
  5. $N/4 = 9$ **7.** $N - 4 = 13$
    number is 36   number is 17

B. **9.** $2N + 6 = 24$ **11.** $1/2N - 6 = 48$
    number is 9   number is 108
  13. $N + N - 4 = 20$ **15.** $2N + 4 = N + 16$
    numbers: 12, 8   number is 12

C. **17.** $\dfrac{2}{3}S = 16$ **19.** $4N - N = 54$
    24 students   numbers: 18, 72
  21. $1/4P = 12$ **23.** $2N + 40 + N$
    48 points $= 580$
                      $180 and $400
  25. $4N - 2N = 28$ **27.** $245 = 80 + 45h$
    numbers: 14, 56   3 2/3 hours
  29. $179 = 47 + 12M$
    $11 per month

Section 11.4

A. **1.** 16   **3.** 71 and 73
  5. 17 and 18   **7.** 46
  9. 30 and 31   **11.** 4

B. **13.** $n + n + 1 + n + 2 = 153$ 50, 51, and 52
  15. $n + n + 2 = 90$ 44 and 46
  17. $n + n + 1 + n + 2 + n + 3 = 178$
                      43, 44, 45, 46
  19. $n + n + 2 = 146$ 72, 73, and 74
  21. $n + n + 4 = 68$ 32, 34, and 36
  23. $n + n + 1 = n + 2$ 1, 2, and 3

25. $n + 1 + 2n = 28$ 9 and 10
27. $n + 4 + 2n = 16$ 4, 6, and 8

Section 11.5

A. **1.** $12 + 21 + 24 + x = 75$ 18 feet
  3. $11 + 20 + x = 48$ 17 meters
  5. $x + x + 18 = 58$ 20 inches
  7. $8x = 176$ 22 feet

B. **9.** $96 = 4s$ 24 meters
  11. $132 = x + x + 6 + x + x + 6$
                      36 in, 30 in
  13. $24 = x + 2x + x + 2x$ 4 yds, 8 yds
  15. $84 = 2x + 6 + x + 2x + 6 + x$
                      12 ft, 30 ft
  17. $x + x + 2x - 7 = 37$ 11 m, 11 m, 15 m
  19. $x + 5 + x + 2x = 41$ 9 in, 14 in, and
                      18 in

Section 11.6

A. **1.** $2N$ **3.** $(0.10)N$, $(0.05)2N$
  5. addition
  7.

Items	Value of Each	No. of Each	Value of Each Kind
nickels	0.05	$2N$	$(0.05)2N$
dimes	0.10	N	$(0.10)N$
			Total Value

B. **9.** $(0.05)2N + (0.10)N = 2.40$ 12 dimes
                        24 nickels
  11. $(0.01)N + (0.25)N = 1.56$ 6 pennies
                        6 quarters
  13. $(0.25)N + (0.10)8N = 2.10$ 2 quarters
                        16 dimes
  15. $(2.50)N + (1.50)2N = 269.50$ 49 adults
                        98 children
  17. $(12.50)N + (38.00)2N = 177.00$
                        2 workbooks
                        4 textbooks
  19. $(0.50)N + (0.25)1/2N = 13.75$ 22 halves
                        11 quarters

Section 11.7 Review Problems

A. **1.** e **3.** g **5.** b **7.** d

B. **9.** l **11.** n **13.** i **15.** j **17.** r

C. **19.** $2n + 7 = 25$ number is 9

  21. $n + n + 1 + n + 2 = 87$ 28, 29, and 30

  23. $1/2n - 4 = 16$ number is 40

  25. $x + 12 + x + x + 12 + x = 52$
                   7 in., 19 in.

  27. $(0.05)N + (0.10)N = 1.95$ 13 nickels
                   13 dimes

  29. $n + 4n = 35$ 7 and 28

  31. $n + n + 4 = 72$ 34, 36, and 38

  33. $n + 5 + 3n = 37$ 13

  35. $x + 6 + x + x + 6 + x = 48$
                   9 in., 15 in.

  37. $n + n + 1 + n + 2 = 60$ 19, 20,
                   and 21

  39. $81 = 3s$ 27 ft., 27 ft.,
                   and 27 ft.

  41. $3/4n - 8 = 10$ number is 24

43. $2x + x + x + 4 = 40$ 18 in.

45. $(0.25)X + (0.10)X = 15.75$ 45 quarters
                 45 dimes

47. $(0.15)X + (0.25)2X = 6.50$ 10 postcards
                 20 letters

Test—Chapter 11

  1. $N + 7$ **2.** $2N$ **3.** $3N - 4$

  4. $1/2Q$ or $Q/2$ **5.** $4R + 5$ **6.** $3/4N + 7$

  7. $N + N + 2$ **8.** $(0.10)N$

  9. $2N - 6 = 20$ **10.** 13

  11. $N + N + 2 + N + 4 = 102$ **12.** 32, 34, 36

  13. $5/6S = 35$ **14.** 42 students

  15. $X + X + 8 + X + X + 8 = 180$

  16. 41 m, 49 m

  17. $(0.25)2X + (1.00)X = 43.50$

  18. 29 dollar bills, 58 quarters

  19. $N + N + 2 = 24$

  20. 11, 12, and 13

Practice Problems—Chapter 12

Section 12.2

A. **1.** -4 **3.** $+26$ **5.** $+276$ **7.** $+5,280$ **9.** $+75$

B.

C. **21.** $<$ **23.** $>$ **25.** $<$ **27.** $>$

D. **29.** 6   **31.** 9   **33.** 0   **35.** 1.4

  37. 8   **39.** 2

E. **41.** 8.54   **43.** 9.85   **45.** 13.95

Section 12.3

A. **1.** -3 **3.** -11 **5.** -7 **7.** $+10$

  9. -5 **11.** 0   **13.** -16 **15.** -6

  17. -10 **19.** $+7$

B. **21.** $(-14) + (+10) = -4$ lost by 4 points

  23. $(-150) + (+250) = +100$ 100 feet above
     sea level

  25. $(+23) + (-40) = -17$ 17 miles east

C. **27.** 8 1/3   **29.** -12

D. **31.** 7.177   **33.** -11.275 **35.** -92.927

Section 12.4

A. **1.** -1 **3.** -3 **5.** -18

  7. $+10$ **9.** -6

B. **11.** $+3$ **13.** $+2$ **15.** $+17$

  17. -23

C. **19.** -12 **21.** $+9$ **23.** -19

  25. -5 **27.** $-3 1/4$ **29.** $-15 1/2$

D. **31.** -11.682 **33.** -82.605 **35.** -17.504

Section 12.5

A. **1.** $+55$ **3.** -48 **5.** $+54$

  7. -36 **9.** 0   **11.** -30

  13. -42 **15.** 0   **17.** $+24$

  19. $+24$

B. **21.** +31   **23.** −1   **25.** −24
   27. −3   **29.** +42
C. **31.** −9   **33.** −64   **35.** −8
   37. +16   **39.** −4   **41.** +108
   43. −49
D. **45.** 13/54   **47.** −5 1/16
   49. −2 13/16
E. **51.** −14.71   **53.** −8,494.26
   55. 2,081.51   **57.** 402.40
   59. −518.69   **61.** 68.81
   63. −16.19

C. **31.** +3   **33.** −15   **35.** −5
   37. −6   **39.** −9   **41.** −3
   43. −3   **45.** −16   **47.** −4
   49. −2   **51.** −4/3   **53.** −9
   55. −16/27   **57.** −6   **59.** 1/50
   61. 3/16   **63.** −1 3/4   **65.** −1 3/5
   67. −7.224   **69.** −1
D. **71.** 0   **73.** +7   **75.** −6
   77. +5   **79.** −36   **81.** +48
   83. +12   **85.** +3/2   **87.** +1/5
   89. −5/13

Section 12.6

A. **1.** −2   **3.** −4   **5.** +4
   7. −2/5   **9.** −12/5   **11.** +9
   13. −5   **15.** −40   **17.** −2
   19. +1
B. **21.** +4/3   **23.** −12/5   **25.** −12
   27. −12   **29.** 36   **31.** −62
   33. 2   **35.** −27   **37.** −1
   39. −7/2

Section 12.7

A. **1.** −1   **3.** +6   **5.** +24
   7. −16   **9.** +22   **11.** +7
   13. +12   **15.** −12   **17.** +45
   19. −32   **21.** −8/3   **23.** +5/3
   25. +12   **27.** +17   **29.** −3/5
B. **31.** −0.825   **33.** −3.25   **35.** +0.25
   37. −8   **39.** 8.675

Section 12.8 Review Problems

A. **1.** distance and direction
   3. less than
   5. +6
   7. add the numbers and use the sign
   11. even   **13.** positive
   15. negative   **17.** see Section 10.7
   19. −2

Test—Chapter 12

1. −3   **2.** 4   **3.** −24   **4.** −7
5. −64   **6.** −28   **7.** −5   **8.** −10
9. 8   **10.** −72   **11.** 41   **12.** −28
13. −6   **14.** 8   **15.** 6   **16.** 1
17. −3   **18.** 1   **19.** 1/5   **20.** 4

Cumulative Review 1–12

1. 33.36   **2.** 19 1/9   **3.** $6x + 11y$
4. −10   **5.** 12   **6.** −3
7. −3   **8.** 1 5/12   **9.** 14.44
10. $-9a$ **11.** 17.6   **12.** +42
13. 289/24 or 12 1/24   **14.** 144/5 or 28 4/5
15. +12   **16.** +4
17. 42   **18.** −7/8
19. −6   **20.** 30
21. 7   **22.** 45/16 or 2 13/16
23. 8   **24.** −3/4
25. −10   **26.** $N − 6$
27. $2N + 7$ **28.** $N(8)$ or $8N$
29. $15/N = −5$ **30.** $C = 5.85N$
31. $Y = +8$
32. $N = 13/5$ or 2 3/5 or 2.6
33. 14.4   **34.** $x > 5$
35. $M = −80/9$ or $−8\,8/9$ **36.** $A = 6$
37. $W = +0.05$ **38.** 20%

B.

39. $x \geq 7$ **40.** $N = 7$ **45.** $-27, -25, -23$ **46.** \$14.63
41. $W = 21$ inches   **42.** $M = \$55.70$ **47.** 62 miles   **48.** 4.3 ounces
43. 5 1/2% or 5.5%   **44.** 63°   **49.** 314 square inches   **50.** 11 yards by 44 yards

Practice Problems—Chapter 13

Section 13.2

A. **1.** $-44 = -44$, yes   **3.** $16 = 16$, yes
       5. $2 = 2$, yes   **7.** $9 \neq 0$, no
       9. $0.2 = 0.2$, yes

B. **11.** 8   **13.** 5
       15. 63   **17.** $3X + 6$
       19. $-5m - 20$ **21.** $6K - 12$
       23. $-8Y + 32$ **25.** $34x - 51$
       27. $16N + 6.72P$ **29.** $-36A + 60B - 24C$
       31. $9A - 6C$ **33.** $-4M - 3N$

C. **35.** $21.645F - 0.039$
       37. $-62.72 + 101.92W - 222.88X$
       39. $3,199,755M + 8,753,805N$

Section 13.3

1. $-1 - 6x$ **3.** $10m - 11$
5. $4C - 2$ **7.** $-5y + 4$
9. $4S + 13$ **11.** $X + 8$
13. $4m + 5$ **15.** $7B - 2$
17. $21 - 2y$ **19.** $5 - 6c$
21. $6m - 18$ **23.** $6 + S$
25. $6 + 2F$ **27.** $24x - 6$
29. $-3 - 5w$ **31.** $4y - 4$
33. $18K - 51$ **35.** $a - 15b$

37. $-5g - 50$ **39.** $8x$
41. $-8c + 48$ **43.** $-12p - 6$
45. $7w - 24$

Section 13.4

A. **1.** $x = 4$ **3.** $C = 11$
        ck: $18 = 18$ ck: $-4 = -4$
       5. $1/5 = p$ **7.** $N = 9$
        ck: $24 = 24$ ck: $38 = 38$
       9. $A = 9$ **11.** $M = 2$
        ck: $12 = 12$ ck: $-12 = -12$
       13. $W = 3$ **15.** $c = 4$
        ck: $8 = 8$ ck: $10 = 10$
       17. $N = 7$ **19.** $a = 32$
        ck: $24 = 24$ ck: $29 = 29$

B. **21.** $16 = R$ **23.** $y = 7$
        ck: $18 = 18$ ck: $44 = 44$
       25. $K = 0$ **27.** No real solution
        ck: $-12 = -12$
       29. $A = -2$ **31.** $x = 2\ 1/2$
        ck: $21 = 21$ ck: $28 = 28$
       33. $-3 = m$ **35.** No real solution
        ck: $15 = 15$
       37. $1 = T$ **39.** $R = -0.5$
        ck: $-3 = -3$ ck: $1.6 = 1.6$
       41. No real solution   **43.** All real numbers

Section 13.5

A. **1.**

3.

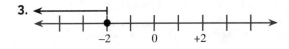

5.

7.

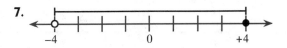

9.

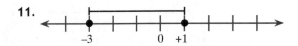

11.

B. **13.** $\{x \mid x \le 7\}$

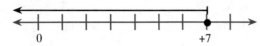

15. $\{x \mid x > 6\}$

17. $\{x \mid x < 4\}$

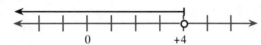

19. $\{x \mid x > 4\}$

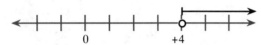

21. $\{x \mid x \ge 1/2\}$

23. $\{x \mid x \ge -9\}$

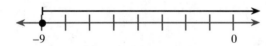

25. $\{x \mid -5 \le x < 3\}$

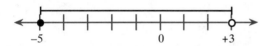

27. $\{x \mid x \ge -3\}$

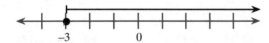

29. $\{x \mid x > -4\,1/2\}$

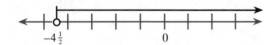

31. $\{x \mid -7/3 < x < 1\}$

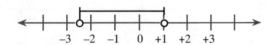

33. $\{x \mid 1 \ge x > -1\}$

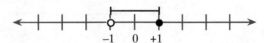

35. $\{x \mid -3 \le x \le 9\}$

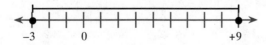

Section 13.6 Review Problems

A. **1.** $56 - 21y$ **3.** $-3b - 18$

   5. $2x + 3$ **7.** $3a + 42$

   9. $26 + r$ **11.** $17n - 5$

   13. $30p + 18$ **15.** $6x - 12y + 15$

   17. 12 **19.** $20 - 26c$

   21. $-13y - 18$ **23.** $-10d + 80$

B. **25.** $x = -7$ **27.** $m = -1$
    ck: $-44 = -44$ ck: $-6 = -6$

   29. $p = 10$ **31.** No real solution
    ck: $45 = 45$

   33. $y = -2$ **35.** $r = 2$
    ck: $-9 = -9$ ck: $7 = 7$

   37. $5 = x$ **39.** $y = 2$
    ck: $-80 = -80$ ck: $24 = 24$

   41. $b = 3$
    ck: $14 = 14$

C. **43.** $\{x \mid x \le 2\}$

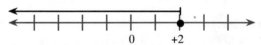

45. $\{x \mid x > 2\}$

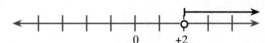

47. $\{x \mid x > 1/7\}$

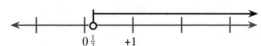

49. $\{x \mid -2 < x < 1\}$

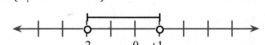

Test—Chapter 13

1. $6x + 12$ **2.** $6 - 2n$

3. $4y - 8$ **4.** $-10R - 14$

5. $x = 10$ **6.** $t = -8$
  ck: $33 = 33$ ck: $13 = 13$

7. $-7 = p$ **8.** $128 = m$
  ck: $24 = 24$ ck: $-48 = -48$

9. No solution   **10.** $k = 3$
  $29 \ne 5$ ck: $-24 = -24$

11. $-9 = T$ **12.** $F = 9$
  ck: $-40 = -40$ ck: $38 = 38$

13. $11 = c$
ck: $-4 = -4$

14. $y = -12$
ck: $-1 = -1$

15. $5 = w$
ck: $-20 = -20$

16. $\{x \mid x > 1\}$

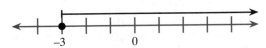

17. $\{x \mid x \geq -3\}$

18. $\{x \mid x \geq 4\}$

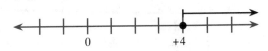

19. $\{x \mid -5 \leq x < 3\}$

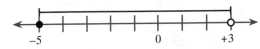

20. $\{x \mid x < -2\}$

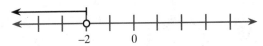

Practice Problems—Chapter 14

Section 14.2

1. $J - 19 + J + 2(J - 19) = 163$
M: \$36, J: \$55, D: \$72

3. $2(n) - 3(n - 5) = -7$
smaller = 17, larger = 22

5. $2T - 123 - T = 627$
World Trade Center: 1377 ft., Prudential: 750 ft.

7. $t + 2 + t + 2(t + 2) = 42$
M: 11 hrs., T: 9 hrs., W: 22 hrs.

9. $h + h + 5 + 2(h + 5) = 90$
Chemistry book cost \$47.50

11. $2(m - 8) + m + (m - 8) = 64$
S: 14, M: 22, L: 28

13. $h + 3 + h + 2(h + 3) = 27$
7 1/2 hours on Tuesday

15. $2N - 14 = 3(N + 3)$
number is -23

17. $3(5N - 1) = 4(N + 2)$
number is 1

19. $6(N - 3) = N - 3$
number is 3

Section 14.3

A.   **1.** $n + n + 1 + n + 2 = -84$ $-29, -28, -27$

    3. $n - 3(n + 2) = -12$ 3 and 5

    5. $4(n) - (n + 4) = 50$ 18, 20, 22

    7. $4(n) + 2(n + 1) = -16$ -3 and -2

    9. $3(n) - (n + 2) = 36$ 19 and 21

B.   **11.** $n - 2(n + 4) = -13$ 5, 7, 9

    13. $n - 2(n + 2) = -23$ largest number is 21

    15. $2(n) = (n + 4) + 7$ 11, 13, 15

17. $n + n + 2 + n + 4 = -18$ $-8, -6, -4$

19. $2(n + 2) = 3(n)$ 4 and 6

21. $2(N + 4) - 3N = -4$ 12, 14, 16

23. $2/3N + 4/5(N + 2) = -38$

     $-27, -25$

Section 14.4

1. $2(x + 8) + 2(x) = 76$
23 feet long, 15 feet wide

3. $2(x - 6) + 2(x) = 88$
25 yards long, 19 yards wide

5. $2(2x + 4) + 2(x - 1) = 96$
30 feet long, 15 feet wide

7. $2(3x - 20) + 2(x) = 440$
160 feet long, 60 feet wide

9. $x + (x - 4) + 2(x - 4) = 28$
10 feet, 6 feet, 12 feet

11. $2(x + 10) + 2(x) = 56$
19 feet long, 9 feet wide

13. $2(3x - 3) + 2(x + 5) = 100$
36 feet long, 12 feet wide

15. $2X + 2(X - 3) = 38$
11 inches, 8 inches

17. $2(2X + 5) + 2X = 40$
5 yards, 15 yards

19. $2X + 2[2(X - 12)] = 66$
15 feet, 9 feet

Section 14.5

1. $0.25(2x + 4) + 0.10(x) = 3.40$
4 dimes, 12 quarters

3. $0.15(x) + 0.25(x - 6) = 8.90$
26 15-cent stamps, 20 25-cent stamps

5. $5(x) + 7[2(x + 6)] + 8(x + 6) = 321$
26 chocolate cakes

7. $1.20(x) + 2.00(40 - x) = 72.00$
10 rolls at $1.20, 30 rolls at $2.00

9. $3.25(x) + 2.25(100 - x) = 2.50(100)$
25 gals. of cranberry, 75 gals. of apple

11. $0.05(x) + 0.10(32 - x) = 2.00$
24 nickels, 8 dimes

13. $0.10(x + 24) + 0.05(5x) + 0.01(x) = 6.72$
36 dimes, 60 nickels, 12 pennies

15. $0.08(300 - x) + 0.05x = 0.06(300)$ OR
$0.08x + 0.05(300 - x) = 0.06(300)$
200 ml of 5%, 100 ml of 8%

17. $0.30x + 0.50(10 - x) = 0.35(10)$ OR
$0.30(10 - x) + 0.50x = 0.35(10)$
7.5 liters of 30%

19. $0.10x + 0.15(50 - x) = 0.12(50)$ OR
$0.10(50 - x) + 0.15x = 0.12(50)$
20 pounds of 15%, 30 pounds of 10%

Section 14.6

1. $7(r) + 7(r + 10) = 658$
42 mph and 52 mph

3. $6(t) = 10(t - 1)$
1 1/2 hours

5. $62(t) + 56(t) = 531$
4 1/2 hours

7. $3(r) + 3(r + 250) = 2,250$
550 mph and 300 mph

9. $3(r) + 3(r + 10) = 330$
Ned: 60 mph, Jim: 50 mph

11. $6(t) = 8(t - 1)$
3 hours

13. $5x + 3x = 24$
3 hours

15. $55x + 65x = 480$
4 hours

17. $4x + 4(x + 20) = 440$
65 miles per hour

19. $60x = 75(x - 1)$
4 hours

Section 14.7

1. $x + 4 - 20 = 2(x - 20)$
Mike: 28 years, Steve: 24 years

3. $56 + x = 3(16 + x)$
in 4 years

5. $x + 32 = 2(x + 8)$
quarter: 40 years old, dime: 16 years old

7. $35 + x = 3(7 + x)$
in 7 years

9. $x + 3 = 2(1/5\ x + 3)$
Becky is 5 years old, Sara is 1 year old

11. $42 - x = 5(14 - x)$
7 years ago

13. $5x + 5 = 3(x + 5)$
Brenda is 25 years old

15. $x + 84 - 20 = 3(x - 20)$
silver dollar: 62 yrs, 5-dollar gold piece: 146 yr

17. $x + 1 = 10(20 - x + 1)$
Hummel, 19 years old; Rockwell, 1 year old

19. $160 - x - 65 = 2(x - 65)$
clock, 75 years old; table, 85 years old

Section 14.8

1. $N - 10 > 8$
minimum integer is 19

3. $3N - 4 > 5 + 2N$
smallest integer is 10

5. $\dfrac{210 + 160 + x}{3} > 175$
more than 155

7. $80 + 72 + 64 + 94 + x \geq 400$
$90 \leq$ grade ≤ 100

9. $2(x + 3) + 2x < 72$
length must be less than 19 1/2 feet

11. $3N - (N + 2) > 54$
29 is the smallest integer

13. $0.29X + 0.20(15 - X) \leq 3.90$
or
$0.29(15 - X) + 0.20X \leq 3.90$
maximum is 10 29-cent stamps

15. $5(35) > 5(20) + 0.15m$
maximum number is less than 500 miles

Section 14.9 Review Problems

A. **1.** $x + 10 + x + 2(x + 10) = 90$
$15, $25, $50

3. $0.45(2x) + 0.30(x) = 2.40$ 4 sn., 2 gums

5. $3(n + 4) + 2n = 97$ 17, 19, 21

7. $2(x) + 2(4.5x) = 440$ 180 yards

9. $3n - (n + 2) = 50$ 26, 27, 28

11. $3n - 4(n + 4) = 0$ $-16, -14, -12$

13. $200(t) = 600(t - 2)$ a) 1 hr., b) 600 miles

15. $2(n) = (n + 4) + 7$ 11, 13, 15

17. $85 - x = 2(75 - x)$ 65 years

19. $3\ 1/2\ x + 15 = 2(x + 15)$ Sue: 35, David: 10

21. $2(r) + 2(2r - 40) = 220$ 60 mph

23. $2(x) + 2(x - 6) = 24$ 9 ft by 3 ft

B. **25.** $3N > 5(N + 4)$
all numbers less than -10

27. $2x + 2(20) \leq 72$
maximum length is 16 feet

Test—Chapter 14

1. $2(3x + 7) + 2x = 174$

2. 67 inches, 20 inches

3. $x + 42 - 20 = 3(x - 20)$

4. chair, 41 years old
table, 83 years old

5. $1.00x + 0.20(50) = 0.50(x + 50)$

6. 30 gallons

7. $6(x) = 5(x + 10)$

8. 60 miles per hour

9. $0.10x + 0.05(2x) + 0.25(x + 4) = 3.25$

10. 5 dimes, 10 nickels, 9 quarters

11. $3x + 4(x - 5) = 204$

12. 32, 27

13. $0.29x + 0.20(28 - x) = 7.40$
or
$0.29(28 - x) + 0.20x = 7.40$

14. 20 29-cent stamps, 8 20-cent stamps

15. $2N + 3(N + 2) = -239$

16. $-49, -47$

17. $35x + 45(x - 1) = 315$

18. 3.5 hours

19. $2x + 2(x + 8) \leq 68$

20. length is less than or equal to 21 yards

Practice Problems—Chapter 15

Section 15.2

A. **1.** $+a$ **3.** $-27g$
   5. $5r^2s$ **7.** $4b^2d^2 + 3b^2d$
   9. $5m^2n$

B. **11.** $-25K$ **13.** $7w^3$
   15. $11wxy^2$ **17.** $-9b^3c$
   19. $45xyz$

C. **21.** $14gh$ **23.** $6x^2$
   25. $-2a^2b^3$ **27.** $11mn$
   29. $7r^2s$ **31.** $272p^3q^2$
   33. $-28a^2b^2c^3$ **35.** $9c^2d^3$

Section 15.3

A. **1.** y^5 **3.** b^4 **5.** a^5 **7.** z^7
   9. 2^5 **11.** $6a^3b^2$ **13.** A^6 **15.** $-y^3$
   17. y^7 **19.** $8x^3$ **21.** $8m^6$
   23. $-\dfrac{8}{27}g^6$ **25.** -64
   27. 64 **29.** -1

B. **31.** $-6a^5b^3$ **33.** $32x^5z^2$
   35. $-a^{15}b^{10}$ **37.** x^6y^7

39. $-64a^3b^3c^3$ **41.** $\dfrac{a^9b^6}{8}$

43. $-6m^4n^3$ **45.** $\dfrac{8}{125}r^6s^3$

47. $-200x^{23}$ **49.** $64a^4b^6$

Section 15.4

A. **1.** x^3 **3.** c **5.** $4d^2$
   7. 1 **9.** $4g^2$ **11.** 3^2 or 9
   13. 1 **15.** $4/9x^2$

B. **17.** $5/8$ **19.** $4xy^2$ **21.** $-6b^2cd$
   23. $1/3h^2$ **25.** $-3a^3$ **27.** $-1/2m$
   29. $8c$

C. **31.** $8a^2b$ **33.** $4m$ **35.** $1/3a^5$
   37. $-1/2x$ **39.** $6ab^3$ **41.** $-3m^2n^2$
   43. $2x/15$

Section 15.5

1. $\dfrac{1}{a^3}$ **3.** $\dfrac{4}{R}$ **5.** $\dfrac{3}{2M}$

7. $\dfrac{1}{x^4}$ **9.** b^4 **11.** $\dfrac{a^2}{4}$

13. $\dfrac{y^2}{4x^4z^6}$ **15.** $3m^2$ **17.** x^3

19. $\dfrac{b^6}{a^4}$ **21.** $\dfrac{6x}{y^4z^2}$ **23.** $\dfrac{4a^7}{b^3c^3}$

25. $\dfrac{-2y^3}{3x^5}$ **27.** $\dfrac{-m^{15}n^3}{8}$ **29.** $\dfrac{b^2}{36a^2}$

31. $\dfrac{1}{32b^{11}}$ **33.** $\dfrac{n}{4m^6}$ **35.** $\dfrac{b^{12}c^{11}}{108a}$

37. $\dfrac{60}{x^3y^3}$ **39.** $\dfrac{1}{x^6y^4}$

Section 15.6

A. **1.** $38x + 6y - 39$ **3.** $28m^2 - 2m - 9$
  5. $7R + S + 3T$

B. **7.** $18a + 3ab - 17b$ **9.** $-8v + 3w - 12x$
  11. $18f^2 - 12f + 8h$

C. **13.** $-4a + 10b$ **15.** $-7m + 7$
  17. $9x^2 + 9x + 3$ **19.** $2R - S - 6T$
  21. $x^2y - 4xy + 5y^2$ **23.** $-b^2 - 11b - 13$
  25. $7y^3 + 7y^2 - 4y + 2$
  27. $-m^3 - m^2 - m - 4$
  29. $6x^3 - x^2 + 4x - 6$
  31. $7p^2q + 2pq + pq^2$ **33.** $6x^2 + 3x - 7$
  35. $-4x - 3y$ **37.** $10y^2 + 9y - 5$
  39. $m^2n^2 - 2m^2n - mn^2 + 7$

Section 15.7

A. **1.** $3a^3 - 12a$
  3. $-56m^3n - 32m^2n^2$
  5. $15x^2 + x - 2$
  7. $-4p^3 - 35p^2 + 17p - 2$
  9. $-3x^4y^3 - 2x^3y^2$
  11. $c^3d^2 + 2c^2d^2 - 4cd - 8d$
  13. $-5x^3 - 10x^2y - 10xy^2$
  15. $y^3 + 4y^2 + 8y + 8$

B. **17.** $a^4 - 4a^2 - 21$
  19. $0.09x^2 - 0.24xy + 0.16y^2$
  21. $x^2 - 6xy + 9y^2$
  23. $g^2 - 16$
  25. $3a^3 - 10a^2b + ab + 8ab^2 - 2b^2$
  27. $-28x^4y^3 + 14x^3y^2 - 21x^2y^3$
  29. $-10m^4n + 8m^3n^2 - 4mn^2$
  31. $9 + 12W + 4W^2$
  33. $y^3 + 3y^2 + 3y + 1$

35. $9y^2 - 1$
37. $x^4 - 9$
39. Answers will vary

Section 15.8

A. **1.** $3b - 1$ **3.** $-2x^3 + 4y^2$
  5. $3m^2 - 2m + 5$ **7.** $-3x^2 - 3/2x - 2$
  9. $7r - 5/r$ **11.** $p - 2q - 3r$
  13. $4vw^3x - 3wx^2$ **15.** $3/2b + 5/2ac^2$
  17. $3 - \dfrac{2b}{a} - a$ **19.** $-x^3y^2 - xy$

B. **21.** $A - 2$ **23.** $n - 2$
  25. $x + 2 - \dfrac{4}{x - 3}$ **27.** $2t + 1$

  29. $8z^2 - 10 + \dfrac{11}{z^2 + 1}$

  31. $-8x^2 + 14x - 22 + \dfrac{48}{x + 2}$

  33. $2x^2 + 3x - 1 + \dfrac{3x}{x^2 - x - 2}$

  35. $a^2 - 3$

Section 15.9

A. **1.** 2.78×10^3 **3.** 7.5×10^{-4}
  5. 6.5×10^{-1} **7.** 7.885×10^3
  9. 4.31×10^{-1} **11.** 6.52×10^{-2}

B. **13.** $6{,}500$ **15.** 0.0157
  17. $90{,}000$ **19.** 0.0007
  21. 0.03144 **23.** $470{,}000$

C. **25.** $(4.6 \times 10^6)(9 \times 10^3) = 4.14 \times 10^{10}$
  27. $(9 \times 10^{-3})(5.1 \times 10^{-4}) = 4.59 \times 10^{-6}$
  29. $(6.4 \times 10^{-3})(8 \times 10^2) = 5.12$
  31. $\dfrac{(6.8 \times 10^3)(4 \times 10^2)}{(4 \times 10^4)(2 \times 10^3)} = 3.4 \times 10^{-2}$
  33. $\dfrac{(2.4 \times 10^{-3})(6 \times 10^3)}{(1.2 \times 10^2)(4 \times 10^4)} = 3 \times 10^{-6}$
  35. $(1.35 \times 10^{-1})(2.7 \times 10^3) = 5 \times 10^{-5}$
  37. $(8 \times 10^{-3})^2 = 6.4 \times 10^{-5}$
  39. $(12 \times 10^3)^2 = 1.44 \times 10^8$

Section 15.10 Review Problems

A. **1.** $19x$ **3.** $-6a$ **5.** $8c^2d + 3c^2$
  7. p^2 **9.** x^3 **11.** $-18m^2n^3$

13. n^4 **15.** t^{15} **17.** $9b^2$
19. $-15a^{10}$ **21.** $2m^2$ **23.** $6a + 3b$
25. $2/b^2$ **27.** $1/b^3$ **29.** $1/4b^2$
B. **31.** $3/4t^5$ **33.** $14x^7y^6$
35. $-8a^3b$ **37.** $p^4q - 3pq^4$
39. $20f^2 + 3fg - 2g^2$ **41.** $3x^5$
43. $2rs - 3r^2s$ **45.** $8a^7b^3 - 11a^3b^6$
47. $3/t^2$ **49.** $a^4c^2/9b^2$
C. **51.** $-2a - 20$ **53.** $-18c^5d^5$
55. $3y^4 + y^3 - 15y^2 + y + 2$
57. $9w^4 + 4w^2 + 2w$ **59.** $25p^2 + 20p + 4$
61. $-2y^2 + 2y - 5$ **63.** $-3p^2 - 13p + 30$
65. $11y - 6$
67. $-2b^2 - 3b - \dfrac{6b}{-6b + 1}$
69. $3x^2z - 5xz + 2yz$
71. $a + 2$ **73.** $\dfrac{-6m}{n^5}$
75. $4x^2 + 4x + 2 + \dfrac{3}{x - 1}$
77. $\dfrac{256y}{x^6}$ **79.** $8 + 12a + 6a^2 + a^3$

D. **81.** 8.7×10^{-3} **83.** $90,000,000$
85. 6.23×10^4 **87.** 0.00065
89. 6.4×10^6 **91.** $9,900$
E. **93.** 5×10^{-6} **95.** 5.58×10^{-5}
97. 1.449×10^4 **99.** 1.6×10^7

Test—Chapter 15

1. $-6a^5b^3$ **2.** $2y^2 - 7y - 4$
3. $-\dfrac{m}{2}$ **4.** $-2a^2c$
5. $1/c^4$ **6.** $4r^4s^2t^6$
7. $\dfrac{a^2}{3bc^3}$ **8.** $\dfrac{2}{c^2}$
9. $-10x + 6y$ **10.** F^6
11. $-\dfrac{6}{a^5b}$ **12.** $x^2 + 6x + 9$
13. $\dfrac{9y^2}{16x^4}$ **14.** $8cd - 2d + 3d^2$
15. $-3x$ **16.** $7m^2n - 11mn + 18$
17. $3a^3 + 4a^2 + a + 10$ **18.** $4y^2 - y + 1 + \dfrac{1}{y + 1}$
19. 6.7×10^{-4} **20.** 1.94×10^1

Practice Problems—Chapter 16

Section 16.2

A. **1.** yes   **3.** no   **4.** yes
5. no   **7.** yes   **9.** yes
B. **11.** $2(x^2 - 2x + 8)$ **13.** $ab(5ab - 12a + 4)$
15. $-12r(r + 1)$ **17.** $0.8m(2m^2 + 8m + 3)$
19. $-a(a^2 - 2a - 6)$
C. **21.** $11(f^3 + 4g^3)$ **23.** $-a(a^2 + a - 8)$
25. $16b^3(4b + 1)$ **27.** Prime
29. $x^2z^2(5x^3y^4 + 7x^2y^2 - 1)$
31. $(a + 4)(a - 4)$
D. **33.** $(x - 7)(ax + 2)$
35. $6x(3w + 1) + w(w + 1)$
37. $c(x + y)(a - b)$
39. $(p + 5)(p + 12q)$

Section 16.3

A. **1.** $x^2 - 4$ **3.** $m^2 + 6m + 9$
5. $4b^2 - 1$ **7.** $x^2 - 4z^2$
9. $1 - 9a^2$ **11.** $9y^2 - 4z^2$

B. **13.** $(x + 7)(x - 7)$ **15.** $(a - 1)(a + 1)$
17. $(k - 11)(k + 11)$ **19.** $(z - 9)(z + 9)$
21. $(c + 2)(c - 2)$ **23.** $(2f + 3)(2f - 3)$
25. $(5T + 6)(5T - 6)$ **27.** $(7J + 8)(7J - 8)$
29. Prime   **31.** $(6 + c)(6 - c)$
33. $(a + 8b)(a - 8b)$ **35.** $(4a + 1)(4a - 1)$
37. $(3bc + 4a)(3bc - 4a)$
39. $(1 + 10x)(1 - 10x)$
C. **41.** $6b(b + 5)(b - 5)$
43. $(x^3 + 9a)(x^3 - 9a)$
45. $3(d^2 + 9)(d + 3)(d - 3)$
47. $5(m + 3n)(m - 3n)$
49. Prime
51. $a(a + 3)(a - 3)$
53. $4(m + 4n)(m - 4n)$
55. $x(5x + 1)(5x - 1)$
57. $2(49 + a^2)$
59. $3(3b + 5)(3b - 5)$

Section 16.4

A.   **1.** $(y + 3)(y + 5)$ **3.** $(t - 15)(t - 1)$

    5. $(a - 5)(a - 2)$ **7.** $(c + 5)(c - 4)$

    9. $(w + 5)(w + 5)$ **11.** $(m - 6)(m - 1)$

  13. $(x - 7)(x - 3)$ **15.** $(r - 1)(r - 1)$

  17. $(b + 5)(b + 4)$ **19.** $(z - 11)(z + 3)$

  21. $(2a + 1)(a + 1)$ **23.** $(3x + 1)(x + 2)$

B. **25.** $(r + 8)(r - 1)$ **27.** $3(v - 5)(v + 4)$

  29. $5(X + 3)(X + 5)$ **31.** $7(t - 3)(t - 3)$

  33. $(m + 3n)(m + 3n)$ **35.** $(w - z)(w - z)$

  37. $(3a + 5b)(a + 2b)$ **39.** $(2x - y)(x - 5y)$

  41. $4a(a + 5)(a - 3)$

  43. $-2Y(2Y + 5)(3Y - 2)$

  45. $(2c + 1)(2c - 3)$ **47.** $(3r - 4)(r + 5)$

  49. $3(3x - 2)(x + 1)$ **51.** $n(7m - 1)(m + 2)$

  53. $(4y - 3)(4y - 3)$ **55.** $3(t - 7)(t - 3)$

  57. $(5 - m)(4 - 3m)$ **59.** $b(3a - 2)(3a - 2)$

Section 16.5

A.   **1.** $(x + 4)^2$ **3.** $(a + 8)^2$

    5. $(2y + 1)^2$ **7.** $(4 - c)^2$

    9. $(5a + 4)(25a^2 - 20a + 16)$

  11. $(6t + 5q)^2$ **13.** $(4 + m)^2$

  15. $(2a - 3b)^2$

B. **17.** $(3 - a)(m + n)$ **19.** $(y - 4)(y^2 - 3)$

  21. $x^2(x - 8)(x + 10)$ **23.** $5(n - 3)^2$

  25. $5(b - 2)(b^2 + 2b + 4)$

  27. $(a + 3)(a + 2)(a - 2)$

  29. $(5a + c)(b - 4)$ **31.** $2(2t + 1)(t + 3)$

  33. $3(m^2 + 4)(m + 2)(m - 2)$

  35. $(a + 2)(b - c)$ **37.** $3y(y - 5)^2$

  39. $x^2y^2(4x + y)(5x - 3y)$

Section 16.6

A.   **1.** $y = -2, y = 5$ **3.** $a = -8, a = 9$

    5. $m = 0, m = 5/3$ **7.** $x = 0, x = 5$

    9. $b = 7, b = -7$ **11.** $y = 0, y = 2$

  13. $c = 0, c = 3, c = -3$

  15. $m = 0, m = -5/7$

  17. $t = 0, t = 1, t = -1$

  19. $x = -3, x = -2$

B. **21.** $n = 5, n = -2$ **23.** $a = -9, a = 8$

  25. $x = -3, x = -7$ **27.** $x = -3/2$

  29. $y = -2/3, y = 3$ **31.** $x = 0, x = 3$

  33. $R = 4, R = -4$ **35.** $x = 2, x = -8$

  37. $y = -7, y = 4$ **39.** $x = -3, x = 9$

  41. $a = 9, a = -8$ **43.** $a = -3/5, a = -2$

  45. $m = -5, m = 3$ **47.** $m = 5, m = -2$

  49. $k = 1/3, k = 4$

Section 16.7 Review Problems

A.   **1.** $3x^3y^2 - 6x^2y^2$ **3.** $y^2 + 3y - 18$

    5. $k^2 - 9$ **7.** $6c^2 + 15c - 9$

    9. $x^2 + 4x + 4$ **11.** $a^2 + 6a + 9$

B. **13.** $3y(y + 1)(y^2 - y + 1)$

  15. $(b - 4)(b + 4)$

  17. $(c - 3)(c - 5)$ **19.** $x^2(x^2 + 1)(x^2 + 1)$

  21. $-t^2(t - 2)$ **23.** $(2d - 1)(2d + 1)$

  25. $m^2n(3n - 5)$ **27.** $(13 - m)(3 - m)$

  29. $(a + 9)(a + 3)$ **31.** $(x + 7)(x + 7)$

C. **33.** $7(5 - 3b)$ **35.** $(m - 9n)(m + 5n)$

  37. $4a(a + 5)(a - 3)$ **39.** $-3y(y + 9)(y - 1)$

  41. $3(m + 4)(m - 4)$

  43. $9xy(xy - 2)(xy - 1)$

  45. $4(a - 4b)(a + 4b)$

  47. $(w^2 + 4z^2)(w + 2z)(w - 2z)$

  49. $6(m^2 - 5n^2)(m^2 + n^2)$

  51. $xyz(x - 6y)(x + 4y)$

D. **53.** $-8m(m - 3)(m^2 + 3m + 9)$

  55. $-6(2a + 1)(a - 1)$

  57. $(x + 7)(x - 7)$ **59.** $10(n + 11)(n + 2)$

  61. $(a + 3)(a + 1)$ **63.** $(x + 8)(x + 3)$

  65. $-4(b - 4)(b + 4)$ **67.** $2(m - 10n)(m + n)$

  69. $(3x + 4)(x - 4)$ **71.** $x^2yz(x^2 + xz + y^2)$

  73. $(x + 1)(xy - 2)$

  75. $(y - 4)(y - 1)(y + 1)$

  77. $8(2g - 5h)^2$ **79.** $(5p + 8)(4p - 3)$

  81. $(3w + 7)(6w - 5)$

E. **83.** $c = 0, c = 6$ **85.** $x = 0, x = 2$

  87. $m = 5/3, m = 3/2$ **89.** $n = 10, n = -4$

  91. $t = 1/4, t = -1/4$

  93. $m = -7/2, m = 9/2$

  95. $x = 7, x = -4$

Test—Chapter 16

1. $12x^3y^2 - 21x^2y^2 + 6xy$ **2.** $4a^2 + 5ac - 6c^2$
3. $2(2R + 5)(R - 4)$ **4.** $c(c - d)(c + d)$
5. $(a + 5)(x + y)$ **6.** $(5 + 4y)^2$
7. $(5t + 1)(3t - 2)$ **8.** $(2m - n)(m + 2n)$
9. $9(p + 5)(p + 2)$ **10.** $4(w - 2)(w + 2)$
11. $(x - 7)(x + 2)$ **12.** Prime
13. $(m + 6n)(m - 4)$ **14.** $(2x + 1)(x + 1)$
15. $3a(a^2 + 7a - 3)$ **16.** $(c - 6)(c + 2)$
17. $(5 + 3Z)(5 - 3Z)$ **18.** $(w - 7)(w - 3)$
19. $s = 6, s = -5$ **20.** $v = -7, v = 4$

Practice Problems—Chapter 17

Section 17.2

A. **1.** positive **3.** positive
  5. II **7.** y-axis
  9. positive **11.** Cartesian coordinate system

B. **13.** $A(+1,5)$ **15.** $C(5,3)$ **17.** $E(0,-2)$
  19. $G(-4,4)$ **21.** $I(-3,-3)$ **23.** $K(5,-1)$

C.

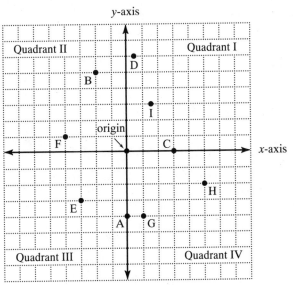

1. $y = -x - 4$

x	y
0	-4
-4	0
-2	-2

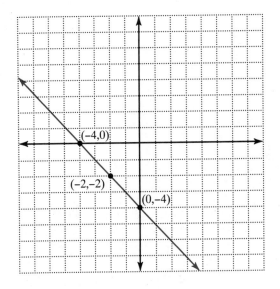

Section 17.3

A. B. and C. The following tables are sample answers and do not contain all possible pairs of numbers.

3. $y = -3x + 5$

x	y
0	5
5/3	0
3	−4

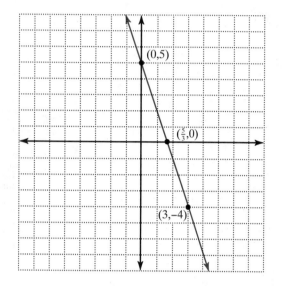

7. $y = -1/2\,x + 0$

x	y
0	0
2	−1
−2	1

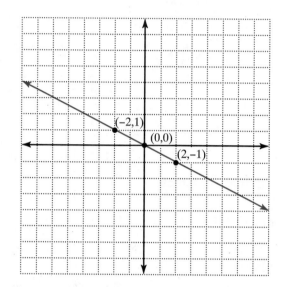

5. $y = x - 3$

x	y
0	−3
3	0
5	2

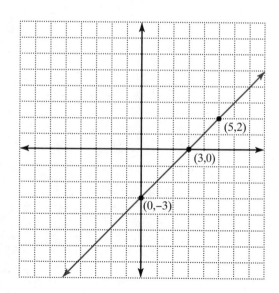

9. $y = 4x - 2$

x	y
0	−2
1/2	0
2	6

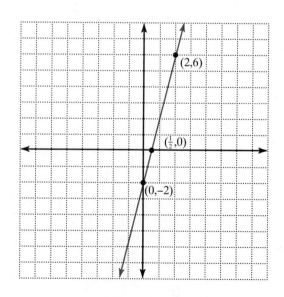

11. $y = 7/2\,x - 3$

x	y
0	−3
2	4
1	1/2

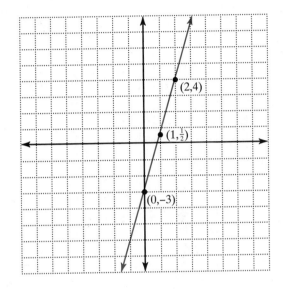

15. $y = 1/2\,x - 4$

x	y
0	−4
8	0
4	−2

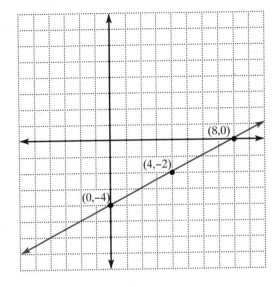

13. $y = -1/3\,x + 0$

x	y
0	0
3	−1
−3	1

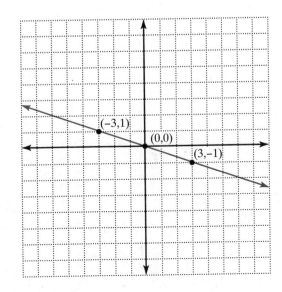

17. $y = 3x - 2$

x	y
0	−2
2/3	0
2	4

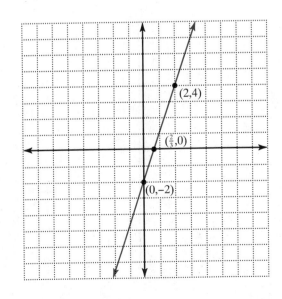

19. $y = 1/3x + 4$

x	y
0	4
3	5
−3	3

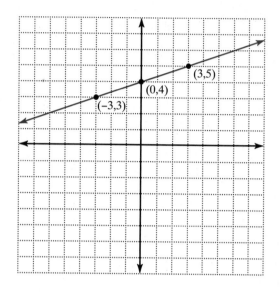

23. $y = 3/2x − 2$

x	y
0	−2
2	1
−2	−5

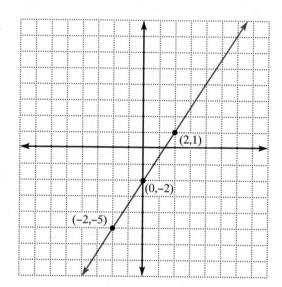

21. $y = −3/4x + 3/2$

x	y
0	3/2
2	0
−2	3

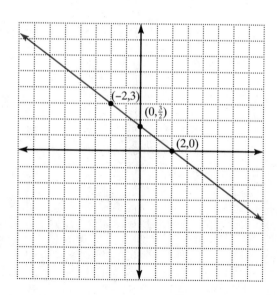

Section 17.4

1. $y = −x − 2$; slope $= −1$; y-intercept $= −2$

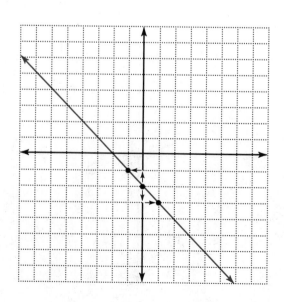

3. $y = 2x + 0$; slope = 2; y-intercept = 0

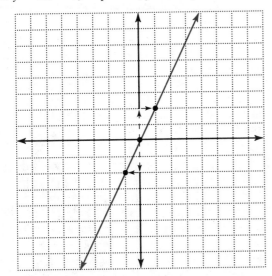

5. $y = x - 1$; slope = +1; y-intercept = −1

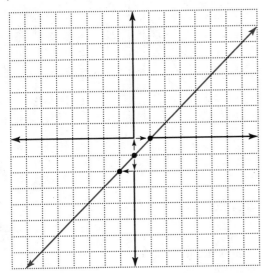

7. $y = -1/3x + 0$; slope = −1/3; y-intercept = 0

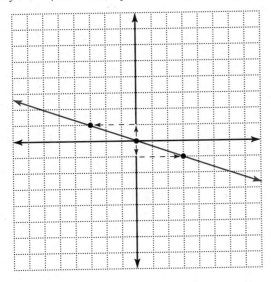

9. $y = -3x - 2$; slope = −3; y-intercept = −2

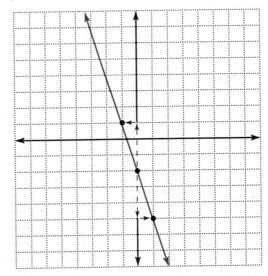

11. $y = 3x + 1$; slope = +3; y-intercept = +1

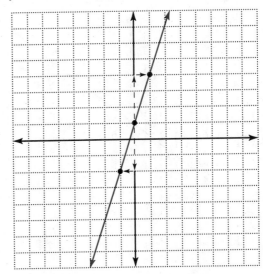

13. $y = 1/2x - 3$; slope = 1/2; y-intercept = −3

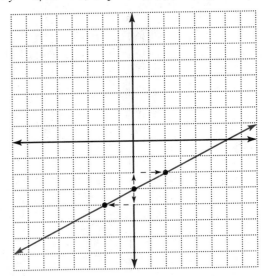

15. $y = -2x - 4$; slope $= -2$; y-intercept $= -4$

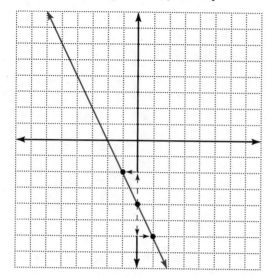

17. $y = 3/4x + 1$; slope $= 3/4$; y-intercept $= 1$

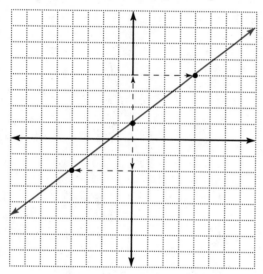

19. $y = -1/2x - 4$; slope $= -1/2$; y-intercept $= -4$

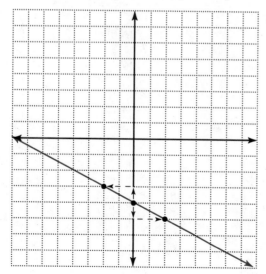

Section 17.5

1. $\dfrac{\text{rise}}{\text{run}} = \dfrac{+3}{+2}$

$m = \dfrac{6 - 3}{4 - 2} = \dfrac{3}{2}$

slant is up,
positive slope

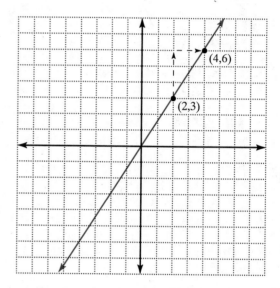

3. $\dfrac{\text{rise}}{\text{run}} = \dfrac{-1}{+1}$

$m = \dfrac{5 - 6}{3 - 2} = \dfrac{-1}{1} = -1$

slant is down,
negative slope

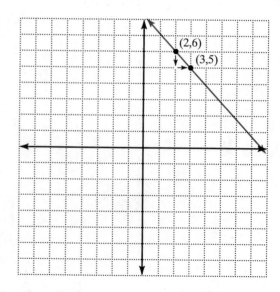

5. $\dfrac{\text{rise}}{\text{run}} = \dfrac{+4}{+3}$

$m = \dfrac{8-4}{3-0} = \dfrac{4}{3}$

slant is up,
positive slope

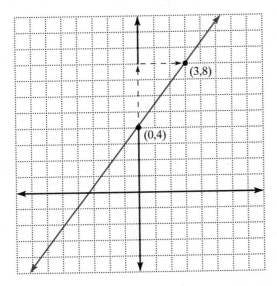

9. $\dfrac{\text{rise}}{\text{run}} = \dfrac{+2}{-3}$

$m = \dfrac{6-4}{2-5} = \dfrac{+2}{-3} = -2/3$

slant is down,
negative slope

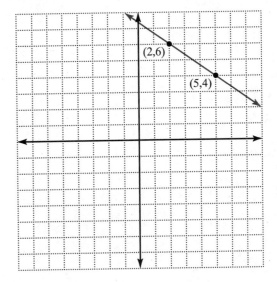

7. $\dfrac{\text{rise}}{\text{run}} = \dfrac{+4}{0}$

$m = \dfrac{8-4}{2-2} = \dfrac{4}{0}$

vertical line,
undefined slope

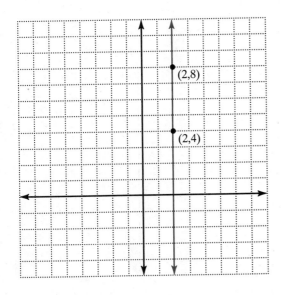

11. $\dfrac{\text{rise}}{\text{run}} = \dfrac{0}{-7}$

$m = \dfrac{4-4}{-4-3} = \dfrac{0}{-7} = 0$

horizontal line,
0 slope

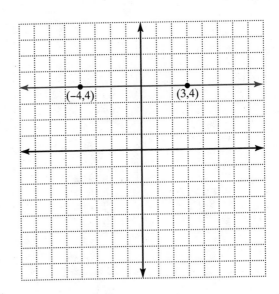

13. $\dfrac{\text{rise}}{\text{run}} = \dfrac{-2}{-2}$

$m = \dfrac{-5 - (-3)}{2 - 4} = \dfrac{-2}{-2} = +1$

slant is up,
positive slope

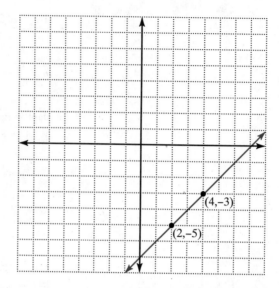

Section 17.6

A. **1.** $y = 3x - 2$ **3.** $y = \dfrac{1}{2}x + 3$ **C.** **21.** $y = 3x - 19$ **23.** $y = -2x - 11$

  5. $y = 0x + 2$ **7.** $y = -5x + 6$ **25.** $2y = x + 3$ **27.** $2y = -x + 1$

  9. $y = -\dfrac{3}{4}x + 2$ **29.** $4y = 3x - 10$

B. **11.** $y = \dfrac{3}{2}x + 3$ **13.** $y = 2x + 12$

  15. $y = \dfrac{5}{3}x$ **17.** $x = 6$

  19. $y = 3$

Section 17.7

1.

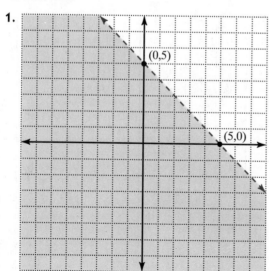

3.

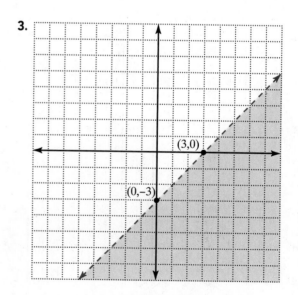

5.

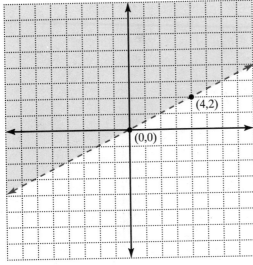

11.

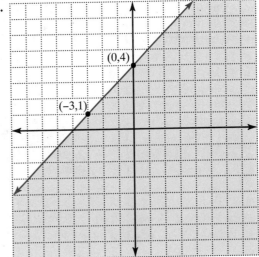

7.

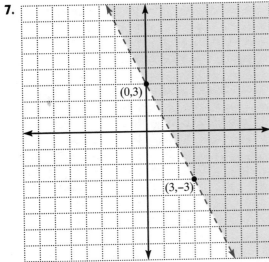

13.

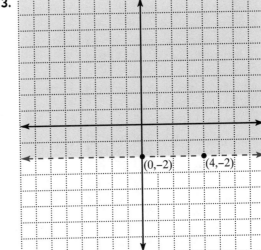

9.

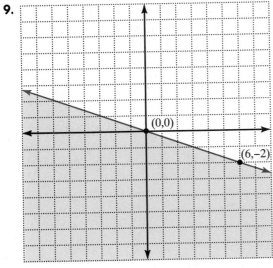

15.

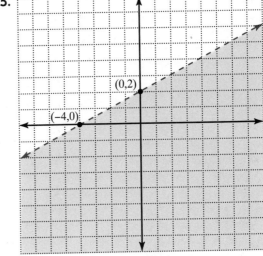

17.

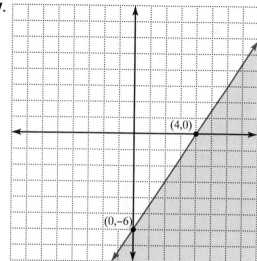

21.

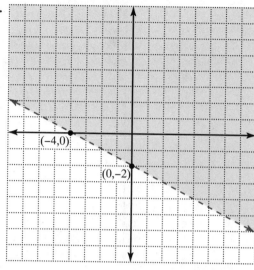

19.

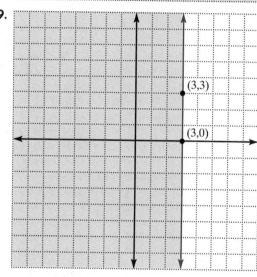

23.

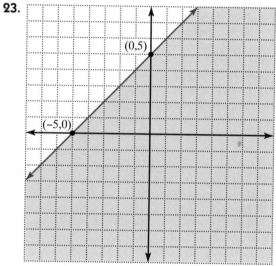

Section 17.8 Review Problems

A. **1.** Quad III   **3.** *y*-axis   **5.** Quad IV   **7.** Quad I   **9.** *x*-axis   **11.** Quad II

B. The following are sample tables and do not contain all possible pairs of numbers.

13. $y = -3x - 2$

slope $= -3$ *y*-intercept $= -2$

x	*y*
0	−2
−2/3	0
1	−5

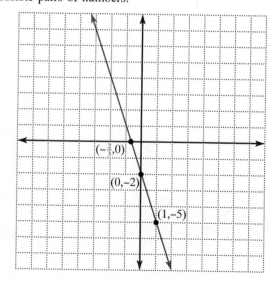

15. $y = 2x - 5$
slope $= 2$ y-intercept $= -5$

x	y
0	-5
2 1/2	0
-1	-7

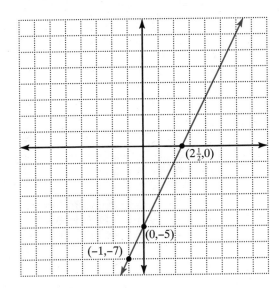

19. can't be done
undefined slope   no y-intercept

x	y
6	0
6	2
6	-2

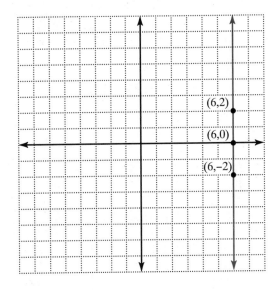

17. $y = x - 3$
slope $= 1$ y-intercept $= -3$

x	y
0	-3
3	0
-1	-4

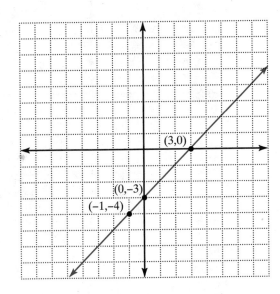

21. $y = -\dfrac{3}{2}x + 3$
slope $= -3/2$ y-intercept $= +3$

x	y
0	3
2	0
4	-3

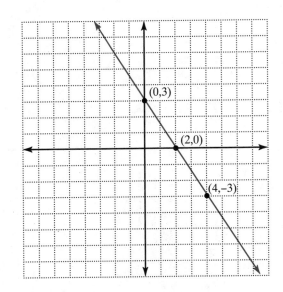

C. **23.** $m = +1$ **25.** $m = +1/2$

    27. $m = 3/0$ **29.** $m = 5/6$
       undefined slope

D. **31.** $y = -2x + 3$ **33.** $y = -3x + 11$

    35. $y = 2x - 3$ **37.** $y = -3x + 6$

    39. $3y = 2x - 3$ **41.** $2y = -x - 9$

    43. $y = -2$ **45.** $y = 3x - 21$

    47. $4y = -2x + 3$ **49.** $7y = 4x - 13$

    51. $3y = 2x + 9$ **53.** $y = -2x - 1$

    55. $3y = -2x - 5$ **57.** $2y = -6x + 1$

    59. $7y = -5x + 1$

65.

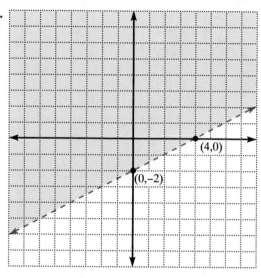

E. **61.**

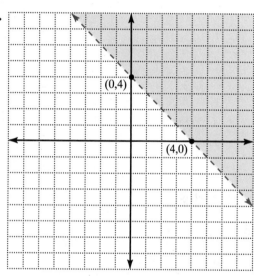

63.

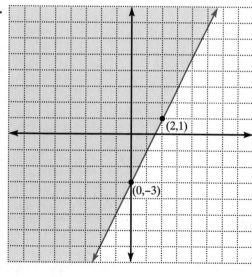

67.

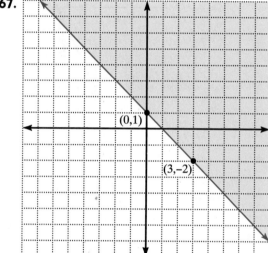

69.

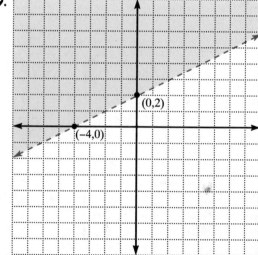

Test—Chapter 17

A.   **1.** y **2.** II

3. origin   4. $-1/2$

5. -2 6. $y = 2x - 4$

7. $-12/5$ 8. undefined

B.   9. $y = -2x + 4$ 10. $3y = 2x - 9$

11. $y = -x + 2$ 12. $x = 4$

13. $y = -2$ 14. $y = -3x + 4$

C.   15.

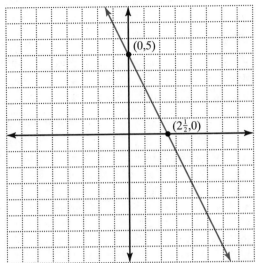

18.

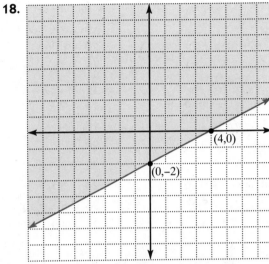

16.

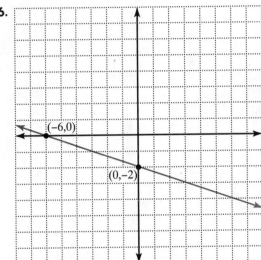

19.

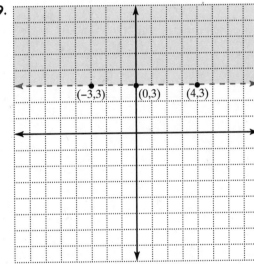

17.

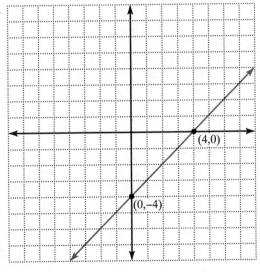

20.

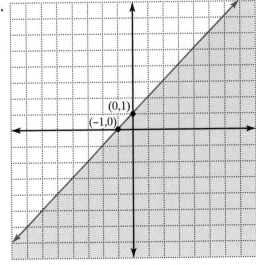

Cumulative Review 1–17

1. 3 67/72

2. 19*a*

3. $5y^2 + 6y + 8$

4. −63

5. 51.69

6. −3

7. 9

8. $-7a - 7ab - 3b$

9. 14*xyz*

10. 11/36

11. 57/8 or 7 1/8

12. −36

13. $6x^4y^3z$

14. 65.4

15. $a^2 + 4ab + 4b^2$

16. $-2ac^2$

17. 3/8

18. 734

19. $-3mn + 2m - n$

20. $X - 6$

21. 5/2

22. 0.21

23. $m = 4/7, b = 3/7$

24. 8/15

25. $(2x - 5y)^2$

26. 3.87×10^1

27. −23

28. 1990 − 1991

29. $1,000,000 more

30. $2,875,000

31. $X = 9/2$ or 4 1/2 or 4.5

32. $-1 \le X \le 1$

33. 106.7

34. $X = -2,$
$X = -4$

35. $Y = -7$

36. $N = 1.8$

37. $Y = \pm 3$

38. 125%

39. $X \ge 2$

40. $T = 1.8$

41. $Y = -1$

42. $h = 7'$

43. number is −1

44. 8 quarters

45. 0.6 inches

46. median: 0.7 inches
range: 1.3 inches

47. 20 miles per hour

48. 16,400 students

49. 5

50. 30 meters by 82 meters

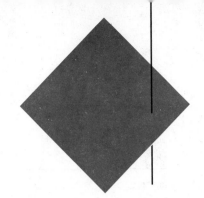

Index